Brown, C H

Structural materials in animals / C. H. Brown. New York : Wiley, [1975]

xv, 448 p. : ill. ; 24 cm.

"A Halsted Press book."
Bibliography: p. 365-421.

1. Morphology (Animals) 2. Histochemistry. 3. Biomechanics. I. Title.

QL799.B76 1975 591.4 75-1293
ISBN 0-470-10641-7 MARC

STRUCTURAL MATERIALS IN ANIMALS

Structural Materials in Animals

C. H. BROWN, PhD

A HALSTED PRESS BOOK

JOHN WILEY & SONS
New York • Toronto

First published in Great Britain in 1975
by Sir Isaac Pitman & Sons Ltd

Published in the U.S.A. and Canada
by Halsted Press
a division of John Wiley & Sons, Inc.,
New York

Library of Congress Cataloging in Publication Data

Brown, C. H.
Structural Materials in Animals

"A Halsted Press Book."
1. Morphology (Animals) 2. Histochemistry. 2. Biomechanics. I. Title.
QL799.B76 1975 591.4 75-1293
ISBN 0-470-10641-7

Printed in Northern Ireland at The Universities Press, Belfast

Contents

PART THREE CHORDATA

Foreword

At the anatomical level the skeletal systems of animals have long been, figuratively as well as literally, the backbone of comparative anatomy, for they provide the most striking anatomical expression of function. Considering this importance, interest in extending the analysis of their properties to progressively lower levels of magnitude developed rather slowly. Such landmarks in the theory of the subject as D'Arcy Thompson's *On Growth and Form* and Sir Wilfred Le Gros Clark's *Tissues of the Body* seemed slow to make an equal impact on research work.

Against this background the discovery by Mark Pryor, thirty years ago, of quinone-tanning in arthropod skeletal structures stood out in relief as dramatic as any in the biology of the time. Not only did it suddenly open up exciting vistas of research on skeletal materials at the chemical level, but it roused interest in all kinds of materials, both external and internal, which until then had scarcely been considered skeletal in any significant sense. This renaissance, largely concerned with invertebrate materials, had also a salutary effect on the rather pedestrian pace of some of the contemporary work on vertebrate materials.

Research on skeletal materials therefore was ready to make good use not only of the current histochemical and biochemical methods but also of the many subsequent techniques, from electron microscopy to the purification and analysis of proteins. As a result the subject has been advancing with increasing rapidity on many fronts, and today we are suddenly aware of a whole new country opened up and asking to be mapped.

That is the aim of this book. It cannot claim to be a map complete in detail, for this is not yet possible, but a working plan for those who wish to explore the country further. It is directed also towards a wider public—those students whose immediate requirement is simply to know the present position in the subject. It is written by a zoologist, primarily for zoologists, but the hope is that the comparative, evolutionary theme will be of equal interest to other biologists for without it no biological subject is fully alive.

With its interest in evolution the book is intended also for palaeontologists, who indeed provided much of the incentive to commission it. Having little available in fossils but skeletal structures their interest in living organisms necessarily centres on the corresponding parts of the body. For them also, it is hoped that the book will provide a base-camp and landmarks for further exploration. In fact, as will be seen, they are already using the standards of extant organisms for some exciting new advances in palaeohistology, palaeochemistry, and geochemistry.

Of course biological evolution is an indivisible entity, and skeletal systems cannot logically, but only from practical necessity, be considered in isolation. Reciprocally students of other biological systems and subjects may find here material which has more than merely parochial interest.

A. E. NEEDHAM

Acknowledgements

After twenty years domesticity it would have been impossible, in three years, to have assembled the material for this book without the help, both indirect and direct, of a great many people to whom I take this opportunity of expressing my most sincere thanks. I have relied heavily on the numerous reviews of subjects related to structural materials in animals and I am more than grateful to their authors for their digestion and summary of material that would have taken many years for me to have digested for myself. References to Dr L. Hyman's *The Invertebrata* fail to indicate the great use that I made of this excellent work for information on the biology of the various phyla that it covers. I owe much to Laurence Picken. *The Organisation of Cells and Other Organisms* was never out of reach and his help and advice over many years I acknowledge with much gratitude. To many others who sacrificed time to discuss with me various aspects of the book, who answered my queries and brought to my notice work of which otherwise I would have remained ignorant I am indeed most grateful, particularly to Dr S. O. Andersen, Dr R. Harkness, Dr W. D. I. Rolfe and Dr K. M. Rudall. Dr H. Y. Elder not only encouraged me by his interest but read and criticized helpfully part of the original manuscript. Alas, tragedy deprived me during the preparation of this book of the wise advice and criticism of Mark Pryor. He it was who first directed my interest towards this subject and whose suggestions have helped to shape many of my ideas. My debt to him is enormous and is more than thankfully acknowledged.

To the editor, Dr A. E. Needham, I owe more than it is possible to determine, not only for encouragement and helpful and kindly criticism, but also for the tremendous care with which he went through the manuscript in its various stages of preparation. But for his advice this book would have been much more incoherent than it is. He saved me from many foolish errors and slips but for those that may remain, I take full responsibility.

My thanks are also due to all those who so generously provided photographs from which the various plates have been compiled: R. W. Cox (Figs. 1.5, 1.8, 1.9); M. R. Dickson (Fig. 7.2); R. W. Drum (Fig. 4.2); H. Y. Elder (Figs. 1.9, 5.4); D. S. Friend (Fig. 2.15); C. Grégoire (Figs. 10.3*a*, *b*, *d*, 10.4); A. V. Grimstone (Figs. 2.27, 2.29); R. H. Hedley (Figs. 2.1, 2.2, 2.3, 2.4, 2.5); J. K. Koehler (Fig. 7.1); A. Nordwig (Fig. 5.5); J. F. Reger (Fig. 2.17); K. M. Rudall and his co-workers (Figs. 10.10,

11.1, 11.2, 11.3); L. T. Threadgold (Fig. 6.1); P. G. Toner (Figs. 3.1, 3.2); J. B. Tucker (Figs. 2.25, 2.26, 2.27); N. Watabe (Fig. 10.2); E. Wisse (Figs. 7.10, 7.11).

The following authors and their publishers have kindly given permission for the use of photographs and figures: The Academic Press and R. W. Drum (Fig. 4.2), J. K. Koehler (Fig. 7.1), E. Wisse (Figs. 7.10, 7.11), and N. Watabe (Fig. 10.2), from the *Journal of Ultrastructure Research;* The Academic Press and M. E. Rawles (Fig. 16.3), from *The Biology and Comparative Physiology of Birds;* The Academic Press and A. Nordwig (Fig. 5.5) from the *Journal of Molecular Biology;* The Academic Press and K. M. Rudall (Figs. 11.2, 11.3), from the *Treatise on Collagen;* and the Academic Press and A. V. Ivanov (Fig. 13.2), from the *Pogonophora.* J. Brill and D. R. Roggen (Fig. 7.6) from *Nematologica.* The Trustees of the British Museum (Natural History) and R. H. Hedley (Figs. 2.1, 2.2, 2.3, 2.4). Cambridge University Press and D. L. Lee (Fig. 7.8), D. W. T. Crompton (Fig. 8.1), from *Parasitology;* and C. Grégoire (Figs. 10.3*a*, *b*, *d*, 10.4) from *Biological Reviews;* Cambridge University Press, the Council of the Marine Biological Association of the United Kingdom and Professor I. Manton (Figs. 2.7, 2.8); Cambridge University Press, the Company of Biologists and J. B. Tucker (Figs. 2.25, 2.26, 2.27). The Carnegie Institute of Washington and P. E. Hare (Figs. 10.5*a*, *b*, 10.6, 10.7). The Clarendon Press, J. G. Blower (Fig. 12.2), J. B. Cowie (Fig. 6.2), and B. D. Watson (Fig. 7.5) from the *Quarterly Journal of Microscopical Sciences.* The Elsevier Publishing Company, K. Simkiss and W. J. Schmidt (Fig. 19.1), and K. M. Rudall (Figs. 12.3, 12.4), from *Comprehensive Biochemistry.* L'Institute Royale des Sciences de Belgique and C. Gregoire (Figs. 10.3*a*, *b*). The *Journal of Protozoolcgy*, and J. F. Reger (Fig. 2.15). E. and S. Livingstone and P. G. Toner (Figs. 3.1, 3.2), from *Cell Structure.* Longmans and G. H. Haggis (Fig. 1.11), from *Introduction to Molecular Biology. Nature*, W. G. Armstrong (Figs. 20.1, 20.2), and L. Pauling (Fig. 17.1). Thomas Nelson and Sons and P. G. 'Espinasse (Figs. 16.2, 16.4), from *A New Dictionary of Birds.* Pergamon Press and E. H. Mercer (Figs. 16.5, 16.6), from *Keratin and Keratinization.* The Rockefeller University Press, D. S. Friend (Fig. 2.15), A. V. Grimstone (Fig. 2.30), and F. L. Renaud (Fig. 2.31), from the *Journal of Cell Biology.* The Royal Society, A. V. Grimstone (Fig. 2.28) and I. Manton (Fig. 2.8). The Royal Society of Edinburgh and L. Auber (Fig. 16.7). The Royal Microscopical Society, R. W. Cox (Figs. 1.5, 1.8, 1.9) and R. H. Hedley (Fig. 2.5). Société Française de Microscopie Électronique and M. R. Dickson (Fig. 7.2). H.M. Stationery Office (Figs. 2.32, 4.1) from the Challenger Reports. John Wiley and Sons (Fig. 1.4) from *Biophysical Science.* The Zoological Society and P. F. A. Maderson (Fig. 16.1). Full references for the source of the various figures is given in the captions to the figures.

Work on this book has been carried out wherever fate chose to deposit me and I am grateful to the University of Michigan, Ann Arbor and Washington State University, Seattle, and particularly to the University of Glasgow for permission to work in their Libraries and for the help I have received from their librarians.

Miss J. Sommerville, Miss J. E. Smith and Mrs C. Macdougall co-operated in typing the manuscript, Mrs McTaggart typed the tables, and Mrs V. Warden typed the references from the card index. I am most grateful to them for the speed with which they completed this part of the work. The Lucy Cavendish Foundation, Cambridge, most generously gave me a grant towards the cost of typing the book.

My husband and children cheerfully accepted a decline in domestic comfort and tolerated periods when I failed to give them the attention that was their due. To them goes my warmest thanks as it was their helpful attitude that allowed the book to be undertaken and finally to be completed.

C. H. B.

AUTHOR'S NOTE

In order to keep the bibliography as up-to-date as possible, an additional list of references has been added at proof stage (see p. 411).

PART ONE

Introduction

CHAPTER 1

The Composition and Properties of Structural Materials

The structural materials to be considered here are those which mainly support and protect animals against forces that tend to deform and disintegrate them. The structural materials may be organic or inorganic. The organic materials are either proteins or carbohydrates or combinations of both of them. The suitability of these organic substances for structural purposes depends largely upon their ability to form long-chain macromolecules. These often form aggregates of parallel molecules described by various names according to the diameter of the aggregation: *protofilaments*, up to 20 Å in diameter may be associated to form *filaments* 50–100 Å wide and these can only be seen in the electron microscope. *Fibrils* are just wide enough to be seen in the light microscope while *fibres* have a diameter of about 1000 Å and over.

A. Mechanical properties of structural materials

The mechanical properties of structural materials that are of importance biologically are: (1) their reactions to forces that tend to stretch, compress, twist or shear them; (2) their resistance to forces that tend to break them; and (3) the changes in these properties within the range of temperature, humidity and ionic variations that are likely to be met with in the environment.

The mechanical properties of a structural material depend on the strength of intramolecular linkages and the strength of the linkages that develop between the various molecules in the material, and these, in turn, depend on the number and nature of the active groups at the surface of the molecules.

No structure is completely rigid, and forces acting on it produce some degree of deformation. After removal of the force the structure may return to exactly its original size and shape, it may undergo only a partial recovery or it may retain the deformation. In the first case it is said to be perfectly elastic, in the second, imperfectly elastic, and in the third, perfectly plastic.

In a material like steel, which consists of evenly spaced molecules, tension results in the molecules being pulled apart and on the release of tension intermolecular forces draw them back to their original positions. In materials made up of flexible, long-chain molecules, the response to tension depends not only on the intermolecular forces in the material but also on the degree of order that exists in the arrangement of the chain molecules. Long-chain macromolecules with active groups in the side chains, when forming a disorganized feltwork, are likely to establish linkages with a large number of adjacent molecules each of which is also linked to numerous other molecules so that the material is equally strong in every direction.

If the molecules at rest are disordered, tension tends to pull them into alignment. When the molecules are linked together this tends to prevent them slipping past each other and puts a limit on the extension which the structure can suffer without breaking. If no new intermolecular linkages develop when the molecules are pulled out straight, then when tension is released they return to their original random arrangement. This is the type of molecular arrangement on which the elasticity of rubber depends.

When the molecules form fibres, if the intramolecular cohesion is weaker than the intermolecular linkages, then the fibre will break when tension is sufficient to overcome the intramolecular cohesion. If, however, the intermolecular linkages are weaker than intramolecular cohesion, then as the fibre is pulled the linkages between the molecules are broken allowing the macromolecules to slide over each other and become pulled apart. The fibre lengthens and finally breaks when sufficient molecules have become separated from each other.

The more highly ordered the arrangement of the molecules, particularly into fibrils, and the more extensively they are linked together, the greater the force needed to deform the structure and the smaller the deformation it is able to undergo. Young's Modulus of Elasticity, the ratio between the tensile or compressive strength (force per unit area) and the strain (relative increase or decrease in length parallel to the direction in which the force is acting), is a measure of the elastic properties of a material.

The strength of a material under tension or compression is measured by the force per unit area which is just sufficient to break a specimen of the material. In these long-chain materials breakage occurs either by rupturing the intermolecular linkages or by breaking the actual chains.

B. Intermolecular linkages

The intermolecular linkages may be roughly divided into three kinds: London—Van der Waals attractive forces, electrovalent linkages, and covalent linkages.

London—Van der Waals forces are those that hold non-polar molecules (e.g. paraffin molecules) together and presumably exist between terminal —CH_3 groups in some protein side chains. They may be regarded as feeble attractions which, considered over any length of time greater than 10^{-20} seconds are electrically neutral but as a result of the development of temporary local concentrations of charge in atoms of adjacent molecules as the electrons revolve in their orbits, they feebly attract each other over short periods of time (Kruyt, 1952; Napper, 1967).

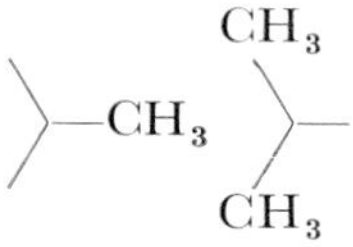

$$\rangle\!-CH_3 \quad \overset{CH_3}{\underset{CH_3}{\rangle}}\!-$$

When a large number of these attractions exist in a material they play a part in holding it together.

Electrovalent linkages depend on electrostatic attractions between positively and negatively charged polar groups as, for example, between ions in salt crystals. The salt link between adjacent $—NH_3^+$ and $—CO\bar{O}$ groups in protein side chains is of this type.

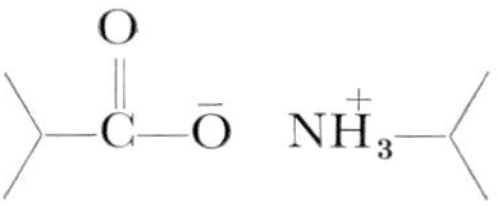

$$\rangle\!-\overset{O}{\overset{\|}{C}}-\bar{O} \quad \overset{+}{N}H_3\!-\!\langle$$

The hydrogen bond may be considered as a special case of the electrovalent bond, resulting from the attraction of two negatively charged atoms to a common hydrogen ion, as for instance between two $—CO\bar{O}$ groups of adjacent molecules,

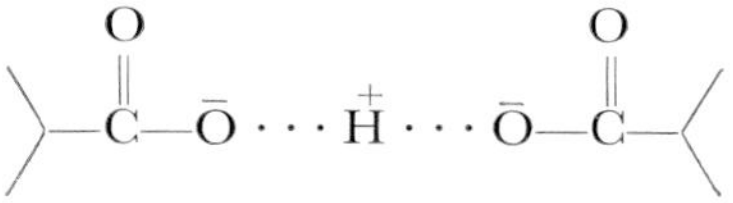

$$\rangle\!-\overset{O}{\overset{\|}{C}}-\bar{O}\cdots\overset{+}{H}\cdots\bar{O}-\overset{O}{\overset{\|}{C}}-\!\langle$$

or between imino and carboxyl groups of adjacent protein backbones.

$$\rangle\bar{N}-\overset{+}{H}\cdots\cdots\cdots\bar{O}=C\langle$$

The strength of electrovalent links is considerably influenced by the ionic content of the medium surrounding them and materials held together only by these linkages maintain constant properties only in a constant environment.

Covalent linkages result from the sharing of electrons by the atoms so bonded. In the disulphur bond of cystine, which may develop by oxidation of two adjacent cysteine —SH groups, two electrons are shared between the two sulphur atoms forming the link.

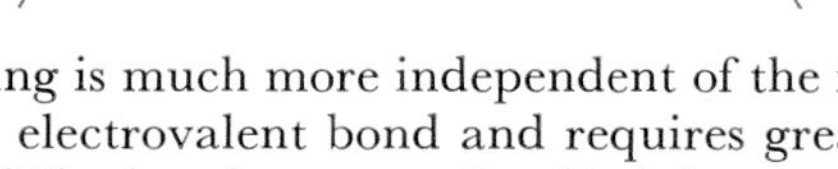

$$\rangle CH-CH_2-S-S-CH_2-CH\langle$$

This sort of bonding is much more independent of the ionic content of the medium than is the electrovalent bond and requires greater expenditure of energy to break it. (The bond energy of a disulphur bond is 63·8 kcal/mol as opposed to approximately 5 kcal/mol for a hydrogen bond.) If, therefore, a material is held together for the most part by covalent bonds, it is very stable.

It is possible to increase the stability of a material by adding to it a substance that establishes more than one linkage, either electrovalent or particularly covalent, between itself and the chain molecules, so increasing the number of linkages present in the system. This is known as tanning from its industrial use in the preparation of leather; here, various tanning agents are applied to skin so as to link up the collagen fibres and stabilize them against the effects of temperature, humidity, and bacterial attack.

One of the natural tanning methods that has been observed in animals involves an orthodihydroxyphenol substance which, on being oxidized to the quinone by a phenolase, establishes linkages with adjacent protein molecules by reaction with free NH_2 groups (Pryor, 1940a; Brunet, 1966).

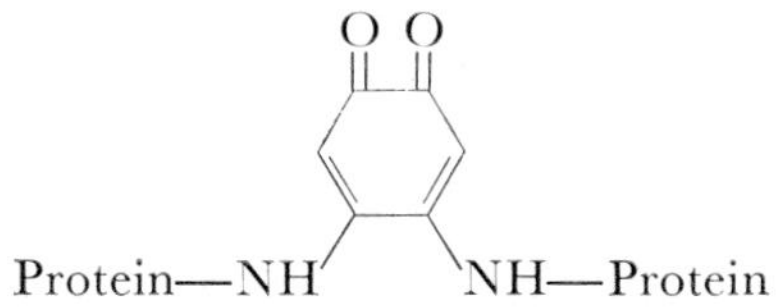

O O

Protein—NH NH—Protein

Skeletal proteins that owe their stability to tanning by orthoquinones are called sclerotins (Pryor, 1962). The quinones produced are often coloured brown or black and when structural materials are so coloured it suggests that quinone tanning may have taken place. Ortho-diphenols are tested for histochemically by their reactions with ferric chloride and ammoniacal silver nitrate (Lison, 1953; see also Smyth, 1953). The detection of a polyphenoloxidase associated with the skeletal material also implies that quinone tanning is present. Sodium hypochlorite will often dissolve quinone-tanned proteins though the chemistry of the reaction is not understood (Brown, 1950b).

Sometimes there is evidence that quinone tanning has taken place, though no free polyphenol precursor can be detected, and in these cases it has been suggested that "autotanning" of the protein may occur, tyrosyl groups within the protein being oxidized by a phenolase to a quinone, which then forms linkages with adjacent protein chains (Brown, 1952).

C. X-ray analysis

When structural material is sufficiently well ordered at the molecular level to give an X-ray diffraction photograph, this, together with a chemical analysis of the material, provides information for determining the atomic arrangement within the molecule. X-ray analysis also shows that materials from different sources may give similar diffraction photographs, so that structural materials may be classified and identified on the basis of their X-ray photographs. The chief organic structural materials in animals are the carbohydrates, cellulose and chitin, and the proteins, collagen, keratin and fibroin, all of which can be distinguished on the basis of their X-ray pictures. Elastin which is also an important structural protein gives no X-ray picture.

D. Organic structural materials

1. *Cellulose*

Cellulose is the most abundant organic compound found in nature. It is the chief constituent of the cell wall of plants, but it also occurs in some protozoa and higher animals.

Chemically, the cellulose molecule is a linear polymer of β-D-glucose forming long, straight chains containing at least 3000 residues.

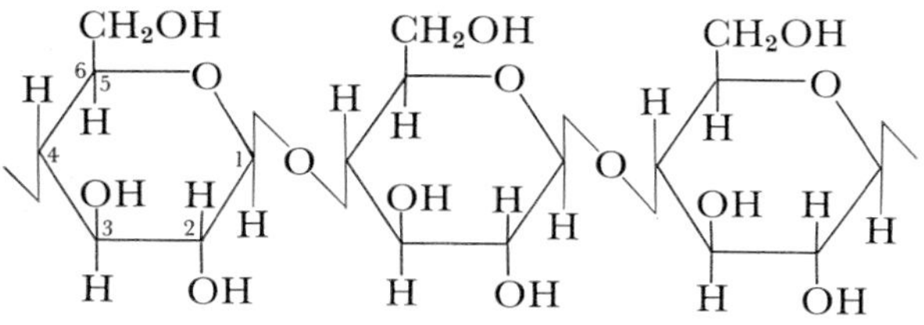

The linear molecules become associated for parts of their length in an ordered, parallel arrangement, held together by hydrogen bonds between the hydroxyl groups of the different chains and by Van der Waals forces, forming crystalline micelles that give well defined X-ray diffraction photographs.

Cellulose is insoluble in water and in any solvents likely to be met in the environment, though it is broken down by an enzyme, cellulase. Its tensile strength increases with its crystallinity though this reduces its elasticity. It is easily identified by its X-ray photograph and its chemical reactions.

2. *Chitin*

Chitin, which also occurs in the cell wall of most fungi is a structural material in many invertebrates, though it has never been found in vertebrates. It is a long chain polymer of *N*-acetylglucosamine, closely resembling cellulose except that the hydroxyl at position 2 in the D-glucose residue in cellulose is replaced, in chitin, by the acetamido ($—NHCOCH_3$) group. The macromolecule contains about 2000 residues.

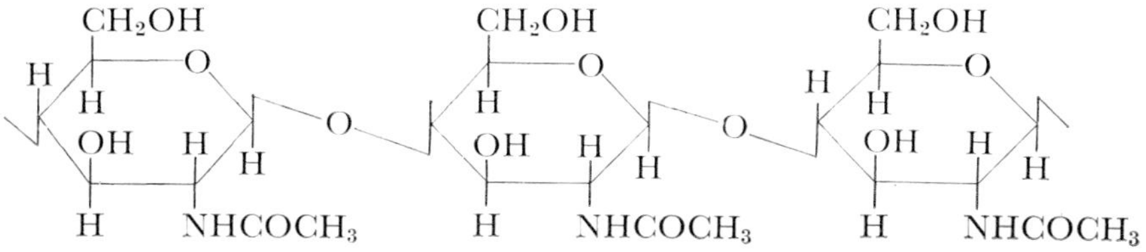

The physical properties of chitin resemble those of cellulose, though it is still more insoluble and unreactive. It is broken down by chitinase which reduces chitin to chitobiose and by chitobiase which splits chitobiose to *N*-acetylglucosamine. Chitin, too, becomes ordered to form crystallites with

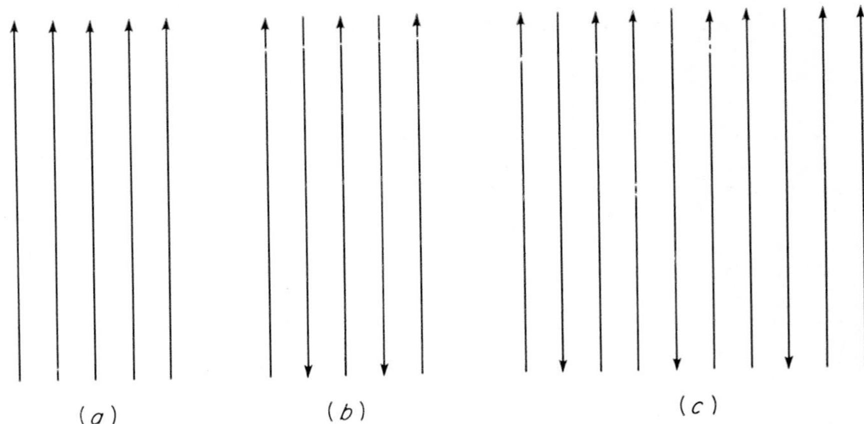

Fig. 1.1. The arrangement of the chain molecules in α-, β-, and γ-chitin (after Rudall, 1965). (*a*) β-chitin (*b*) α-chitin (*c*) γ-chitin.

hydrogen bonds between the acetamide side groups (Darmon and Rudall, 1950). In structural materials chitin is almost invariably mixed with other substances, generally proteins, to which it becomes firmly bound by covalent links (Hackman, 1960). From X-ray studies chitin has been found to exist in three different crystalline forms: α-, β-, and γ. α-Chitin is the most commonly occurring form. Rudall (1965) suggested for the three forms the molecular arrangements shown in Fig. 1.1. The chain molecules of chitin are asymmetrical and when they are organized into fibres all the molecules may have an exactly similar orientation, as in β-chitin (parallel arrangement), or some may have their orientation reversed through 180° as in α- and γ-chitin (antiparallel arrangement). β-Chitin can be converted to α-chitin by treatment with 6NHCl which brings about a contraction of the material to about half its length. Rudall suggests that it causes the molecules to become folded on themselves (Fig. 1.2). γ-Chitin may result from the chains undergoing a double bend (Fig. 1.3).

3. *Mucus and allied substances*

There are also a variety of complex polysaccharides more or less related to cellulose and chitin, which though they are composed of long-chain molecules, do not always have the mechanical properties usually associated with structural materials, for they generally form plastic masses with little strength.

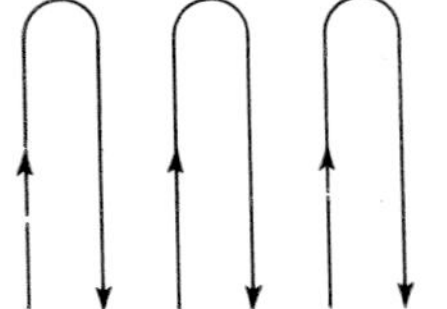

Fig. 1.2. Conversion of β-chitin to α-chitin (after Rudall, 1965).

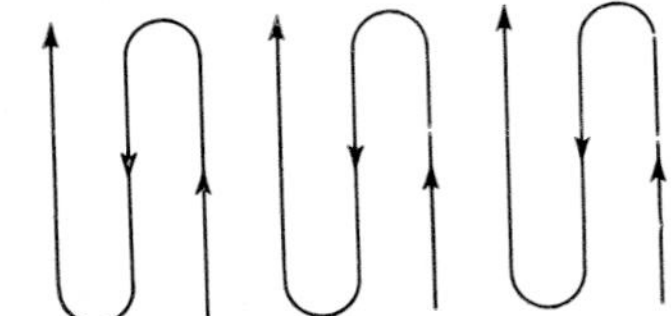

Fig. 1.3. Double-folding of β-chitin to produce γ-chitin (after Rudall, 1965).

But since these polysaccharides may be associated with structural proteins and may in part determine the properties of the complex structure of which they form part, they warrant consideration. As mucins they are also components of mucus, which is a viscous secretion over the surface of cells or epithelia, and as such is an excellent lubricant. It is often found at places in or at the surface of animals where friction is likely to develop. It is able also to absorb and hold considerable quantities of water so that it may maintain about an animal covered by mucus, a layer of water that protects the animal from desiccation. Mucus also has antifreeze properties (Needham, 1965) which may give protection against damage by low temperatures. It may also be used as a cement to bind together foreign particles to form cases or tubes in which animals may live. Mucus is used, particularly at the surface of the lower invertebrates, for purposes which typical structural materials fulfil in the higher invertebrates and vertebrates. Mucus deserves brief consideration, therefore, not only for the uses to which it is put, but also to see if some typical structural materials represent a chemical evolution from the components of mucus.

In the majority of mucins, the polysaccharide is associated with protein or polypeptide. In mucoproteins, protein is the dominant component; in mucopolysaccharides the polysaccharide is dominant. Although in the majority of cases the polysaccharide contains amino sugars, not all viscous secretions do so; often the sugar is sulphated or has the terminal CH_2OH oxidized to carboxyl to produce acid mucopolysaccharides (Kent, 1964; Mathews, 1965; Jackson and Bailey, 1968).

The following are the main mucopolysaccharides that have been isolated from mammalian skeletal tissues:

(1) Hyaluronic acid, with repeating units of D-glucuronic acid + 2-acetamido-2-deoxy-D-glucose.

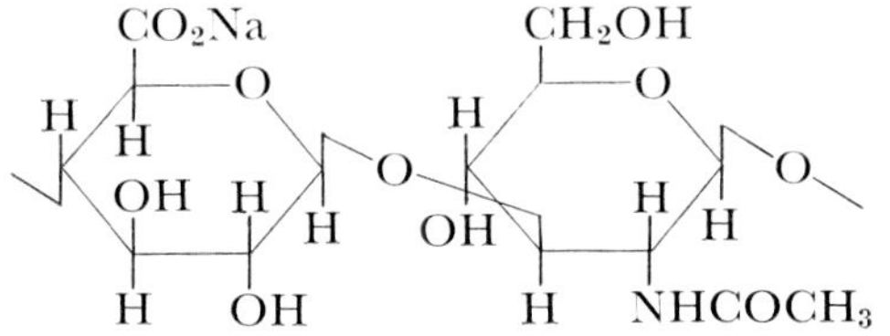

CO_2Na CH_2OH
O O
H H
H
O O
OH H H
H OH H
H OH H $NHCOCH_3$

(2) Chondroitin, with repeating units of D-glucuronic acid + 2-acetamido-2-deoxy-D-galactose.

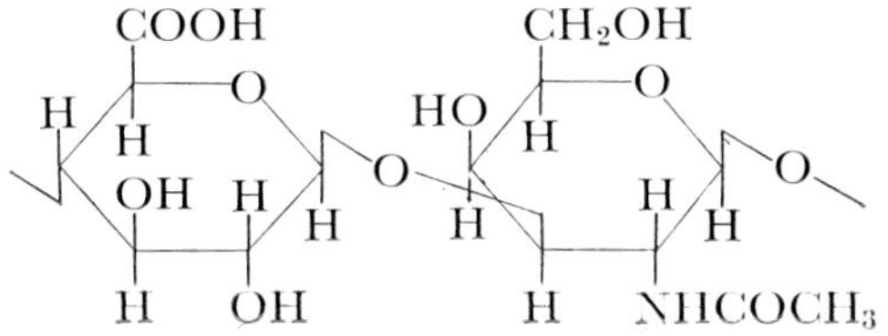

COOH CH_2OH
O O
H HO
H H
O O
OH H H
H H H
H OH H $NHCOCH_3$

(3) Chondroitin 4-sulphate or chondroitin sulphate A (CS-A) with repeating units of D-glucuronic acid + 2-acetamido-2-deoxy-4-O-sulpho-D-galactose.

OSO_3Na
CO_2Na CH_2OH
O O
H H H
O O
OH H H H H
H H
H OH H NHAc

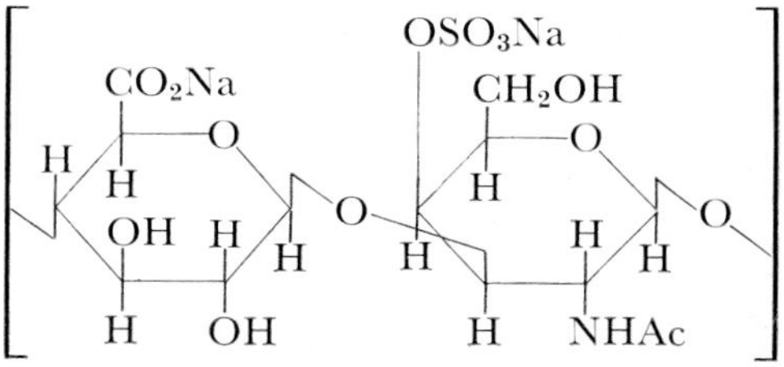

(4) Chondroitin 6-sulphate or chondroitin sulphate C (CS-C) with repeating units of D-glucuronic acid + 2-acetamido-2-deoxy-6-O-sulpho-D-galactose.

COONa CH_2OSO_3Na
O O
H H OH H
O O
OH H H H H
H H
H OH H NHAc

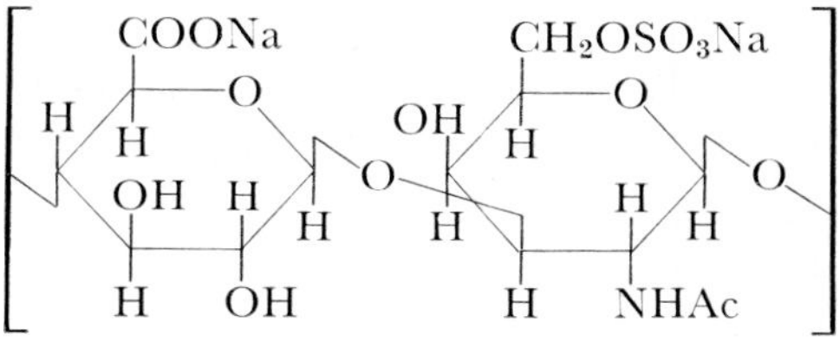

(5) Dermatan sulphate or chondroitin sulphate B (DS) with repeating units of L-iduronic acid + 2-acetamido-2-deoxy-4-O-sulpho-D-galactose.

OSO_3Na
H CH_2OH
O O
H CO_2Na H
O O
OH H H H H
H H
H OH H NHAc

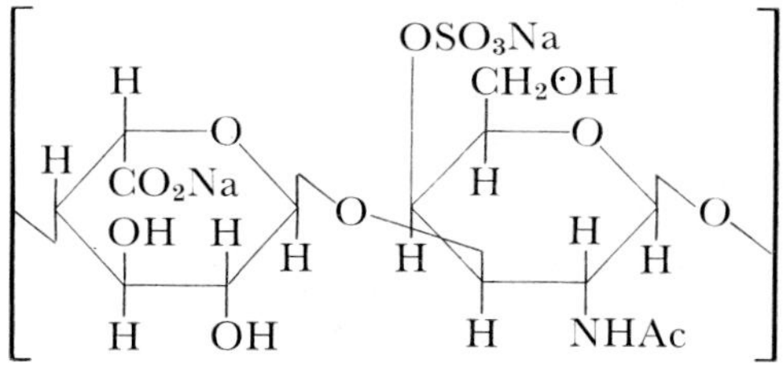

(6) Keratan sulphate (KS) with repeating units of D-galactose + 2-acetamido-2-deoxy-6-O-sulpho-D-galactose.

CH_2OH CH_2OSO_3Na
O O
HO H H H
O
O H H H OH H H
H
H OH H NHAc

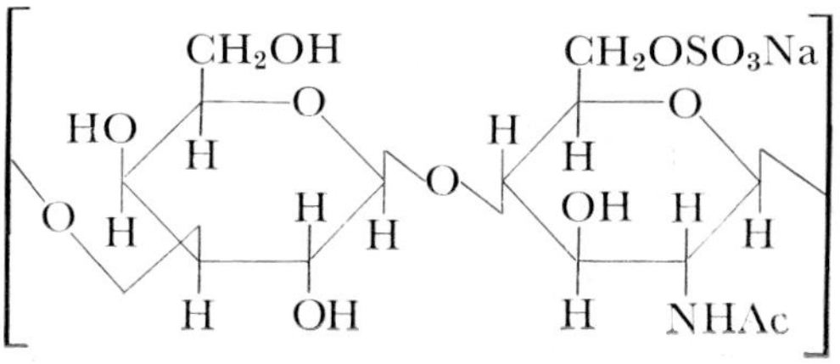

Mucopolysaccharides with some repeating units containing two or more ester sulphates are described as sulphated, e.g. sulphated chondroitin sulphate C (SCS-C) or sulphated keratosulphate (SKS).

Little is known as yet about the chemistry of invertebrate mucopolysaccharides though they often react in the same way as mammalian mucopolysaccharides to specific polysaccharide dyes. Some mucopolysaccharides can be digested by the enzyme hyaluronidase.

4. *Collagen*

Collagen has been detected in almost all classes of multicellular animal from sponges to mammals. In the very few classes in which its occurrence has not been recorded this is likely to be because it has not been looked for. It is a fair assumption that collagen occurs in all multicellular animals. Collagen occurs as extracellular fibrils or fibres which in the electron microscope usually have a banded structure of period approximately 640 Å. Smith, J. W. (1965) has shown, however, that banding is unlikely to be detectable in fibrils less than 100 Å in diameter. In young collagen fibrils the banding period often is only 210 Å (Randall, 1953). The arrangement of the fibres in various tissues shows a relationship to the stresses to which the tissues are subjected. In any particular tissue the fibres are of rather uniform diameter, but they vary in diameter from tissue to tissue, from about 100 Å to 2000 Å or more. Their length is not known, but is certainly considerable, possibly even of the order of millimetres; nor is it known how each fibre ends (Harkness, 1968).

Though it might be rational to consider first the properties of collagen as it is found in the sponges, the most primitive of multicellular animals and subsequently consider the modfications that occur at various levels of animal organization, this unfortunately is not possible. The bulk of work on collagen has been carried out on mammalian material and information on invertebrate collagens is very incomplete. Mammalian collagen must therefore be used as the standard of comparison for other collagens.

(*i*) *Mechanical properties*. Measurements of the mechanical properties of animal skeletal materials have so far been carried out mainly *in vitro* with little or no consideration of the conditions under which the material functions in the animal. It is not possible to measure the tensile strength of a single fibre of collagen so measurements are usually made on tendons where the collagen fibres are all of approximately the same diameter and orientated parallel to each other. On the basis of weight of material, tendon is about as strong as steel, but on the basis of tensile strength per unit cross-sectional area, the strength of collagen is low compared with that of steel, though collagen is one of the strongest of animal fibres (Table 1.1). The tendons of smaller animals may be weaker per unit cross-sectional area than those of larger animals. Their strength, however, is more than enough to withstand the maximum isometric tension that the muscles attached to them can develop. Rupture does occasionally occur but only under exceptional conditions of rapidly applied tensions (Harkness, 1968; Elliot, 1965). Collagen has a high modulus of elasticity (140 kg/mm^2). High tension leads to about 10 per cent

Table 1.1 Mechanical properties of structural materials

Material	Young's Modulus kg/mm²	Tensile strength kg/mm²	% Extension at break	Reference
Cellulose	600–1000	50–60	6–8	Locke, 1964
Chitin dry	4500	58	1·3	Locke, 1964
Purified chitin dry		9·5		Thor and Henderson, 1940
Purified chitin wet		1·8		Thor and Henderson, 1940
Mesogloea	0·0003			Alexander, 1962, 1964
Insect cuticle	960	9·6	2–3	Jensen and Weis-Fogh, 1962
Collagen	140	56	10	Locke, 1964
Cartilage		1·4		Koch, 1917
Bone	1800	10·5		Currey, 1962
Elastin	0·61	1·02	127	Bergel, 1961 Burton, 1954
Resilin	0·064	0·3–0·4	300	Weis-Fogh, 1961a
Abductin	1·27			Kelly and Rice, 1967
Silk fibroin	700–1600	35–60	10–35	Locke, 1964 Lucas *et al.*, 1958
Wool keratin	100–300	15–20	30–40	Locke, 1964
Steel	20,000	100		Kaye and Laby, 1948

extension before the fibres rupture. If extension is not greater than 2 per cent, the fibres return to their original length, but greater extensions lead to permanent set and beyond 10 per cent the fibres break.

(*ii*) *Chemical properties*. Although traditionally collagen is defined as a protein that, on heating in water, dissolves to give gelatin, it is now much more precisely defined by its characteristic wide angle X-ray diffraction picture with a principal axial repeat of 2·8–2·9 Å and an equatorial periodicity of 11 Å, or more, depending on the state of hydration of the collagen (Rudall, 1968).

Although slight chemical differences do occur between collagens from different mammals, and from different sites within the body in any one species, the variations are small (Table 1.2) and do not affect the wide angle X-ray photograph, though differences between collagens from different sources may become apparent in small angle X-ray photographs (Bear, 1952).

To aid comparison of the chemical compositions of the different structural proteins the amino acids are always tabulated in the same order. Amino acids have molecular weights in the same general range so that in analyses proportions on a weight basis are not grossly different from those on a molar basis; further, their average molecular weight is not far from 100 so that the actual figures (ignoring decimal points) are not greatly different. It is therefore possible to make meaningful comparisons between protein analyses given in different forms, e.g. residues per 1000 residues, grammes per 1000 grammes, moles per 1000 moles, so that it has not been thought worth while to convert all the analyses to the same form.

The amino acid composition of mammalian collagen is highly characteristic and completely different from that of any other type of protein. Glycine, the simplest amino acid, accounts for almost exactly 1 residue in three. Alanine, the next simplest, makes up one residue in nine. The imino acids, proline and hydroxyproline together account for two residues in nine so that collectively glycine, alanine, proline and hydroxyproline account for two out of three residues. The remaining third of the residues is made up of varying amounts of fourteen of the other amino acids commonly occurring in proteins, only cystine and tryptophan being absent (Eastoe, 1967).

There is a slight preponderance of basic over acidic side-chain groups so that collagen is a basic protein. The total content of residues with hydroxyl groups is nearly 1 in 6 and together with acidic and basic side chains greatly exceeds the number of lipophilic side chains, so that collagen is strongly hydrophilic (Eastoe, 1967).

Hydroxyproline and hydroxylysine were thought to be confined to collagen and the related protein elastin, but recently free and protein-bound hydroxyproline have been found in various plants (Hulme and Arthington, 1952; Radhakrishnan and Girri, 1954; Piez *et al.*, 1956; Lamport and Northcote, 1960). Viswanatha and Irreverre (1960) found hydroxylysine in trypsin and

Table 1.2 Composition of Mammalian Collagen. Number of residues of each amino acid per 1000 total residues

	Man			Rat	Ox
	Skin[a]	Uterus[b]	Bone[c]	Tendon[d]	Tendon[e]
Alanine	114·5	95·9	113·5	99·3	97·8
Glycine	324·4	337·1	319·0	351·0	336·5
Valine	24·5	21·6	23·6	22·5	21·5
Leucine	24·8	24.5	25·5	22·2	27·3
Isoleucine	10·4	10·8	13·3	13·2	14·5
Proline	125·1	108·2	123·4	123·0	144·2
Phenylalanine	12·6	12·3	13·9	14·3	15·2
Tyrosine	3·5	2·2	4·5	5·4	4·8
Serine	36·9	41·9	35·9	27·8	29·5
Threonine	18·3	20·6	18·4	19·1	18·9
Cystine	tr.	tr.	—	—	—
Methionine	7·0	5·8	5·3	5·8	3·6
Arginine	49·0	49·0	47·1	46·5	45·4
Histidine	5·4	5·4	5·8	3·3	6·5
Lysine	26·6	24·9	28·0	35·6	22·4
Aspartic acid	47·2	52·5	47·0	47·1	48·0
Glutamic acid	77·7	73·6	72·2	73·7	71·4
Hydroxyproline	90·9	108·6	100·2	90·4	83·4
Hydroxylysine	5·9	5·1	3·5	—	9·3

[a] Fleischmajer and Fishman, 1965.
[b] Harding, 1963.
[c] Eastoe and Leach, 1958.
[d] Neuman, 1949.
[e] Nordwig *et al.*, 1961.

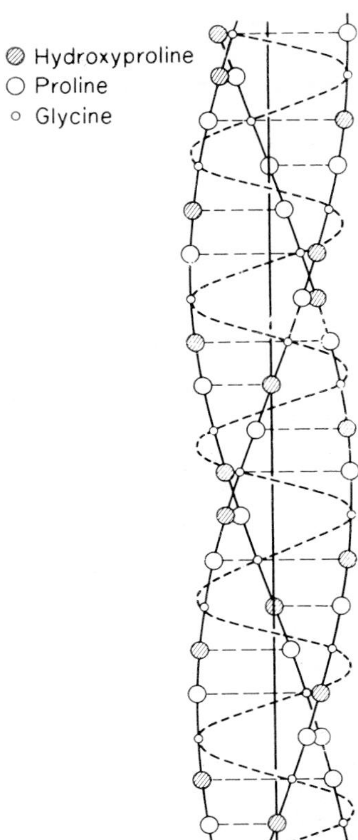

Fig. 1.4. Schematic picture of the helical co-coiling of three extended peptide chains in the collagen molecule (from Schmitt, 1959).

chymotrypsin and it has also been found occasionally in wool. The chemical composition of gelatin is very similar to that of the collagen from which it is derived.

The unit of collagen structure is now held to be the tropocollagen molecule which is found in solutions of collagen. It has a molecular weight of about 300 000, a length of 2800 Å and a diameter of 11–14 Å. It is the diameter of the molecule which is responsible for the 11 Å equatorial periodicity in the X-ray diffraction photograph. Each molecule is formed from three helically coiled polypeptide chains which are twisted together to give a coiled coil configuration (Fig. 1.4) with 2·8–2·9 Å between the residues; this is responsible for the axial repeat in the X-ray photograph (Rich and Crick, 1961).

It is possible, by denaturation, to separate the tropocollagen molecules into their component chains, the α-chains, each composed of 1000 residues. In mammalian collagen there are generally believed to be two similar α_1 chains and a differing α_2 chain per molecule. Although the chemical composition of the α_1 and α_2 chains is of the same general pattern as that of whole collagen, they differ from one another in detail, particularly with regard to particular amino acids (Table 1.3). Bornstein and Piez (1965) have evidence that all three α-chains may be different and this is certainly so in collagen from the cod (Piez, 1964).

Petruska and Hodge (1964) suggested that the two α_1 chains were made up of five identical subunits linked end to end by non-peptide linkages while the α_2 chain is made up of seven subunits. Petruska (see Hodge, 1967) has now suggested that calf skin collagen, at least, may have a 5:5:6 rather than a 5:5:7 subunit structure and that within the α_1-chains there may be a sequence of type *a b a b a* where *a* and *b* are similar but not identical subunits while the α_2-chain has a sequence of *c d c d c d* subunits.

When first formed the α-chains probably are linked together only by intramolecular hydrogen bonds and in native collagen they are also linked by similar bonds to adjacent molecules. There is evidence however, that, with time, intra and intermolecular covalent bonds may develop which make the collagen more resistant to solution in acids and to disruption by denaturation than it would be if only electrovalent bonds were present.

On denaturation of collagen solutions the hydrogen bonds are broken and most of the α-chains separate but any covalent bonds that have been formed are not broken and some of the chains remain linked together. Two chains linked together are called β-units; γ-units contain three linked chains. β-units may contain two α_1 chains (β_{11}-units) or an α_1- and an α_2-chain (β_{12}-units) or two α_2-chains (β_{22}-units). This latter could only result from

Table 1.3 Composition of subunit components of human skin collagen. Number of residues of each amino acid per 1000 total residues

	α_1[a]	α_2[a]
Alanine	115	105
Glycine	333	337
Valine	20·5	33·3
Leucine	19·5	30·1
Isoleucine	6·6	14·8
Proline	135	120
Phenylalanine	12·3	11·7
Tyrosine	2·1	4·6
Serine	36·8	35·1
Threonine	16·5	19·2
Methionine	4·9	5·2
Arginine	50	51
Histidine	3·0	9·7
Lysine	30·0	21·6
Aspartic acid	43	47
Glutamic acid	77	68
3-hydroxyproline	0·8	0·9
4-hydroxyproline	91	82
Hydroxylysine	4·4	7·6

[a] Bornstein and Piez, 1964.

the survival of an intermolecular linkage. Similarly, γ-units may be made of α_1-chains, α_2-chains or combinations of α_1 and α_2-chains, some of the combinations only being possible through the survival of intermolecular linkages.

The exact nature of the covalent bonds and how they come to be formed is not certain. One possible covalent bond may result from the conversion of some of the lysine residues to aldehydes and then two of these aldehyde groups in separate chains combining to form a cross link through an aldol-type condensation (Piez, 1967). Such a reaction is a form of autotanning. Since these covalent linkages develop gradually, the properties of collagen within an individual change gradually with age (see Sinex, 1968, for a consideration of collagen changes and ageing).

Collagen fibres suddenly shrink to about one-third their original length when slowly heated in water. The temperature, T_S, at which this shrinkage takes place is the temperature at which the three chain collagen molecule becomes converted to a random-coiled gelatin structure. In mammalian collagen T_S is 60–65°C, although it varies slightly with the age of the material; T_S from rat tail tendon collagen being $62{\cdot}5 \pm 0{\cdot}77$ for material from 60-day-old rats, but $65{\cdot}5 \pm 0{\cdot}25$°C in that from 950-day-old rats (Chvapil and Jenšovský, 1963). Shrunken collagen loses its typical X-ray pattern, and its banded appearance in the electron microscope, and is elastic, rather like rubber. Gustavson (1953) suggested that hydroxyproline formed both intra and extra-molecular hydrogen bonds with backbone carboxyl groups and that these played an important part in stabilizing the collagen structure. He found a relationship between the T_S of a collagen and its hydroxyproline content. However, Piez and Gross (1960) showed that in fish collagens, with lower T_S than mammalian collagens and fewer hydroxyproline residues, the number of hydroxy groups present was not lower because the deficiency of hydroxyproline was compensated for by extra serine and threonine residues. They found a much closer relationship between T_S and total imino acid content (proline plus hydroxyproline) (Figs. 5.6(*a*)). It is now thought that the planar pyrrolidine ring, which occurs in both types of residue, controls the geometry of the collagen structure. Free rotation about the C—N bond in the back bone portion of these residues cannot take place since the C—N unit is incorporated into the ring and regions of the collagen chain containing imino residues therefore tend to maintain the triple helix structure against distorting forces such as heat. The temperature, T_D, at which vertebrate collagen in solution undergoes denaturation to free polypeptide chains is also related to total imino acid content and is 25–30°C lower than T_S (Fig. 5.6(*b*)). In mammalian collagen, T_D is approximately 39°C (Burge and Hynes, 1959). There is evidence of a relationship between T_D of an animal's collagen and the normal environmental temperature of the collagen.

Hodge *et al.* (1960) and Rubin *et al.* (1963, 1965) have obtained from tropocollagen, by treatment with proteolytic enzymes, groups of peptides that they have called teleopeptides that differ from collagen in being tyrosine rich and lacking hydroxyproline and hydroxylysine. Hodge *et al.* suggest that the teleopeptides extend beyond the triple helical body of the tropocollagen molecule as randomly coiled appendages. They may contain the loci of intramolecular covalent linkages since their removal considerably reduces the number of β units found in solutions of denatured tropocollagen and this is consistent with the suggestion that tyrosine may play a part in cross-linkages (Bensusan, 1965). The detachment of teleopeptides also reduces the capacity of tropocollagen to reaggregate to form native-type fibres; they are, therefore, of importance in fibrogenesis and may be implicated in the region of overlap (see p. 20) between the molecules (Bailey, 1968). Various other components seem regularly to be associated with collagen. Some nucleic acid (Pouradier and Accry, 1962) and reactive aldehydic substances (Pouradier and Venet, 1952*a* and *b*; Landucci *et al.*, 1958) appear to be firmly fixed to it. All collagens contain hexoses, e.g. vertebrate collagens have 0·5–1·5 per cent, as glucose and galactose, and these appear to be an integral part of the molecule, where they may be involved in ester linkages with acid groups of the protein chains.

Trypsin and pepsin, though they split off the teleopeptides, do not otherwise break up the collagen chains. Collagen is degraded by collagenase obtained from the bacterium, *Clostridium histolyticum*: this breaks linkages along the length of the chain. A new collagen-digesting enzyme has recently been obtained from the epidermis overlying the tail fin, skin, and gills of tadpoles undergoing metamorphosis (Gross and Lapiere, 1962; Eisen and Gross, 1965). It is quite different from *Clostridium* collagenase in its action, as it splits all the chains of the triple helix at one particular place so that the tropocollagen molecule is broken into two unequal pieces, one 25 per cent, the other 75 per cent of the original length. Both pieces retain the triple helix structure. This breakage may sufficiently lower the temperature at which the fragments are denatured so that within the metamorphosing tadpole this happens spontaneously and the individual chains are then digested by other proteolytic enzymes. The rapid reduction in collagen in the mammalian uterus after parturition, with the loss of half the collagen content in about a day, may also be brought about by a special enzyme (Harkness, 1964).

Although most of the amino acids necessary for collagen formation can be manufactured by the animal, some of the residues have to be obtained, preformed, from the food. There is evidence that reduced food intake can affect the properties of collagen. Deyl *et al.* (1968) found that food restriction in rats reduces the number of β_{11} and β_{12} units found in denatured solutions of their collagen. This is believed to be due to a reduction in the number of carbonyl compounds with free carboxyl groups. This reduces the degree of cross-linking between α_1 chains and between α_1 and α_2 chains, and this in turn affects the properties of the collagen.

It has long been known that vitamin C is essential for the formation of collagen and is therefore of great importance in growth and the healing of wounds. Other vitamins also may influence aspects of collagen metabolism but the subject is not fully understood (Gould, 1968).

Although collagen in the adult is relatively inert, as one would expect a skeletal material to be, metabolically it does undergo a slow turnover. This varies with the different tissues, liver collagen having a half-life of 20–30 days and that of muscle 50–60 days, while that of aorta and peripheral nerves is of extreme metabolic stability. In growing animals collagen is continually being resorbed and reformed, particularly in bone. This allows the achievement of the correct size and shape of the bone in relation to the size of the animal.

(*iii*) *Molecular organization.* Electron micrographs of dyed native collagen fibres show a pattern of regularly repeating light and dark bands with a repeat period, D, of approximately 640 Å. This banding is believed to be due to the arrangement of the tropocollagen molecules in the fibres. Within these bands a finer, asymmetrical but repeating, banded pattern can be seen (Fig. 1.5).

Tropocollagen molecules in solution can be reprecipitated to form various structures according to the conditions of precipitation. They may form filaments that are indistinguishable in their banding pattern from native collagen, or they may form differently banded filaments, the main bands having a period of 2800 Å (fibrous long spacing: F.L.S.) and a pattern of fine banding that differs from that of native collagen in forming a pattern that is symmetrical about a mid point between the bands. Other conditions lead to the precipitation of isolated units 2800 Å long (segment long spacing: S.L.S.), with a fine banding pattern that corresponds with that in native collagen. Tropocollagen molecules are approximately 2800 Å long and S.L.S. is believed to result from lateral aggregation of parallel tropocollagen molecules (Fig. 1.6(*a*)). In this arrangement the molecules appear unable to link together end to end to form filaments. In F.S.L. the molecules are thought to have an anti-parallel arrangement (Fig. 1.6(*b*)) and are able to link together to form filaments. In native collagen it has been suggested that the molecules have a staggered arrangement. Each molecule is taken as being approximately $4{\cdot}4D$ long ($D = 640$ Å), and is separated from the molecule directly in front by $0{\cdot}6D$, giving a distance of $5D$ from the beginning of one molecule to the beginning of the next. Each sequence overlaps the adjacent sequence by a distance D, but since there is a gap of $0{\cdot}6D$ between the members of a sequence the head of each molecule overlaps the tail of the adjacent molecule by $0{\cdot}4D$ (Fig. 1.6(*c*)). From this diagram it can be seen that the fibre is divided up into regions of overlap with a periodicity of D (640 Å) which are believed to form the dark staining regions of the fibres and regions containing the gaps between the molecules which produce the light staining bands (Hodge *et al.*, 1965),

The asymmetrical pattern of fine banding is thought to be caused by regions of polar amino acids with large side chains which dye with electron-dense dyes alternating with regions containing a preponderance of imino acids with shorter side chains which do not take up the dye. This results from the summation of the polar and non-polar groups along the five similar subunits in the two α_1-chains with those in the seven similar subunits in the α_2-chain. These form a pattern that can be divided into 4·4 approximately similar segments. This basic banding pattern is clearly seen in S.L.S. where the pattern is in register across all the laterally associated molecules. That

Fig. 1.5. Native rat tail collagen negatively stained with potassium phosphotungstate. The alternating *A* bands (light) and *B* bands (dark) have a periodicity of approximately 640 Å. The tropocollagen macromolecules are distinguishable as thread-like structures which are approximately 15 Å in diameter and which cross each other in both the *A* and *B* bands. Marker represents 1000 Å (Grant *et al.*, 1967).

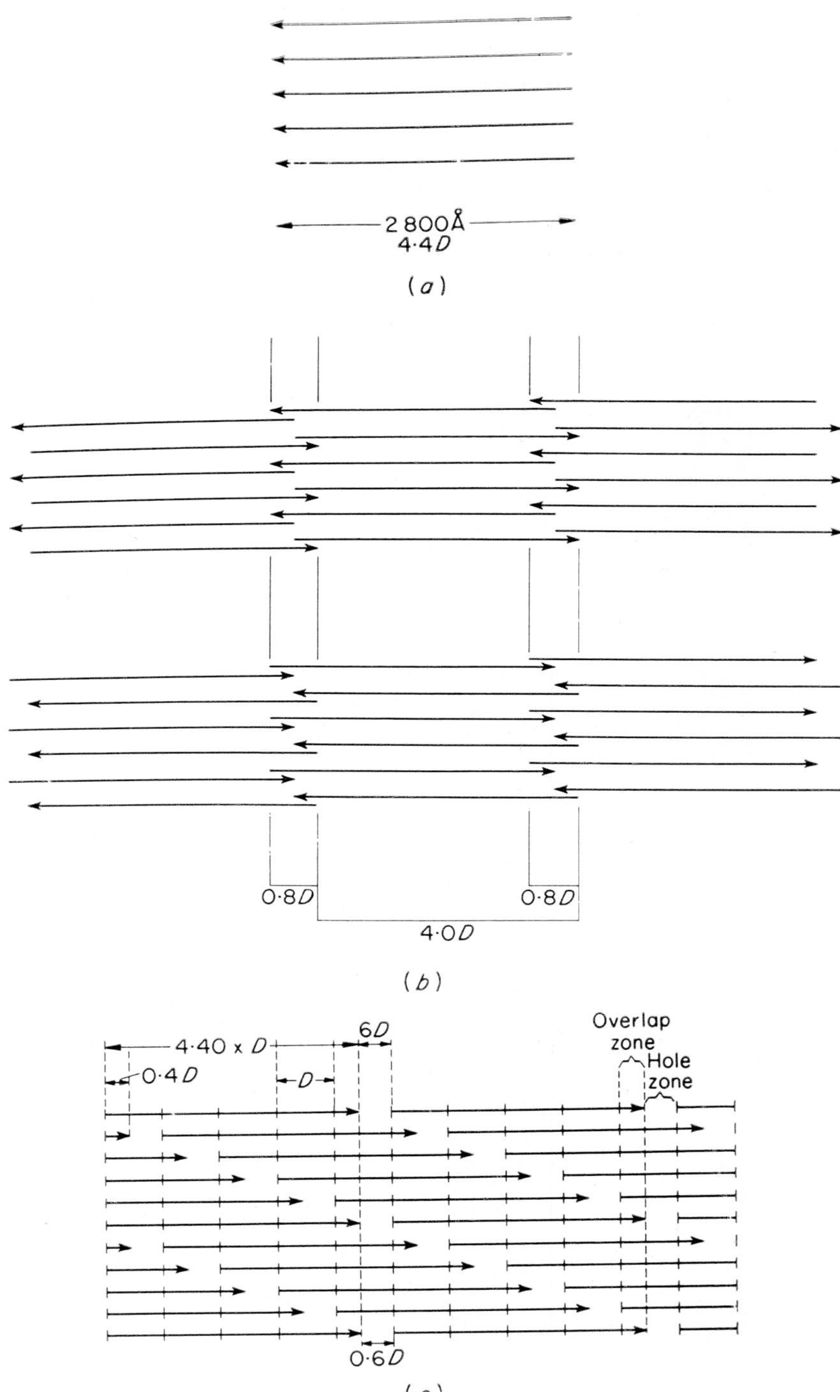

Fig. 1.6. Diagrams to show: (*a*) Arrangement of tropocollagen molecules in S.L.S. (*b*) Arrangement of tropocollagen molecules in F.L.S. (*c*) Arrangement of tropocollagen molecules in native type collagen fibrils $D = 640$ Å. Length of tropocollagen molecule $= 4{\cdot}4D = 2800$ Å. (After Hodge *et al.*, 1965.)

each segment of the pattern is essentially similar is shown by the capacity of S.L.S., under certain conditions, to form as outgrowths from native collagen filaments. The outgrowths all have the same orientation and are positioned with each segment aligned with a segment of the collagen filament and with their banding in register with that of the inter period banding of the collagen filament (Fig. 1.7). In the staggered arrangement of the molecules in the native collagen fibre each segment is in register with similar but differently positioned segments in the adjacent molecules, the banding patterns coincide and produce the fine banding pattern in the fibre (see Hodge (1967) for a full account). Stretching collagen fibres to their maximum extension produces about 9 per cent elongation in the 640 Å spacing but produces only a slight increase in the residue spacing in X-ray photographs, indicating that stretching results in molecular rearrangement rather than in extension of the molecular chains (Cowan *et al.*, 1955).

Although this suggested staggered arrangement of tropocollagen molecules has been generally accepted for the last decade or more, recently Grant *et al.* (1967) and Cox *et al.* (1967) have proposed an alternative hypothesis. They suggest that each molecule has five main bonding regions 280 Å apart separated by four regions where intermolecular attachments are weak. The aggregation of molecules to form a fibre is essentially a random process with a bonding region on one molecule able to link with any of the bonding zones in an adjacent molecule (Fig. 1.8), the bonding regions being responsible for light bands in phosphotungstate stained fibres while in the regions of reduced linkage, the stain can penetrate and produce dark bands. McGavin (1964) and Ross and Bendit (1961) also have offered a different explanation from that of Hodge *et al.* (1965).

Structures proposed for collagen are only two-dimensional and little consideration has been given to three-dimensional arrangements. X-ray photography suggests that within a fibre the molecules are either arranged in a series of circles of increasing diameter around three central molecules or more probably in a spiral as if formed by the rolling up of a sheet of molecules (Ramachandran, 1967).

Collagen fibres, in animal tissues, are always embedded in a ground substance containing acid mucopolysaccharide. Fitton-Jackson (1965) has assembled evidence for an organized structure in the ground substance. How far there is chemical bonding between the ground substance and the collagen fibres is not known, though particular condensations of ground substance round the fibres is sometimes seen. The chemistry and properties of the ground substance vary in the different tissues, from the soft, pliable connective tissues round organs to the firm tissues of cartilage.

Both collagen and ground substance are secreted by the same mesodermal cells, the fibroblasts. There is, however, increasing evidence that epidermal and even endodermal cells also can produce collagen. Fitton-Jackson (1968) has considered various aspects of the morphogenesis of collagen.

Collagen is identified histologically by its reactions with dyes or in the electron microscope by its banded appearance. Histochemically, its presence can be confirmed by its digestion from sections by collagenase. Where

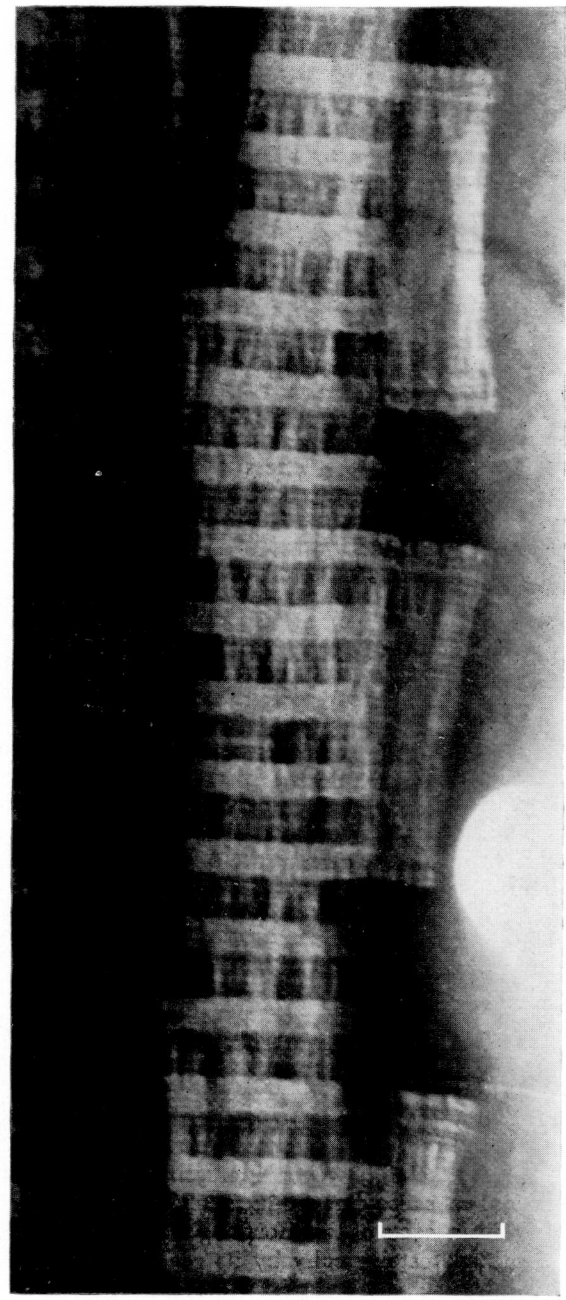

Fig. 1.7. Reconstructed native collagen fibrils with S.L.S. crystallites deposited on them, negatively stained with potassium phosphotungstate. Marker represents 1000 Å (Cox *et al.*, 1967).

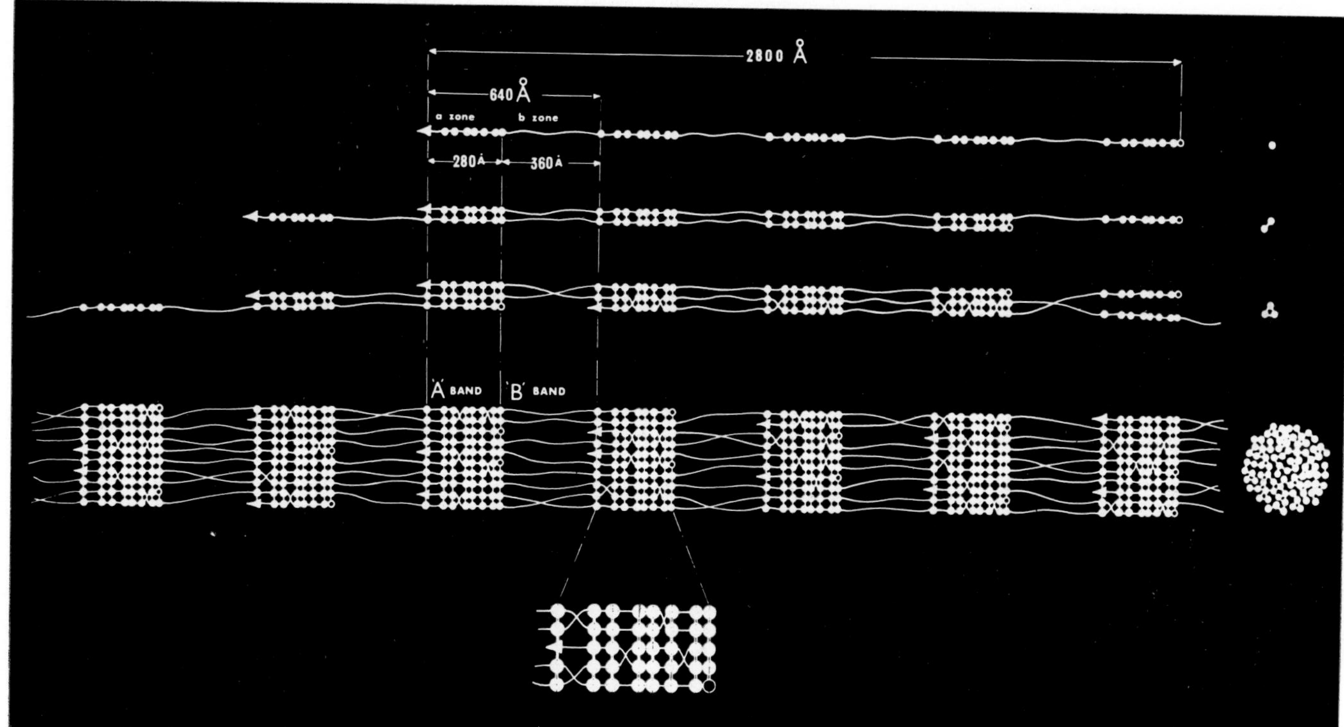

Fig. 1.8. Diagram to illustrate the formation as suggested by Cox *et al.* (1967) of a native collagen fibril with 640 Å periodicity from tropocollagen molecules 2800 Å in length. The tropocollagen molecule is divided into five main bonding or *a* zones (approximately 280 Å) separated by four main non-bonding or *b* zones (approximately 360 Å). Some of the bonding rates within the main bonding zones are represented by asymmetrically arranged dots—by virtue of this asymmetry, the tropocollagen molecules are polarized (from Cox *et al.*, 1967).

sufficient material is obtainable, X-ray analysis is used to identify the material with certainty.

5. *Elastin*

Collagen fibres are differently arranged in different tissues. If they are in a loose feltwork the tissue can undergo a limited deformation which tends to remain until new stresses again alter it. In vertebrates a return to the resting arrangement of the tissue is aided by the presence in these tissues of extracellular elastic fibres which, on the removal of any stress, contract and pull the tissue back into its original shape. Similar elastic material is particularly abundant in parts of the heart, in the walls of arteries and in tracheae and bronchi that undergo rhythmic contractions and in the *ligamentum nuchae* which helps herbivorous cattle to hold up their heads.

In the light microscope elastic fibres are highly refringent and slightly yellow and they frequently branch and fuse to form networks. In the larger arteries the elastic material is present as a fenestrated membrane composed of superimposed, irregularly perforated sheets each 2–3 microns thick. Early electron microscopy studies suggested that all elastic structures were formed

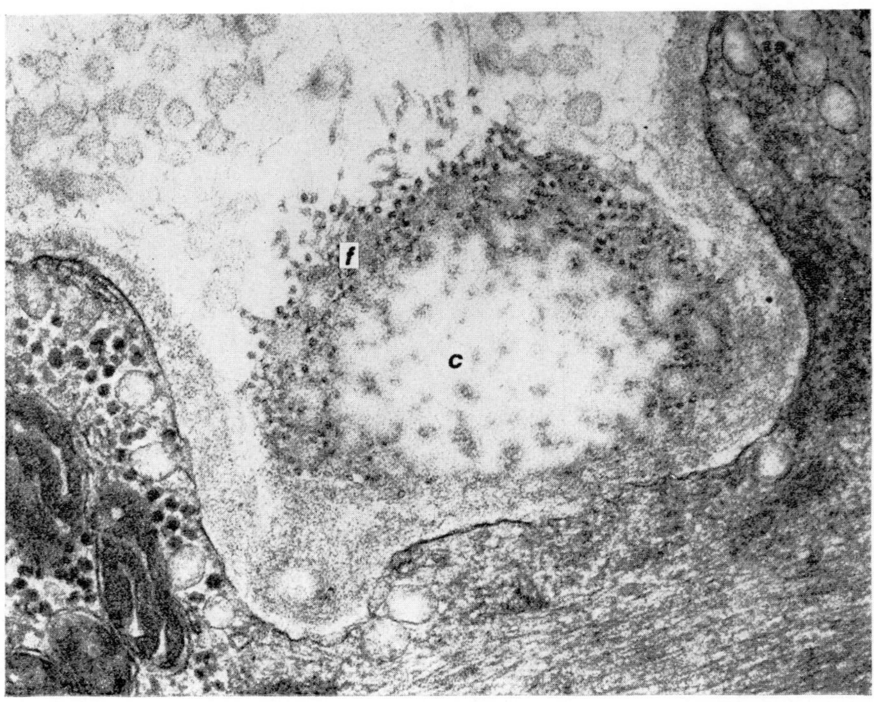

Fig. 1.9. T.S. of rabbit ear artery elastin. *c* = amorphous core of elastin fibre, *f* = peripheral fibres. (H. Y. Elder.) [×68 700]

of fine homogeneous filaments 30–100 Å in diameter but more recent work (Greenlee *et al.*, 1966) shows that they are formed of a core of amorphous material surrounded by tubular appearing microfibrils about 110 Å in diameter (Fig. 1.9). In development the microfibrils form an aggregate structure before any amorphous component is visible (Greenlee *et al.*, 1966; Ross, 1968) and this suggests that the fibrils may play a role in determining the shape of elastic structures. The elastic component is held to be the amorphous core. Whether or not the peripheral fibrils are also elastic is not yet known, although if they are not they must accommodate in some other way the changes in length of the core.

(*i*) *Physical and mechanical properties.* Mammalian elastic material can be extended 100–150 per cent of its resting length and show complete recovery. Although it is less strong than collagen (Table 1.1) collagen fibres associated with elastic fibres in various tissues prevent them from being stretched beyond their elastic limit. While unstretched elastic fibres are isotropic in polarized light, they become positively birefringent with respect to their axes when fully stretched (Schmidt, 1924). X-rays give no indication of internal structure, neither in stretched nor in unstretched material. This lack of crystallinity in mammalian elastic material and its elastic behaviour give every indication that it has a rubber-like organization. Weis-Fogh (1965) pointed out the difficulties in constructing an isotropic, elastic material from fibrils and this newer picture of the amorphous nature of the elastic core is therefore more satisfactory.

(*ii*) *Chemical composition.* Mammalian elastic material differs from collagen in its high resistance to acids and alkalis. Formerly it was thought to contain a single protein, elastin, which resembled collagen in that about one third of all the residues were glycine but it contained more alanine, valine, and proline, only a very little hydroxyproline, and no hydroxylysine.

Ross and Bornstein (1969) managed to separate the two components of elastic material from the ligamentum nuchae of foetal calves. The amino acid composition of the amorphous core is virtually identical with that previously found for elastin (Table 1.4). The peripheral fibrils, which can be separated from the core either by treatment with proteolytic enzymes or by reduction of disulphide bonds with dithioerythritol in 5 M guanidine, are rich in polar, hydroxy, and sulphur-containing amino acids, and contain less glycine, valine, and proline than the amorphous core.

Analyses of elastin from human aorta and from various sites in the ox, purified by autoclaving, differ from those of similar material purified by extraction with 0·1 N NaOH (Gotte *et al.*, 1963, 1965). If the same method of purification is used there is little difference chemically between elastins from different sites in the same animal (Table 1.5). Gotte and his co-workers suggest that elastin is composed of protein molecules in a glycoprotein cement which is removed by the alkaline treatment. A true rubber-like material retains its elasticity even under conditions of dehydration, its elasticity being entirely dependent on its single-phase molecular structure which forms a self-lubricating system under stress. Elastin becomes brittle when dehydrated and must therefore be a diphasic system with the elastin

Table 1.4 Amino acid composition of the elastic fibre and its components. Residues per 1000 total residues. (Ross and Bornstein, 1969.)

		Amorphous component		Microfibrils	
	Elastic fibre	After enzymatic digestion	After DTE[a]	Enzymatic digest	Alkylated[b]
Alanine	212	223	223	82·6	58·9
Glycine	305	324	324	142	120
Valine	130	140	135	69·7	54·1
Leucine	60·3	67·8	61·1	65·5	57·2
Isoleucine	29·6	26·2	25·5	43·8	45·2
Proline	90·2	110	120	73·5	70·4
Phenylalanine	34·4	30·9	30·1	32·8	32·1
Tyrosine	9·4	8·2	7·1	27·6	30·0
Serine	16·1	9·4	9·9	52·8	58·9
Threonine	13·8	8·6	8·9	47·3	55·1
Cystine $\frac{1}{2}$	10·2	5·0	4·1	56·3	8·0
Methionine	0·7	—	—	13·0	15·8
Arginine	11·1	5·2	5·4	42·3	45·2
Histidine	3·7	0·7	0·6	11·5	14·2
Lysine	11·9	7·6	7·4	36·7	36·9
Aspartic acid	16·1	5·4	6·4	92·5	114
Glutamic acid	24·6	14·7	15·0	98·3	111
Hydroxyproline	16·4	11·3	10·7	1·7	n.c.
Hydroxylysine	—	—	—	—	—
Isodesmosine/4	3·9	4·1	3·3	—	—
Desmosine/4	5·5	6·5	4·6	—	—

[a] Obtained as a residue after dissolving the microfibrils with dithioerythritol.
[b] Dissolved by reduction and alkylation.

molecules embedded in a lubricant whose properties depend upon its state of hydration. This lubricant is probably the glycoprotein cement found by Gotte *et al.*

Thomas *et al.* (1963) suggest that the network of elastin protein chains is linked together by two hitherto undescribed amino acids, desmosine and isodesmosine. These substances are isomers consisting of a pyridinium ring with four side chains each terminating in a carboxyl group and an α-amino group.

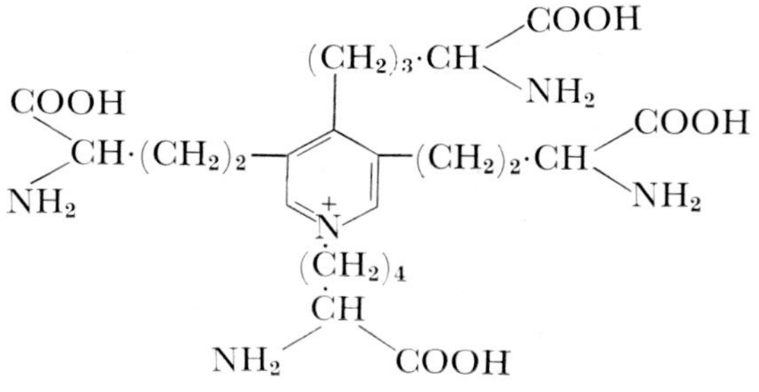

COOH
$(CH_2)_3 \cdot CH$
NH_2
COOH
COOH
$CH \cdot (CH_2)_2$
$(CH_2)_2 \cdot CH$
NH_2
NH_2
N^+
$(CH_2)_4$
CH
NH_2 COOH

I-Desmosine

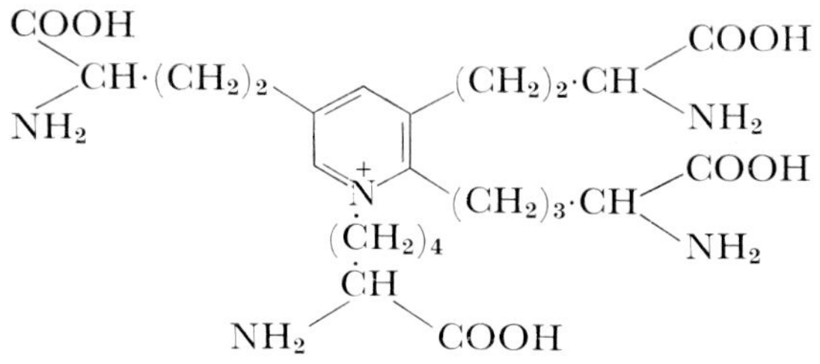

COOH   COOH
$CH\cdot(CH_2)_2$ $(CH_2)_2\cdot CH$
NH_2 NH_2
$\overset{+}{N}$ COOH
$(CH_2)_3\cdot CH$
$(CH_2)_4$ NH_2
CH
NH_2 COOH

II-Isodesmosine

Their structure implies that both these substances could form links with up to four independent peptide chains. From the work of Miller *et al.* (1964) it is suggested that desmosine is synthesized from four lysine residues, probably already situated in different protein chains. It is a slow reaction recalling the gradual formation of covalent linkages in collagen and again can be looked on as a form of autotanning. Desmosine and isodesmosine do not occur in the microfibrils.

Very small quantities of carbohydrates, the exact amounts depending upon the method of purification, are obtained from elastin and appear to be intimately bound to the protein (Ayer, 1964). Lipid also is present. The carbohydrates may come from the cement. Elastin structures appear to be surrounded by an outer sheath of glycolipoprotein and it may be this sheath material that contributes some of these carbohydrate and lipid components.

Table 1.5 Amino acid composition of bovine elastin from various sites. Grammes of amino acid per 1000 g of dry, ash-free protein. (Gotte *et al.*, 1963.)

	Ligamentum nuchae		Aorta		Ear cartilage	
	Auto-claved	Extracted with NaOH	Auto-claved	Extracted with NaOH	Auto-claved	Extracted with NaOH
Alanine	213	228	199	226	168	203
Glycine	267	268	239	268	229	258
Valine	177	182	160	—	134	160
Leucine	90	88	88	90	99	94
Isoleucine	38	35	40	38	33	30
Proline	135	152	—	140	137	143
Phenylalanine	62	55	60	—	59	56
Tyrosine	15	16	23	18	33	26
Serine	09	09	11	10	16	09
Threonine	11	10	16	10	16	10
Cystine ($\frac{1}{2}$)	—	—	—	—	—	—
Methionine	tr.	tr.	03	tr.	tr.	tr.
Arginine	13	08	17	08	25	13
Histidine	01	tr	03	01	05	01
Lysine	05	06	12	11	18	05
Aspartic acid	11	07	20	10	35	21
Glutamic acid	24	24	34	24	64	46
Hydroxyproline	16	15	15	18	19	18
Tryptophan	01	01	05	01	03	03

Elastin is broken down by a large number of proteolytic enzymes. Trypsin and pepsin do not digest it but a number of enzymes that do can be prepared from the pancreas and from bacteria. All these enzymes have a wide peptide bond specificity, and so attack a wide range of proteins; they are only resisted by proteins like collagen that have a highly individual structural organization.

Within the body, elastin undergoes a very slow turnover; when tissues are damaged the damaged elastin is resorbed and later replaced, though the mechanism for this is unknown.

Histologically, elastin is recognized by the way it dyes with orcein and Weigert's resorcin-fuchsin dye, though it also takes up other dyes. It can be distinguished from collagen histochemically by its resistance to digestion by collagenase, and in the electron microscope by its lack of banding.

There is evidence that elastin is secreted by the same fibroblasts that secrete collagen and ground substance. The elastin is probably secreted as soluble precursor molecules that aggregate to produce the definitive elastin structure.

6. *Keratin*

Keratin is the intracellular structural protein of the skin of amphibia and higher vertebrates and of feathers and mammalian hair. It is distinguished by its high content of cystine which forms disulphur bonds between the molecules. The detailed chemical composition of keratin will be dealt with in Chapter 17. The X-ray picture of keratin has been interpreted as resulting from the coiling of the chain molecules into the so-called α-helix configuration. In this arrangement, the chain molecule is twisted into a helix with $3\frac{2}{3}$ amino acid residues to each turn, or eleven residues in three turns of the helix. It is held together in this configuration by the positive charges of the N—H groups along the backbone forming hydrogen bonds with negatively charged backbone oxygen atoms in the next turn of the helix (Fig. 1.10). The X-ray photograph gives evidence of a pattern repeating every 5·0–5·5 Å along the structure and this corresponds to successive turns of the helix. There is also evidence for a repeat pattern at 1·5 Å and this is the distance between successive amino acid residues. Other proteins of different chemical composition such as extracellular silk from hymenoptera also give an α-type X-ray photograph since the α-helix represents a stable arrangement for a variety of amino acid sequences. Keratin can be digested by a specific enzyme, keratinase. Histochemical detection of cystine and the solution of the protein by thioglycollate solution are used as indications of the presence of keratin.

7. *Silk fibroin*

If an α-helix, as in hair, is stretched, the molecules may uncoil to produce β-type fully extended molecules or as in silk-worm fibroin, the amino acid content and arrangement may determine that the protein normally exists in this extended condition (Figs. 1.10 and 1.11(*a*)). Because of the zig-zag form of the protein backbone, with the side chains of the amino acid residues

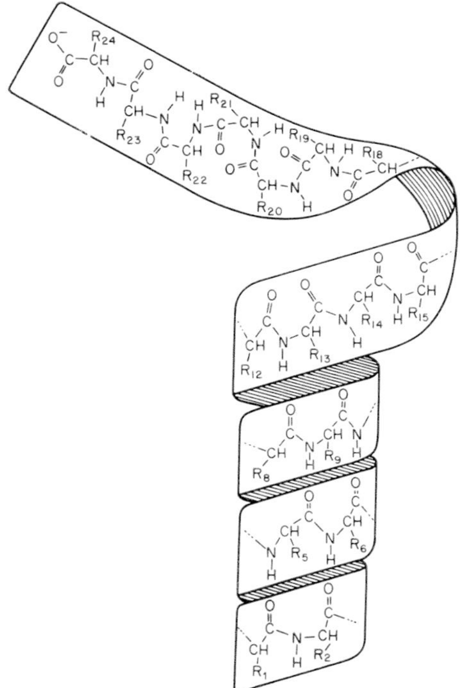

Fig. 1.10. Drawing of a backbone of a peptide chain curled into a right-handed α-helix. The lower part of the chain forms two turns of an α-helix, the upper part of the chain is folded in a random way with the chain fully extended into the β-form at the extreme upper end (from Haggis *et al.*, 1964).

extended first on one side of the chain and then on the other, one repeat pattern of a fully extended protein chain contains two amino acid residues; this shows in the X-ray photograph as a spacing of 6·7 Å. The chains aggregate laterally to form pleated sheets held together by interchain hydrogen bonds which hold the chains 4·65 Å apart. The side chains project above and below the pleated sheets and determine the spacing between the sheets, which varies with the proportions of the different side chains present (Fig. 1.11(*c*)).

E. Inorganic components

Inorganic salts that aid in the support and protection of animals are generally salts of calcium, though salts of magnesium, strontium, aluminium, iron, and silica also may occur. In invertebrates, the most common calcium salt is the carbonate, which may occur in a variety of forms that can be distinguished by their X-ray photographs, solubility, and density. Calcite is the least soluble and most stable form and *in vitro* experiments show that it is the form in which calcium carbonate crystallizes under normal conditions of temperature and pressure. Aragonite only forms in the laboratory at

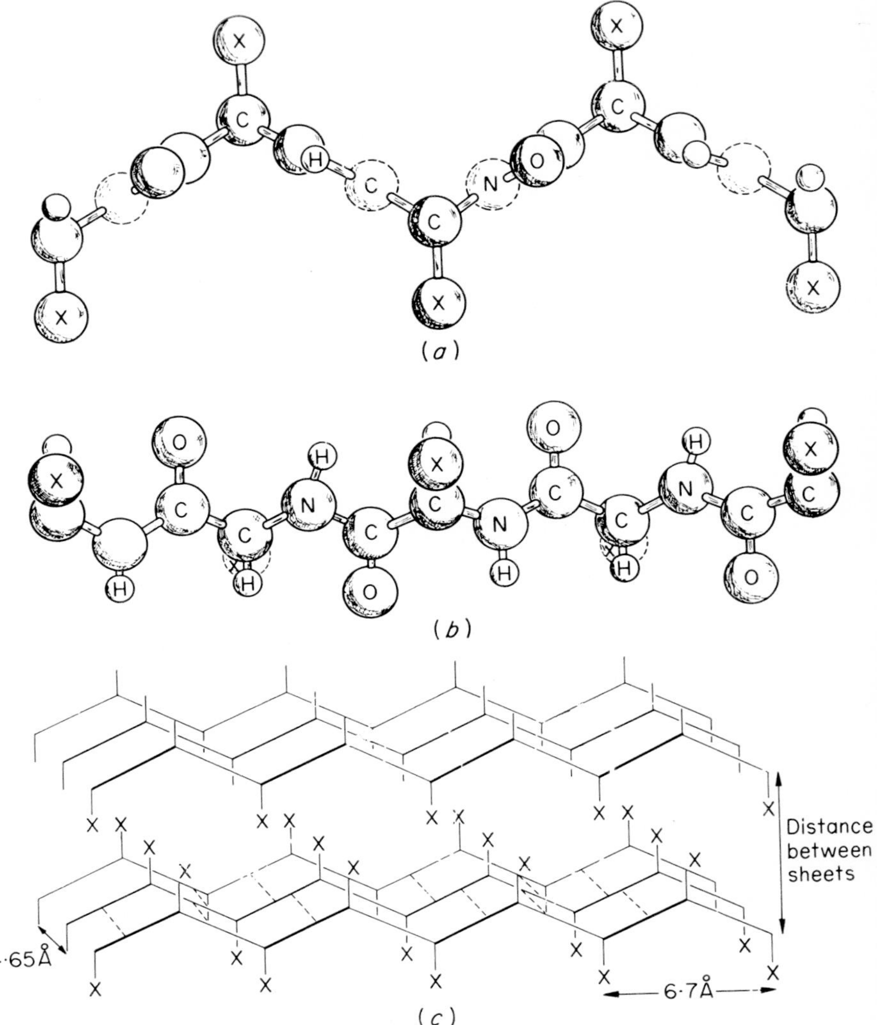

Fig. 1.11. (*a*) Diagram of a β-type extended protein chain (silk fibroin). (*b*) The same chain molecule rotated 90° so that the side chains lie above and below the plane of the page ((*a*) and (*b*) redrawn from Mercer, 1961). O = oxygen; N = nitrogen; X = side chains; C = carbon; H = hydrogen. (*c*) Diagram of β-chains associated to form two pleated sheets with the side chains extending above and below each sheet. The dotted lines between the chains represent hydrogen bonds between the chains.

temperatures considerably above those that can possibly exist in living matter, yet animals are able to deposit aragonite in their tissues. Occasionally, another crystalline form, vaterite, is found, or the calcium carbonate may be amorphous. With time, *in vitro*, all forms of calcium carbonate become converted to calcite, yet animals that deposit aragonite are able to hold it stable during their life time. How aragonite is deposited and held stable is not known (Prenant, 1927; Wilbur and Watabe, 1963; Lowenstam, 1954).

To consider only the chemistry of biological structural materials relegates them to the province of the chemist and physicist. The zoologist is also interested in how they are made and shaped into the structures that support and protect animals and how the intrinsic properties of the materials determine the behaviour of these structures. He is also interested in whether any evolutionary trends are detectable in structural materials and whether their properties have, in any way, conditioned the course of evolution. This has necessitated a taxonomic approach to the subject, considering the nature of the structural materials and their deployment in each phylum of animals. The classification system used is that of Rothschild (1965).

PART TWO

Invertebrata

CHAPTER 2
Protozoa

Protozoa are small, unicellular, and live in water or in the body fluids or cells of their hosts and since in aquatic animals weight is of little significance they need only limited support, though they do require protection against the environment and predators.

The electron microscope has revolutionized knowledge of protozoan anatomy, not only by adding details of fine structure to organelles already known to be present, but also by showing up structures whose existence was not previously suspected. Unfortunately, much of this work remains at the purely descriptive level and while we do now know where the various structural systems and materials are situated little is known about their behaviour and virtually nothing about their chemistry.

A. Surface structures

Cells, whether protozoan or metazoan, are bounded by a three-ply membrane, consisting of a pair of electron-dense layers each 2–3 nm thick separated by a less dense layer also 2–3 nm thick. The appearance of this layer is taken as evidence in favour of Davson and Danielli's (1943) hypothesis that the plasma membrane is composed of a biomolecular layer of lipid with protein adsorbed onto its inner and outer surfaces. This membrane controls the permeability and sensitivity of the cell to substances in its environment and the escape of substances from the cytoplasm. From studies, mainly on erythrocytes and echinoderm eggs, the membrane appears reversibly extensible and this elasticity must contribute to determining the shape of the cell, tending to round it up unless opposed by some structure or activity within or at the surface of the cell. Outside the plasma membrane there is often a layer of diffuse material believed to be mucopolysaccharide.

1. *Rhizopoda*

A typical plasma membrane, often with a diffuse outer layer, forms the outermost surface of a large proportion of Protozoa. In some there is no further apparent specialization of the surface either outside or beneath the plasma membrane. Such a simple plasma membrane is invariably found in those Rhizopoda that exhibit amoeboid locomotion and take in food by phagocytosis at any place on the body surface. Older studies of some of the large amoebae such as *Pelomyxa illinoisensis* and *P. carolinensis* suggested that a distinct pellicle covered these forms (Kudo, 1951) but this has not been confirmed by electron microscopy.

Amoeboid movement and the production of pseudopodia cause continual changes in the shape of these organisms. Since cytoplasm is relatively incompressible, these changes of shape must be accomplished without change in volume and must result in changes of surface area. How the surface structure is able to achieve this is not known. What opposes the tendency for the plasma membrane to round up the organism also is not known though the gel-like ectoplasmic layer so common in rhizopods in some way may be responsible. When *Amoeba proteus* is caused to shrink by osmosis, the cytoplasm, including the ectoplasm, shrinks away from the plasma membrane which then becomes spherical (O'Neil, 1964). Heliozoa have definite internal structures associated with the formation and maintenance of their finer pseudopodia.

The diffuse layer outside the plasma membrane in *Amoeba* is organized into hair-like processes 5–15 nm in diameter and up to 150 nm long (Mercer, 1959) and it is thought that these play an important part in phagocytosis by trapping and holding the prey while it is being engulfed (Lehmann *et al.*, 1956). Bell (1962) suggests that the mucus layer may also play a fundamental part in the contraction and expansion of the plasma membrane resulting from surface area changes.

Some Rhizopods secrete around themselves shells which always have openings through which pseudopodia can be extended. In the Testacea the shell is generally urn or vase-shaped with a definite mouth. *Cochliopodium*, *Chlamydophrys* and *Pamphagus* have thin membranous shells of unknown chemical composition (Hyman, 1940). In *Gromia oviformis* (Hedley and Wakefield, 1969) the shell is made up of four components (Fig. 2.1). The outer wall, which is perforated at intervals, is finely fibrillar and proteinaceous and also has a high content of bound ferric iron (Hedley, 1960). This outer wall is bounded externally by an electron-dense layer with a high polysaccharide content. Beneath the outer wall is a complex of honeycombed membranes arranged parallel to the outer shell. Each membrane is composed of an hexagonal array of cylinders 10–20 nm in diameter with walls 2–2·5 nm thick. Each cylinder is connected to the six surrounding cylinders by a septum in the middle of which is an electron-dense spot (Fig. 2.2). In cross section, these membranes appear as banded ribbons, the spacing of the banding depending on which part of the cylinders have been cut through. These membranes form an interleaving stack of up to fifteen layers (Fig. 2.3). Beneath the honeycombed membranes lies the plasma membrane of the cytoplasm. In the outer wall, honeycombed membranes and plasma membrane may all be in close contact with each other, but more commonly they are separated from each other by a finely fibrillar substance that is morphologically similar to the material comprising the outer wall and filling the wall canals and may be a precursor of the outer wall substance. The following amino acids have been obtained from the outer wall and its associated honeycombed membranes: alanine, arginine, aspartic acid, glutamic acid,

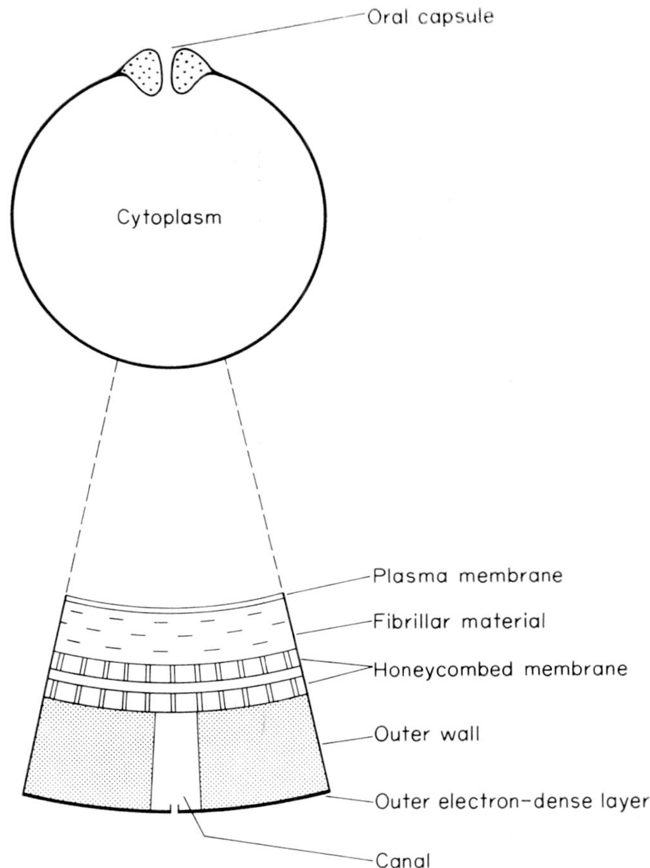

Fig. 2.1. Diagram of the shell of *Gromia oviformis* showing the position of the oral capsule and details of wall-structure (from Hedley and Wakefield, 1969).

glycine, hydroxyproline, leucine/isoleucine, lysine, ornithine, phenylalanine, proline and valine. The occurrence of hydroxyproline suggests the presence of collagen but no fibres morphologically resembling those of collagen can be detected in the shell (although the absence of banded fibres does not preclude the presence of collagen: cf. nematode and annelid cuticle). Hedley and Wakefield consider it as highly unlikely that the honeycombed membranes represent an atypical form of collagen and conclude that the hydroxyproline probably originates from a non-collagenous hydroxyproline protein such as are now known to exist (see p. 13). The function of the canals through the outer wall of the shell is not known. They cannot serve for the extrusion of pseudopodia as they are covered over by the honeycombed membranes.

Fig. 2.2. Tangential section of a honeycombed membrane of *Gromia oviformis* (from Hedley and Wakefield, 1969). [×121 000]

They may possibly serve for the escape of gametes when the parent is partly buried in mud with the oral region firmly attached to the substratum, but even this could only be accomplished by the breakdown of the honeycombed membranes.

The mouth of the shell is filled by the extracellular oral capsule composed of outer wall material enclosing a mass of microtubules (Fig. 2.4) which are mainly circumoral in arrangement. Microtubules are of very common occurrence in cytoplasm. In longitudinal sections they appear as fibrils, although a substructure has been detected in them (Behnke and Zealander, 1967). In cross section, microtubules appear as tubes with dense walls and less dense cores. However, the microtubules of the oral capsule of *Gromia* differ from all other microtubules so far described in being extracellular; other described microtubules have all been intracellular. Also, the oral

covered only by a plasma membrane (Parke and Manton, 1967). The theca in *Platymonas* is formed not of cellulose but contains galactose and a uronic acid together with a trace of arabinose and amino acids (Lewin, 1958). In *Pedinomonas tuberculata* no scales are seen on the body surface but underneath the plasma membrane are tubercles (Manton and Parke, 1960) which X-ray analysis suggests may contain cellulose, calcite, and quartz (Brandenberger and Frey Wyssling, 1947).

Some Chrysomonadina are described as being naked (Fritsch, 1935) and bounded only by a plasma membrane, e.g. *Chromulina psammobia* (Roullier and Fauré-Fremiet, 1958) but others, e.g. *Chrysochromulina* sp. (Parke *et al.*, 1955, 1959, see Pitelka, 1963 for other references), are covered by one or more layers of scales though in this order scales do not extend over the flagella. In electron micrographs of the majority of the species no cementing material can be detected and the scales are easily dislodged, though in *Prymnesium parvum* material appears to cement the scales to the cell surface and they remain more firmly in position (Parke *et al.*, 1955).

Many chrysomonads have calcified surface scales or coccoliths and in *Coccolithus pelagicus*, *Hymenomonas roseola*, and *Cricosphaera carterae*, Manton and Leedale, 1969; Manton and Peterfi, 1969 have shown that the organic portion of the calcified outer scales is essentially similar to the uncalcified outer scales in such species as *Chrysochromulina chiton* (Fig. 2.8(*a*) and (*b*)). In *Coccolithus pelagicus* Parke and Adams (1960) describe the calcified scales as being embedded in a two-layered membrane, the inner one containing cellulose. In Manton's sections (Manton and Peterfi, 1969) no such membrane is seen but there lies between the scales and the cell membrane a columnar deposit (Fig. 2.8(*a*)) of unknown chemical composition which might represent some form of cement. Calcified and uncalcified scales and the columnar material are produced in Golgi cisternae (Fig. 2.9). The scales of *Chrysochromulina chiton* are fibrillar, the fibrils at the centre of the scale being arranged radially while those of the rim are wound round the circumference several times. The fibrils are carbohydrate, the main components of which are ribose and galactose. X-ray photography indicates that the material either forms a crooked chain or is composed of a chain of mixed sugars (Green and Jennings, 1967).

The calcium of the coccoliths is in the form of calcite although, under conditions of nitrogen starvation, aragonite and vaterite are deposited instead (Watabe and Wilbur, 1963). Light promotes the deposition of calcium. This may be through photosynthesis utilizing carbonic acid and so increasing the pH of the cytoplasm, which favours the deposition of calcium carbonate (cf. calcification in coral, p. 102), although there are indications that the effect of light may be more complex than this (Paasche, 1968).

Isenberg *et al.* (1965, 1966) obtained a protein–polysaccharide complex from decalcified coccoliths of *Hymenomonas*. The protein (Table 2.1) contains hydroxyproline and relatively little sulphydryl and aromatic amino acid, and although much less glycine and much more aspartic and glutamic acid is present than in collagen they concluded that the matrix contains collagen and resembles the matrix of vertebrate bone in which calcification takes place.

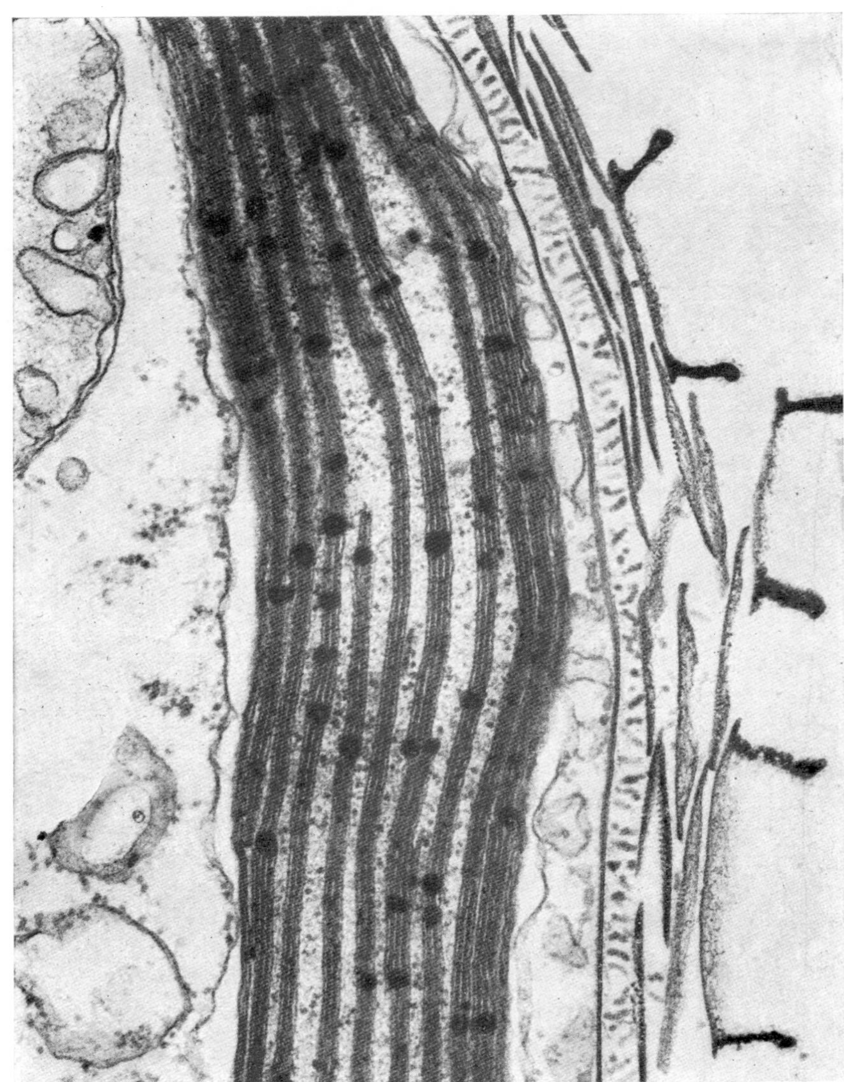

Fig. 2.8. (*a*) T.S. of *Hymenomonas roseola* showing outer layer of scales calcified round rim. Inner layer of flat, uncalcified scales. Columnar deposit between inner layer of scales and the cell plasma membrane. [×60 000]. (Manton and Peterfi, 1969). (*b*) T.S. of *Chrysochromulina chiton* showing outer and inner layers of scales. [×40 000]. (Manton and Leedale, 1969.)

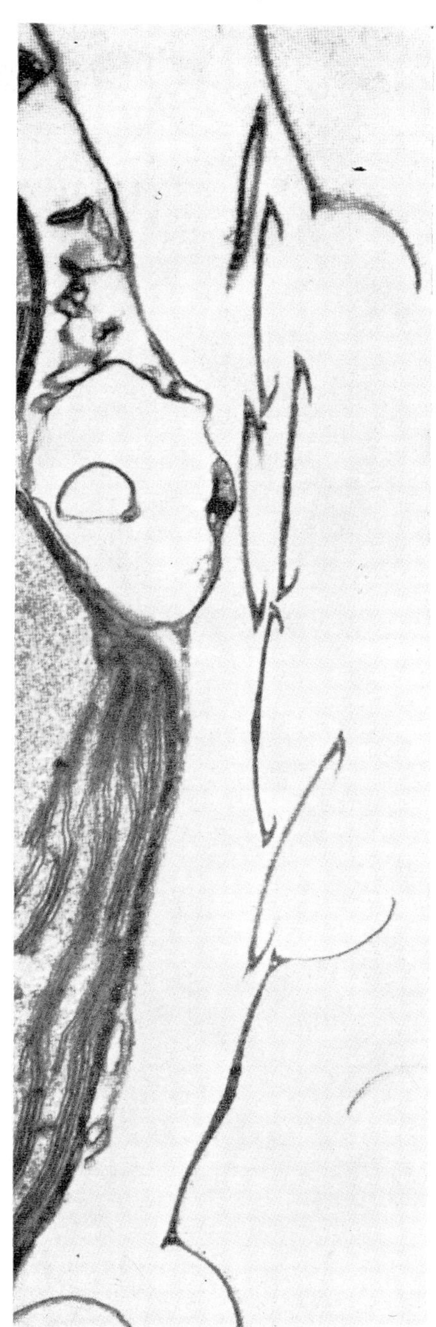

Fig. 2.8. (*continued*)

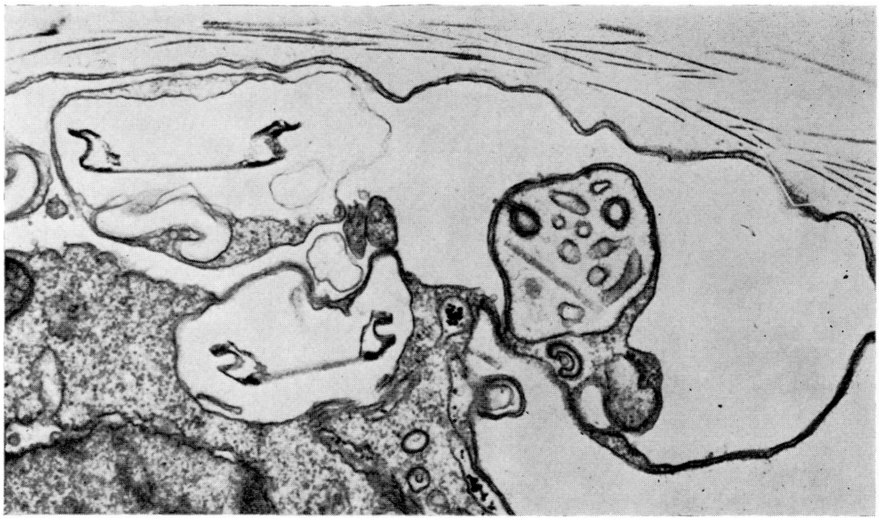

Fig. 2.9. Section near the surface of a cell of *Cricosphaera carterae* showing two coccoliths in Golgi vesicles ready for liberation. Micrograph D. 8186 × 20 000. (Manton and Leedale, 1969.)

Table 2.1 Amino acid composition of the protein extracted with distilled water from coccoliths. Residues per 1000 total residues.

Amino acid	No. of residues
Alanine	87
Glycine	96
Valine	44
Leucine	47
Isoleucine	27
Proline	44
Phenylalanine	29
Tyrosine	17
Serine	76
Threonine	48
Methionine	11
Arginine	44
Histidine	8
Lysine	40
Aspartic acid	138
Glutamic acid	153
Hydroxyproline	69
Hydroxylysine	—
Cysteine	—
Cystine	4

It is surprising that these scale-covered phytomonads are often able to produce pseudopodia which they extend between the surface plates, and Parke *et al.* (1955, 1959) have evidence of phagocytosis taking place in some species. Some chrysochromulinids about 4 μm in diameter ingest particles of up to 2·5 μm in diameter. A covering of separate scales would be no hindrance to the expansion of the surface to accommodate the increase in volume resulting from the intake of such large pieces of food.

The Euglenoidea have developed surface specializations that lie beneath the plasma membrane. In the light microscope their surface is seen to have helical striations. In transverse sections the electron microscope shows the striations to be due to a series of electron-dense strips that start as a whorl at the posterior end, bifurcate a few times before passing helically over the surface of the organism, and then come together again as they reach the canal leading to the anterior reservoir where they curve over and continue down the canal; here the strips appear to fuse together but do not extend to cover the reservoir into which the contractile vacuole discharges (Leedale, 1967). Underlying and parallel to the strips are microtubules which vary in number and distribution in the different species. Parallel with the strips are rows of mucus-producing bodies which discharge through small canals opening in the troughs between the strips.

In forms such as *Distigma* that show considerable changes of shape, there are numerous narrow strips, about 200 Å thick, and in transverse section they vary from very convex outwards to almost flat, presumably depending on the shape of the organism at time of fixation (Fig. 2.10(*a*)). In *Euglena stellata*, with less changeable shape, the strips are thicker and more angular, with fewer underlying microtubules (Mignot, 1965, 1966). *Euglena spirogyra* (Leedale, 1967) *Euglena acus*, and the colourless *Cyclidiopsis acus* (Mignot, 1966), all with almost rigid surfaces, have strips that appear designed to resist surface distensions. Each strip has two projections on one side, one of which runs under the adjacent strip while the other runs back beneath the parent strip to meet the projection from the strip on the far side (Fig. 2.10(*b*)). In *E. acus* and *C. acus* the strips run longitudinally rather than spirally. In these three species the strips are connected to each other by fibres while in *Hyalophacus ocellatus* a further system of fibres unite the underlying projections (Fig. 2.10(*c*)). In *Euglena spirogyra* Leedale describes a ridge and groove articulation between the strips within the troughs and, having observed limited sideways movement between the strips, suggests that this may be lubricated by mucus discharged into the groove. How such movement can take place between strips joined together anteriorly and posteriorly and connected by the plasma membrane is not certain.

Calycimonas physaloides and *Petalomonas steinii*, again more or less rigid forms, have only seven or eight strips which are thick and concave outwards with up to eighty underlying microtubules (Fig. 2.10(*d*)). In *Menoidium bibacillatum* (Fig. 2.10(*e*)) there is beneath the plasma membrane a continuous smooth layer surrounding the organism with a few underlying microtubules (Mignot, 1966).

Barras and Stone (1965) found that the surface material in *Euglena gracilis*

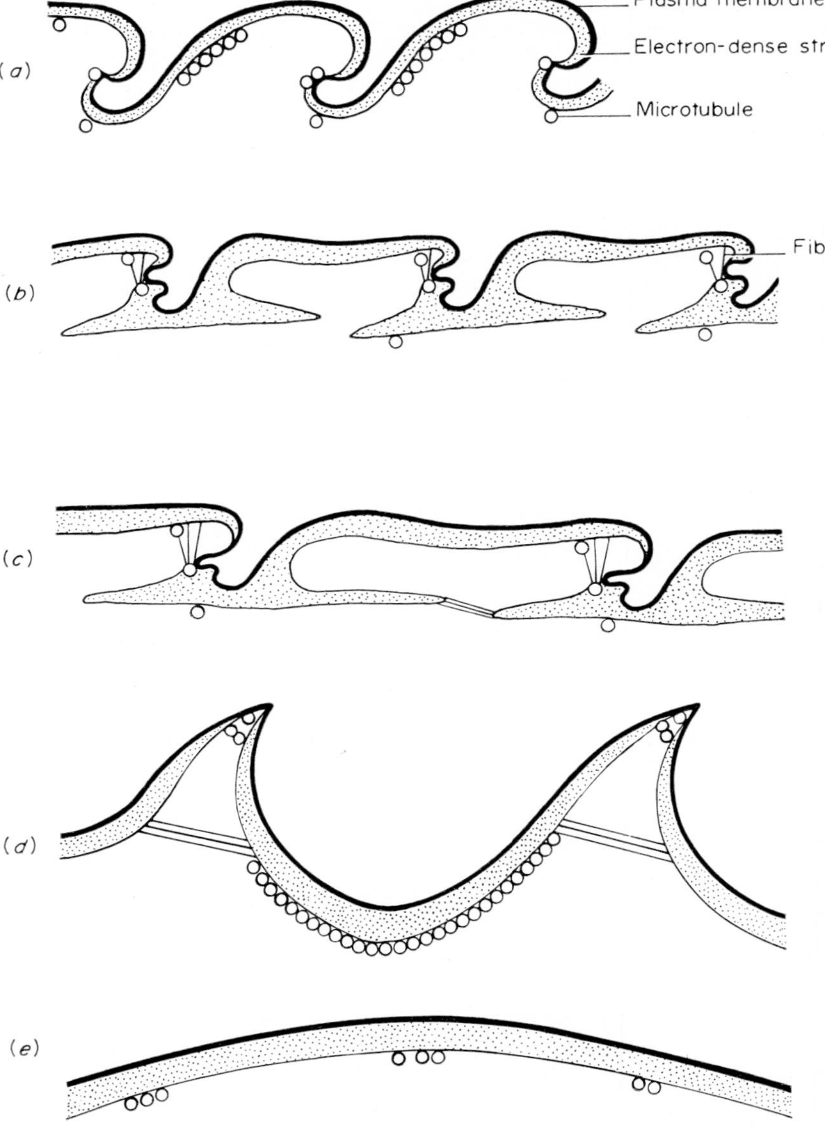

Fig. 2.10. Transverse sections of the cuticle of various Euglenoids. (*a*) *Distigma proteus*; (*b*) *Euglena acus*; (*c*) *Hyalophacus ocellatus*; (*d*) *Calycimonas physaloides*; (*e*) *Menoidium bibacillatum*. (After Mignot, 1966.)

contained 80 per cent protein, 11 per cent lipid, and between 6·4 and 17 per cent carbohydrate depending on the method of analysis. Kirk (1964) found the same species lysed by pepsin and trypsin suggesting that these enzymes destroyed an essential protein surface structure, but Northrop (1926) found *Euglena* able to survive in solutions of trypsin and pepsin as can most animals.

Some Euglenoids secrete, outside the plasma membrane, loose thecae or shells with an apical pore through which the locomotory flagellum emerges. Others have a more vase or cup-shaped envelope into which the organism can retract. The chemical nature of these structures is not known.

Although some dinoflagellates are described as naked (Fritsch, 1935), and the classification of the family differentiates between armoured and unarmoured types, Leadbeater and Dodge (1966) believe that eventually all dinoflagellates will be found, by electron microscopy, to have a theca or surface structure at least as definite as that which they have described in *Woloszynskia micra*. Here two outer membranes, of the plasma-membrane type, overlie two inner triple membranes and are fused with the inner membranes around the edges of hexagonal areas. The two inner membranes are perforated by plug-like structures which also perforate an underlying triple membrane beneath which are longitudinally arranged microtubules (Fig. 2.11). *Amphidium elegans* has a similar type of membrane-bound surface with an outer fringe of projecting fine fibrils believed to be of mucus (Grell and Wohlfarth-Botterman, 1957). In *Nematodinium armatum* (Mornin and Francis, 1967), there is an outer plasma-membrane-type membrane with

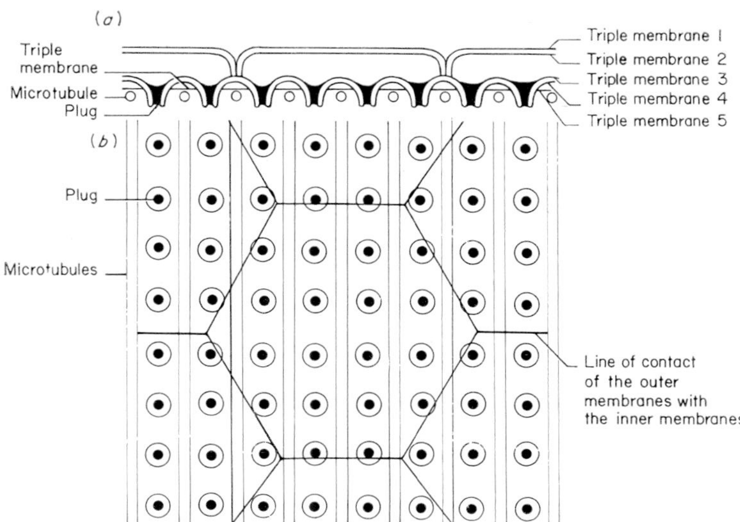

Fig. 2.11. Diagrammatic representation of the surface of *Woloszynskia micra*. (*a*) transverse section; (*b*) surface view. (After Leadbeater and Dodge, 1966.)

underlying membrane-bound vesicles. These rest on an inner membrane beneath which is a layer of electron-dense material bounded at the inner surface by another membrane, in all a pin passing through the surface of *Nematodinium* would go through five membranes (Fig. 2.12). Beneath the inner surface of the outer part of the vesicle membrane is a layer of material that shows radial stripes. In the dinoflagellate symbiotic in the anemone, *Anemonia sulcata*, Taylor (1968) has described a sequence of surface differentiations which may serve to relate the different structures so far described in the other species. When first formed by binary fission, the daughter cells of the symbionts are surrounded by two plasma membranes, though later

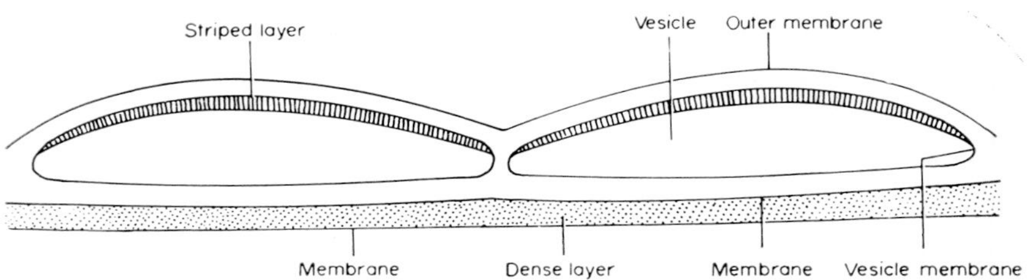

Fig. 2.12. Diagrammatic representation of a transverse section of the surface of *Nematodinium armatum*. (Constructed from the photograph of Mornin and Francis, 1967.)

these increase to five membranes. Under certain conditions that may be associated with ageing, the two membranes immediately beneath the outermost membrane become folded and constricted to enclose a series of vesicles, and later still the symbionts become separated from the host cytoplasm by a capsule whose wall is formed from a layer of homogeneous material separating two finely fibrillar surface layers. From histochemical tests, Taylor obtained a cellulose reaction from the surface of the symbionts presumably from this capsule.

In the freshwater dinoflagellate symbiont of *Hydra* (Park *et al.*, 1967) the surface is covered only by a single plasma membrane. The multi-layered surface of the *Anemonia* symbiont would presumably hinder metabolic exchange between it and its host and suggests a different host—symbiont relationship from that established in *Hydra*.

In the armoured dinoflagellates the theca, lying outside the plasma membrane, is composed of two or more valves or plates which can generally be separated by treatment with warm, dilute potash. The plates, often very thick and variously ornamented, are generally pierced by pores (Dodge, 1965) through which pseudopodia may be extended for the capture and digestion of food. In *Ceratium hirundenella* the valves are homogeneous in texture (Fott and Ludvik, 1956). In *Aureodinium pigmentosum* they lie between two sets of of double membranes (Dodge, 1967).

Growth is said to take place by interpolation of material at the sutures between the plates. When division takes place sometimes one daughter-cell inherits all the parental plates, while the other has to grow a new set, sometimes the daughters may share the plates or the parental theca may be abandoned and both daughters grow a new set. The work of Magnin (1907) showed these plates contain cellulose. Silica may occasionally be found in the plates.

The subclass Zoomastigina are a polyphylogenetic assemblage of flagellates that lack chloroplasts and are generally thought not to have clear structural relationships with the Phytomastigina. Within the order Protomonadina, the Choanoflagellata are sedentary, single, or colonial forms that produce, around the single flagellum, a collar of cytoplasmic microvilli (Fjerdingstad, 1961) that serves to filter food particles from the current of water produced by the flagellum, and pass them, trapped in mucus, to a deep invagination of the cell surface lying outside the collar. Though no specialization of the cell surface has been observed, they often secrete a hyaline, sometimes silicified cup or vase-shaped envelope that not only serves for protection but may play a part in the mechanism of feeding. Hyman (1940) points to a similarity between the microvilli of chanoflagellates and the fine axopods of heliozoans, though the microvilli do not, like axopods, contain internal axial filaments. She suggests that choanoflagellates have evolved from heliozoans amongst which the Helioflagidae can pass into a flagellate condition either losing or retaining the axopods. Lackey (1940) described a naked, chloroplast-containing chrysomonad, *Stylochromonas minuta*, with a protoplasmic collar round the flagellum but could find no evidence that this organism was holozoic.

Another sessile, protomonadinid group, the Bikosoecidae, attached to the substratum by a stalk, often surround themselves with an envelope; both stalk (Pringsheim, 1946) and envelope (Robinow, 1956) contain iron salts. The stalk is extruded as a thin, fragile, colourless, sticky mass which, with time, becomes thick and brown. Pringsheim says there is little evidence of organic material in the stalk which he found is formed from a mixture of brown manganic compounds within an almost colourless ferric hydroxide gel. The form of the envelope when present is almost identical with that of the chrysomonad *Stokesiella* (Petersen and Hansen, 1960) and is formed, at least in the lower part, from a spirally wound membranous band (Fig. 2.13). In feeding, particles are driven down onto the cell surface, most frequently near the base of the flagellum where the protoplasm continually changes shape forming short amoeboid processes. Picken (1941) suggests that these processes round the flagellum indicate a possible relationship with the choanoflagellates.

Also amongst the protomonadinids are the trypanosomes and bodonids, two orders that show little affinity with the two preceding groups but do show certain relationships with each other. Trypanosomes are parasites of vertebrates, invertebrates, and plants and may show, in some species, an alternation between a vertebrate and an invertebrate host. Different genera and different life-cycle stages vary in size and body-shape, but typically they

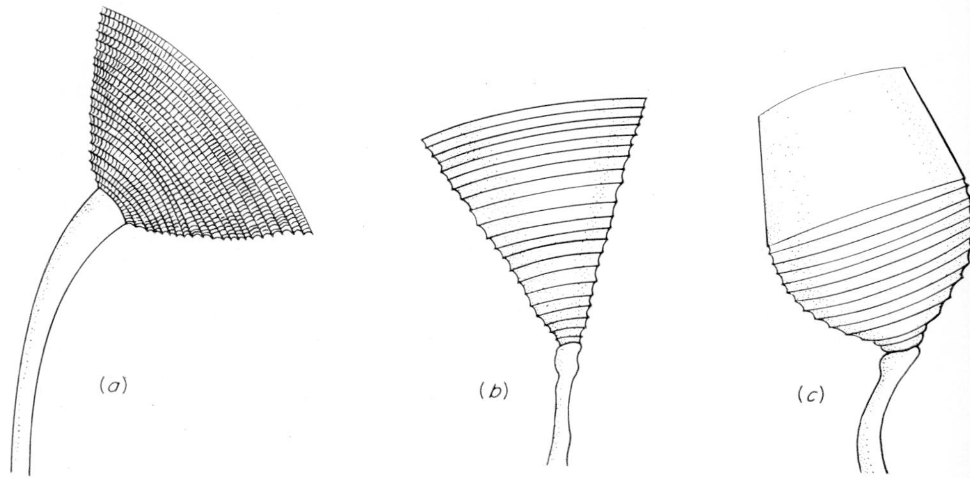

Fig. 2.13. (*a*) *Stokesiella dissimilis* drawn from the photograph by Petersen and Hansen (1960). (*b*) *Bikosoeca petiolata* drawn from the photograph by Robinow (1956). (*c*) *Bikosoeca* drawn from the photograph by Robinow (1956).

have an elongated form, pointed at one or both ends. In all species investigated there is a single layer of evenly spaced microtubules underlying the plasma membrane (see Pitelka, 1963, for references). These microtubules are formed from protein for they are destroyed by trypsin. The microtubules spiral round the body and converge, with a reduction in numbers by fusion, at the anterior and posterior ends. Trypanosomes are believed to be saprophytic but Steinert and Novikoff (1960) have seen in the leech stage of *Trypanosoma mega* a cytostome which may enable it to take in, by phagocytosis, haemoglobin from broken red blood corpuscles of the frog on which the leech feeds, the particles of which are too large to be absorbed through the general plasma membrane.

Bodonids are parastic or free living. The surface of *Crytobia helicis*, parasitic in the garden snail, is very similar to that in trypanosomes (Pyne, 1959, 1960). In the free-living *Bodo saltans* most of the body is covered only by a plasma membrane but there is a complex arrangement of the microtubules underlying the membrane at the anterior end (Pitelka, 1961, 1963).

Within the Metamonadina, the Polymastigina are generally parasitic in the intestines of vertebrates, leeches, or termites, although there are a few free-living species. They are generally oval or elongate with two to eight flagella, one of which may be recurrent and form an undulating membrane (Anderson and Beams, 1961; Grimstone, 1959). In trichomonads, believed to be the most primitive forms, the surface is covered only by a plasma membrane though in *Saccinobaculus* it is curiously folded (Grimstone, 1963). The organisms can bend and twist though limitations to change of shape

are generally considered as responsible (see Kümmel, 1957; Beams *et al.*, 1959; Jarosch, 1959) though jet-propulsion by the ejection of mucus has also been suggested.

Generally, it has been assumed that sporozoa are saprophytic, and the perforations in the inner membrane in *Pyxinoides balani* and *Plasmodium* sp. may be associated with this. Rudzinska and Trager (1957) found evidence for phagocytosis in erythrocyte stages of *Plasmodium lophurae*. Aikawa *et al.*, (1966) have shown that phagocytosis takes place, not over the general body surface, but only through a specialized cytostome, surrounded by pellicular rings. Cytostomes have also been observed in *Lankestrella* (Garnham *et al.*, (1962) and in intercellular stages of *Eimeria stiedae*, and *E. bovis* (Scholtyseck *et al.*, 1966).

The systematic position of *Sarcocystis* and *Toxoplasma* is a matter of doubt. Chessin (1965) believes that in spite of the possession by these forms of conoid, apical rings, and a cytostome, the differences between their life cycles and that of the Sporozoa prevent them from being placed in this class, although life cycles often are very different in nearly related parasites.

4. *Cilata*

The ciliates are a monophyletic group, unique in their possession of two types of nuclei. Also they all possess, at some stage of their life cycle, cilia which are used as locomotor and food collecting organelles. Though cilia are structurally similar to flagella, their arrangement and the system of fibres connecting their basal bodies distinguish them from the multiflagellate Mastigophora. The majority of ciliates have a permanent body shape, a cytopharynx, and never develop pseudopodia.

It is impossible in a book of this size to consider all the different surface structures that have been described in ciliates and only sufficient examples will be described briefly to indicate the range of structures found.

Paramecium, completely covered by longitudinal rows of cilia which with their basal apparatus are collectively called *kineties*, has an outer plasma membrane that is continuous over the body surface and over the cilia. Beneath this, over the whole body surface, is a regular mosaic of large vesicles or alveoli, the units of which correspond with the ciliary units. The base of each cilium (or in some species and in some body regions, each pair of cilia) is surrounded by a pair of kidney-shaped alveoli. The closely apposed membranes of adjoining alveoli are the sites in which silver is deposited in Von Gelei's silver impregnation technique (Gelei, 1936) and this serves to outline the limits of the alveoli (Fig. 2.2C(*a*)). Below the alveoli are the basal bodies of the cilia from which there originates, at the right-hand, anterior region a striated, tapering fibril, the kinetodesmal fibril, that runs anteriorly, past four or five of the basal bodies immediately in front in the same kinety, and becomes loosely twisted with the kinetodesmal fibrils from these adjacent basal bodies (Fig. 2.20(*b*)). This system of anteriorly directed fibres, originating on the right-hand side of the basal bodies, is known as the *kinetodesmal system* and is found in many ciliates. Below the basal bodies is a network of fine fibres forming a roughly polygonal system (Ehret and

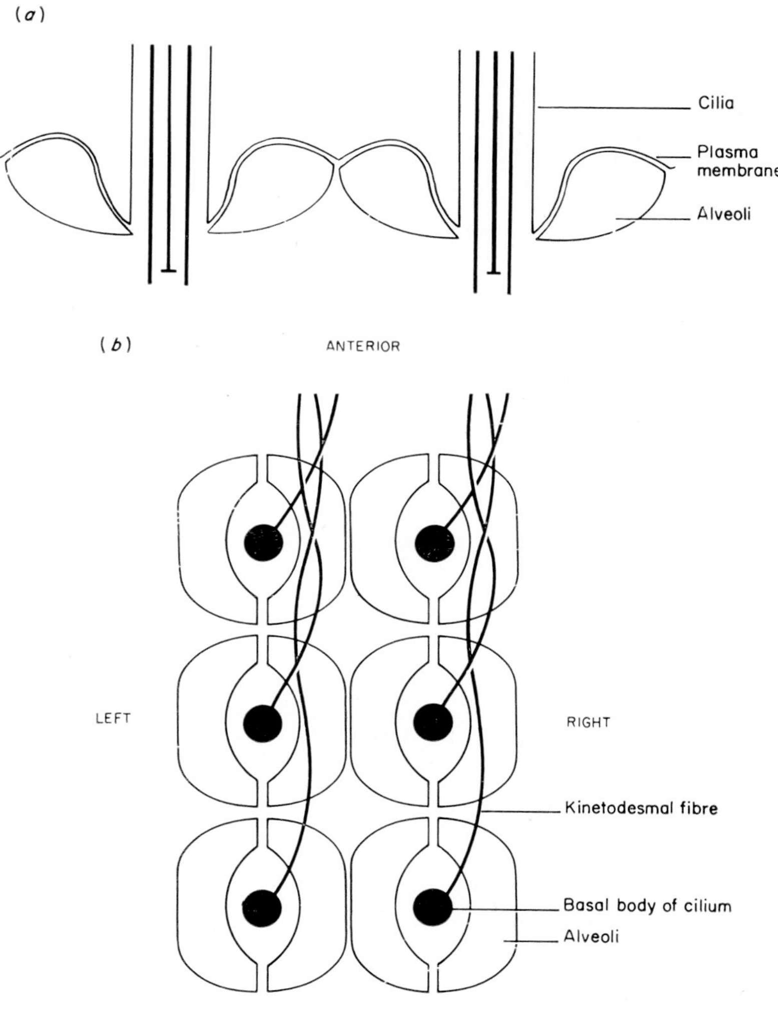

Fig. 2.20. Diagrammatic representation of surface structures of *Paramecium* (*a*) Transverse section through centre of alveoli (after Grimstone, 1961). (*b*) Surface view showing shape of alveoli and the underlying kinetodesmal fibres.

Powers, 1959; see Pitelka, 1963, for further references and further details). This whole system of surface structures is called the pellicle.

Colpidium has in addition to the kinetodesmal system, three systems of microtubular fibres, two of which are attached to the basal bodies of the cilia and run upwards towards the surface while the third, composed of overlapping short lengths of microtubules, forms longitudinal fibres between the rows of cilia. In this species the alveoli are much flattened (Fig. 2.21).

Whether or not the fibres attached to the basal bodies play a physiological role in co-ordinating ciliary activity (and there is no visible connection between fibres from different basal bodies in *Paramecium*) they must almost certainly play a part in strengthening the surface of the organism and in

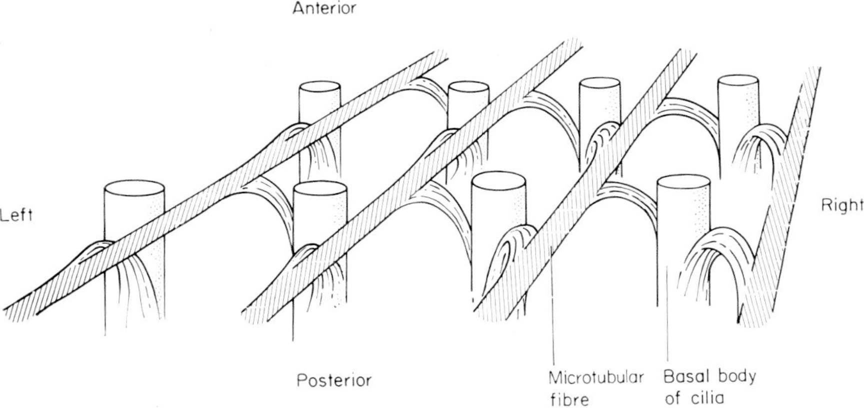

Fig. 2.21. Diagrammatic representation of the microtubular fibre system associated with the basal bodies in *Colpidium* constructed from the description of Pitelka (1963).

maintaining its shape. They may also represent a type of ciliary root system adapted to prevent the beating of cilia from causing oscillations of the basal bodies and distortion of the surface layers.

In other ciliates with general ciliation, like *Stentor* and *Nyctotherus*, the plasma membrane has beneath it not alveoli but an inner membrane, although in *Stentor* the inner membrane sometimes appears as composed of two closely apposed membranes such as might result from the collapse of membrane-bounded vesicles (Randall and Fitton-Jackson, 1958). *Nyctotherus* has numerous vesicles underlying the inner membrane (Paulin, 1967). In *Hyalophysa chattoni* the pellicle between the kineties is composed of three layers, an outer plasma membrane, a homogeneous, dense, middle layer and an inner layer, 50 nm thick, of regularly arranged short fibrils orientated perpendicular to the pellicle. In the region of the kineties the pellicle is much thinner (Bradbury, 1966).

Those ciliates that lack general ciliation also lack the complicated fibrillar systems normally associated with the basal bodies of cilia. *Euplotes patella* with a few specialized surface cilia merely has two systems of microtubules at right angles to each other beneath an inner membrane (Roth, 1956). The ophryoscolecids are covered by a plasma membrane. Below this plasma membrane is a thick, homogeneous layer containing glycoprotein; beneath this is a layer of bundles of longitudinal microtubules, and below this, a layer of transversely arranged fine fibrils (Noirot-Timothée, 1960).

The Suctoria, which as adults lack cilia, feed by capturing and sucking out small organisms. Some species (e.g. *Ephelota*) have, outside the plasma membrane, an alveolar layer containing mucopolysaccharide. Internal to the plasma membrane is a homogeneous protein layer, the epiplastic membrane, which contains embedded in it the basal bodies of the cilia possessed by the organism in its larval stage. This layer is much reduced in the tentacles (Rouiller *et al.*, 1956). In *Tokophyra infusionum* the outer layer is homogeneous, not alveolar, and there is no epiplastic membrane (Rudzinska, 1965). The tentacles can stretch out and contract, so that the surface layers are probably elastic, though in Rudzinka's photographs the outer layer is often wrinkled. The method by which suctorians suck out their prey is not known. Kitching (1952) observed, in *Podophyra*, a wrinkling of the body surface immediately after capture of prey and suggested that an increase in body surface, triggered off by food capture, produced a negative pressure sucking the food into the body; as the food flows in, the wrinkles are smoothed out. Extra output of the contractile vacuole is believed to produce this negative pressure.

Pyxicola nolandi, a sedentary ciliate is surrounded by a theca and attached to the substratum by a stalk-like prolongation of the theca. The stalk and theca, when first formed, are colourless and slightly elastic (Finley and Bacon, 1965), but become brown and brittle with age, the colour-change suggesting either the presence of iron or possibly quinone tanning. In the strangely shaped stalk of *Epistylis helicostylum* (Fig. 2.22) which also becomes brown with age, Vavra (1962) suggests that chitin may be present as it stains intensely with congo red, though this is not considered a diagnostic stain for chitin. In *Chilodochona* the firm elastic stalk, composed of protein fibrils, is

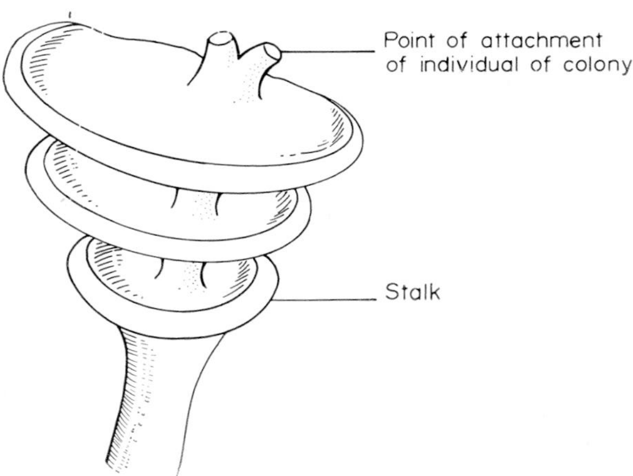

Fig. 2.22. The three-tiered stalk of *Epistylis helicostylum* (after Vavra, 1962).

secreted by glandular vesicles at the base of the organism (Fauré-Fremiet *et al.*, 1956a).

Some ciliates, such as *Blepharisma*, undergo a process of moulting in which the outer surface is lost but the cilia remain attached to the organism (Nadler, 1929). What the process involves in terms of ultrastructure is not yet known.

However rigid the surface of some protozoa may seem, it probably has at least a degree of elasticity. This is a requisite in all protozoa to permit some increase in volume under osmotic intake (Kitching, 1967) and through the intake of food, particularly in those predaceous forms that feed on organisms almost as large as themselves.

5. *Conclusions*

The various surface specializations found in protozoans almost certainly play a part in determining and maintaining the permanent shape that is often characteristic of these organisms. When *Paramecium* is burst and the cytoplasm squeezed out, the empty pellicle holds the original shape of the animal. This is generally true also for *Euglena spirogyra*, a species with a relatively rigid body form. However, if *E. spirogyra* is dried slowly, very often it rounds up as if some framework maintaining the shape against the elasticity of the pellicle has collapsed.

If the morphological similarity of inner triple membranes with the plasma membrane indicates similar physical properties, they could not, by themselves, provide a rigid framework. Nor do the frequently occurring alveoli seem obviously designed to fulfil this role, though collectively they may produce a not easily distorted or compressed layer which would have skeletal possibilities.

The frequent occurrence of microtubules beneath plasma membranes suggests that they may be important skeletal elements (Puytorac, 1965), but they are also present in forms that undergo considerable changes of shape and have, in fact, been held responsible for these changes. What is known about the chemistry of microtubules in flagella and cilia supports the idea that they form part of the mechanism of movement (see p. 69). How far changes of shape result from a mechanism causing the tubules to bend and how far it is also accompanied by alterations in their length is not known. That some microtubules are able to change their length is suggested by their presence around the contractile vacuole in *Paramecium*. However, contraction here need not be an active process but may result from inherent elasticity in these fibres and the release of tension upon them. The microtubules in cilia appear not to contract. Maybe microtubules in different sites are different functional entities.

B. Internal structures

If an internal framework to determine the shape of protozoans does exist the electron microscope cannot yet detect it. On the other hand, there do exist, within protozoa, various structures that are definitely skeletal, either supporting the whole body, or limited regions which are concerned with feeding or fixation of the organism to the substratum.

1. *Microtubules*

Recently, in *Clastroderma debaryanum* and *Physarummelleum* (Rhizopoda), McManus and Roth (1967) have detected in the cytoplasm a dense body with microtubules radiating from it. Nachmias (1964) found two types of fibres in *Chaos chaos*. Whether the microtubules in these rhizopods are concerned with support or movement is not known.

Helizoa produce stiff, radiating pseudopodia called axopods in which there is a core of microtubules. These microtubules extend deep into the cytoplasm and end near the nucleus (Anderson and Beams, 1960; Wohlfarth-Botterman and Kruger, 1954). The axopods, while generally appearing rigid, may bend and be retracted (Barrett, 1958). High hydrostatic pressure makes the microtubules distintegrate and the axopods retract but on release of pressure the axopods reform about the regenerating microtubules, suggesting that they play a part not only in the support but also in the formation of the axopods (Tilney *et al.*, 1966). Although a similarity between axopods and the microvillous collar of choanoflagellates has been suggested, microtubules have not been seen in these microvilli. Nor have microtubules been observed in the very fine pseudopodia of the Radiolaria, though they occur in the pseudopodia of the foraminiferan *Shepheardella taeniformis* (Hedley *et al.*, 1967).

Rudzinska (1965) has described, in the tentacles of the suctorian *Tokophyra infusionum*, two distinct passages separated by a system of microtubules: an inner circle of seven groups of four microtubules is surrounded by seven groups of three microtubules (Fig. 2.33(*a*)). Cytoplasm from the prey passes down the inner tube, while between the rings of tubules and the plasma membrane bounding the tentacles, complex, missile-shaped bodies (Fig. 2.23(*b*)), possibly containing enzymes or narcotics, pass out along the tentacle. They pierce the plasma membrane at the end of the tentacle, where it is in

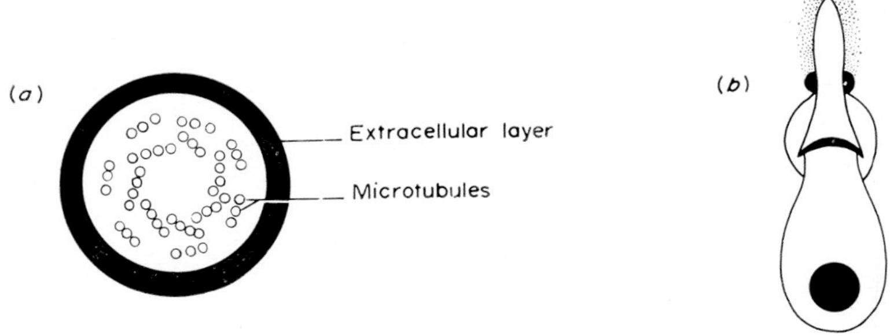

Fig. 2.23. (*a*) Transverse section of tentacle of *Tokophyra infusionum*. (*b*) Missile-shaped bodies possibly containing enzymes or narcotics. (After Rudzinska, 1965.)

contact with the captured organism, but these bodies do not appear to pass out into the prey. The cytoplasm of the prey, being sucked in by some unexplained mechanism, may force the suctorian cytoplasm with its missile-shaped bodies up the outer tube, the microtubules serving only to separate the two streams; alternatively the microtubules may, in some way, be actively concerned in these movements (Porter, 1966). *Ephelota* has two distinct types of tentacles; the sucking tentacles are very similar to those of *Tokophyra*, except that radially arranged lamellae extend into the inner tube. The other tentacles, used for capturing prey, are divided into longitudinal compartments and in the centre of each is a thick protein rod (Rouiller *et al.*, 1956).

Internal microtubules may also be organized into compact structures. The pharyngeal rods of *Peranema*, each of which consists of about 100 parallel, hexagonally packed microtubules (Roth, 1959), are concerned in capturing and swallowing food (Chen, 1950). A very complex feeding apparatus mainly composed of microtubules (Figs. 2.24, 2.25, 2.26) has been described in the ciliate, *Nassula*, by Tucker (1968).

The rostral apparatus (Fig. 2.27) in *Trichonympha* and *Pseudotrichonympha* (Grimstone and Gibbons, 1966) and *Lophomonas* (Beams *et al.*, 1960) formed from microtubules, banded fibrils, and lamellae probably provides support for the large number of flagella in these species. The considerable specific variation that occurs in these structures might indicate that they perform different functions in the different species but Grimstone and Gibbons suggest that for organisms such as these, living in a highly stable, homogeneous environment, protected from predators, passed regularly to new hosts, and having an unlimited food supply, variations in the structure of organelles need not necessarily indicate differences of function but may be the consequence of less vigorous elimination of genetic novelty.

Trichomonads may also possess a long fibrous structure, the axostyle, running through the length of the body (Fig. 2.28), sometimes arching over the basal bodies of the flagella at the front end, but not in direct contact with them, and sometimes projecting at the posterior end. In *Trichomonas* and *Foaina* it is a flexible tubular structure limited by an envelope of microtubules and filled with spherical granules of varying size (Anderson and Beams, 1959, 1961; Grassé, 1956). In *Pyrsonympha* (Grassé, 1956), *Oxymonas*, (Fig. 2.29) *Saccinobaculus* and *Notila* (Grimstone and Cleveland, 1965) it is composed of sheets of longitudinal microtubules and by its active undulations brings about locomotion in these forms.

Microtubules are also found in flagella and cilia; their arrangement into an outer ring of nine filaments, each of which is composed of two microtubules, surrounding two inner filaments each consisting of a single microtubule, is now too well known to need further consideration here (see Grimstone, 1962, 1966; Pitelka, 1963). Whatever the relationship of the microtubules to the contractile process there is some evidence that they do not themselves contract. As cilia bend, the framework of microtubules shears so that the distal ends of the tubules no longer remain level, those on the inside of the bend extending beyond the others (Satir, 1965). Even so, it is doubtful if they are merely skeletal structures (Gibbons and Grimstone, 1966) for

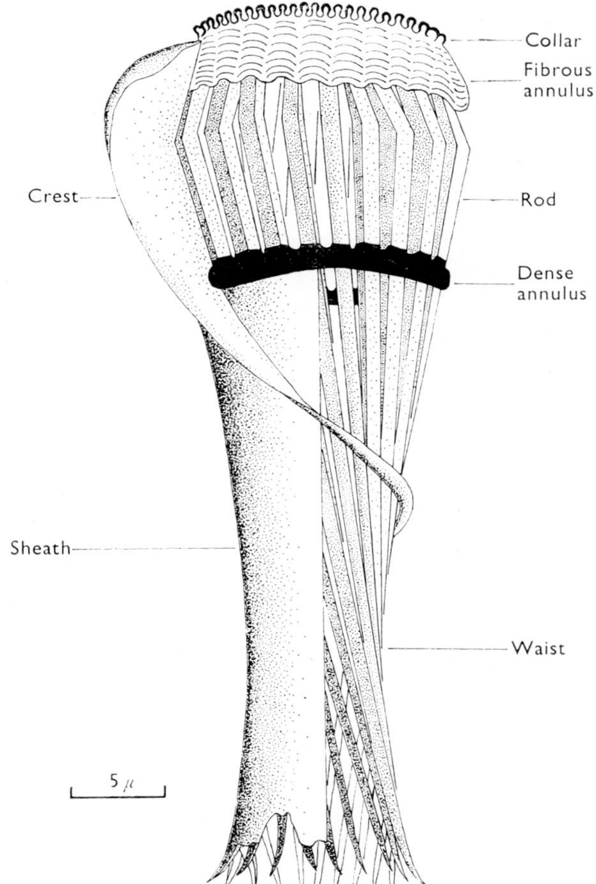

Fig. 2.24. Diagrammatic reconstruction of a cytopharyngeal basket in *Nassala*. The sheath encircles the palisade of rods below the dense annulus but has been omitted on the right side of the diagram to show arrangement of rods inside it. A crest is attached to each rod above the dense annulus and spirals around the outer surface of the sheath at the lower levels; for clarity only one crest has been illustrated (Tucker, 1968).

ATPase activity, recalling that found in muscles, has been detected in cilia (Gibbons, 1965). Chemical analysis of the outer filaments of *Tetrahymena* cilia shows they are made from a protein very similar chemically to actin from muscle (Fig. 2.30), though differing from it in its reactions to colchicine and in its immunological reactions (Renaud *et al.*, 1968). These facts suggest that these microtubules in cilia and flagella are an essential component of the mechanism generating movement. How far these chemical findings for the outer microtubules of cilia apply to all microtubules is not known.

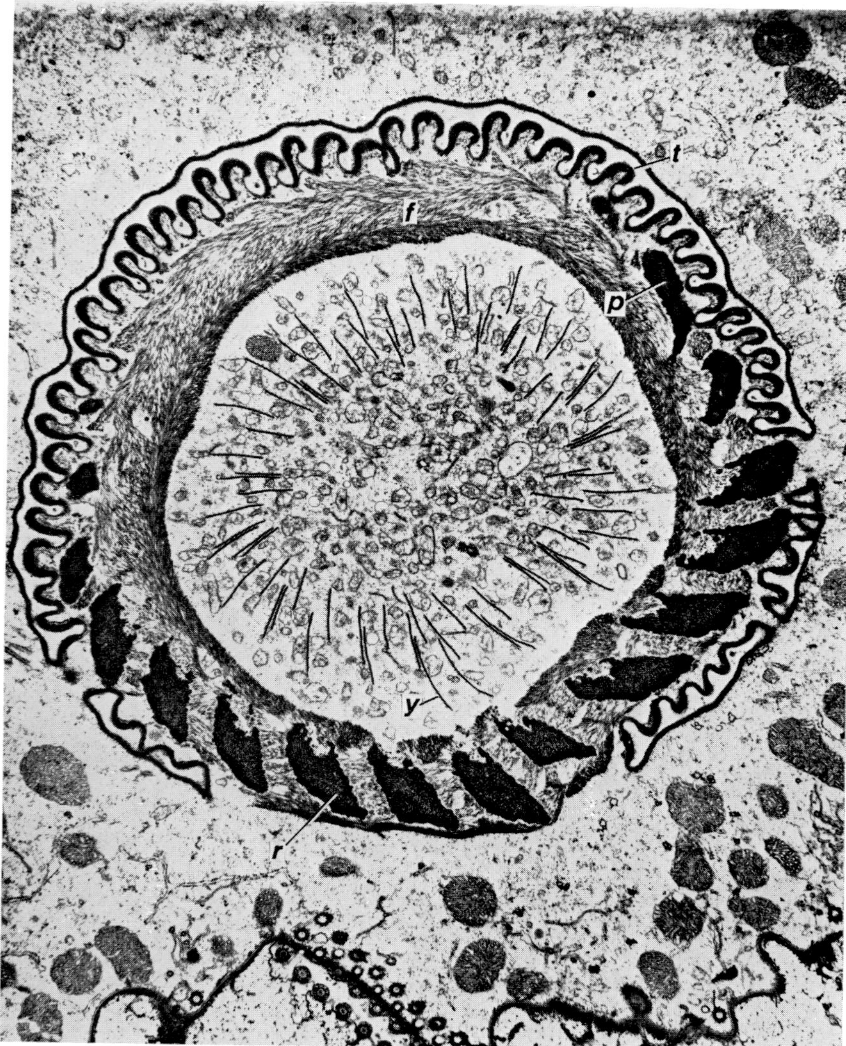

Fig. 2.25. Cross section of a cytopharyngeal basket of *Nassala* just below its top at the level of the fibrous annulus (*f*), cutting through the tops of some of the rods (*r*), the outer ends of some of the thickened corrugations of the collar (*t*), radially arranged subcytostomal lamellae (*y*), and some of the caps (*p*), on the tops of the rods (Tucker, 1968). [×8 140]

Long filamentous appendages called haptonemes, superficially resembling flagella, but differing from them in fine structure and functions, have been found in *Chrysochromulina coccolithus*, *Prymnesium*, and *Platychrysis*. They are anchoring structures which coil into a helix. They contain six to seven longitudinal single microtubular filaments though at their bases the number of

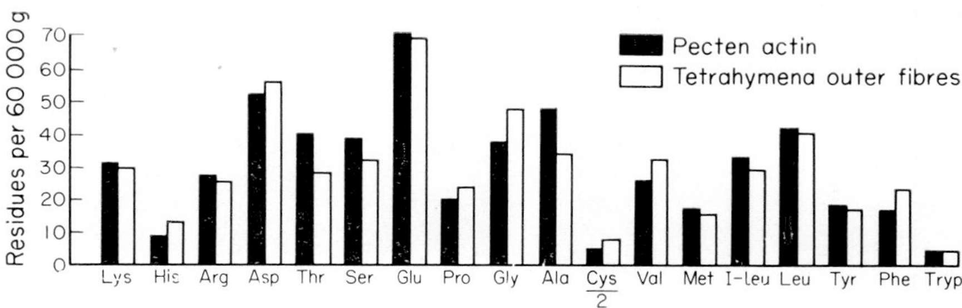

Fig. 2.30. Amino acid composition of outer filament protein from cilia compared with that of actin from *Pecten* muscle (Renaud *et al.*, 1968).

2. *Contractile elements*

The contractile element in *Vorticella* stalk is a bundle of very fine filaments 2–4 nm in width and of indefinite length. In cross section, according to Sotelo and Trujillo-Cenóz (1959), the structure appears alveolar somewhat resembling the sarcoplasmic reticulum of vertebrate striated muscle. Similar myoneme structures occur also at the surface of the body of contractile peritrichs. In *Opercularis* Fauré-Fremiet *et al.* (1956b) described regularly spaced transverse fibres beneath the surface membrane which they believed opposed the contractions of the longitudinal myonemes. In the contracted state the surface layer of these animals is often wrinkled or regularly annulated as if below a certain length the elasticity of the surface alone can no longer accommodate the decrease in length of the animal. The heterotrich ciliate, *Stentor*, also has contractile elements, called M-bands by Randall and Fitton-Jackson (1958), somewhat similar to the myonemes of peritrichs. When fully extended *Stentor* may have as much as four times its volume when contracted. This increased volume is achieved by taking in water.

3. *Internal fibrillar systems*

Fibrils or rhizoplasts run from the basal bodies of some flagella and cilia for varying depths into the cytoplasm and in some flagellates become attached to the nuclear membrane. In holotrich and heterotrich ciliates, e.g. *Stentor*, cirri and membranelles formed from aggregates of cilia often have very complex rooting systems (Randall and Fitton-Jackson, 1958). These may be mainly anchoring in function.

The *parabasal fibres* and *costa* of trichomonads are both connected to the basal bodies of the flagella. They are, like some kinetodesmal fibres and rhizoplasts, formed of cross-striated protein fibres but the periodicity of that of the costa differs from that of the parabasal body. These cross striations suggest that the protein may be collagen but Simpson and White (1964) failed to digest them with collagenase. In *Trichomonas gigantea* and *T. termopsidis* the costa, an undulating rod extending through the length of the organism, has an extremely complex ultrastructure. Its undulating movements, aided by the four flagella and the undulating membrane, propel these relatively large flagellates (*T. gigantea* is 200 μm and *T. termopsidis*, 60 μm long) through the dense contents of the termite stomach (Amos, 1968) though the costa must also support the body and limit the changes of shape these animals can undergo.

Fibrous systems separating the endoplasm from the ectoplasm in various ciliates may serve a skeletal function. In *Isotricha*, *Metaradiophyra*, and some Ophryoscolecidae, all relatively large organisms, two layers of very fine fibres are orientated at right angles to each other. From the inner layer, in *Isotricha*, tracts of fibres pass inwards to form a network round the nucleus (Noirot-Timothée, 1958). In *Metaradiophyra* the outer layer has connections with the basal bodies of the cilia (Puytorac, 1959). In *Gregarina rigida* (Sporozoa) an ectoplasmic fibre layer connects with the fibrous membrane dividing the anterior "protomerite" from the posterior "deutomerite" (Beams *et al.*, 1959).

4. *Organic skeletal structures*

The so-called skeletal plates of ophryoscelecids are made of neutral polysaccharide and Noirot-Timothée (1960) thinks that they are more likely to be food reserves than skeletal in function.

One of the most complex structures yet found in a protozoan occurs in the ciliate, *Trichodinopsis paradoxa*. This has a complex skeletal disc made up of overlapping plates supporting the base. Above this is a corona of very regularly arranged slender rods radiating outwards from the centre, all of the parts being made of protein. There is also a complex fibre system believed to be contractile and three elastic fibres supporting the infundubulum leading to the mouth (Fauré-Fremiet *et al.*, 1956c).

Except for some Acantharia, radiolarians possess a central capsule separating endoplasm from ectoplasm. It is a perforate structure allowing continuity between the two regions. It is generally colourless or slightly brown. Chemically, it is very resistant. Material lying in the ectoplasm around the central capsule in *Thalassicola* gives positive reactions in tests for the presence of orthodiphenols (Brown, 1950b), suggesting that quinone-tanning may be present. The majority of radiolarians are pelagic and while some occur at considerable depths they are generally planktonic. They are helped to float by the calymma, a thick, bubbly layer of gelatinous material below the ectoplasm. When the bubbles collapse the organism sinks, and when the bubbles reform, the organisms rise again.

5. *Inorganic skeletal structures*

Internal inorganic skeletons are developed in Heliozoa and Radiolaria. Heliozoa may have silicious spicules or plates radiating out from the endoplasm. Radiolaria present a wonderful array of skeletal patterns (Fig. 2.31). In Acantharia, the spicules, formed from strontium sulphate or from calcium aluminium silicate, may unite to form a lattice. In the remaining genera the skeleton is composed of silica.

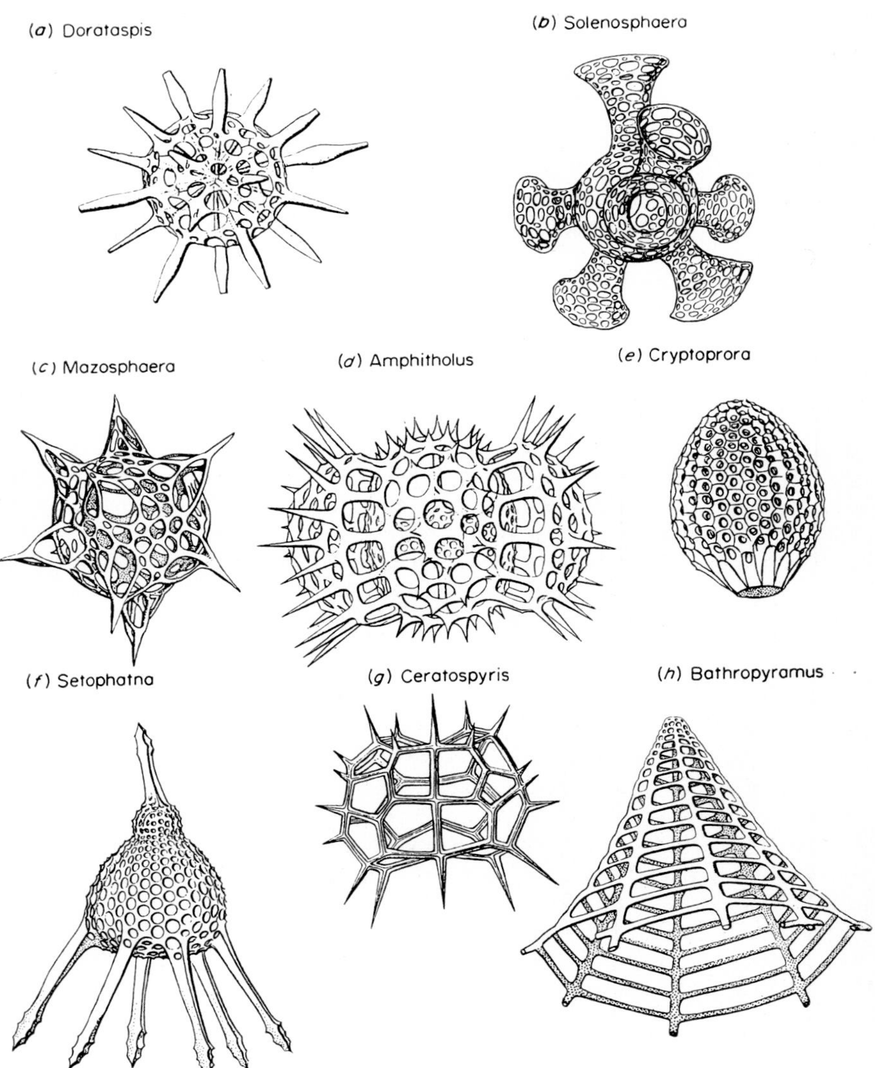

Fig. 2.31. Radiolarian skeletons from Haekel (1886).

In the deeper parts of the ocean where calcareous shells dissolve, the bottom mud consists largely of the silicious skeletons of radiolarians together with those of sponges and diatoms. Radiolarian deposits are also important in the formation of flints and chert.

Spirostomum ambiguum, a benthonic ciliate, sometimes produces, under the surface layer, a system of articulating plates of hydroxyapatite (an alkaline calcium phosphate) which appears to form in some specific relationship to to the surface network of fibres. Pautard (1959a and b) suggests that these plates serve both as a support when the organism burrows into the silt and also as a mineral bank from which phosphate can be withdrawn. Calcium phosphate is also deposited at the surface of other ciliates, e.g. *Coleps hirtus* (Fauré-Fremiet *et al.*, 1948).

C. Trichocysts

Ejectile trichocysts occur in various flagellates and ciliates and somewhat similar bodies occur also in the Cnidosporidia. Though the trichocysts differ widely in form and apparent mode of functioning in the different species, Grimstone (1966) suggests that they may all be evolved from the much simpler muciferous bodies that occur in both flagellates and ciliates just below the cell surface and discharge their contents to the exterior (see Grimstone, 1966, for references). In *Paramecium* the undischarged trichocyst appears as a carrot-shaped mass of amorphous protein which, on discharge, is converted into a long, tapering shaft with a conspicuous crossbanding of about 550 Å periodicity. The banding resembles that of collagen but Jackus's (1945) investigations failed to establish that this was its nature.

D. Cysts

Freshwater and parasitic protozoa, but rarely marine protozoans, may secrete about themselves temporary cysts of various sorts. Many Rhizopoda, Mastigophora, and Ciliata encyst to survive unfavourable conditions, often at the same time achieving dispersal by various agencies. Some dinoflagellates and ciliates encyst while digesting a large meal. Parasites are often transmitted in cysts to new hosts. Cysts occur regularly in the life of sporozoans around pairing gamonts and sporulating zygotes.

The cysts are often two or three-layered, the layers differing chemically. In *Endamoeba histolytica*, *Endolimax nana*, *Councilmania lafleuri*, and *C. dissimilis* (Rhizopoda) and in *Giardia lambila* (Zoomastigina) the cyst walls all contain protein and possibly polysaccharides; they are very resistant to acids, alkalis and digestive enzymes (Kofoid *et al.*, 1931). MacLennan (1937) found protein in the cyst wall of the ciliate, *Ichthyophthirius*, though in this case it is very soft and offers no resistance to drying out or to noxious substances. Chitin has been found in the cysts of the rhizopod, *Pelomyxa illinoisensis* (Sachs, 1956), the ciliate *Vorticella* (Stein, 1854), and the cnidosporidian, *Nosema* (Dissanaike and Canning, 1957). A carbohydrate allied to glycogen occurs in the cysts of the ciliate, *Colpoda cucullus* (Goodey, 1913) and one allied to cellulose in those of *Diplosalis acuta*, a dinoflagellate (Sebestyen, 1935). In the rhizopod, *Acanthamoeba*, a thick layer containing cellulose and lipoprotein underlies a thin layer that is mainly protein and there is an inorganic component, part of which is phosphatic (Neff *et al.*, 1962, 1964; Neff and Benton, 1962). In *Naegleria gruberi*, a rhizomastiginid, the cyst is probably mucopolysaccharide (Schuster, 1963).

Many cysts, particularly of sporozoans, appear brown and in the oocysts of several coccidians, Monné and Honig (1954b) found evidence of quinone tanning in the outer layer of the cysts; the inner layer is composed of lipid firmly bound to protein. While the chemical resistance of the oocysts of the gregarines, *Lepistophilia* and *Colepistmatophila*, suggests quinone tanning, Ratnayake (1960) could find only circumstantial evidence for this. Wasielewski (1924) describes granules flattened against the surface of unfertilized coccidian eggs which, after fertilization, are extruded and play an important part in the formation of the cyst (cf. egg shell formation in trematode eggs, page 119).

Holophytic flagellates are able to assume a palmella state in which individuals lose their flagella, round up, and secrete a jelly-like substance which may unite numerous individuals in large colonial masses; in these, feeding and division continue to take place. *Volvox* and related species represent a permanent, highly organized palmella state in which flagella are retained. The method of secretion of the jelly and its chemical composition have not been determined.

F. Conclusions

Little is yet known as to how protozoans organize materials to form their complex surface and intracellular structures. Grimstone (1966) has considered some of the questions that must be answered in attempting to find out. A prerequisite is a knowledge of the anatomy of the structure and its chemical composition. The observation with the electron microscope of the gradual accumulation of material to form a structure should be possible with patience. The knowledge that scales and coccoliths of Prasinophycae and Chrysomonodina are formed in golgi cisternae holds out prospects that further details of the development of these structures will soon be available. From such a descriptive basis the more fundamental problems of the influences directing development will be opened up to attack.

This capacity of cytoplasm to produce and operate, within its substance, local structures of remarkable complexity, more concerned with support and mechanical activities than with the metabolism of the cell, is mainly confined to protozoans. In metazoans, structures playing similar roles in the life of the animal are generally built up of whole cells as are scales, feathers, and hairs of vertebrates, or of extracellular materials as in the cuticles of invertebrates.

The absence of any surface specializations similar to those in ciliates and flagellates and the frequent performance of phagocytosis by metazoan cells suggests that, whether or not the organisms from which metazoans evolved were flagellated, they were probably amoeboid and lacked any surface specializations. The cells of multicellular plants are bounded by cell walls, and microtubules are often found at the cell surface, and it is generally assumed that the plants evolved from unicellular organisms with specialized surfaces.

No protozoan has yet been proved definitely to produce collagen but if collagen can only be produced extracellularly, from soluble precursors, then this is hardly surprising. In those protozoa, particularly flagellates and ciliates, that tend to draw a current of water over the body-surface, the precursors most likely would be dispersed before they could aggregate into fibres. Collagen might be expected to occur in protozoans in places where other extracellular structures are present that could prevent the tropocollagen molecules from diffusing away. The outer shell in the rhizopod, *Gromia*, in the foraminiferan, *Haliphysema*, and the surface layer in which the coccolith scales are embedded in *Hymenomonas*, might provide the necessary barriers to the dispersal of the tropocollagen molecules in these three organisms, in which there is some evidence that collagen does occur.

CHAPTER 3

Multicellular Animals

Multicellular animals or Metazoan have two common structural capacities: (1) the capacity of the cells to adhere to each other and to non-cellular components of the body, so allowing multicellular bodies to be formed; (2) the capacity of at least some of the cells to produce collagen which is organized into a framework to support the cells.

A. Cell adhesion

What causes metazoan cells to adhere to each other is as yet uncertain (Curtis, 1967). The electron microscope shows the plasma membranes of adjacent cells often to be convoluted and interlocking. Whether the gap of 100–2C0 Å, which almost always separates the two adjacent plasma membranes, exists in life is a matter of dispute, though its reality is generally accepted. The presence, in this gap, of a mucopolysaccharide cement has been suggested as the cause of adhesion, but there is now evidence that the contents of the gap have, in life, a very low viscosity. In certain cells, particularly nerve cells, no gap is seen between the membranes, at least over certain small regions of the opposed surfaces.

Special structures, the desomosomes, have also been implicated in cell adhesion. Desmosomes are densely staining patches formed below the plasma membranes and opposite each other in two adjacent cells. In the 100–200 Å gap that persists in the region of the desmosomes a number of dense lines parallel to the plasma membranes are seen (Fig. 3.1) though what they represent structurally is not known; they offer no resistance to the penetration of ferritin particles into the region. When cells are pulled away from each other they sometimes remain attached in the region of their desmosomes, and in sections of mammalian skin the cells of the "prickle" layer thus adhere to each other.

Curtis (1967) has suggested that cell adhesion results from the forces of repulsion between the cells, caused by similar changes on their surfaces, being balanced at a certain distance of separation by London—Van der Waals forces of attraction. Any tendency of the cells to move closer together is prevented by the increase in repulsive forces that results from decreasing the distance of separation and any tendency to move apart is countered by the relatively smaller decrease with distance in the forces of attraction than

B. Internal skeletal structures

Though sponges weigh very little in relation to water, the size of some of them is such that they would collapse under their own weight if their body was not supported, as it is in the majority of species, by a skeleton. This is formed from inorganic spicules, organic fibres, or both.

1. *Inorganic skeletons*

The Calcarea, which are in general of small size, have calcium carbonate spicules, mainly in the form of calcite though aragonite was found by Lister (1900) in *Asteroderma willeyana*. Varying amounts of magnesium carbonate is mixed with the calcium carbonate and Fox and Ramage (1931) found strontium in *Clathrina*.

The spicules are birefringent and behave optically as if cut from a single homogeneous crystal. Each spicule has an optical axis (the trigonal axis of symmetry of the calcite crystal) which has been shown for various species to be orientated with respect to the sponge anatomy (Ebner, 1887; Bidder, 1898). Jones (1954a, b; 1955a) in a series of papers has suggested various intrinsic and extrinsic factors which together operate to produce this orientation. If calcareous sponges are grown in calcium-free sea water they are unable to form spicules and spicules already formed may disappear.

Hexactinellida, deep sea forms, have silicious spicules that are basically 6-rayed, the rays being mutually at 90° to each other. Silicious spicules also occur in tetraxonid and monaxonid Desmospongiae but they are never 6-rayed. The silica is in the form of amorphous hydrated silica, the degree of hydration varying in different species (Vinogradov, 1953). Jørgensen (1944) found, in addition, magnesium, potassium, and sodium oxides. How sponges manage to extract and concentrate silica from sea water is not known, but Vinogradov suggests that part of it may come from diatoms which are often found in large numbers in sponges. Jørgensen cultured *Spongilla lacustris* in water of varying silica content and found that once the concentration was above a certain minimum of about 0·02 mM SiO_2 the number of spicules per unit volume of developing sponge is independent of the silica content. At a concentration above 1·28 mM SiO_2 the production of large spicules is retarded. The length of the spicules is not affected by concentration but the higher the concentration the thicker the spicules. When solid silica in various forms is added to the culture, it has no effect on the size of the spicules.

The form of the inorganic skeletons in various sponges is very fully described in the relevant volumes of the Challenger Reports. While spicules detached from the animal often appear bizarre in shape, when viewed in position their shape is seen as highly relevant to its function of supporting and protecting the body (Fig. 4.1).

It has long been held that in Calcarea each simple spicule originates intracellularly but grows as an extracellular structure through the co-operation of two cells, probably derived from the division of the spicule mother-cell. An organic axial filament forms between the two cells, round which the

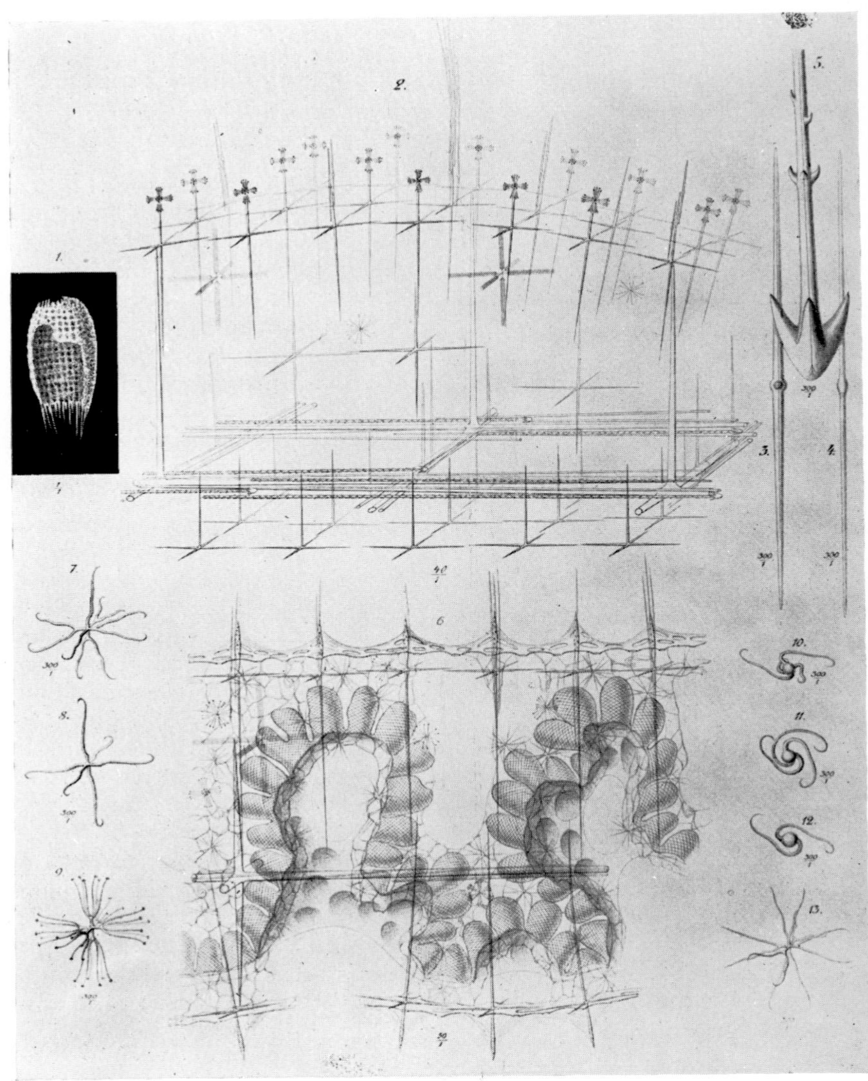

Fig. 4.1. To show the shape of some sponge spicules and their arrangement in the body wall of two sponges, *Euplectella nodosa* and *Holascus stellatus* (from Schulze, 1887).

inorganic material of the spicule is deposited. One of the pair of cells acts as the "founder", establishing the length and shape of the spicule, while the "thickener" later passes over the spicule thickening it up. Triradiate spicules are formed by the association of three pairs of cells whose initial axial filaments join up centrally, producing a three-rayed primordium. When the spicule is complete the cells move away into the mesogloea (Minchin, 1908).

In mature spicules a dark core can often be seen and this is assumed to be the axial filament now encased in the spicule. In electron micrographs of decalcified spicules of *Scypha*, Travis *et al.* (1967) described what they took to be the axial filament. Jones (1967) however, failed to find any evidence of axial filaments in electron micrographs of transverse sections of spicules. Further, he presents evidence that where they have been described both in the developing and in the mature spicule they are artefacts or possibly only a core of impure calcium carbonate at the centre of the spicule. He also questions the intracellular origin of the spicule and suggests that it forms in a space between the two apposed spicule-forming cells, the shape of the space determining the arrangement of the crystals and the shape of the spicule.

In Hexactinellida the six-rayed spicules are said to form in one piece in multinucleate cells. In Demospongia some form of axial filament is almost certainly present (Schulze, 1925; Schröder, 1936). In *Mycale conturenii* the axial filament, formed from spirally twisted filaments 70–100 Å in diameter develops between two large vacuoles in the spicule-forming cell (Levi, 1963). Travis *et al.* (1967) do not mention an axial filament in electron micrographs of demineralized spicules of *Euplectella*. In *Heteromyenia* (Fig. 4.2) it consists

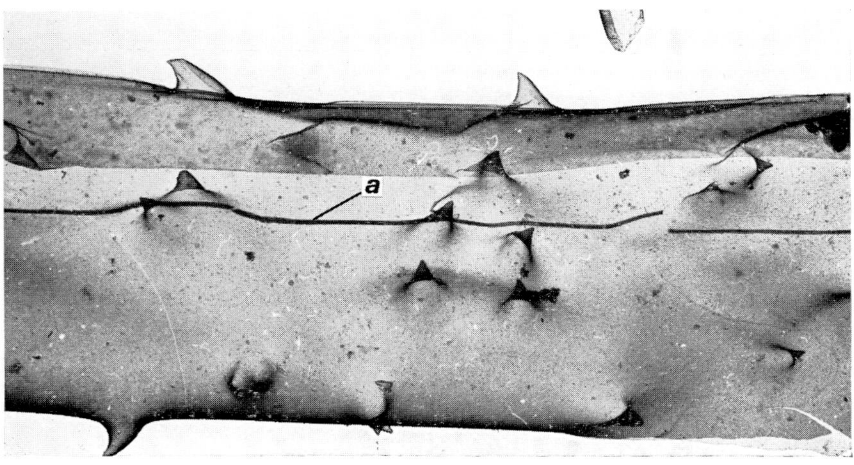

Fig. 4.2. Replica of portion of a spicule of *Heteromyenia* showing axial filament, *a* (Drum 1968). [×4250]

of a carbohydrate, presumably polysaccharide (Drum, 1968), and a polysaccharide surface appears essential for the deposition of silica in a variety of other organisms (Drum, 1968).

Both calcareous and silicious spicules are surrounded by an organic sheath. In *Scypha* it is formed of unstructured, relatively amorphous material in which fibrils about 150 Å thick could be seen (Travis *et al.*, 1967). Associated with the sheaths are fibrils identified as collagen by their axial repeating pattern of approximately 625 Å. While these authors assumed that the collagen fibrils came either from the sheath or from inside the spicule they could be mesogloeal fibrils attached to the sheath to anchor the spicule in position (see Jones, 1967). No X-ray diffraction photograph could be obtained from the sheath, other than that of quartz, that could not be removed by any treatment. In *Clathrina coriacea* Jones (1956) describes the sheath as contracting to considerably less than the size of the spicule after decalcification. An extensible, if not elastic, sheath would seem necessary to allow for the growth of the spicule. Jones found the sheath took up dyes that are known to dye mammalian elastin but chemical analysis of *Scypha* sheaths (Travis *et al.*, 1967) resembles neither that of elastin (Table 4.1) nor that of mammalian collagen.

In *Euplectella* the sheath is composed of a network or lattice of filaments 8–25 Å in diameter associated with amorphous material and a few fibrils with a periodic banding of 520 Å are also present. Although a little hydroxyproline is present, the amino acid composition of the material (Table 4.1)

Table 4.1 Amino acid composition of the organic material from demineralized sponge spicules compared with that of elastin. Number of residues per 1000 total residues

Amino acid	Class Calcarea *Scypha*[a]	Class Hexactinellida *Euplectella*[a]	Elastin[b]
Alanine	86	72	223
Glycine	221	132	332
Valine	45	41	141
Leucine	44	66	64
Isoleucine	27	27	27
Proline	71	73	109
Phenylalanine	16	34	35
Tyrosine	10	25	7
Serine	86	84	8
Threonine	53	57	8
Cystine (half)	17	5·3	—
Methionine	4	12	—
Arginine	38	32	7
Histidine	18	55	—
Lysine	42	41	4
Aspartic acid	112	119	7
Glutamic acid	98	97	15
Hydroxyproline	—	11	—
Hydroxylysine	6·6	1·5	—
Hexosamines	Moderate	High	

[a] Travis *et al.* (1967).
[b] Andersen and Weis-Fogh (1964).

is quite different from that of mammalian collagen. A considerable quantity of hexosamine also is present. X-ray diffraction photographs indicate the presence of a cellulose-like material (Travis *et al.*, 1967) and this may come from the polysaccharide axial filament described by Drum.

While spicules support the body of sponges, Jones (1956) thinks that they may be equally important as protective devices, making sponges unpalatable, though fishes in captivity have been seen to eat them. Calcareous spicules also may buffer the mesogloea against a fall in pH which, Jones has shown, leads to an alteration in its properties. The fine spicules that protrude from the surface may keep the surface clear of detritus and ectoparasites.

2. *Organic skeletons*

In Demospongia, the Tetraxonida have only siliceous spicules, the Monaxonida have spicules associated with organic fibrous material, and the Keratosa have fibrous skeletons. The Keratosa probably evolved from Monaxonida by the loss of spicules and, since they are mainly large sponges, this sort of skeleton must have advantages for the support of large sponge bodies, probably being more pliable and less brittle than a purely inorganic skeleton. The material of the organic skeleton is called spongin.

Spongia graminae has a spongin skeleton of main fibres 0·1 mm thick; smaller fibres form a dense network with regular meshes, on average 0·15 mm wide, formed of primary fibres, 0·027 mm in diameter and secondary fibres 0·01 mm thick. From this sponge Gross *et al.* (1956) obtained two morphologically distinct spongin fibres: spongin *A* and *B*. Spongin *A* is in unbranched fibres approximately 200 Å thick with a 625–650 Å periodic structure. They describe these fibres as forming round the body a cuticle of parallel, closely packed fibre bundles in which are interspersed the remains of cells. In the mesogloea there is a widely ramifying reticular network of similar fibres that join the cuticular fibres without interruption. The spongin *A* fibrils are too fine to form the visible skeleton but build a submicroscopic network in the mesogloea. While the outer layer of fibres may function as a cuticle, their connection with the mesogloeal network suggests that they are a peripheral condensation of mesogloeally produced fibres rather than a secretion by the epidermal cells, as is a true cuticle.

Spongin *B*, in *Spongia graminae*, is in thick branched fibres formed from bundles of thin, unbranched filaments less than 100 Å wide. These occasionally show a faint periodic structure but their diameter is below that at which collagen periodic structures are clearly seen (Smith, 1965). Gross *et al.*, also obtained spongin *A* and *B* from *Microciona proliferata* and *Haliclona oculata*. It is the thick fibres of spongin *B* that form the visible skeletal system.

Although spongin, since it was known to contain sulphur, was thought to resemble the keratin of hair, Marks *et al.* (1949) and Gross *et al.* (1956) have obtained X-ray diffraction photographs from spongin indicating that it is a collagen; Keratosa is therefore a misnomer. Spongin *A* gives a clearly defined photograph with indications of the presence also of a non-collagenous material. Spongin *B* gave a less well-orientated photograph. That collagen fibrils similar to spongin *A* occur in the mesogloea of Calcarea and Hexactinellida is indicated by the work of Travis *et al.* (1967).

The amino acid composition of both spongins *A* and *B* (Table 4.2) approaches that of mammalian collagen with glycine accounting for nearly a third of the residues. Both spongins also contain appreciable quantities of

Table 4.2 Amino acid composition of spongin *A* and spongin *B* from *Spongia graminae*. Number of residues per 1000 total residues. (Gross *et al.*, 1956.)

	Spongin *A*	Spongin *B*
Alanine	56	93
Glycine	315	323
Valine	29	24
Leucine	28	24
Isoleucine	24	17
Proline	78	73
Phenylalanine	9·3	10
Tyrosine	4·7	4·0
Serine	38	24
Threonine	43	21
Cystine	3·3	6·0
Methionine	4·7	3·1
Arginine	47	43
Histidine	3·9	3·7
Lysine	9	24
Aspartic acid	92	97
Glutamic acid	95	86
Hydroxyproline	108	94
Hydroxylysine	12	24

cystine. Sugars are firmly bound to both spongin *A* and *B*, and in much higher quantities than in mammalian collagen (Table 4.3), including arabinose which is not present in mammalian material (Gross *et al.*, 1956). Roche (1952) found considerable variation in the composition of spongin from various species but Gross *et al.* think that his material may not have been satisfactorily purified.

Neither T_S nor T_D have been measured in these collagens. Since spongins *A* and *B* both occur in the same animal they must be subjected to the same

Table 4.3 Sugar content of spongin *A* and spongin *B* relative to that of cow hide collagen. Grammes per 1000 g dry weight

	Cow hide	Spongin *A*	Spongin *B*	Amorphous material (Mesogloea)
Hexosamine	0·5	20	3	30–60
	Glucosamine	Glucosamine	Glucosamine	Glucosamine
	Galactosamine	Galactosamine	Galactosamine	Galactosamine
Uronic acid	0	8	3	1·7–3
Hexose	4	105	50	110–160
	Glucose	Glucose	Glucose	Glucose
	Galactose	Galactose	Galactose	Galactose
	Mannose	Mannose	Mannose	Mannose
	Fucose	Fucose		Fucose
Pentose		28	4	27–43
		Arabinose	Arabinose	Arabinose

environmental temperature yet spongin *A* has a total imino acid content of 186/1000 residues while spongin *B* has only 167/1000 residues. The relationship between environmental temperature of the collagen, T_D, and the total imino acid content cannot therefore hold precisely for both of these collagens.

Spongin *B* fibres are much more chemically resistant than mammalian collagen, being undissolved by acids and alkalis and undigested by collagenase. This resistance might be due to disulphur bonds formed by the cystine. However, the brown to black colour of Keratosa species and the decided blackening that developed in the macerated material during the preparation of spongin *A* and *B* samples in the experiments of Gross *et al.* suggest the possibility of quinone-tanning. Histochemical tests for the presence of quinones gave very indefinite results (Brown, unpublished results). That spongin contains iodine has been known since the work of Fyffe (1819). Akerman and Muller (1941) obtained di-iodotyrosine and di-bromotyrosine from bath sponges and Low (1951) found mono (3-) bromo- and iodotyrosines as well. Tong and Chaikoff (1961a) demonstrated that quinones can activate iodine utilization in tissues and there is evidence from other animals that skeletal materials which are quinone-tanned may bind halogens in appreciable amounts (Roche *et al.*, 1963; Tong and Chaikoff, 1961b; Barrington and Thorpe, 1968), so that the iodine content of spongin may indicate the presence of quinone-tanning.

The formation of spongin *B* fibres was first described by Schulze and has since been described by Loisel (1898), Wilson and Penny (1930), and Tuzet (1932). In *Reniera elegans*, Tuzet found that the cells responsible became arranged in a row, and rod-like units from within each cell joined up to produce the fibre (Fig. 4.3). In *Reniera simulans*, only one cell is concerned in the initial production of a fibre. Herlant-Meewis (1948) failed to find any cells associated with the production of fibres in any of the large number of sponges she investigated and she concluded that they formed from material secreted into the mesenchyme. She observed the intracellular bodies seen by Tuzet and others but never saw them become linked up to form a continuous structure. She concluded that they formed an additional articulated skeleton of unknown chemical composition. Pavans de Ceccaty and Thiney (1963) and Tuzet and Connes (1964) have suggested that in some sponges lophocyte-like cells produce collagen fibrils in a tuft at one end of the cell. Simpson (1963) found in *Microciona prolifera* that the cell which produces the siliceous spicules subsequently produces a cap of spongin *B* at either end of the spicule. Should it be true that spongin *B* fibres are produced intracellularly, then their formation differs from that of mammalian collagen.

Various other fibrillar systems in sponges which presumably are skeletal in function, have been described. Minchin (1900) found in *Clathrina* cells which pass into the gastral cavity between the collar cells and give rise to a cellular network ramifying through the whole gastral lumen. The strands of the network are composed of cells placed end to end and the axis of each strand contains a fibre which stains like the sheath of spicules. Sollas (1888), in the tetractinellid *Dragmastra*, describes connective tissue cells with fibrous walls and axial filaments. Topsent (1893) described in *Clathrina coralloides*,

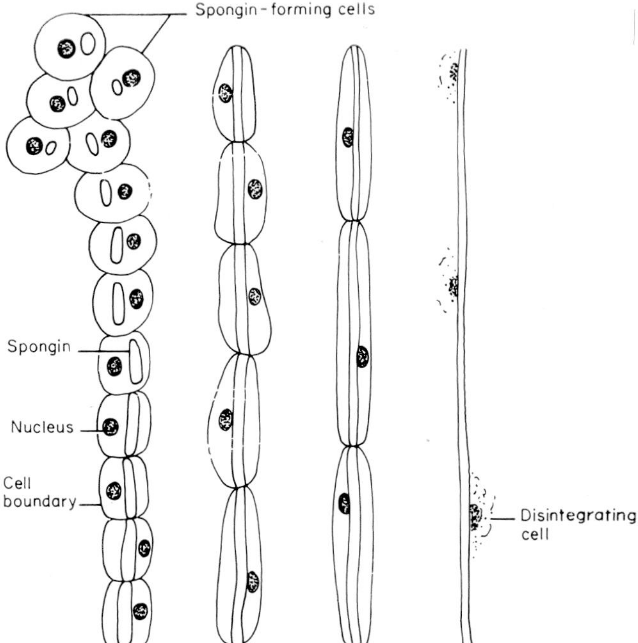

Fig. 4.3. Intracellular formation of spongin *B* fibrils in sponges (after Loisel, 1898).

Echinoclathrina seriata, and *Microciona*, elastic fibres which contract into a spiral when they are severed (cf. elastic fibres in the coelenterates, *Lucernaria* and *Pelagia*). Herlant-Meewis (1948) found fibrous systems that dyed with orcein which she therefore concluded were formed from elastin in a large number of sponges, but Gross *et al.* failed to find elastin in the three sponges they investigated and Jones (1961) found little elasticity in the mesogloea of *Leucosolenia*. The mesogloea in this species is tough and can only be torn by suddenly applied tension. If subjected to gradually increasing tension the mesogloea slowly yields and on removal of tension there is very little recoil. In *Reniera*, Jones (1956) found that it was the ends of mesenchymal cells that became dyed by orcein and Hyman (1940) has described layers of long, fibre-shaped cells in the regions in which Meewis claimed to have found elastin fibres. Fauré-Fremiet (1931) found non-spongin fibres in *Ficulina ficus*. Jones himself has suggested, in the Calcarea, a system of spiralling fibrils in a sub-epidermal membrane which could be responsible for orientation of the spicules.

Sponges are generally fixed to the substratum. In Hexactinellida, and some Tetraxonida living on mud, anchorage is by special spicules, but in the Calcarea, Monaxonida, and Keratosa anchoring roots are formed, and in some types producing spongin a special plate of spongin may be formed sticking the sponge to rock. Brøndsted and Carlsen (1951) have described, in the area of a developing sponge in contact with the substratum, the development of a series of fibres firmly attaching the young sponge to the surface on which it is growing.

The firmness of the mesogloea shows considerable variation not only in different species, but also in different regions of the same species (Laubenfels, 1932). In some species that lack either an organic or an inorganic skeleton the mesogloea may be as stiff as mammalian cartilage, in others it is so fluid that it is deformed by the currents of water passing through the animal. In *Leucosolenia*, Jones (1956) has described the effect that changes in pH have on the stiffness of the mesogloea.

The composition of the ground substance of the mesogloea of *Spongia graminae* has been analyzed by Gross *et al.* (1956). The method of analysis did not lead to the exact identification of the hexosamines present but the overall picture (Table 4.3) is somewhat similar to that of the ground substance of the mammalian connective tissue.

In *Leucosolenia*, Jones (1956) describes the mesogloea as secreted by the choanocytes or by amoebocytes closely applied to them. In *Microciona proliferata*, Simpson (1963) says it is secreted by special rhabdiferous cells. How far this represents a difference from the mammalian condition where the same cells secrete both collagen and ground substance, is not clear, as sponge cells are highly plastic, changing shape and performing different functions at different times.

C. Gemmulae

Sponges may form, within themselves, reproductive bodies called gemmulae. These are aggregations of amoebocytes which become covered by a two-layered membrane. In *Spongilla*, a fresh water sponge, this membrane is said to contain chitin (Wester, 1910; Kunike, 1925). Though Rasmont (1956) found the membrane composed of spongin-like protein, Jeuniaux (1963) again found 1–3 per cent chitin in the gemmulae of *Ephydatia mulleri* and *Spongilla lacustris*. There is evidence that this chitin is firmly linked to a protein.

CHAPTER 5

Coelenterata

Coelenterates are basically radially symmetrical animals showing considerable variation in size from microscopic forms to jelly fish that may weigh half a ton. The body-wall consists of epidermis and endodermis separated from each other by mesogloea. The cellular components of the layers differ considerably from those in the sponges. Coelenterates have a well-developed muscle system which makes possible considerable body-movement and changes of shape that sponges cannot achieve. This requires the body to possess firm anchorages for the muscles and a body-architecture that permits and exploits the muscle activity.

A. External skeletal structures

It is thought that primitively the whole of the epidermis was ciliated and contained nematocysts, and therefore probably lacked a cuticle. Certainly, the absence of a cuticle from some of the largest forms shows that the integrity of these animals is not dependent on an outer membrane holding them together. Essentially they are held together by the structures or forces that hold the individual cells together in the epidermis and endodermis (Grimstone *et al.*, 1959; Hess, 1961) and which anchor the epidermis and endodermis to the mesogloea; much depends, therefore, on the properties of the mesogloea which is the basic skeletal structure. In some classes, however, additional skeletal structures are found.

1. *Hydrozoa*

In the polyp stage many colonial Hydrozoa are covered by an extracellular organic secretion which may be closely applied to the ectoderm as a cuticle or separated from it to form a "periderm" or "perisarc".

In the Gymnoblastea the periderm, where present, stops below the hydranths and gonophores. Rees (1957) suggests that the earliest sedentary hydroids resembled the Corymorphines, solitary forms such as *Euphysa*, *Hypolytus*, *Amalthaca*, and *Acaulis*, living in mud and sand, with the periderm forming a loose sheath round the stalk and pierced by filaments anchoring the animal to the substratum. These sheaths may be discarded and replaced. They can hardly give support but may protect the animal from abrasion by mud or sand. In *Corymorpha nutans* the egg secretes a membrane which Rees (1957) describes as very elastic but which from his description should more correctly be called highly extensible. When the egg settles it pushes out membrane-covered processes from which the cytoplasm then retracts leaving behind the membranous tubules which shrivel and form the first anchoring filaments.

In *Myriothela cocksi* the obvious dark periderm is continuous with a colourless cuticle extending over the hydranth. *Myriothela penola*, which lacks an obvious periderm, is covered completely by a thin cuticle. In these

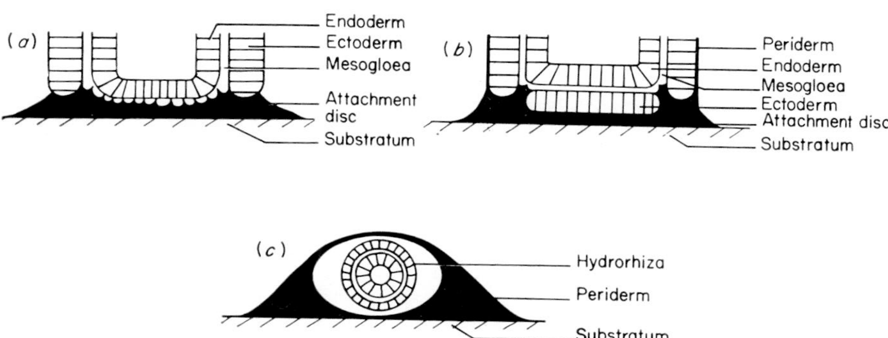

Fig. 5.1. Methods of attachment of Gymnoblastea to the substratum. Diagrams of longitudinal sections through the attachment filaments of (*a*) *Myriothela capensis* and (*b*) *Myriothela cocksi*. (*c*) Transverse section through a hydrorhiza stolon of *Obelia* (after Manton 1940).

two species, the attachment filaments resemble ordinary tentacles, the tips of which become cemented to the substratum. The cement is secreted by globule-filled cells which send processes up into the mesogloea and down through the cuticle, to make contact with the substratum. In *M. penola* and *M. capensis* these cells then disintegrate and the globules run together and harden to form a disc adhering to the substratum and firmly attached to the mesogloea (Fig. 5.1(*a*)). This firm attachment may be correlated with the comparatively large size of the animal (up to 850 mm) and the absence of a periderm. In *M. cocksi* (Fig. 5.1(*b*)) which is smaller, the epidermis at the tip of the attachment tentacle remains intact except around a narrow ring where the attachment disc sends an extension into the mesogloea (Manton, 1940, 1941). In this species, tentacles at the base of the blastostyles, very similar to the attachment tentacles, become attached to the membrane of fertilized eggs. The epidermis in the region of attachment completely disappears and the mesogloea comes into close association with the egg, possibly serving to pass food to it.

In most of the remaining attached forms the base of the animal or the hydrorhiza of the colony becomes cemented to the substratum (Fig. 5.1(*c*))

As soon as the actinula larva of *Tubularia larynx* is liberated, its aboral end becomes covered by a membrane which forms the attachment disc when the larva settles.

While in many colonial forms such as *Pennaria* the stems with their surrounding periderm remain distinct, in *Hydractinia* the stems fuse to form a network of endodermal tubes embedded in mesogloea and sandwiched between an upper and lower layer of epidermis, both of which initially secrete a periderm. However, the upper periderm disappears and the lower layer is continuously increased till it forms a thick mat attaching the colony to the snail shell on which *Hydractinia* normally lives. In the Solanderiidae the thick mat of skeletal material, with the stems of the colony passing through it, appears to be in the mesogloea but Vervoort (1966) has shown that the mesogloea is always separated from the periderm by a layer of epidermis. In *Millepora* and *Stylasterina* a somewhat similarly disposed organic skeleton becomes impregnated with aragonite.

Hydra, classified with the Gymnoblastea, but highly specialized and living in fresh water, has a very thin cuticle (Hess, 1961), which Schulze (1871) describes as attached to the epidermis by numerous small bulbous protruberances. The basal cells contain spherical granules 1–1·5 μm in diameter, with complex internal structure, which are secreted to cement the animal to the substratum (Philpott *et al.*, 1966).

In Calyptoblastea the periderm is extended around the hydranths and gonophores as hydrothecae and gonothecae. In *Campanulina*, *Lovenella*, and the Sertulariidae the hydrothecae have lids of one to several pieces. The hydranth of *Obelia* is anchored at its base to the hydrothecae by desmocytes, dead cells containing hardened material, which stretch from the epidermis into the periderm.

The medusoid stages of Gymnoblastea and Calyptoblastea have never been described as covered by a cuticle, though mucus sometimes covers the surface. Nor have cuticles been found in Limnomedusae, Trachymedusae, or Narcomedusae, in which the polyp generation is absent or reduced.

Siphonophora are polymorphic, swimming or floating, colonial forms that contain both medusoid and polypoid structures. *Physalia physalis* has a cuticle over the float, gastrozooids, tentacles, ampullae, palpons, and gonodendra but not over the gonophores, nectophores, or the jelly polyps, though it is possible that the cuticle over these structures is too thin to have been detected by the means employed by Mackie (1960). The cuticle over the float turns in at the apical pore and is continuous with the lining of the air sac. Over the general body surface the cuticle is friable and splits and flakes away. It is anchored to the epidermal cells by funnel-like structures ending in bulbous swellings which may represent the remains of cells that secreted the cuticle, and recall the desmocytes in *Obelia*. The epidermis of the body surface has gland cells which may be responsible for replacing the worn cuticle but gland cells are missing from the epithelium of the air sac where the lining, protected from wear, does not need replacing. In the hydromedusan, *Phialidium gregarium*, Bonner (1955) found that all cells of the epidermis are concerned with cuticle production.

1. *Composition and structure*

By far the most abundant component of mesogloea is water and in some medusae it accounts for 96 per cent of the material, the rest being made up of 1 per cent organic material and 3 per cent inorganic salts. In an anemone, such as *Metridium*, the organic content is considerably more than this. The organic material is made up of fibres, interstitial material, and sometimes cells (G. Chapman, 1966).

While in some forms the fibres in the mesogloea appear to have no regular arrangement, in *Calliactis* and *Metridium* they are in a geodesic or cross-lattice arrangement towards the inner and outer surfaces, while in between they have a more radial arrangement and this seems well adapted to the marked changes of shape from tall and thin to short and fat that these anemones can achieve through muscular action, and from expulsion or intake of water. In *Metridium* immediately beneath the epidermis and endodermis the mesogloeal fibres are very fine and form a three-dimensional network or "basement membrane", the fibrils of which become directly attached to the cell membranes of the muscle processes (Grimstone *et al.*, 1958). A similar basement membrane has been seen in *Amphinema* (D. M. Chapman *et al.*, 1962) and *Hippopodius* (Mackie and Mackie, 1967). In *Hippopodius* Mackie and Mackie suggest that in some way it may be concerned with the reversible opacity that this animal develops when disturbed, resulting from the production of small opaque granules in this part of the mesogloea. The nearly related *Chelophyes*, which does not become opaque, has no basement membrane.

In schyphozoans, fibres stretch across the mesogloea from epidermis to endodermis and if they are damaged and broken they become convoluted (G. Chapman, 1953a). Bouillon and Vandermeerssche (1957) claimed that the fibres did not dye like collagen but like mammalian elastin, though this was not confirmed by G. Chapman (1959). Elder and Owen (1967) found in the stauromedusan, *Lucernaria*, and in *Pelagia*, fibres forming a network under the endodermis and which at fairly regular intervals give rise to thick fibres that pass through the mesogloea and branch to form a similar network under the epidermis. In sections they appear convoluted (Fig. 5.4), like G. Chapman's damaged fibres in medusae. Initially these fibres do not dye with mammalian elastin dyes but do so after treatment with potassium permanganate solution. They can be distinguished from collagen by their resistance to acid and their digestion by elastin digesting enzymes. They show complete recovery after being stretched to twice their resting length. Elder and Owen failed to find similar fibres in anemones, but Mackie and Mackie (1967) have found them in *Hippopodius*, a siphonophore.

The Octocorallia generally have calcite spicules in the mesogloea formed, at least in the Alcyonacea, within special cells, as may occur in sponges. In the Xeniidae the spicules are minute oval discs, but in all the other forms they are elongated spindles or rods, sometimes ornamented with warts. In Octocorallia, Gorgonacea, Pennutulacea, and Antipatharia the horny material, "gorgonin", forms a central, axial rod running through the stems and branches of the colonies. The rod generally has a spongy core and a

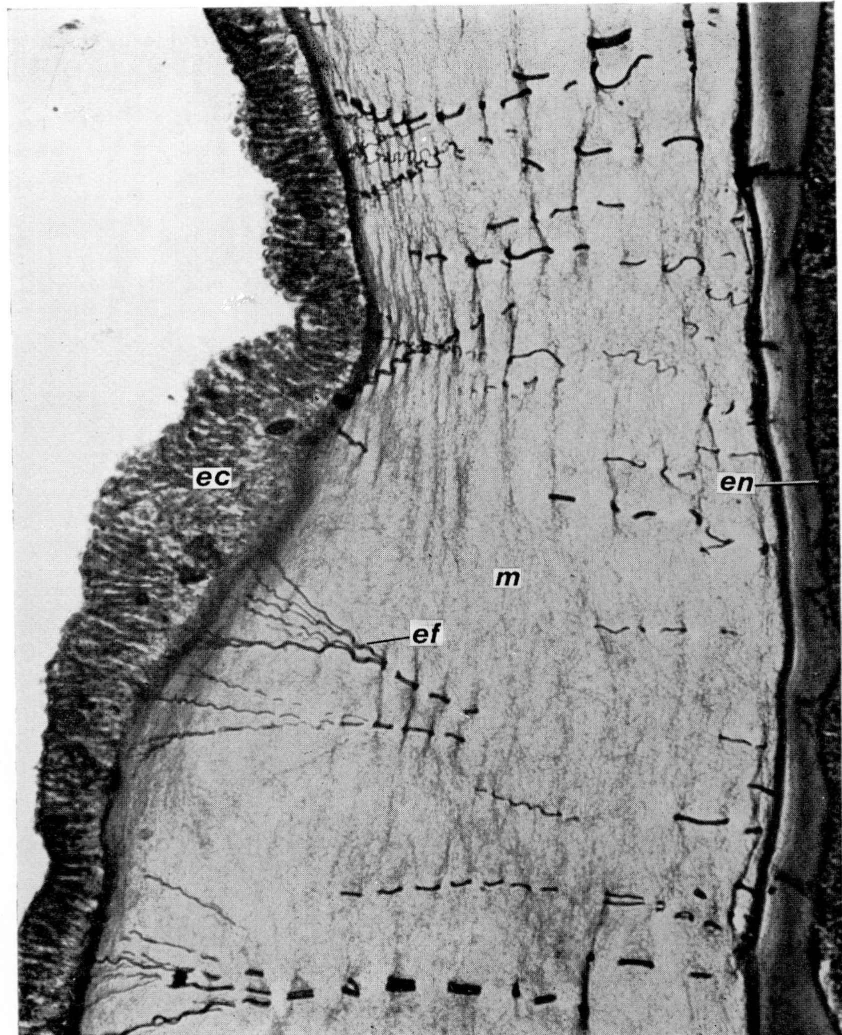

Fig. 5.4. L.S. of the bell of *Lucernaria* showing mesogloeal elastic fibres extending from the endoderm on the right to the ectoderm. *ec*, ectoderm; *en*, endoderm; *ef*, elastic fibres; *m*, mesogloea. (H. Y. Elder.) [×1500]

lamellar cortex and it may, in addition, be impregnated with calcite. The axial rod is always covered by an epithelium, which some claim is epidermal, from the base of the colony, and this would make the axial rod essentially an external skeletal structure. Kukenthal (1916) says, however, that the epithelium is always formed from cells within the mesogloea. In *Corallium*, a gorgonacean, the red coral of commerce, the axial skeleton is formed from calcite spicules united by a calcareous cement; the gorgonin is restricted to fibres in the mesogloea. The red colour is due to iron salts.

In other Octocorallia, e.g. the organ-pipe coral, *Tubipora*, the spicules fuse to form skeletal tubes and transverse platforms. *Heliopora* has a massive skeleton of crystalline fibres of aragonite fused into laminae. The blue colour of the skeleton is due to the presence of bile salts (Tixier, 1945).

2. *Mechanical properties*

Mesogloea is surprisingly strong for a material that may only contain 1 per cent organic matter, a concentration lower than that of the gelatin required to form a firm jelly. In *Metridium* and *Calliactis* Alexander (1962) found that under constant stress the mesogloea stretched very slowly but finally, after many hours, reached a constant length. Elastic recovery was very slow, but more or less complete. The very slow stretching shows that the viscosity of the mesogloea is high though its modulus of elasticity is low (about 0·0003 kg/mm^2) so it is much harder to stretch the mesogloea quickly than to stretch it slowly. This gives the body wall a resistance to sudden knocks, but allows the animal to inflate under the small water pressure produced by the cilia driving water into the enteron. On the other hand, the mesogloea can permit rapid contraction of the muscles. Mesogloea from the jelly fish, *Cyanea* and *Chrysaora*, in similar experiments continued stretching until a very low rate of strain was obtained; elastic recovery was incomplete suggesting some permanent alteration of the mesogloea under prolonged stress (Alexander, 1964). The differences in behaviour between anemone and jelly-fish mesogloea must depend in part upon differences in their fibre content. In swimming medusae, rapid elastic recovery of the mesogloea, deformed by contraction of the exclusively subumbrella muscular system, is the only force to oppose the muscles and bring about their relaxation, and this may be mainly dependant on the elastic fibres that Elder and Owen find suitably situated in various medusae. The mesogloea of *Hydra*, which is a thin lamella containing collagen fibres but in which no elastic fibres have been reported, can be stretched to more than twice its original length and show complete elastic recovery (Shostak *et al.*, 1965).

It is impossible to account for the elastic properties of mesogloea in these forms, in which elastic fibres have not as yet been described, in terms of the deformation of a network of inextensible collagen fibres in a viscous matrix, even if the collagen fibres are free to slip by each other to permit the extension. There would still be no force to produce the elastic recoil. Either elastic fibres are present but have not yet been detected, or the ground substance must be elastic, or the collagen fibres themselves must have elastic properties not found in mammalian collagen. The presence of an ordered structure in the collagen fibres (see below) makes this unlikely.

When the epidermis and endodermis are removed the mesogloea remains intact, maintaining the shape of the animal, and there is little doubt that the mesogloea has a skeletal function not in the traditional sense of the word (σκελετός = dried up) of providing a hard bony framework, but a pliable elastic skeleton to which the cells of the epidermis and endodermis adhere and which can, because of its physical properties, alter its shape in response to muscle contractions and hold the body together against environmental forces.

3. *Chemical composition*

That collagen occurs in the mesogloea has been known since Astbury and Bell (1939) took X-ray photographs of material from a "jelly-fish". Since that time, further X-ray studies have confirmed this observation in Alcyonacea (Champetier and Fauré-Fremiet, 1942), *Alcyonium* (Rudall, 1955), *Physalia*, *Velella*, and *Hippopodius* (Rudall, 1956). However, Bouillon and Vandermeerssche (1956) obtained only the pattern of an amorphous material from *Limnocnida*, *Pelagia*, and *Aurelia*.

Although the fibres often show a periodic banding in the electron microscope the spacing is variable (see Table 5.1). Although in *Pelagia* the periodic spacing (660 Å) approaches that of mammalian collagen, in the majority of species it lies between 200 and 250 Å. The 420–460 Å period in *Mimetridium cryptum* (Batham, 1960) might possibly be divisible into a fundamental period of 200–220 Å. In *Hippopodius* the fibrils are so thin (50–80 Å) that they would not be expected to show banding, and in *Physalia*, where also no banding has been detected, the fibrils have been described as very fine (Piez and Gross, 1959). It is possible to precipitate mammalian tropocollagen as filaments with a 210 Å periodic structure by increasing the sodium chloride concentration of the medium from 0·2 M, at which filaments with the 640 Å banding are formed, to 0·35 M (Gross, 1956). The salt content of the mesogloea could profitably be investigated from this point of view. *Limnocnida*, a freshwater form, with possibly a different salt concentration in the mesogloea, lacks periodic banding in the mesogloeal fibres (Bouillon, 1956).

Chemical analysis of the collagen component of the mesogloea has been carried out on various medusae (G. Chapman, 1953), *Pelagia* (G. Chapman, 1959), *Metridium* (Gross *et al.*, 1958), the float wall of *Physalia* (Piez and Gross, 1959), and *Actinia equina* (Nordwig and Hayduk, 1969). In the *Metridium* analysis the material also contained mesogloeal cells and in *Physalia*, epidermal and endodermal cells. In spite of these sources of inaccuracy, the analyses show a collagen-like protein (Table 5.2) in which glycine forms nearly a third of the residues and with a high content of hydroxyproline, though not as high as in mammalian collagen. It differs from mammalian collagen in containing more tyrosine and in the presence of cysteine, though Nordwig and Hayduk (1969) do not report cysteine from *Actinia* collagen. Also, much greater amounts of hexose and pentose are firmly bound to coelenterate collagen than to mammalian collagen.

In spite of apparent chemical differences, Nordwig and Hayduk (1967) obtained, from *Actinia equina*, SLS with a pattern of light and dark banding that exactly matches those of SLS from calf-skin collagen (Fig. 5.5). While this does not necessarily imply complete chemical similarity between the two, it does strongly suggest a basic similarity of organization.

G. Chapman (1966) found that mesogloea of *Metridium* contracted when heated at 80–90°C, considerably above T_S for mammalian collagen, although

Table 5.1 Collagen fibre thickness and banding periodicity

Source	Thickness (Å)	Band period (Å)	Reference
Spongin *A*	200	625–650	Gross *et al.*, 1965
Spongin *B*	100	faint banding	Gross *et al.*, 1956
Scypha	150	625	Travis *et al.*, 1967
	2000–3000	625	Travis *et al.*, 1967
Euplectella	—	540	Travis *et al.*, 1967
Physalia	very fine	no banding	Piez and Gross, 1959
Metridium	200–300	200	Piez and Gross, 1959
Metridium senile	100	220–250	Grimstone *et al.*, 1958
Mimetridium cryptum	100	420–460	Batham, 1960
Actinia equina	—	650	Nordwig and Hayduk, 1969
Limnocnida	—	none	Bouillon, 1956
Pelagia	—	660	Chapman, 1959
Euphysa	up to 10 000	260	Mackie and Mackie, 1967
Hippopodius	50–80	none	Mackie and Mackie, 1967
Polycelis nigra	100	200	Skaer, 1961
Fasciola hepatica	—	no banding	Threadgold and Gallagher, 1966
Fasciola hepatica	—	640	Nordwig and Hayduk, 1969
Nippostrongylus	—	640(?)	Lee, 1965
Ascaris cuticle	—	none	Various authors
Ascaris body collagen	—	none	Rudall, 1968a
Lipobranchius	200	500	Elder, 1966
Earthworm	100–500	280–640	Rudall, 1968a
	v. fine filaments	shortband period	Rudall, 1968a
Leech		300	Bradbury and Meek, 1958
Achatina	500	210–300	Plummer, 1966
Mya	—	500–600	Elder and Owen, 1967
Busycon	—	600–650	Person and Philpot, 1963
Mytilus	—	500–625	Travis *et al.*, 1967
Helix aspersa	300–500	580	Meek, 1966
Thyone	150–2000	650	Piex and Gross, 1959
	very thin	300–400	Piex and Gross, 1959
Strongylocentrotus	—	667	Travis *et al.*, 1967
Lytechinus	—	625	Travis *et al.*, 1967
Leucophaea (insect)	150–1000	570–610	Harper *et al.*, 1967
Galleria mellonella	~140	~170	Pipa and Woolever, 1965
Peripatopsis	2000	640	Robson, 1964
Limulus		500	Dumont *et al.*, 1964
Vertebrate material			
Cattle vitreous humor	60–150	200	Propst and Leb, 1964
Eye zonula	100–250	300–400	Propst and Leb, 1964
ciliary body	2000	540–600	Propst and Leb, 1964
Vertebrate collagen	100–2000	640	

Table 5.2 Amino acid composition of coelenterate collagens. Number of residues per 1000 total residues

Amino acid	*Metridium*[a] Body wall	*Physalia*[a] Float	*Actinia*[b] *equina*
Alanine	113	66	66
Glycine	308	307	309
Valine	34	26	31
Leucine	37	31	34
Isoleucine	23	22	23
Proline	63	63	75
Phenylalanine	12	11	8
Tyrosine	7·9	5·6	4
Serine	54	47	38
Threonine	39	33	38
Methionine	8·8	5·8	5
Arginine	57	54	62
Histidine	5·1	1·9	3
Lysine	27	27	19
Aspartic acid	81	83	79
Glutamic acid	95	104	88
Hydroxyproline	49	61	92
Hydroxylysine	25	30	28
Cysteine	3·2	1·6	—
Total imino acid content	112	124	167

[a] Piez and Gross, 1959.
[b] Nordwig and Hayduk, 1969.

only containing 112/1000 imino acid residues compared with 215/1000 for calf-skin. In *Actinia equina* there are 168/1000 imino residues and Nordwig and Hayduk (1969) found T_D for this collagen to be 26·6°C compared with 39·0°C in calf skin collagen, a value that conforms better but not very well, with the relationship established between the total number of imino acid residues and T_D (*see* Fig. 5.6(*b*)). This species tolerates a wide range of temperature as it occurs from the Arctic to the Equator.

The decalcified axial rod of the pennatulacean, *Balticina*, gives a typical collagen X-ray photograph but only shows a low angle periodic structure if the skeleton is decalcified in the cold. Decalcification at room temperature destroys this structure, possibly through heat generated by the decalcification (Marks *et al.*, 1949). G. Chapman (1966) also obtained collagen-like X-ray photographs from pennatulacean skeletons.

No full chemical analysis of gorgonin has yet been made. Roche (1952) carried out partial analyses on samples from a variety of species with special reference to their cystine, tyrosine, and halogen content. Reports of high values for cystine content in the older literature have not been confirmed: the content is usually less than 3·5 per cent. The tyrosine content is high and the higher the tyrosine, the higher the halogen content. The halogens are present as mono- and di-iodo-, and bromotyrosines, and their significance is not known. They may result from halogen-binding by quinones (see

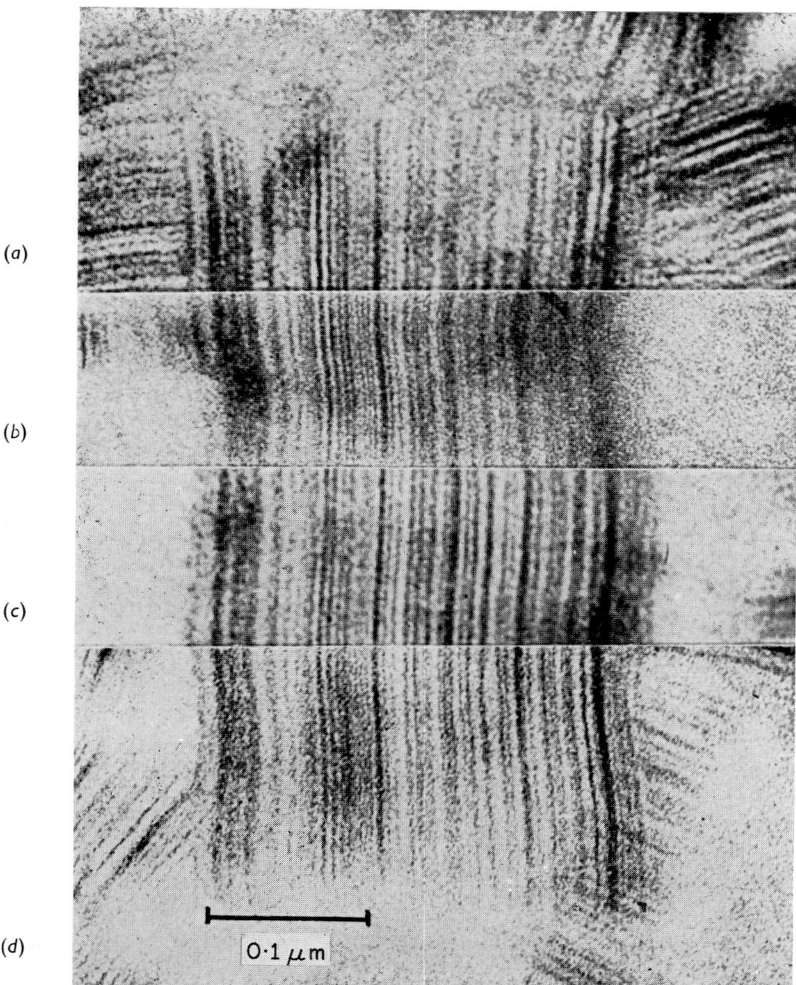

Fig. 5.5. Comparison of SLS of collagen from: (*a*) Mesogloea of *Actinia equina*; (*b*) *Fasciola hepatica;* (*c*) Carp swim bladder; (*d*) Calf skin. Segment length 2800 Å. Positively stained with phosphotungstic acid solution (0·5 per cent) and subsequently with uranyl acetate (1 per cent) (Nordwig and Hayduk, 1969).

Barrington and Thorpe, 1968) though quinone-tanning has not yet been detected in gorgonin.

As for the matrix in which the fibres are embedded, nothing is known about it chemically except that it is faintly dyed by carbohydrate dyes and therefore almost certainly contains mucopolysaccharide, which must be responsible for absorbing the enormous amount of water that is found in the mesogloea. In *Metridium senile*, Grimstone *et al.* (1958) showed that the collagen fibres were surrounded by a definite sheath of amorphous material which may be mucopolysaccharide.

Mesogloea must be secreted, at least in those forms that lack mesogloeal cells, by the cells of the epidermis or endodermis. The formation of mesogloea in *Actinia*, *Calliactis*, and *Sagartia* has been briefly considered by Leghissa and Mazzi (1959).

C. Nematocysts

Nematocysts that occur in all coelenterates are formed in cells called cnidoblasts. The walls of the nematocyst consist largely of protein which in nematocysts of *Corynactis*, Brown (1950b) showed to contain cystine. This has now been found in the walls of a variety of other species (see Picken and Skaer, 1966 for references).

In *Diadumene* the nematocyst capsule, but not the thread, is dissolved by sodium thioglycollate which is known to break disulphur bonds (Yanagita and Wada, 1954). This suggests that the protein of the wall may be keratin-like. Lenhoff *et al.* (1957) found approximately 22 per cent hydroxyproline in discharged nematocysts of *Hydra* and suggested that the wall was made of collagen, but in the nematocysts of *Metridium*, Phillips (1956) failed to find hydroxyproline. Lane Dodge (1958) found the most abundant amino acids in *Physalia* nematocyst walls were proline and alanine and they failed to detect tyrosine, though tyrosine has been found in other species. The exact nature of the nematocyst capsule wall remains uncertain.

D. Cyst and egg shells

Various coelenterates form resting bodies covered by a membrane which in its formation by ectoderm and its chemical composition resembles periderm. Some Scyphistomae that have limited powers of movement produce "podocysts". A mass of cells that forms in the mesogloea near the mouth makes its way to the aboral attachment region and becomes covered by a membrane incorporating the original attachment disc. The scyphistoma

---

Fig. 5.6(*a*)—Key.

●	Internal collagen vertebrates
×	Internal collagen invertebrates
⊗	Cuticular collagen
A	*Ascaris lumbricoides* (Mason and Rice, 1963)
A′	*Ascaris lumbricoides* (Rigby, 1968)
Al	Allolobophora sp.
C	Carp
Co	Cod
Cr	Crocodile
D	*Digaster longmani*
H	Human
Ha	*Helix aspersa*
He	*Helicella virgata*
Ho	*Holothuria*
L	*Lumbricus* sp.
Lf	Lungfish
M	*Melarapha unifasciata*
Ma	*Macracanthorhynchus hirudinaceus*
N	*Nodilittorina pyrimidalis*
P	*Pheretima megascolidioides*
Pk	Pike
Py	Python
R	Rat
S	Shark
St	Sturgeon
Xe	*Xenopus*

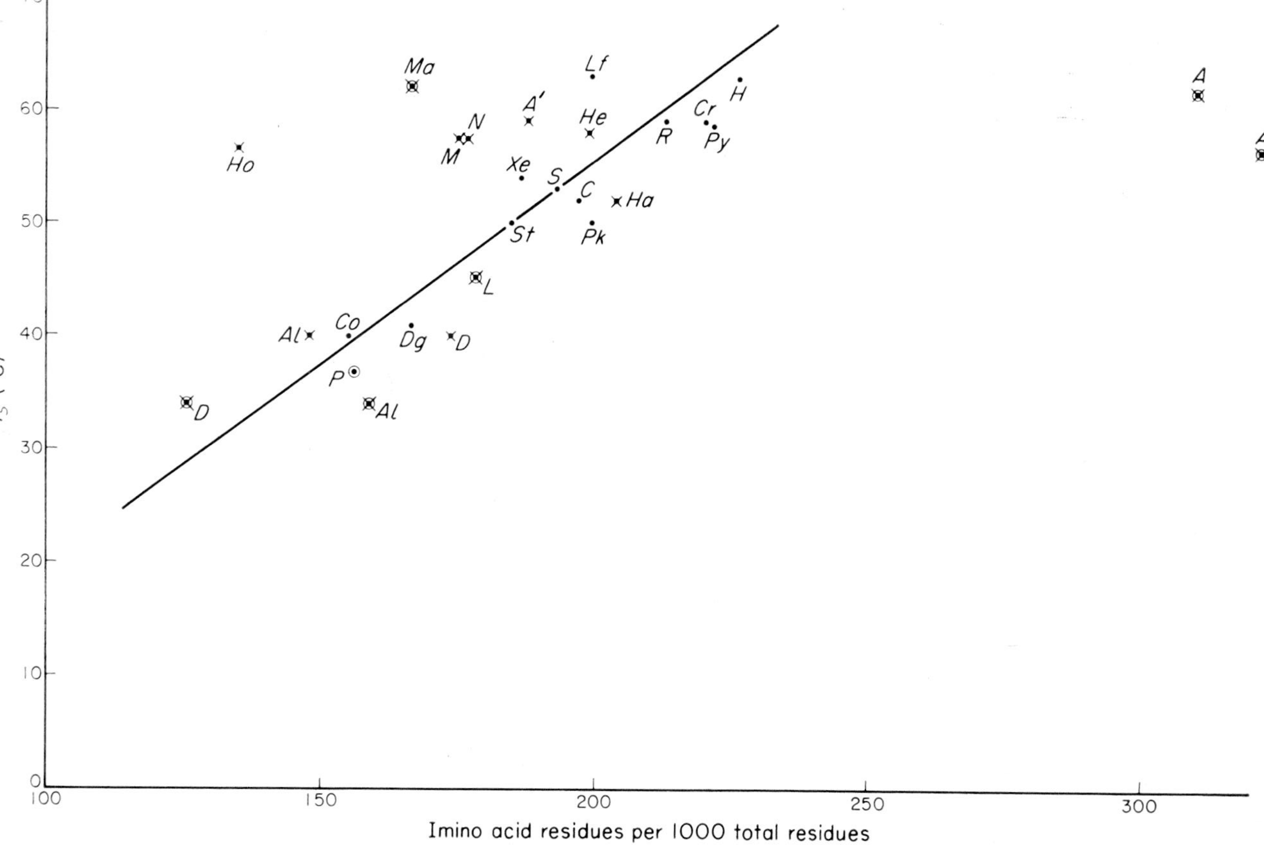

Fig. 5.6(*a*). Relationship between imino acid content and T_s of vertebrate and invertebrate collagens.

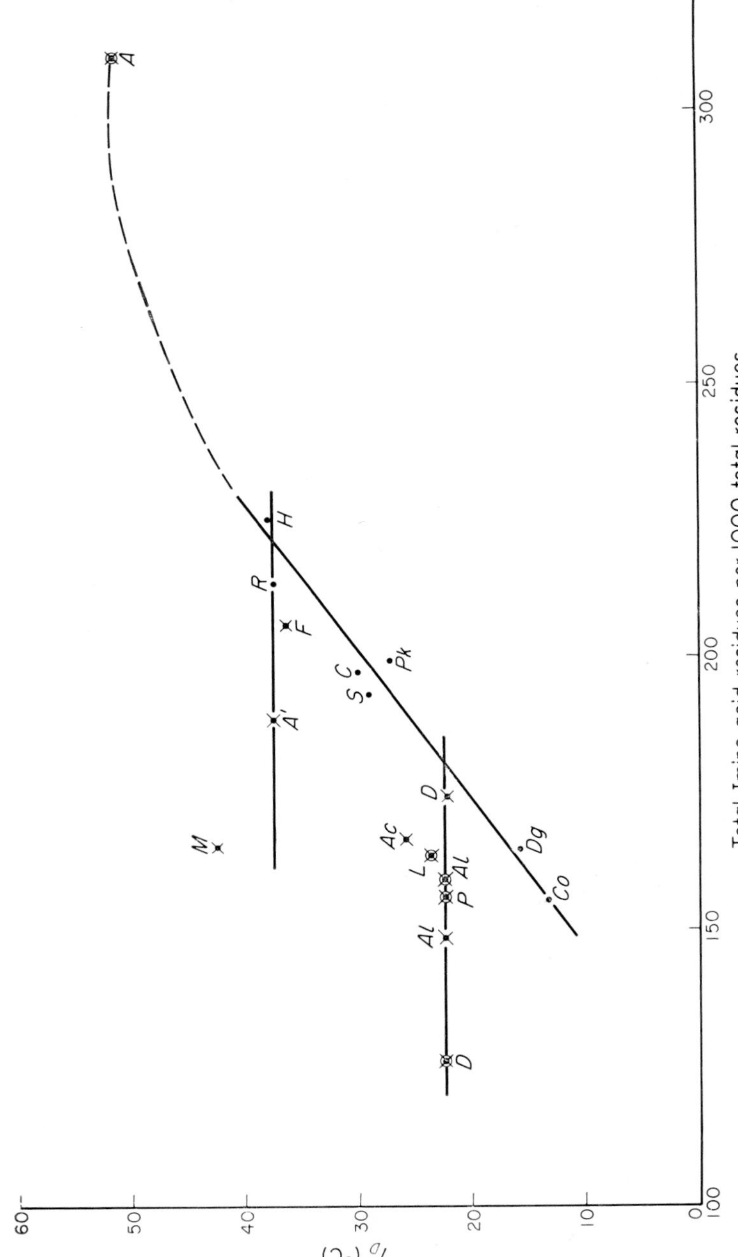

Fig. 5.6(*b*). Relationship between imino acid content and T_D of vertebrate and invertebrate collagens.

detaches itself and moves away leaving the podocyst attached to the substratum where it eventually germinates to produce a new scyphistoma (D. M. Chapman, 1966, 1968). Similar sorts of bodies are also formed in *Chrysaora* (Tcheou-Tai-Chuin, 1930); *Rhizostoma* (Paspaleff, 1938a); *Cyanea* (Verway, 1960) and in the hydroid *Ostroumovia* (Paspaleff, 1938b).

Some eggs become covered by a shell after fertilization, as in *Corymorpha*, which however, develops without delay. A membrane covers the eggs of *Myriothelia cocksi* attaching it to claspers during development (Manton, 1941). In the pelagic tubularian, *Margelopsis*, eggs produced in autumn, unlike eggs produced earlier in the year, do not undergo direct development but form cysts similar in shape to podocysts. The egg of *Hydra* also encysts. Encystment is not common in marine animals, but may be related to seasonal supplies of food.

Fig. 5.6(*b*)—Key.

●	Internal collagen vertebrates
×	Internal collagen invertebrates
⊠	Cuticle collagen
A	*Ascaris lumbricoides* (Josse and Harrington 1964)
A′	*Ascaris lumbricoides* (Rigby 1968)
Ac	*Actinea*
C	Carp
Co	Cod
D	*Digaster longmani*
Dg	Dogfish
F	*Fasciola hepatica*
H	Human
L	*Lumbricus* sp.
Ma	*Macracanthorhynchus hirudinaceus*
P	*Pheretima megascolidioides*
Pk	Pike
R	Rat
S	Shark

CHAPTER 6

Platyhelminthes and Nemertea

1. PLATYHELMINTHES

Platyhelminthes, or flat worms, are at the tissue to organ grade of construction. They lack a coelom and the space between the organs is filled by parenchyma, differing from the mesogloea of sponges and coelenterates in being largely composed of cells. There are three classes of Platyhelminthes: turbellarians are mainly free-living, in the sea, fresh water or damp terrestrial habitats; trematodes and cestodes are, without exception, parasitic. They are mostly internal parasites with complicated life histories, generally involving two or more different hosts but often with certain free-living stages confined to water. Platyhelminthes are usually small, 1 mm to 2 cm in length, although a few turbellarians may reach a length of 50 cm and cestodes a metre or more.

A. External skeletal structures

In turbellarians the surface is covered by a ciliated epidermis though in creeping forms the cilia tend to be restricted to the ventral surface and in some commensals and sand living forms, cilia may be completely lacking. Skaer (1965) has shown that in *Polycelis nigra*, although in the embryo the primary epidermis is formed from flattened blastomeres and is ectodermal in origin, it is subsequently replaced by cells that migrate outwards from the parenchyma and therefore are dermal. The nuclei of the epidermal cells may remain in the parenchyma and are connected with the epidermal layer by a neck of cytoplasm that passes through the muscle layers.

Mucus cells, the cell body lying in the parenchyma and discharging through a canal to the surface, cover the animal with a secretion of mucus which is important in locomotion, in protection against desiccation, and possibly against injurious substances, and may be unpalatable to predators. Mucus may also be used in making resistant cysts (Hyman, 1951a). Beneath the epidermis are layers of circular, oblique and longitudinal muscle and these together with the epidermis make up the body wall of the animal.

Formerly, it was thought that trematodes and cestodes were covered externally by an extracellular cuticle, but electron microscope studies, initially by Threadgold (1963a, b, 1967) (for further references, see Lee, 1966b), have shown that the surface layer in both these classes is a syncytial

of the shell is related to the part that the egg plays in the life cycle of the animal.

Acoela are exclusively marine and Polycladida have only a single freshwater species; they produce thin-shelled eggs. The remaining Turbellaria, with marine, freshwater, and terrestrial species, generally produce eggs with thicker shells. It is as an egg surrounded by a thick, resistant shell that some species survive unfavourable conditions. Many rhabdocoels of the family Typhloplanidae produce thin-shelled "summer" eggs and thicker shelled "winter" eggs.

The eggs of the parastic trematodes are generally laid and develop in water, protected by a thick shell, though some develop in the uterus where sometimes a shell is present, as in the Haplosporidae and Zoogonidae, or may be lacking, as in *Gyrodactylus*.

The pseudophyllid cestodes also lay their eggs in water, where the young develops within the shell, and hatches as a free-swimming coracidium which is subsequently eaten by the intermediate host. The eggs of Tetraphyllidea and Proteocephaloidea, surrounded by a thin membrane, develop in the uterus and are ready to hatch when they are laid in water. Cyclophyllidea have no aquatic stage in the life cycle and the eggs, surrounded by a thin membrane, develop in the uterus and remain there till liberated by the decay of the parent proglottid.

1. *Shell formation*

In Acoela the shell appears to be secreted by the egg itself. In most other Platyhelminthes it is secreted by cells from the vitelline gland that become associated with the egg in the ootype. Formerly, it was thought that the "shell" gland or Mehlis' gland in trematodes and cestodes was responsible for secreting the shell and although its secretions play some part in egg shell formation, only in the cyclophyllidea has it been suggested that they contribute the bulk of the shell.

The egg shells are often brown and in turbellarians (Valli, 1950; Nurse, 1950; Valcurone, 1953), trematodes (Valli, 1950; Stephenson, 1947; Smyth, 1954), and pseudophyllid cestodes (Smyth, 1956; Smyth and Clegg, 1959) shells have all been shown to contain quinone-tanned proteins. The chemistry of tetraphyllid and protocephaloid shells is not known. In the thin shells of cyclophyllid eggs no quinone-tanned protein has been found. The exact nature of these shells is in doubt. In *Moniezia expansa*, *Echinococcus* (Rybicka, 1964, 1966), and *Mesocestoides corti* (Ogren, 1956) it is described as formed from acid mucopolysaccharide, secreted by the vitelline cells, but Hoy and Clegg (see Rybicka, 1966) suggest it is a lipoprotein secreted by the Mehlis' gland. In *Multiceps smythi* Johri (1957) found acid mucopolysaccharide in Mehlis' gland but not in the shell.

There is good evidence for the presence, in the vitelline cells of a variety of trematodes, of a phenolase capable of oxidizing *o*-diphenols to quinones (see Clegg and Smyth, 1968 for references). There is histochemical evidence for the presence of *o*-diphenols in the vitelline cells of *Fasciolahep atica* and *Schistosoma;* however, the material does not appear to be a free phenol but a constituent of the protein from which the egg shell is formed. Rainsforth (see Clegg and Smyth, 1968) has prepared from this material a purified protein giving phenolic reactions. Burton (1963) found that radioactively labelled tyrosine fed to frogs infected with the lung fluke, *Haematoloechus medioplexus*, rapidly appears in globules in the vitelline cells of the fluke and can be detected in the egg shells. Probably the *o*-diphenol is formed by oxidation of tyrosine, therefore tanning of the shell by a quinone, formed initially from tyrosine already part of the protein, would be a form of auto-tanning. The egg shell of *Fasciola hepatica* contains a protein with 28 per cent glycine, 20 per cent serine, 19 per cent glutamic acid, 18 per cent aspartic acid and small amounts of lysine, histidine, arginine, and alanine (Wilson, 1967). Only a trace of tyrosine is present but a spot in the chromatographic analysis was identified as an amino-phenol representing a hydrolysis product of a modified tyrosine molecule, probably resulting from the tanning reaction. This protein resembles no other structural protein yet described.

It is surprising that the substrate and phenolase occur together in the vitelline cells. In trematodes like *Fasciola hepatica*, living in an environment with a very low oxygen content (Brandt, 1952), this lack of oxygen may prevent premature tanning from taking place. When *Fasciola* is cultured *in vitro* under aerobic conditions the vitelline glands turn brown from *in situ* tanning of the globules (Smyth and Clegg, 1959). For species like *Haptomelia*, that lives in the lungs of frogs in a well aerated environment, some other explanation is necessary and possibly some inhibitor system is present, or the substrate needs activation, as in insects.

In both trematodes and cestodes Mehlis' gland is believed to secrete a water-soluble lipoprotein (Clegg, 1965; Hoy and Clegg, *see* Clegg and Smyth, 1968), though Johri (1957) found acid mucopolysaccharide in the Mehlis' gland of *Multiceps smythi*. Dawes (1940) suggested that Mehlis' gland secretions form, round the egg, a thin membrane on to which the vitelline cells secrete the globules that form the shell. Clegg (1965) found a thin membrane of Mehlis gland material on both the inner and the outer surface of *Fasciola* egg shells. There is also evidence of a second type of gland cell in Mehlis' gland which Löser (1965) suggests produces a surface-active agent that assists the running together of the egg shell globules.

The egg shell of *Fasciola* is permeable to water and to a number of compounds with molecular weights below 150 (Rowan, 1962). The lipoprotein membranes may be responsible for the semi-permeability. The hardened shell may not only give protection against mechanical damage but also prevent the egg from bursting should its osmotic pressure become raised.

F. Embryophores

Cyclophyllids, during development in the uterus, lose their egg shell and become enveloped in membranes formed mainly by the embryo itself. It is generally possible to distinguish an inner embryophore and an outer membrane which, although it is generally formed by the embryo, has been described as formed by uterine secretions in *Syncoelium spathulatum* (Coil and

Kuntz, 1963) and uterine secretions may contribute to the outer membrane in *Moniezia expansa* (Hoy and Clegg, see Rybicka, 1966).

Dipylidium has a hardened outer membrane and a thin embryophore. In *Dilepis undula* Orgen (1959) claimed the outer membrane contains quinone-tanned protein but Rybicka (1966) was unable to confirm this. In *Hymenolepis diminuta* (Ogren and Magill, 1962) the outer membrane is composed of a protein containing tyrosine and tryptophan and in the outer membrane of *Moniezia expansa* Clegg and Smyth (1968) found a disulphur-bonded protein.

In some species it is the embryophore which is the main hardened membrane. In the family Taeniidae it is composed of roughly hexagonal columns cemented together (Silverman, 1954b; Lee *et al.*, 1959). Morseth (1965) studied the formation of these columns in *Taenia pisiformis*. Deposition of the columns begins against the inner surface of the outer membranes, material being deposited round "circular bodies" that remain visible in the finished embryophore.

In *Taenia hydatigena*, *T. pisiformis*, and *T. ovis*, Moreth (1966) found the embryophore contained sulphur and has almost the same infra-red absorption pattern as mammalian keratin. Johri (1956) found disulphur bonded protein in the embryophore of *Multiceps smythi* and no quinone-tanning or chitin were present. In *Taeniarhynchus saginatus* Pavlova (1963) claimed to find quinone-tanned protein in the columns and a lipoprotein cementing the columns together.

The embryophores are believed to be intracellular formations. Rybicka (1966) describes them as forming "from the hardening of a cytoplasmic layer" of a syncytium formed from the outer cells of the embryo and in some species the remains of degenerating nuclei are visible in the embryophore. This intracellular origin of the embryophore material, together with the presence in it of disulphur bonds, increases its resemblance to vertebrate keratin. At the end of development the embryo becomes surrounded by a further, thin membrane, the oncospheral membrane that separates the embryo from the embryophore. In *Taenia pisiformis* and *T. saginata*, Silverman (1954b) found this membrane contained lipids.

Cyclophyllid embryos never leave the uterus while the parent parasite remains in its host. Eventually the whole gravid parental proglottid is released from the host and the embryo, within its protective membranes, is released only when the proglottid disintegrates. In some, the uterus may divide up into fibrous sacs or "uterine capsules", each containing several eggs, or the eggs may pass into the parenchyma and become surrounded by a fibrous coat, presumably of collagen fibres, to form "para uterine organs".

The embryo, dormant in the hardened embryophore or outer capsule membrane and possibly also in a "uterine capsule" or "para uterine organ", waiting to be eaten by its intermediate host, is well protected against mechanical damage and bacterial attack. The presence of disulphur bonding in the protective embryophores around cyclophyllid embryos which are deposited on land, compared with the predominance of quinone-tanning in the shells of trematode and cestode eggs that are laid in water, might suggest that disulphur bonded proteins are better able to give protection under terrestrial conditions by being more resistant to the passage of water than are quinone-tanned proteins. Certainly, the egg shell of *Fasciola hepatica* is permeable to water (Rowan, 1962) but Silverman (1965) found that *Taenia saginata* embryos within the embryophore died in 14 days in the absence of surface moisture; although the membranes may reduce the rate at which water is lost it is likely that resistance to desiccation, where it exists, depends not only on the protective membranes but also on the capacity of the living tissue to withstand water loss.

G. Hatching of eggs and embryophores

Factors leading to hatching from eggs and embryophores might be expected to throw additional light on the composition of the egg shells and embryo membranes. Nothing is known about hatching in turbellarians. Hatching in trematodes and cestodes must be carefully regulated so that it takes place under conditions that give maximal opportunity for infection of the host. In trematodes hatching in water there is generally some stimulus, such as light, that induces hatching, probably by stimulating the larva to produce a hatching enzyme as for instance in *Fasciola hepatica* (Rowan, 1956). The specificity of the enzyme has not been determined and it probably attacks not the general substance of the shell but the material sealing the operculum to the shell (see Gönnert, 1962 for a description of the formation of the operculum). Monogenean eggs generally lack an operculum, and the hatching enzymes in this family would warrant careful study as they may be the only enzymes able to disrupt quinone-tanned proteins.

The hatching of cyclophyllid embryos from their enveloping membranes takes place within the gut of the intermediate host. Digestive enzymes of the host, enzymes produced by the embryo itself, and pressures set up by osmosis, may all play a part and these may be aided by movements of the embryo itself. In various Hymenolepidae and Dilepididae that have insect intermediate hosts, the hardened envelope is broken by the insect's mandibles (Smyth, 1969). Hatching of *Taenia* is generally brought about by the digestive enzymes of the intermediate host. Silverman (1954a) found the cement holding the columns together in the embryophore was digested by trypsin in *T. pisiformis* and by pepsin in *T. saginata*. The released embryo is still enclosed by the lipid oncospheral membrane from which it escapes on treatment with pancreatin and bile salts; these stimulate the embryo actively to break out of the membrane. Meymarian (1961) found the cement in *Echinococcus granulosus* undigested by proteolytic enzymes but dissolved away by alkaline solutions.

Perhaps, then, the use of quinone-tanned proteins for the shells of eggs that hatch in water and of disulphur bonded proteins for the membranes of embryos that hatch out in the gut of their host reflects some necessary difference in the hatching mechanisms of the embryos in these two different environments. Alternatively, perhaps they represent different metabolic capacities between the cyclophyllids and the other cestodes and trematodes, or between the vitelline glands that produce the quinone-tanned protein and the ectodermal layer of the embryo that produces an embryophore.

H. Cysts

Both trematodes and cestodes generally make the passage from the intermediate to the definitive host in an encysted state, either in the tissues of the intermediate host or encysted on vegetation that is eaten by the host.

Where the larva encysts within an intermediate host, nothing is known about the structure or composition of the cyst wall. Sometimes the host tissues are believed to contribute to the formation of the cyst.

The cyst that cercarian larvae of *Fasciola hepatica* form on blades of grass is of surprising complexity (Dixon, 1965; Dixon and Mercer, 1964). It has an outer and inner wall, both of which are complex in structure. The outer wall has two layers, an outer of quinone-tanned protein and an inner fibrous layer made up of strands of acid mucopolysaccharide mixed with strands of mucoprotein. The inner wall has an outer mucopolysaccharide layer probably made up of three sub-layers: an outer muco-(or glyco-)protein, a middle acid mucopolysaccaride and an inner, neutral mucopolysaccharide layer. The innermost layer is made up of sheets of disulphur bonded protein in a matrix containing protein and lipids, but over a small area this layer gives way to a neutral mucopolysaccharide plug. The cyst is formed by cystogenous glands in the larva. The innermost layer is secreted by cells containing rods about 0·75 μm $\times$ 6 μm formed of a tightly rolled, continuous sheet of material. The rods are arranged in bundles, each containing some hundreds of rods. Each rod unrolls when extruded from the cell to produce a sheet of material 6 μm $\times$ 30 μm, which associates with similar sheets to form the laminae of the inner cyst layer.

Notocotylus urbanensis also encysts on vegetation and the cyst wall has five layers, largely composed of protein (Herber, 1950). The cyst wall of *Ascocotyle* is resistant to trypsin, ficin, and papain, and contains hydroxyproline (Singh and Lewert, 1959). In *Posthodiplostomum minimum* the cyst wall which is formed by the host fish, is a single layer containing a protein—carbohydrate complex chemically similar to vertebrate connective tissue matrix though no collagen was found in it (Bogitsh, 1962; Lynch and Bogitsh, 1962).

Dixon (1965) studied *in vitro* the excystment of *Fasciola hepatica.* Treatment with pepsin, trypsin, and "pancreatin" gave uncertain results but the larvae were activated in 30 min in 10 per cent sheep bile plus a reducing agent such as sodium dithionite, and hatched out of the cyst on transfer to Eales saline. The cyst wall is not digested away by this treatment which probably activates the larva to break out of the cyst. Perhaps the ventral plug is important in allowing the activating medium to reach the larva. In both the hatching of embryos, and excystment, bile slats play an important and possibly a highly specific part; both Silverman (1954b) and Smyth (1962) have suggested that they are an important factor in controlling host specificity.

Variations in structure and composition of egg shells and cysts within the Platyhelminthes probably are related to differences in the environment, in life cycles, and in habits of the parasite and its hosts, but it is not yet possible to determine what these relationships are.

II. NEMERTEA

The nemertines, mainly marine, bottom-living forms, are long, thin, and either cylindrical or dorsoventrally flattened, though less flattened than platyhelminths, and generally less than 20 cm in length, though some species may measure several metres when fully extended. Their bodies are soft, slimy, and very extensible.

A. External skeletal structures

The surface of the body is covered by a ciliated epidermis and no cuticle is present. The proboscis of hoplonemertines is armed with a stylet of unknown chemical composition though it does not appear to contain chitin (Jeuniaux, 1963).

B. Internal skeletal structures

Control and maintenance of body shape depends on a collagenous dermis underlying the epidermis in all nemertines. In heteronemertines the dermis is very thick. In *Amphiporus lactifloreus* it consists of five or six layers of fibres arranged in alternating right- and left-handed geodesic helices making a lattice whose constituent parallelograms have side lengths of 5–6 μm. The fibres are attached to each other where they cross so that no slipping relative to each other can occur (Cowey, 1952). The same arrangement of fibres has also been found in other nemertines (Clark and Cowey, 1958).

In *Amphiporus* two systems of skeletal fibres also occur in the layer of circular muscle beneath the dermis, one set of fibres running parallel to the muscle fibres at intervals of 10 μm while the other set of fibres runs radially from the basement membrane to the myoseptum underlying the circular muscle layer. This septum is similar in structure to a single layer of the dermis. The myoseptum is, in turn, connected to a series of longitudinal radial membranes on either side of which are arranged the longitudinal muscles (Fig. 6.2). The gut, gonads, and proboscis also are covered by membranes similar to the myoseptum (Cowey, 1952). The spaces between the body organs are filled by a gelatinous material containing a few cells.

The helical arrangement of collagen fibres beneath an epidermis has already been commented on in *Metridium* (Chapman, 1959) and some turbellarians (Clark and Cowey, 1958) though this arrangement was not confirmed in *Polycelis nigra* by Skaer (1961). Such an arrangement of virtually inextensible fibres nevertheless allows changes of shape to take place in response to contraction of circular and longitudinal muscles, by bringing about a change in the pitch of the helices. Since the fibres are of fixed length, this must be accompanied by an alteration in the diameter of the helices tending to produce a change in the volume contained by them. In *Metridium* changes in volume, to a certain extent, can be accommodated by changes in the volume of water contained in the enteron. In other animals in which mouth and anus are normally held closed, the volume enclosed by the helices

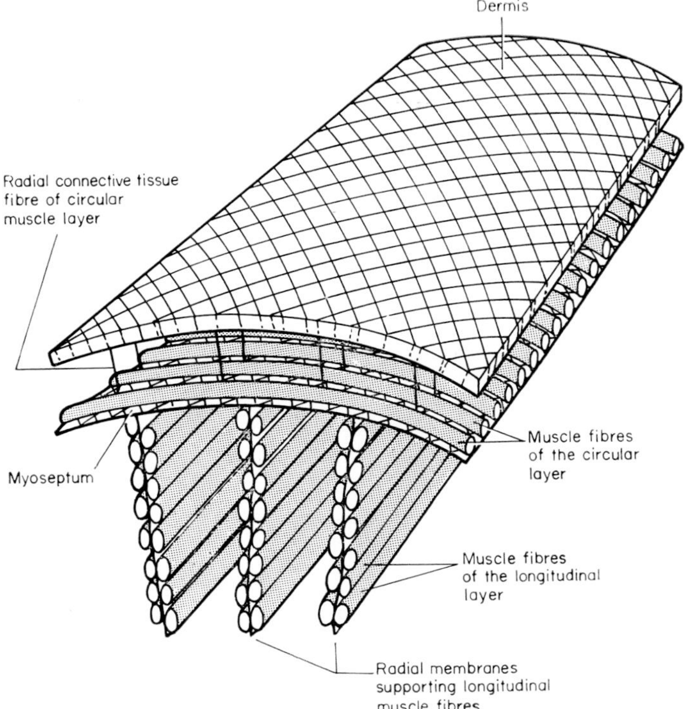

Fig. 6.2. Diagram illustrating the structure of the body wall in *Amphiporus lactifloreus* (after Cowey, 1952).

must remain constant and this can only occur if the cross-sectional shape of the animal is able to alter. In such an animal as a nemertine, lengthening causes the cross section to become more elliptical and the animal becomes flattened, contraction results in a more circular cross section. These changes in shape are accompanied by changes in surface area of the animal.

From a consideration of the arrangement of the fibres in the dermis it is possible to calculate the extension that theoretically a nemertine should be able to undergo (Cowey, 1952). *Amphiporus* can achieve extensions almost as great as those theoretically possible to it, but *Cerebratulus* is considerably less extensible than it should be. Such factors as longitudinally arranged connective tissue fibres and the arrangement and anchoring points of the muscles, and limited deformability of organs may restrict extension in some nemertines. In other animal phyla internal skeletons and external cuticles which may restrict changes in surface area will also restrict extension.

In nemertines mucus, of unknown chemical composition, is used for purposes that in some other phyla are fulfilled by hardened organic or inorganic structures. Besides providing a slime track for locomotion and possibly protecting the animal from predators by its noxious character, it may also be used to make tubes in which the animal lives. Several animals may surround themselves with a common mucus sheath into which eggs and sperm are shed. Mucus sheaths are also formed round starving animals and regenerating fragments, and in freshwater forms may provide protection against desiccation. Eggs are sometimes laid in gelatinous masses, the jelly being produced by epidermal glands. Within these masses the eggs are separate or in small groups.

1–4 mm in length but some parasites of vertebrates may be as long as 40 cm. The diameter of the body in relation to length shows considerable variation, even within a single family. In some the diameter is only $\frac{1}{50}$ of the length; at the other extreme it is as high as $\frac{1}{8}$. The females of heteroderids are short, fat and lemon-shaped.

A. External skeletal structures

The external surface is covered by an extracellular cuticle which is also continued inwards to cover the surface of the oesophagus and rectum.

Although a few species of *Enoplus*, *Mononchus*, and *Dorylaimus*, amongst adult free living forms, and *Acanthocheilus*, amongst the parasites, have smooth cuticles, the majority have them transversely annulated. In Chromodoridae, Steiner and Hoeppli (1926) have correlated the type of annulation with the movements made in capturing food and with habit, some of the animals anchor themselves by twining round, others by hooking on to, seaweed. In the tylenchids, a family of plant parasites, it is possible to correlate the degree of cross striation, and the form of the annuli, with the relative thickness of the worm. The thicker the worm, the more pronounced the striations (Chitwood and Chitwood, 1950). Epsilonematidae, which move like geometrid caterpillars, have the body bent into a shape resembling the greek letter epsilon. The dorsal portion of the annuli is expanded and subdivided to allow for this body shape (Steiner, 1928).

1. *Structure*

It is generally possible to distinguish in the cuticle three more or less distinct layers, an outer cortex, a middle matrix, and an inner basal layer, and each of these layers may be subdivided. They vary considerably in relative thickness. The basal layer rests on the epidermis which is assumed to be responsible for secreting the cuticle. The outer surface of the cuticle is generally covered by an epicuticle, usually only detectable in electron micrographs. Occasionally the cuticle, as in *Ascaris lumbricoides*, is pierced by canaliculae.

In some free-living nematodes annulations at the surface are due in part to transverse structures in the matrix, but in the large parasites of vertebrates it is the cortex that is annulated. Transverse grooving may affect only the outer layer of the cortex, or the inner layers may be involved as well (Fig. 7.3). The outer layers of the cortex generally appear dense and homogeneous while the inner layers are often fibrous. In *Meloidogyne javanica* Bird and Rogers (1965) describe the outer layer of the cortex as formed from very fine radial fibrils.

The matrix layer shows considerable variation in structure not only in different species but also sometimes along the length of an individual. Chitwood and Chitwood (1950) and Inglis (1964a and b) have given the fullest accounts of the matrix layer. It generally contains radially arranged fibres or structures. In the marine, free-living Chromadorinae, generally considered the most primitive nematodes (see Inglis, 1965 for evolution of nematodes), the radial structures are generally columnar. In *Cyatholaimus*

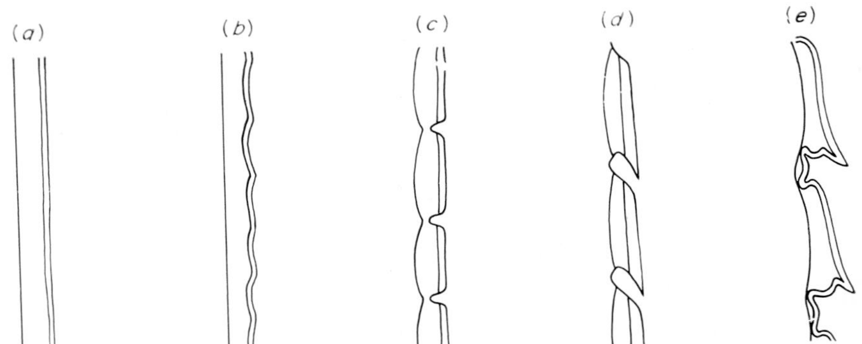

Fig. 7.3. Diagram of longitudinal sections of the cortical layer of nematode cuticles to show different types of annulation. (*a*) *Acanthocheilus;* (*b*) *Anisakis physeteris;* (*c*) *Ophidascaris filaria;* (*d*) *Crossphorus collaris;* (*e*) *Anisakis simplex*. Based on Inglis (1964a and b).

the columns have no regular arrangement (Fig. 7.4(*a*)); in *Mesonchiumnini* they are in transverse rows that correspond with annulations at the surface (Fig. 7.4(*b*)). In *Euchromadora* (Watson, 1965a), *Acanthochus duplicatus* (Wright and Hope, 1968), and the anterior part of *Longicyatholaimus* (Inglis, 1964a) the columns have become greatly elaborated (Fig. 7.4(*c*) and 7.5), but the derivation of these from simple columns is indicated by the gradual change to simple columns in the posterior region of *Longicyatholaimus*. In *Sigmophora*, *Metachromadora*, and *Desmodora* the matrix contains not columns but a longitudinal series of bands or rings encircling the animal (Fig. 7.4(*d*) and (*e*)). In surface view the columnar structures in such a form as *Cyatholaimus* appear tubular, and Inglis believes them to be canals leading from the basal layer to the cortex. He believes that all matrix layer structures can be considered as modifications of this type of canal system.

In the free-living Monohysteridae and Axonolaimidae, believed to be closely related to the Chromadorinae, the matrix layer is like that of *Metachromadora*. Enoploidea, free-living, mainly marine forms, represent another line of evolution from the Chromodorinae. They have, directly beneath the cortical layer, two layers of spiralling fibres, the direction of spiralling being reversed from one layer to the other (Chitwood and Chitwood, 1950; Timm, 1953; Inglis, 1946a). Below these two layers the matrix layer has radially arranged fibres at the inner and outer margins but they do not span the whole width of the layer.

Dorylaimids and mermithids, which are free-living as adults, are closely related to enoplids and in the light microscope appear to have a similar cuticular structure (Fig. 7.4(*f*) and (*g*)). In the dorylaimid, *Xiphinema index*, (Wright, 1965; Roggen *et al.*, 1967) electron microscopy has shown up a very complex structure (Fig. 7.6). A layer (3) of regularly spaced, longitudinal ribbons, roughly rectangular in cross section, lie outside the two spiral layers (4, 5) and at each cuticular grove are connected to an outer triple membrane (1) by a series of thin, radial filaments which pass through the cortex (2) and end in tiny, rod-shaped, electron-dense structures immediately beneath the

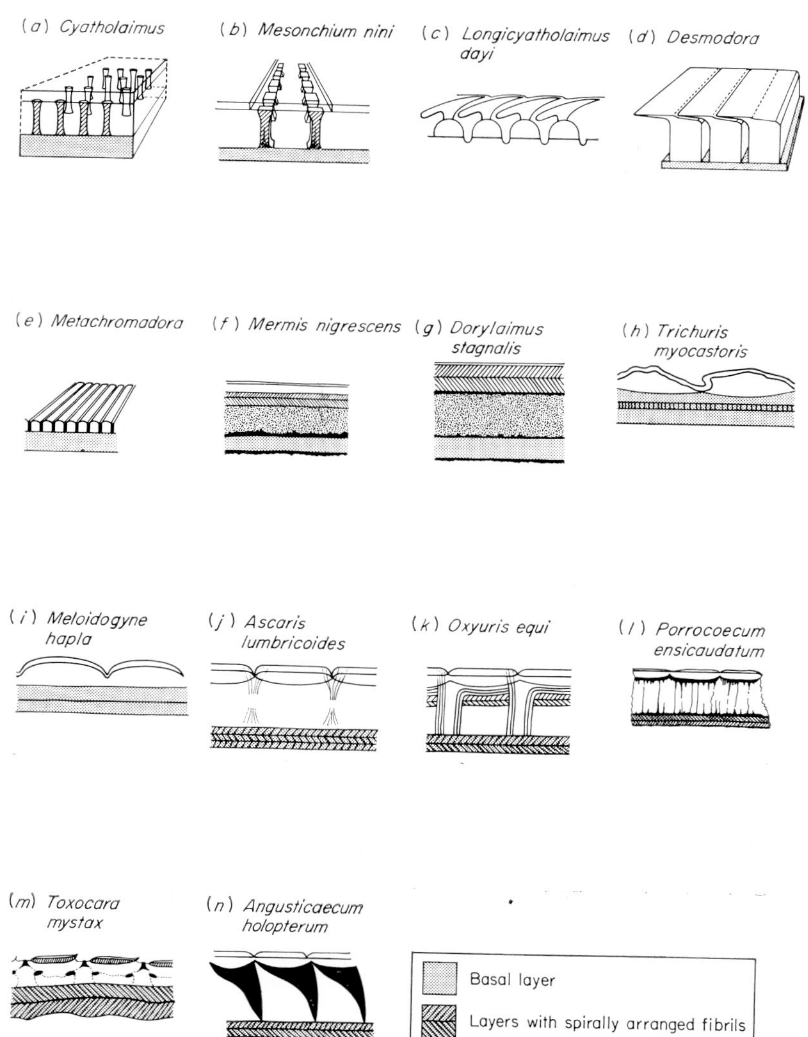

Fig. 7.4. Longitudinal sections of nematode cuticles to show various types of matrix-layer structure. Mainly after Inglis (1964a and b).

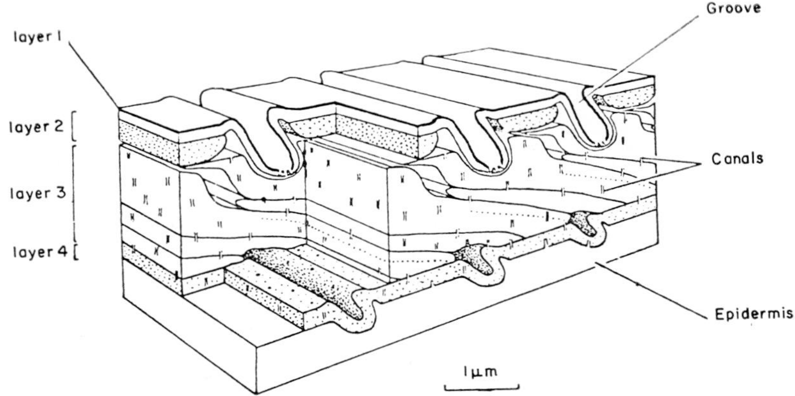

Fig. 7.5. Three dimensional reconstruction of the cuticle of *Euchromadora*. 1, outer cortex; 2, inner cortex; 3, matrix layer; 4, basal layer. (Watson, 1965a.)

bounding membrane. Beneath the spiral fibre layers there is a layer (6) of longitudinal fibres. There is no division of the inner part of the cuticle into matrix and basal layers but only a series of lamellae (7), the number probably depending on the age of the worm.

Parasitic Trichuroidea have affinities with the dorylaimids. The cuticle of *Capillaria hepatica* and *Trichuris myocastoris* (Wright, 1968) and *Trichuris vulpis* (Sheffield, 1963), however, is morphologically very simple. It is a reticulum of rather low density which in *C. hepatica* has a very irregular contour forming crests and valleys. In *T. myocastoris* it is broken up into blocks corresponding with the annuli (Fig. 7.4(*h*)).

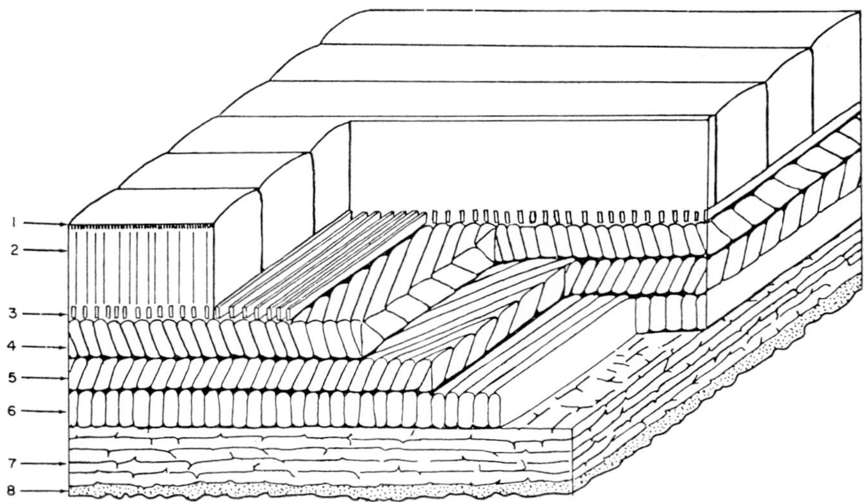

Fig. 7.6. Schematic drawing representing the structure of the cuticle in *Xiphinema index* (*see* text). (After Roggen *et al.*, 1967.)

The Rhabdoidea, mainly free-living or plant parasites, are believed to be related on one hand to the chromadorids and on the other to the parasitic oxyuroids and strongyloids. In the majority the matrix layer is homogeneous or fibrous, e.g. in *Rhabditis pellio* (Beams and Sekhon, 1967), the male of *Heterodera glycines*, and *Hoplolaimus tylenchiformis* (Hirschmann, 1960) and in *Meloidogyne javanica* (Bird and Rogers, 1965). In *Meloidogyne hapla* (Fig. 7.4(*i*)) the matrix is subdivided into two dense layers.

Strongyloids, oxyuroids, and ascaroids, which are believed to have evolved independently from the rhabdoids, show a marked similarity in the organization of their matrix layer, which is more complex than has yet been found in rhabdoids.

The cuticle of *Ascaris lumbricoides* has been described by Chitwood and Chitwood (1950), Bird and Deutsch (1957), Inglis (1964a), and Watson (1965b). It is very similar to that of *Parascaris equorum* described by Hinz (1963) and of *Strongylus equinus* (Bird, 1958). In the matrix, under each transverse cortical grooves, there are bundles of radial fibres. At the inner and outer ends of the bundles the fibres splay out and externally connect with tube-like structures in the cortex (Fig. 7.4(*j*)). In young adults the fibre bundles extend right across the matrix layer but in fully grown adults the bundles break and the inner and outer ends become separated by homogeneous matrix material. Bird (1958a) and Watson (1956b) both describe the fibres as tubular but Hinz did not find them so in *Parascaris*. In *Oxyuris equi* there are two rings of bundles of fibres per groove, fibres from the bundles of one ring penetrate the cortical layer while those from the other ring turn and run longitudinally just below the cortical layer towards the next cortical groove behind (Fig. 7.4(*k*)). Immediately beneath these longitudinal fibres there are two layers of spirally running fibres (Bird, 1958a). From Bird's photograph they do not form a continuous layer but are interrupted at each cortical groove.

In some ascaroids (Inglis, 1964b) the radial fibres are not aggregated into bundles and again, as in *Porrocoecum* (Fig. 7.4(*l*)), appear to have been broken so that homogeneous matrix separates the two ends. In a number of species regularly arranged circular fibres are also present, e.g. in *Toxocara mystax* (Fig. 7.4(*m*)). In *Angusticaecum holopterum* (Fig. 7.4(*n*)) these circular fibres have become organized into wedge-shaped structures. In *Aspicularis tetraptera* (Anya, 1966a) the matrix contains longitudinal fibres arranged in waves (Fig. 7.7). In *Nippostrongylus brasiliensis* Lee (1965a) has described longitudinal ridges in the matrix which are formed by a series of approximately triangular plates or struts, each held in place by fibrils that connect them to the cortex and the basal layer (Fig. 7.8). These fibrils in the electron microscope appear banded like collagen fibrils. In the development of definite radial structures the cuticles of these nematodes appear more closely related to the cuticle of chromadorinids. In some of the filarioids, which are distantly related to the ascaroids, the matrix is a thick fibrous layer, as in *Setaria* (Kagei, 1960); in others, e.g. *Litomosides carinii*, the cuticle is a single homogeneous layer (Armstrong and Hawkins, 1964).

The basal layer varies considerably in relative thickness. In the majority of

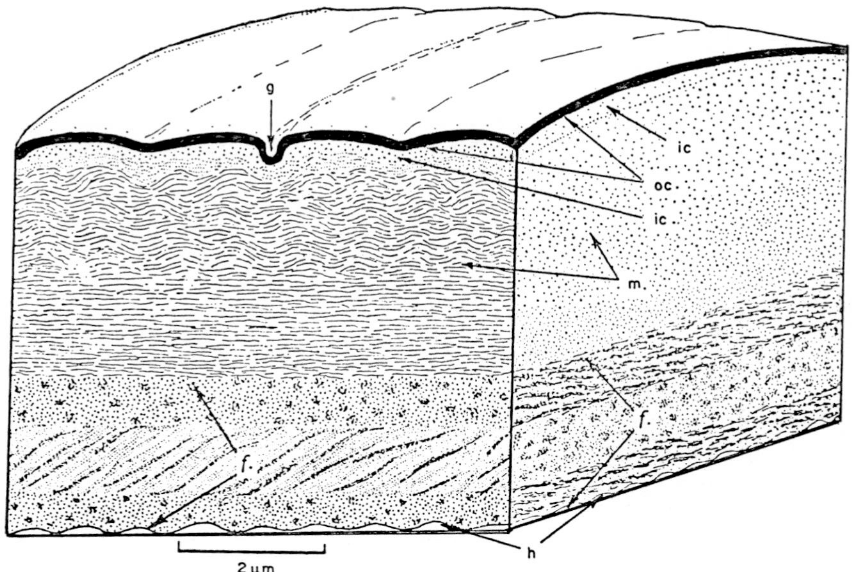

Fig. 7.7. Diagrammatic, three-dimensional drawing of the cuticle of *Aspicularis tetraptera*. *g*, grooves of the cuticle; *ic*, inner cortex; *oc*, outer cortex; *m*, matrix layer; *f*, basal fibre layer; *h*, "hypodermis" (epidermis). (Any, 1966a.)

chromadorinids, enoplids, dorylaimids, mermithids, monohysterids, and axonolaimids, the basal layer shows no structure except where its inner surface may be modified as a basal membrane. The basal layer of *Capillaria hepatica* (Wright, 1968) contains fine transversely orientated structures. In rhabdoids, a similar fine radial structure is found in the basal layer of most species examined (*Rhabditis strongyloides:* Peebles, 1957; *Turbatrix aceti:* Watson, *see* Lee, 1966b; *Meloidogyne javanica*, Bird and Rogers, 1965; see also Wright's (1968) comments on Bird and Rogers description of the cuticle; *Ditylenchus dipsaci:* Yuen, 1967). In the majority of large nematodes parasitic in vertebrates the basal layer is divided into two or three layers of fibres that spiral round the body. *Ascaris lumbricoides* has three layers, the fibres in the outer and inner layers are parallel while the fibres in the middle layer are at an angle of approximately 135° to those in the outer and inner layers. (Picken *et al.*, 1947). In *Aspicularis tetraptera* the fibres in the outer and inner layers do not appear well orientated though the fibres in the middle layer take an approximately spiral course. Jamuar (1966) found in the outer basal layer of *Nippostrongylus brasiliensis* some fibres with a periodic banding of 640 Å and Hinz found banded fibres in *Parascaris equorum*.

Where the cuticle lining the oesophagus has been examined (Lee, 1966b) it is simpler in structure than that covering the body. At the head end of many nematodes the cuticle is modified to form specialized structures such as canals supporting the style used to penetrate plant tissues (e.g. in *Xiphinema*

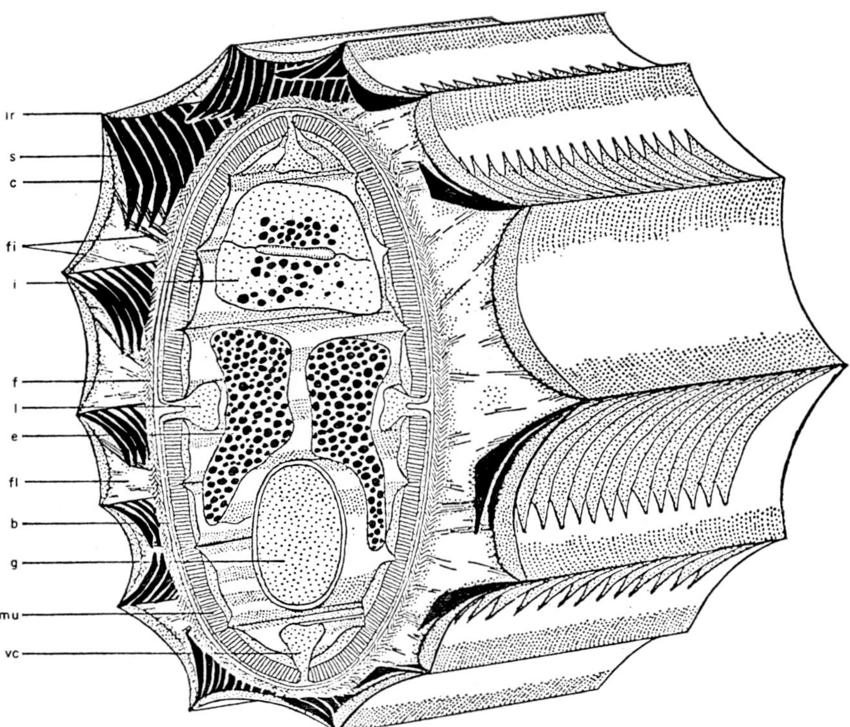

Fig. 7.8. Stereogram of a thick section taken from the middle region of an adult *Nippostrongylus brasiliensis* showing general anatomy and the arrangement of the various layers in the cuticle. *a*, Attachment of muscle to cuticle; *b*, basement lamella; *c*, cortex; *e*, excretory gland; *f*, fibre layer of cuticle; *fi*, fibrils of collagen; I, fluid filled layer of cuticle; *g*, gonad; *i*, intestine; *l*, lateral cord; *lr*, longitudinal ridge of cuticle; *m*, mitochondria; *mu*, muscle of body wall; *s*, skeletal strut; *vc*, ventral cord. (Lee, 1965.)

index, Roggen *et al.*, 1967). In others it may form a structure (Fig. 7.9) supporting the whole head, as in *Ditylenchus dipsaci* (Yuen, 1967).

Nematodes moult their cuticle four times during development; the mechanism of moulting is little understood but Rogers (1962) and Rogers and Somerville (1963) suggest it may be under neurosecretory control. When *Trichinella* moults (Lee, 1966) almost all the old cuticle is shed, little

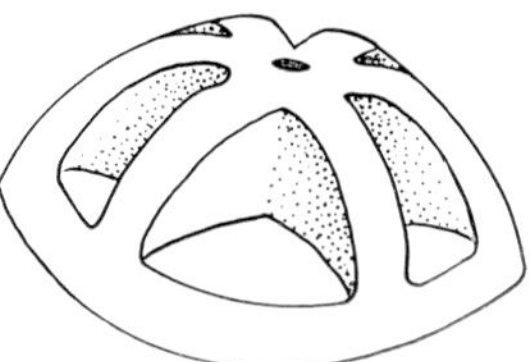

Fig. 7.9. Schematic diagram of the head section of *Ditylenchus dipsaci* (after Yuen, 1967).

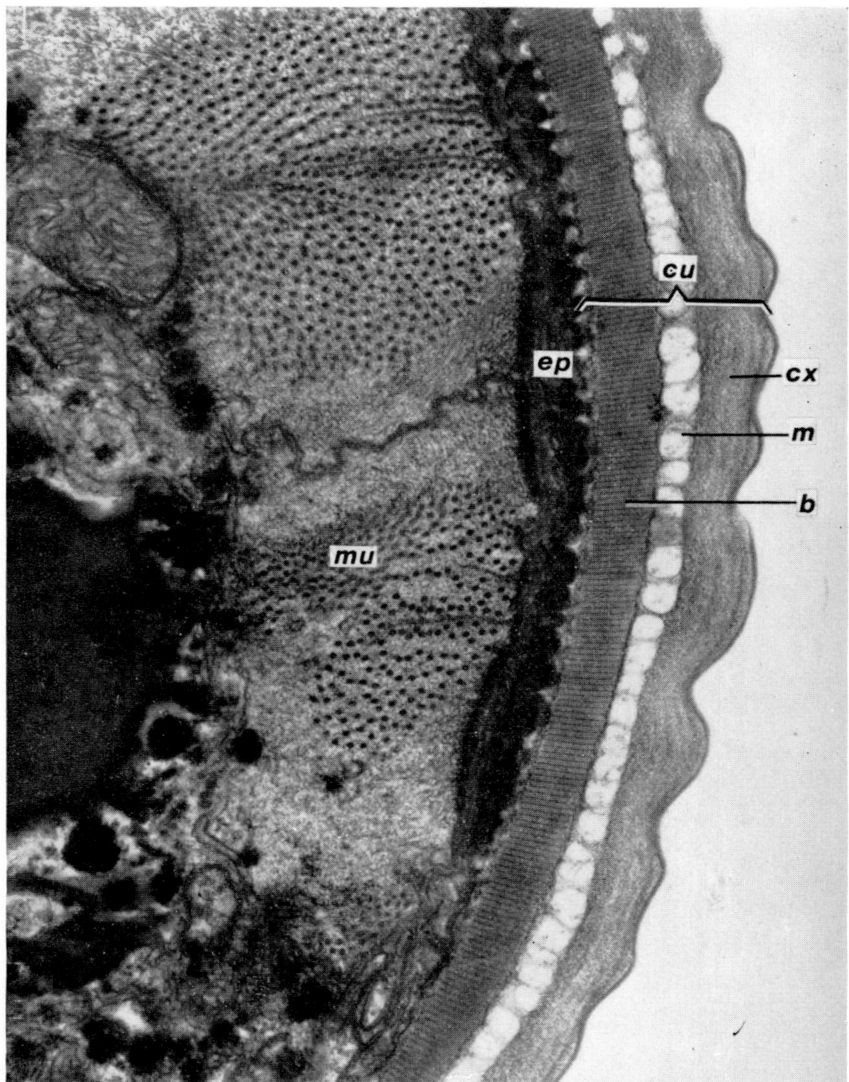

Fig. 7.10. Transverse section of the cuticle of second stage larva of *Heterodera rostochiensis* *cu*, cuticle composed of *cx*, cortex; *m*, matrix layer; *b*, basal layer; *ep*, epidermis; *mu*, muscle cell. (Wisse and Daems, 1968.) [×26 000]

being resorbed. In *Meloidogyna javanica* (Bird and Rogers, 1965) almost all the old cuticle is resorbed, only the outer layer of the cortex being shed. The new cuticle forms beneath the old before it is shed and is convoluted to allow for rapid expansion of the body after moulting has taken place (cf. moulting in arthropods.)

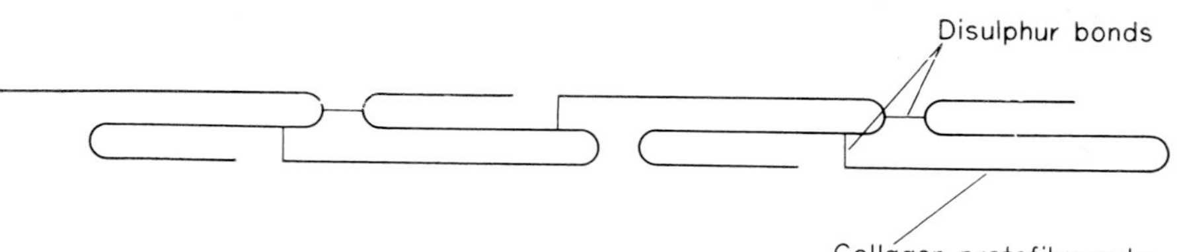

Fig. 7.12. The proposed model for the arrangement of polypeptide chains in *Ascaris* cuticle showing the disulphide cross linkages between the subunits. Regions of chain overlap within a subunit and are thought to be in the triple helix pattern as in vertebrate collagen (after McBride and Harrington, 1967).

nematode cuticle fibres suggest that the organization of the molecules into fibres differs from that in mammalian collagen.

Carbohydrates have been found in Ascaris cuticle (Chitwood, 1936; Fairbairn, 1957; Fairbairn and Passey, 1957), and large quantities of non-glycogen carbohydrate in the cuticle of *Trichuris myocastoris* (Wright, 1968). Monné (1959) found mucoid material in various nematode cuticles and Anya (1966) identified hyaluronic acid and chondroitin sulphate-containing mucopolysaccharides in *Aspicularis*. Wright (1968) suggests that the carbohydrate in *Trichuris* may be concerned in providing a permeability barrier in the cuticle. There is evidence that the cuticle is metabolically active, since ribonucleic acid, ascorbic acid, adenosine triphosphate, and acid phosphatases have been found in the cuticle of *Aspicularis syphacia* and *Ascaris* (Anya, 1966a, e) and esterases and haemoglobin in that of *Nippostrongylus* (Lee, 1965); some of the carbohydrate may be a source of energy for growth and for the control of diffusion through this cuticle.

Various bristles and stylets occur in nematodes and the spicules in the male may be necessary to open the vulva of the female against the high internal pressure of the body fluid. Nothing, is known of the chemistry of these structures or their method of formation.

B. Internal skeletal structures

The space between the body wall and the viscera is filled with fluid, and also contains fibrous material. In some cases the fibres appear to be processes of cells. In the ascaroids and other parasitic worms there is often a single cell dorsal to the pharynx which gives rise to a fenestrated fibrillar membrane covering the viscera and muscle layers. Though no banding can be detected in the fibrils, Rudall (1955) obtained from this membrane in *Ascaris lumbricoidēs* a typical collagen X-ray photograph. Dawson (1960) found that a membrane surrounding the longitudinal muscles was destroyed by bacterial collagenase. Rudall (1968a) has analysed the intestinal membrane of *Ascaris* for glycine, proline, and hydroxyproline, though he was doubtful about the purity of his sample. This collagen contained fewer glycine residues (242) and proline residues (86) than does the cuticle collagen, and the content of both these amino acids falls far below that of mammalian collagen. It contains more hydroxyproline (86 residues) than does the cuticle collagen, and in the content of this amino acid approaches that of mammalian collagen. Neither T_S nor T_D have been measured. Rigby (1968) has analysed an internal body-wall collagen from *Ascaris lumbricoides* (Table 7.1). This collagen resembles that from the intestinal membrane in having less glycine than mammalian collagen. It has less hydroxyproline than that from both mammals and the intestinal membrane but more tyrosine and glutamic acid than mammalian material. Nematode collagen has a total imino acid content of 188/1000 residues, well below that of collagen from that of its host, the pig, with 226/1000 residues. The T_S (59°C) and T_D (38°C) of *Ascaris* collagen are virtually the same as the T_S (58°C) and T_D (37°C) of pig collagen, and in both collagens T_D is related to the environmental temperature which is that of the body temperature of the pig (37°C). In *Ascaris* collagen therefore while T_D is related to environmental temperature as in mammalian collagens, T_S and T_D are both much higher than would be expected if its stability had the same relationship to total imino acid content as does pig collagen. Some different factor must in part determine the stability of *Ascaris* collagen. Disulphur bonds, which are important in the organization of *Ascaris* cuticle collagen, cannot contribute to the stability of the internal collagen as it contains no cystine. The relatively high content of tyrosine might be significant. Elder and Owen (1967) failed to find any elastic fibres in nematodes.

The nematodes, together with the other classes that make up the phylum, Aschelminthes, are the first metazoan phylum not to have the gut and reproductive organs supported by a solid mesogloea or parenchyma. The collagenous connective tissue sheath around organs is a common feature of animals with some form of body cavity. It supports the organs and protects them from frictional wear and tear due to movements of body fluid, resulting from movements of the animal.

C. Egg shells

Nematode eggs are fertilized internally and become enveloped in membranes. The eggs generally undergo some development in the uterus. In some species the eggs then are passed out from the host and hatch, when conditions are favourable, into free-living larvae. In these species the egg shell is generally thin, though in some, e.g. *Nematodirus*, that over-winter as an egg, it is thick. In many species no free larval stages are produced, the egg enclosed in a thick shell waits till it is eaten by the appropriate host before it hatches.

Eggs of ascaroids and oxyuroids are covered by three membranes. The inner and middle membranes are secreted by the egg itself while the outer membrane is believed to be formed from secretions of the oviduct (Anya, 1964). In *Aspicularis tetraptera* the outermost layer is formed from lipoprotein, the middle layer contains chitin, lipid, and protein and the innermost layer again contains lipid and protein. Chitin has been found in the egg shells of various ascaroids (Fauré-Fremiet, 1913; Chitwood, 1938; Monné, 1955), of other oxyuroids (Jacobs and Jones, 1939, Crites, 1958), of *Distophyma renale* and *Ditylenchus dipsaci* (Chitwood, 1938), and of *Heterodera rostochiensis*

(Tracey, 1958). However, Monné and Honig (1954a) and Monné (1955) failed to find it in the egg shells of *Capillaria vulpis*, *Trichuris*, *Echinuria uncinata*, or in any of the strongyloids, and Mansard (1954) doubted if it occurs in *Subulura distans* and *S. dispar*. In ascaroids the chitin is generally in the form of thin lamellae between lamellae of protein (Monné and Borg, 1954; Monné, 1962).

Monné (1959, 1962) reported quinone-tanning of the protein in the shells of ascaroid and oxyuroid eggs and also in those of *Trichuris vulpis*, *Capillaria hepatica*, and *Echinuria uncinata*. Though the outer layer of *Aspicularis* egg shell is brown, Anya (1964) could find no polyphenols or polyphenol oxidases associated with the production of the shell. If quinone-tanning does recur in

Table 7.2 Amino acid composition of hydrolysates of the egg shells of *Heterodera rostochiensis*. Grammes per 1000 g. (Clarke *et al.*, 1967.)

	6 hours hydrolysis	12 hours hydrolysis
Alanine	28	23
Glycine	98	85
Valine	15	19
Leucine	19	16
Isoleucine	13	14
Proline	342	383
Phenylalanine	18	15
Tyrosine	31	21
Serine	83	74
Threonine	19	27
Cystine	26	—
Methionine	06	26
Arginine	22	17
Histidine	17	23
Lysine	35	28
Aspartic acid	114	109
Glutamic acid	52	60
Hydroxyproline	47	52
Hydroxylysine		
Total imino content	389	435

these shells Anya suggests that it must result from some non-enzymatic oxidative deamination of the tyrosine present in the shell protein (Hackmann, 1958), producing a form of autotanning. Cystine has also been found in ascaroid shells (Monné, 1962) and in *Aspicularis* shells (Anya, 1964), and may contribute linkages, stabilizing the shell material.

A chemical analysis of the egg shell of *Heterodera rostochiensis* by Clarke *et al.* (1967) gives 59 per cent protein, 9 per cent chitin, 7 per cent lipid, and 3 per cent polyphenols. The presence of hydroxyproline (Table 7.2) suggests that some of the protein may be collagen, though glycine is not abundant. X-ray photography gives no indication of crystalline substances in the shell.

The inner membrane in *Ascaris* is composed of ascarosides which are glycosides of three long-chain alcohols ("ascaryl alcohols"). The monose "ascarylose", 3,6-dideoxy-L-arabinose is peculiar to this material (Fauré-Fremiet, 1913, see also Jezyk and Fairbairn, 1967). Whether this is the case in all nematodes is not known. The relative impermeability of nematode eggs has been attributed to the properties of this inner membrane. During uterine development of *Enterobius vermicularis* the inner membrane is permeable to water and salts but later it becomes impermeable and further development can take place in media that are toxic to the free larva. In *Ascaris lumbricoides* embryonic development takes place outside the host, moisture is necessary for development and drying the eggs kills the embryos, but the egg membranes are able to keep substances like mercuric chloride. Generally, the inner membrane is thin but in some oxyuroids it is thick and fibrous. In *Heterodera rostochiensis* while the egg is in water it is permeable to water, but should the egg become dried, this appears to make the shell relatively impermeable to water, preventing the embryo from becoming desiccated (Ellenby, 1968). Eggs of *Ascaris* are laid in a transparent, colourless jelly. A jelly layer surrounds the eggs of *Toxascaris leonina* when they are in the uterus, but is missing when the eggs are laid. The eggs of some endoparasitic rhabdoids have a gelatinous coat secreted, in *Meloidogyne* by rectal glands, in *Tylenchulus* by the excretory cell, and by cells of the uterine wall in *Heterodera cruciferae*. In *Meloidogyne* it contains carbohydrate and quinone-tanned protein and appears morphologically homogeneous (Bird and Rogers, 1965). These jelly coats may facilitate the extrusion of the eggs.

Although the eggs of nematodes are subjected to the same sort of hazards as are the eggs of trematodes and cestodes, the shells of nematode eggs are morphologically and chemically more complex, and it is reasonable to assume that each layer has a particular physiological significance, though the particular properties and functions of the different layers have not yet been determined. Nor has it yet been discovered if the complex shells of nematodes are more efficient in protecting the embryo than the simpler shells of cestodes and trematodes. Hatching of nematode eggs results from the same sort of influence as does the hatching of Platyhelminthe eggs (W. P. Rogers, 1958; Anya, 1966b).

CHAPTER 8
Acanthocephala

Acanthocephalids are parasites that may reach a length of 0·5 m but which are mainly between 1·5 mm and 2·5 cm. They have a complex life history involving an arthropod intermediate host, a reptilian or amphibian transport host and a final host which is either a bird or a mammal.

A. External skeletal structures

The surface of *Polymorphus minutus* is covered by a layer of slightly granular mucopolysaccharide and Crompton (1963) has called this the "epicuticle". A similar layer has also been found in *Macracanthorhynchus hirudinaccus* but Stanach *et al.* (1966) could find no trace of it in *Pomphorhynchus laevis*. Beneath the epicuticle there is, in *Polymorphus*, a thin homogeneous membrane that looks like an extracellular structure and Crompton and Lee (1965) have called it the "cuticle" but it is bounded at the outer surface by a three-layered structure, closely resembling a plasma membrane, whereas at the various interfaces between the cuticle and the underlying striped layer, and between the striped layer and the syncytial epidermis, no plasma membrane is seen, so that the extracellular nature of the cuticle is in doubt. The striped layer is much thicker than the cuticle and is crossed by radial canals which continue through the cuticle to open beneath the epicuticle, by pores that are smaller in diameter than the canals. The plasma-membrane-like membrane turns inwards to line the whole length of the canals; these are separated from each other by homogeneous material which makes up the substance of the striped layer. The striped layer is in contact with the cytoplasm of the epidermis in which there is an outer region containing numerous fibrils arranged tangentially to the surface and an inner region in which the fibrils are fewer, and are arranged radially (Fig. 8.1). The cuticle of *Moniliformis dubius* has been investigated by Nicholas and Hynes (1963) and Nicholas and Mercer (1965). There is little difference between it and that of *Polymorphus* and *Pomphorhynchus* but these authors interpret the photographs somewhat differently. A similar surface structure is also found in *Acanthocephalus ranae* (Hammond, 1967).

Crompton (1963) describes the cuticle in *Polymorphus* as composed of a lipoprotein with disulphur and sulphydryl groups, the total sulphur content

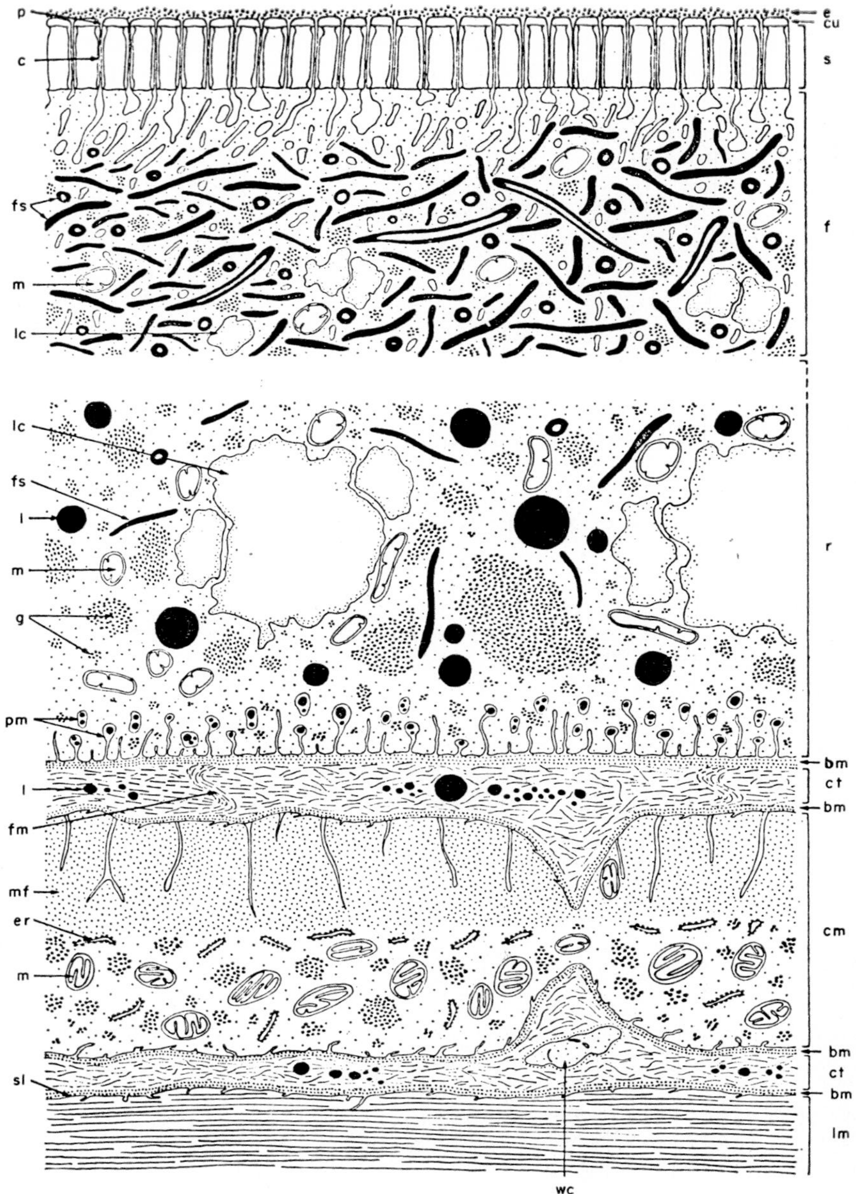

Fig. 8.1. A diagrammatic representation of the metasomal body wall of *Polymorphus minutus*. *bm*, Basement membrane; *c*, canal; *cm*, circular muscle; *ct*, connective tissue; *cu*, cuticle; *e*, epicuticle; *er*, endoplasmic reticulum; *f*, felt layer; *fm*, fibres attaching muscle to body wall; *fs*, fibrous strands; *g*, glycogen; *l*, lipid; *lc*, lacuna channel; *lm*, longitudinal muscle; *m*, mitochondrion; *mf*, myofilaments; *p*, pore; *pm*, folded plasma membrane; *r*, radial layer; *s*, striped layer; *sl*, sarcolemma; *wc*, wandering cell. (From Crompton and Lee, 1965.)

being about 4 per cent. Monné (1959) claims to have found quinone-tanning in the cuticles of *Polymorphus botulus* and *P. boschadis*. From the dyes that they take up Crompton suggests that the fibrils in the epidermis are chemically similar to the cuticle. In the anterior region the canals through the cuticle contain electron-dense material which Crompton and Lee (1965) suggest is lipid being absorbed into the animal but Hammond (1968a, b) suggests that it is lipid being secreted and not absorbed.

The relationship of acanthocephalids to other parasitic groups is uncertain; whether or not the layer beneath the epicuticle is extracellular, the overall picture of the surface layers resembles neither that of the nematodes nor that of the Platyhelminthes, though if all the surface structures are intracellular they might represent specializations of a microvillous surface-layer such as is found in cestodes which are, like the acanthocephalids, without mouth or gut. On the other hand, the acanthocephalan surface structures might be derived from something similar to the surface structures of a rotifer like *Asplanchna sieboldi*, by the surface invaginations increasing in number, becoming regularly arranged, and extending down through the fibrous layer.

In *Macracanthorhynchus hirudinaceus* the hooks which occur on the eversible proboscis, and which serve to anchor the parasite to the tissues of its host, do not dissolve in KOH but Brandt (1940) could not demonstrate chitin in them.

B. Internal skeletal structures

The epidermis of *Polymorphus* rests on a dermis which is markedly fibrillar, and fibres pass from it to the membranes enveloping the layer of circular muscles lying beneath it. An internal collagen from *Macracanthorhynchus hirudinaceus* (Table 7.1) somewhat resembles the internal collagen of *Ascaris* in its amino acid composition (Rigby, 1968). Though its total imino acid content (167/1000 residues) is below that of *Ascaris*, its T_S (62°C) and T_D (42°C) are approximately similar to those of its host, the pig, with a total imino acid content of 226/1000 residues. Again in this internal collagen while T_D is related to environmental temperature T_S and T_D do not show the expected relationship to total imino acid content, and its stability must be determined in part by some other factor. Acanthocephalids have well developed layers of circular and longitudinal muscles in the body wall which suggests that the surface structures may not limit the shape of these animals, and that the basement membrane, together with the muscle layers, are responsible for this. Alternatively, the adult, lacking a gut and being relatively immobile, living attached to the intestinal wall of its host, may require muscular contractions of the body wall to help circulate food and metabolic products through the body.

C. Egg shells

The eggs which are fertilized internally are initially covered by a membrane of unknown origin inside which the fertilization membrane forms. Development is internal and a membrane, presumbaly secreted by the embryo, is formed inside the fertilization membrane. Also a shell forms between the outer and the fertilization membrane so that the larva (or acanthor), when it leaves the parent's body, is covered by four membranes. In forms with a terrestrial intermediate host, the shell is generally thick, often brown and striated or granular at the surface. In forms with aquatic hosts the shell is poorly developed (Hyman, 1951b). Chitin occurs in the innermost membrane of *Macracanthorhyncus hirudinaceus* (Brandt, 1940), and in *Acanthocephala jacksoni* and *Echinorhynchus gadi* the chitin is associated with protein (West, 1963). Brandt found a carbohydrate in the fertilization membrane of both these species and West identified this carbohydrate as cellulose. Monné and Honig (1954c) found a keratin-like protein and weak quinone-tanning in the egg shells of *Polymorphus botalus* and *P. minutus*.

In *P. minutus* the embryo, surrounded by the egg shell, is passed out of the host and eaten by *Gammarus*, the intermediate host, where possibly a rise in CO_2 tension stimulates the parasite to activity. It breaks out of the shell and passes into the tissues of the host where it develops into a cystacanth with a well developed surface structure. It becomes surrounded by a thin membrane which is probably a product of the host's cellular reaction to the parasite. This species does not have an obligate transport host and reaches its final host when a duck eats an infected *Gammarus*. The parasite passes unharmed through the gizzard, probably protected by its surface structures, to take up its final position in the duck's intestine.

CHAPTER 9

Endoprocta and Lophophorate Phyla

Although Endoprocta have a superficial resemblance to some of the Lophophorate phyla they are not closely related to them. Endoprocta lack a coelom whereas all the lophorate phyla are coelomate.

I. ENDOPROCTA (KAMPTOZOA)

The Endoprocta, a phylum of small, pseudocoelomate animals not exceeding 0·5 mm in length are mainly sedentary and either solitary or colonial; except for one freshwater genus, *Urnatella*, all are marine. The animals appear superficially like coelenterates, with a circle of ciliated tentacles on what is called the ventral surface, surrounding the "vestibule" into which open the mouth and anus. Peripherally the tentacles are surrounded by a ridge that can be drawn over them when they are curled down over the vestibule. The body or calyx is attached to the substratum or to other members of the colony by a stalk arising from the dorsal surface.

A. External skeletal structures

Except for the tentacles and vestibule, the body surface is covered with a cuticle which may be thick and layered over the stalk and which over the dorsal surface may be thickened into shields, as in *Chitaspis* and *Loxosomatoides*; the shields may be ornamented by spines or by polygonal markings. Wester (1910) obtained evidence for chitin in the cuticle of *Pedicellina* but Kunike (1925) could not find any in that of *Bartensia*.

B. Egg shells

The eggs, which are fertilized internally, become covered with a secretion of the eosinophil glands during their passage through the gonoduct. This secretion forms a loose membrane round the egg and becomes drawn out into a stalk which attaches the egg to the embryophore developed in the vestibule, anterior to the anus.

II. POLYZOA (ECTOPROCTA)

The three lophorate phyla, the Polyzoa, Phoronida, and Brachiopoda are all sedentary and possess a tentacle-bearing structure, the lophophore. The tentacles are ciliated and used to create a current of water bringing food to the animal and carrying away waste products.

The Polyzoa, microscopic, coelomate animals are colonial and generally firmly attached to the substratum. They are fastened into exoskeletal cases or zoecia with openings through which the lophophore can be extruded and retracted. While the simpler Polyzoa superficially resemble the endoprocts, in all the Polyzoa the anus lies outside the crown of tentacles; this can be withdrawn into the body, an impossibility in the endoprocts since their body cavity is filled with gelatinous parenchyma.

A. External skeletal structures

In the simplest marine Gymnolaemata, with colonies having stoloniferous growth, the zoecium is little more than an investing cuticle, though in some species it forms round the body a more definite tubular or vase-shaped structure. Such a growth-form is too delicate to permit the development of large colonies. Where large colonies are formed stoloniferous growth has been abandoned, individual zoecia become more or less box-shaped, and are in direct contact with neighbouring zoecia except on the ventral surface, connections between the individuals of the colony being maintained through pores in the zoecium wall. The walls of the zoecium also tend to become thickened and sometimes hardened by deposition of calcite between the cuticle and the epidermis. However, hardening of all the walls would present difficulties in the accommodation of volume changes resulting from the extrusion and retraction of the tentacles. Usually the ventral wall remains wholly or partly flexible (Fig. 9.1(*a*)). Because the ventral wall remains thin and pliable it is presumably more liable to damage than the rest of the zoecium and is often protected by spines. In the Cribrilinidae these calcified spines arch over and fuse to form a shield above the ventral wall. In some cheilostomes a calcified shelf grows forward beneath the ventral wall, though the shelf is always perforated by muscles responsible for operating the ventral wall for volume control (Fig. 9.1(*b*)). In the ascophoran cheilostomes the ventral wall is completely calcified and below it a new structure, the "compensation sac" which can be filled with sea water, permits necessary volume changes (Fig. 9.1(*c*)). In stenostomes (cyclostomes), again with completely calcified ventral walls, the method of extruding the tentacles has been described by Borg (1926).

Schneider (1963) has described, in *Bugula*, the deposition of calcite initially as crystal spherites and later in bands of thread-like crystals with a rather uniform parallel orientation, the neighbouring bands separated by a zig-zag junction. According to Hyman (1958) the zoecium of *Bugula neritina* is devoid of calcium.

Carnose ctenosomes may form colonies more than 1 m high and here support for the colony is provided by a gelatinous material of unknown

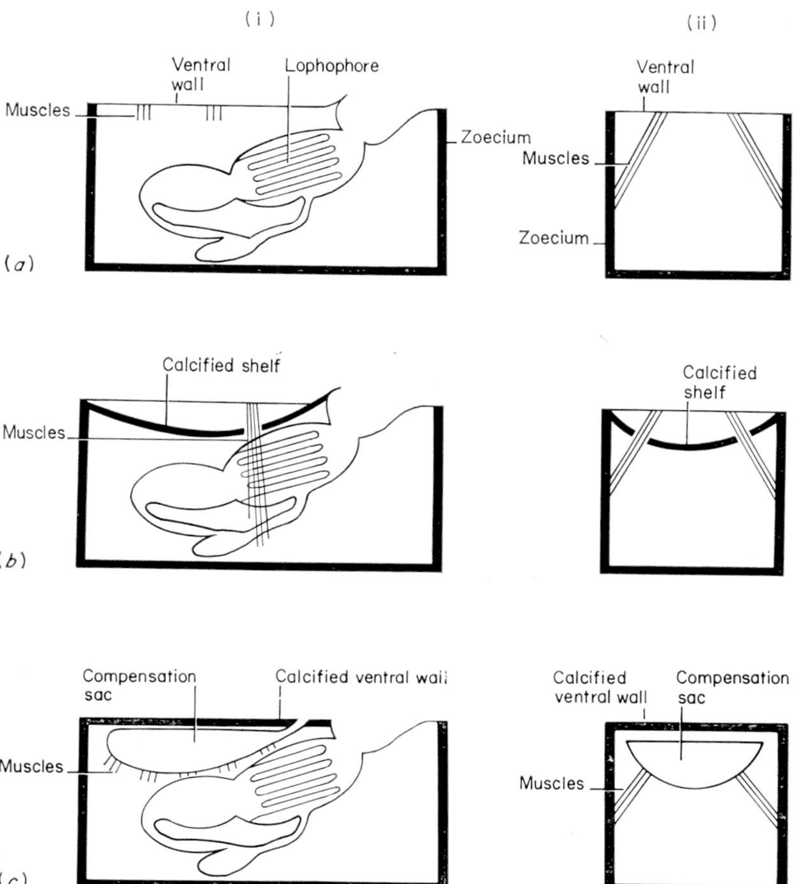

Fig. 9.1. Diagrams to show the different methods of providing for the extrusion of the lophophore in cheilostomes: (i) longitudinal and (ii) transverse sections through zooid. (*a*) cheilostome with flexible ventral wall; (*b*) cheilostome with calcified shelf (cryptocyst) beneath the ventral wall; (*c*) ascophoran cheilostome with calcified ventral wall, and compensation sac. (After Hyman, 1959.)

chemical composition in which the zoecia are embedded. In freshwater Phylactolaemata, the zoecium is either membraneous, sometimes incorporating foreign particles, or gelatinous, but never calcified.

The bulk of the organic part of the exoskeleton is protein. In *Bugula* about 90 per cent is protein but up to 10 per cent chitin may be present (Fischer and Nebel, 1955). The most extensive investigation of the distribution of chitin in ectoprocts has been carried out by Hyman (1958, 1966). All the forms she tested, except the cyclostomes, whether they were calcified or not, were found to contain chitin, though the exoskeleton of *Thalamoporella gothica* gave rather indefinite results. The gelatinous material surrounding *Pectinatella* (Phylactolaemata) also contains 1·25 per cent chitin (Kraepelin, 1887); no chitin could be found in the cyclostomes.

Roche *et al.* (1962, 1964) have found that *Bugula neritina* actively takes up radioactive iodine from sea water and incorporates it into the organic part of the exoskeleton in the form of 3-monoiodotyrosine and 3,5-diiodotyrosine. This, for reasons already given, suggests that it would be worth looking for quinone-tanning in these structures.

Schneider (1963) has described how growth takes place in colonies of *Bugula*, by the formation of cuticle-covered buds, in which certain cells then soften the cuticle and add more material to it so that increase in size can take place, with the formation of a new individual.

B. Egg shells and cysts

Marine polyzoan eggs are fertilized in the coelom and are devoid of egg shells. They develop internally and produce a ciliated larva provided with a small bivalve shell which is said to contain chitin.

Freshwater Phylactolaemata produce reproductive bodies called statocysts. During development a two layered epidermis is formed round them. The outer layer secretes between itself and the inner epidermal layer, a hardened shell which sometimes bears spines or hooks. The outer epidermis also produces a band of tall columnar cells round the shell. The walls of these cells harden, the contents disappear, and the cells become filled with air to form a floatation device. Chitin occurs in the statocyst coat (Hyman, 1958) and in the float wall (Lerner, 1954), though little chitin was found in the hooks of *Pectinatella.*

III. PHORONIDA

Phoronids are worm-like, coelomate, non-segmented animals about 6–200 mm in length with a terminal, bicornuate crown of tentacles. They are all marine and sedentary, living in tubes.

A. External skeletal structures

It is uncertain whether the tube is secreted by a limited area or by the whole surface of the animal. When first formed it is soft, transparent, and sticky and may take up foreign particles, but it sets hard on contact with sea water. These tubes have been found to contain chitin (Hyman, 1958). The body is covered by a layer of cuticle which is best developed on the outer surface of the tentacles but diminishes along the trunk till it is no longer detectable. No chitin could be found in the cuticle (Hyman, 1958).

B. Internal skeletal structures

Underlying the epidermis is a basement membrane which plays an important skeletal function. It is very thick at the base of the tentacles and extends as a very thick membrane up into the tentacles, making them stiff with little power of movement. The tentacles generally lack muscles though both circular and longitudinal muscles are present in the rest of the body wall.

C. Egg shells

The eggs, after internal or external fertilization, develop into free-living larvae without at any time being surrounded by an egg shell.

IV. BRACHIOPODA

The brachiopods are marine coelomate animals 5–80 mm in length, although some extinct forms reached a length of 375 mm. They superficially resemble lamellibranch molluscs, as they have bivalve shells, though in brachiopods the valves are dorsal and ventral (Fig. 9.2). The two valves, lined anteriorly by projections of the body-wall called the mantle, enclose the body and an anterior space in which the lophophore is situated. The animals are sedentary either with the ventral valve cemented to the substratum or with a stalk emerging from the shell at the posterior end; this often ends in adhesive filaments, anchoring the animal to shells or rocks or, as in *Lingula*, is embedded in mud at the bottom of the tube in which the animal lives and down which it can retreat by coiling the stalk. There are two classes of brachiopods: (1) the Inarticulata, with valves held together only by muscles and without a skeletal support for the lophophore; (2) the Articulata which have a tooth and socket valve-articulation, and an internal support for the lophophore.

A. External skeletal structures

The valves of most brachiopods are mainly mineral. In the Inarticulata, except the Craniidae, they contain calcium phosphate. In *Lingula* it is in the form of calcium fluoroapatite which gives the same X-ray photograph as human dentine (McConnel, 1963). The Articulata, together with the Craniidae, amongst the Inarticulata have calcite shells.

The valves are covered externally by an organic layer, the periostracum, which is often thick in the Inarticulata. In the Articulata it is generally thin and in the form of a honeycomb network with strengthening bars enclosing polygonal spaces or less commonly as in *Terebratulina* a diamond pattern. In the phosphatic shells organic matter may be uniformly distributed through the mineralized part as in the discinids where the shell is made up of lamellae diagonal to the valve surface. In lingulids organic and phosphatic layers alternate and are parallel to the shell surface. Calcite shells also have a matrix of organic matter and often have two distinct layers, the outer layer generally has a fine fibrous crystal structure, the inner layer has long prismatic crystals arranged slightly obliquely to the valve surface (Williams, 1956, 1965). Beyer (1886) described the inner surface of the mantle as covered by cuticle, but Prenant (1928) describes it as ciliated. The stalks, where present, are covered by a well developed cuticle which is very extensible.

The shell is secreted by the epidermis of the posterior body wall and the external epidermis of the mantle. The mantle edge in articulate brachiopods is bilobed, the inner surface of the outer lobe secretes the periostracum, and the outer face of the outer lobe secretes the outer calcite layer, while the inner

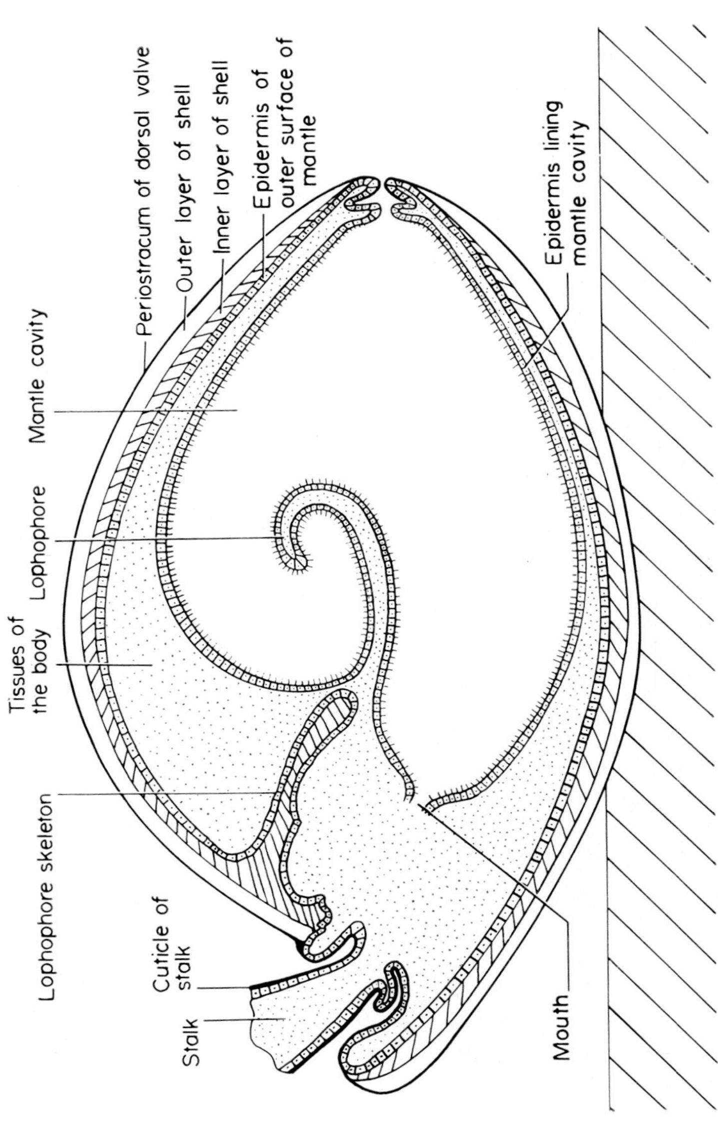

Fig. 9.2. Diagram of a sagittal longitudinal section of terebratuloid articulate brachiopod to show the relationship of the shell to the mantle and body (modified from Williams 1956).

calcite layer is secreted by the whole of the outer surface of the mantle and posterior body wall (Williams, 1966). Each prism of the inner layer is secreted by a single epidermal cell. In articulates the epidermis forms minute papillae which penetrate through the substance of the shell to the outer surface. The mantle round the edge of the valves is often armed with setae or bristles produced in deep epidermal invaginations set along a groove.

The lophophore in articulates is supported by a calcareous skeleton which Hyman (1959) regarded as formed in the connective tissue close under the epidermis, on the proximal side of the lophophore, but Williams (1956, 1965) describes it as secreted by a local ingrowth of the epidermis lining the dorsal valve of the shell; according to him, the lophophore skeleton is a local specialization of the shell, and since it is firmly attached to the valve this seems reasonable.

Krukenberg (1885a, b), Wester (1910), and Schmeideberg (1882) all found chitin in the shell of *Lingula*, occurring in both the periostracum and the shell matrix (Jope, 1965). Rudall (1955, 1965), by X-ray analysis, found chitin in the γ-form in both the stalk cuticle and the setae. The cuticle of the stalk shows an additional well defined meridional arc at 2·86 Å which could be a collagen spacing. Hyman (1958) found chitin in the cuticle of *Lingula* but concluded it was absent from the setae, as they disintegrated completely in KOH, though subsequently (Hyman, 1966) she proved the presence of chitin in the setae of both *Lingula* and *Discinisca*. She found chitin lacking in the shells of articulates and craniids though it is present in the stalk cuticle of *Terebratulina traversa* and *Laqueous californicus*. Jope (1965) confirmed the absence of chitin from articulate and craniid shells, though in *T. traversa* and *L. californicus* 0·4–0·5 per cent hexosamine is present.

Protein also is present in the periostracum and shell matrix. The periostracum of *Lingula* contains 67 per cent protein and 10·6 per cent chitin. The protein contains a high proportion of alanine, arginine, and proline and a small proportion of hydroxyproline. Hydroxyproline is also present in the periostracum of *Discinisca lamellosa*, though it is absent from the periostracum of articulates (Jope, 1965), some of which have a high proportion of tyrosine, phenylalanine, proline, and glycine (Table 9.1); hydroxyproline is absent also from the periostracum of craniids.

The amino acid composition of the shell matrix differs from that of the periostracum and shows considerable variation, not only between the articulates and inarticulates, but also between the different families within these orders (Table 9.2). Hydroxyproline occurs again in the matrix of inarticulates, excepting craniids (Jope, 1967a). The occurrence of hydroxyproline infers the presence of collagen. Should collagen fibres be proved to occur in the matrix of inarticulate shells this would be an additional point of similarity with vertebrate bone, where collagen fibres are believed to be essential for the deposition of calcium phosphate crystals. Perhaps it is the absence of collagen from the matrix of articulate and craniid shells which determines that calcite rather than calcium phosphate is present in them.

The brown colour of the periostracum of *Lingula* might suggest the presence of quinone-tanning but Jope (1965) says that it is due to the presence of iron,

Table 9.1 Amino acid composition of the periostracum of brachiopods: Residues per 1000 total residues. (Jope, 1967a.)

		Inarticulata		Articulata	
	Lingula sp	*Discinisca lamellosa*	*Crania anomala*	*Laqueus californicus*	*Terebratalia transversa*
Alanine	132	231	81	16	41
Glycine	72	291	165	315	324
Valine	54	22	39	16	23
Leucine } Isoleucine }	86	9	51	55	19
Proline	45	6	9	33	11
Phenylalanine	24	2	20	46	7
Tyrosine	31	2	60	78	12
Serine	89	47	93	97	70
Threonine	45	54	74	32	37
Methionine	—	—	—	—	—
Arginine	73	33	5	12	39
Histidine	55	71	67	55	113
Lysine	27	4	7	8	7
Aspartic acid	156	166	226	149	235
Glutamic acid	91	122	104	75	59
Hydroxyproline	21	12	—	—	—
Hydroxylysine	—	—	—	—	—
Cystine	—	—	—	3	3
Tryptophan	—	—	—	9	—

probably in the form of ferric hydroxide, which may form as much 10 per cent of the material present. The periostracum of *Laqueus* also contains iron as ferric hydroxide but in much smaller amounts.

B. Internal skeletal structures

Within the stalk of *Lingula* there is a fibrous dermis which Rudall (1955) showed to contain collagen. The stalk contains muscles and can be curled into a spiral, drawing the animal down into its tube. The stalks of Articulata are short and thick, and lack muscles. The core of the stalk is almost completely filled with connective tissue and muscle, virtually obliterating the coelomic cavity. In the lophophore, besides the calcified skeleton, there is firm connective tissue resembling vertebrate cartilage in texture and microscopic appearance, and this cartilage-like material extends up into the tentacles. The calcified skeleton and connective tissue make the lophophore stiff and capable of little movement; protrusion of the lophophore is impossible in most articulates. The internal tissues of the mantle, the posterior body wall, and the general connective tissues of the lophophore of some articulates also contain calcite spicules.

C. Egg shells

The eggs are fertilized either internally or externally and no egg shells have been described round the egg.

Table 9.2 Amino acid composition of the matrix of brachiopod shells. Residues per 1000 total residues. (Jope, 1967a.)

	Inarticulata				
	Lingulida			Acrotretida	
	Lingula			*Discina*	*Crania*
	Outer layer	Middle layer	Inner layer	Whole shell	Whole shell
Alanine	215	156	166	188	61
Glycine	153	156	130	22	141
Valine	53	37	71	19	50
Leucine	35	48	46	15	45
Isoleucine	21	28	28	10	30
Proline	67	71	77	55	49
Phenylalanine	21	23	26	7	22
Tyrosine	25	22	27	1	2
Serine	53	63	56	35	140
Threonine	34	40	39	27	58
Cystine	1	6	3	7	15
Methionine	5	5	9	1	10
Arginine	—	—	—	—	—
Histidine	—	—	—	—	—
Lysine	—	—	—	—	—
Aspartic acid	109	113	103	70	185
Glutamic acid	58	61	59	49	88
Hydroxyproline	43	63	53	137	—
Hydroxylysine	—	1	—	1	—
Tryptophan	5	—	—	—	1

	Articulata							
	Terebratulida							Rhynchonellida
	Laqueous	*Neothyris*			*Terebratella*	*Terebratalia*	*Macandrevia*	*Notosaria*
	Whole shell	Whole shell	Outer shell	Inner shell	Whole shell	Whole shell	Whole shell	Whole shell
Alanine	44	126	117	143	110	74	71	86
Glycine	199	268	247	317	226	234	196	332
Valine	65	56	60	42	60	78	78	31
Leucine	46	42	50	30	48	43	56	23
Isoleucine	45	38	42	26	47	37	43	14
Proline	51	51	50	46	73	58	60	88
Phenylalanine	39	35	40	30	35	34	30	83
Tyrosine	63	—	—	1	—	—	—	2
Serine	81	66	71	64	70	90	91	69
Threonine	51	37	39	27	40	51	53	27
Cystine	53	24	11	47	12	tr.	—	2
Methionine	18	13	13	9	11	14	16	9
Arginine	—	—	—	—	—	—	—	—
Histidine	—	—	—	—	—	—	—	—
Lysine	—	—	—	—	—	—	—	—
Aspartic acid	82	100	101	96	105	105	123	119
Glutamic acid	61	57	65	46	64	65	86	46
Hydroxyproline	—	—	—	—	—	—	—	—
Hydroxylysine	—	—	—	—	—	—	—	—
Tryptophan	—	—	—	—	—	—	—	—

CHAPTER 10

Mollusca

Molluscs are coelomate animals the majority of which show no traces of segmentation, though the recently discovered *Neopilina*, which is definitely a mollusc, is pseudometamerically segmented. Most molluscs have an external shell, but in the squids the shell has become internal and in some other genera it has been completely lost. Mollusc shape varies considerably and so does their size for while some clams and snails may be less than 1 mm, the giant squids may reach a length of 40 m, excluding tentacles and weigh 2–3 tonnes.

A. External skeletal structures

The worm-like, marine Aplacophora, with neither shell, mantle, nor foot, may present the most primitive extant molluscan organization (Hyman, 1967; Beedham and Trueman, 1968). They show a considerable range of size, 1–300 mm in length, though most are less than 50 mm.

The entire body, except the ciliated ventral groove, is covered by cuticle, generally thick, in parts up to 100 μm. It is penetrated by papillae from the epidermal cells which extend to the cuticle surface where they form and discharge vesicles, which may be excretory.

The cuticle also contains, in *Proneomenia aglaopheniae*, hollow, actinular, calcified spines or spicules, each borne internally on a thin cup-like organic base, which in turn is attached to an epithelial papilla. They scarcely extend into the outer, detritus-covered surface of the cuticle.

Chemically, the cuticle is composed of a glycoprotein-complex with a high acid mucopolysaccharide and low protein content (Beedham and Trueman, 1968). Although in *Proneomenia aglaopheniae* Beedham and Trueman failed to find conclusive evidence of chitin in the cuticle, Hyman (1966) reported chitin "in the body wall" of *Solenogaster*. There are indications of quinone-tanning in the organic bases of the spicules.

Glycoprotein polymers are found in the matrix of almost all tissues that calcify but generally the protein content is fairly high, and in the matrix of molluscan shells protein is the dominant organic component. Beedham and Trueman suggest that calcification and shell-formation in the higher molluscs has resulted from the addition of more protein to the primitive mucoid cuticle and that this protein subsequently became tanned, as in the periostracum of bivalves and gastropods (Trueman, 1949; Brown, 1952; Beedham, 1958; Abolins-Krogis, 1963a, b), making it a semi-conductor which Digby (1968) suggests facilitates calcifications (see p. 182). The calcified spines probably should not be considered as a primitive attempt to calcify the cuticle but as special structures peculiar to the aplacophorans, serving a protective, supporting and in certain regions a sensory, role (Beedham and Trueman, 1967).

Presumably the aplacophoran cuticle keeps pace with the growth of the animal by continual addition of new material from the epidermis beneath, and this would also make good wear and tear, and allow for the healing of wounds. No process of moulting has been observed in these animals.

The Polyplacophora or chitons are a class of marine molluscs, more or less oblong in shape, flattened and mostly 1–50 mm in length, though the longest species may reach a length of 33 cm and a breadth of 15 cm. They resemble the Aplacophora in some features but have a shell composed of eight overlapping, dorsal plates surrounded by a lateral extension of the body-wall, the mantle or girdle, in which the edges of the plates are embedded. Polyplacophorans also have gills in the mantle cavity and a ventral, flat, muscular foot.

The shell-valves consist of two layers of aragonite (Clarke and Wheeler, 1922); the outer, the *tegmentum*, contains pigment and a high proportion of organic matter and the surface is generally sculptured in complex patterns of ridges or spots. The inner layer, the *articulamentum*, is mainly inorganic

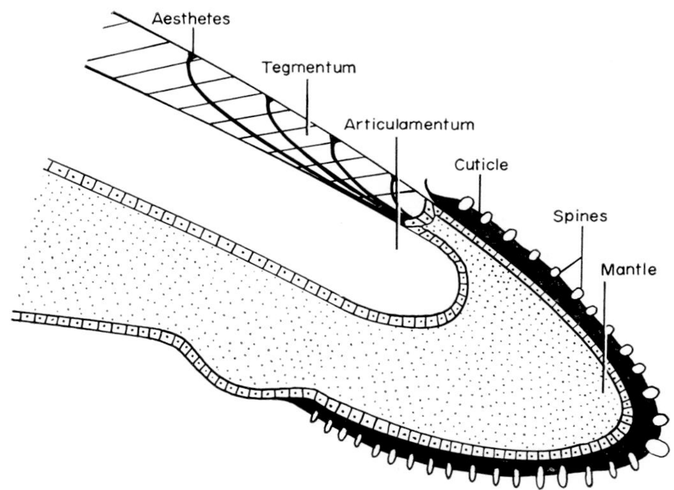

Fig. 10.1. Diagram of a transverse section through the edge of a polyplacophoran shell valve showing the relationship of the valve to the mantle edge (after Beedham and Trueman, 1967).

(Fig. 10.1). Beneath the articulamentum, an additional layer, the *hypostracum* (Bergenhayn, 1930), commonly associated with the point of insertion of the shell muscles, is generally discernible. It is probably a subdivision of the articulamentum (Beedham and Trueman, 1967).

Both dorsal and ventral surfaces of the mantle are covered by cuticle which also covers the body wall between the valves and extends as a transient layer over the shell valves but is soon worn away. The inner part of the cuticle is differentiated into a more or less distinct layer. Calcareous spicules of a variety of shapes occur embedded in the cuticle of the mantle, more numerous on the dorsal than on the ventral surface, but are lacking in the cuticle between the valves. The epidermis of the pallial groove and foot lack cuticle, and the foot is ciliated.

Cuticle and spines are secreted by the epidermis of the mantle. The tegmentum of each valve is secreted by the underlying epidermis cells with which the tegmentum is in contact where it overlaps the articulamentum around the edge of the valve so that this part of the shell can grow only round the edge. Strands of epidermal cells pass through the tegmentum in this region (Fig. 10.1) to reach the valve surface where they form sense organs, the aesthetes, generally covered over with a cup of clear material. In some chitons the aesthetes form eyes, with the cornea believed to be formed from overlying modified tegmentum. The articulamentum is secreted by the epidermal cells of the dorsal body wall directly underlying it, so that it can continually increase in thickness.

Cuticular spines, which generally project beyond the cuticle, are enclosed in an organic sheath and rest in specialized cuticular cups. In some chitons these cup-like structures, lacking spines, elongate and project beyond the cuticle as bristles.

The cuticle contains a mucopolysaccharide or mucoprotein, part of which at least is in the form of an acid mucopolysaccharide. Chitin and some calcium carbonate also are present and there are indications of quinone-tanned protein in the inner layer (Beedham and Trueman, 1967). Jeuniaux (1963) found chitin in the bristles of *Acanthochites discrepans*.

Hyman (1967), with others, suggests that the mantle cuticle in Polyplacophora is continued into the shell to form the matrix of the tegmentum and that in this class there is no homologue of the periostracum which covers the shell in bivalves and gastropods. Because Beedham and Trueman (1967) found that the cuticle initially covered the valves of the polyplacophoran shell and because the cuticle, like the periostracum of bivalves and gastropods, contains quinone-tanned protein, they suggest that the cuticle is homologous with the periostracum of other molluscs. They further suggest that the tegmentum and articulamentum of the polyplacophoran shell are homologous with the prismatic and nacreous layers of the bivalve shell.

Due, no doubt, to the shell and cuticle-covered mantle, the movements of chitons are limited and sluggish. Waves of muscular contraction pass over the foot producing a slow locomotion. Chitons are able to roll up with the dorsal shell valves on the outside protecting the soft, more vulnerable ventral parts of the body.

Although the remaining classes of molluscs, the Monoplacophora, Gastropoda, Bivalvia, and Scaphopoda, differ widely in the gross structure of their shells, the fine structure is basically the same. Where external shells are present in the cephalopods, as in *Nautilus,* the shell structure is similar to that of other external shells. Where internal shells are present in the remaining living cephalopods, the shells are much changed, both in architecture and fine structure, from typical external shells. *Sepia, Spirula, Architeuthis,* and *Loligo* show a progressive reduction of the internal shell (Appellof, 1893; Denton and Gilpin Brown, 1961; Mutvei, 1964) and the octapods have no shell at all, except for the female of *Argonauta;* this has a neoform coiled shell secreted, not by the mantle as in other molluscs, but by the two dorsal arms, and it is used to carry the egg-mass.

Typically there is an outer organic layer, the periostracum, though this may be missing in some gastropod shells, or replaced by lobes of the mantle, as in cypraeids, and it is often partly rubbed away, particularly in burrowing forms. It varies considerably from a thin, apparently structureless membrane in *Tellina* (Trueman, 1942) to a complex three layered structure in *Mytilus* edulis (Brown, 1952; Dunachie, 1963) and a four layered structure in *Solemya parkinsoni* (Beedham and Owen, 1965). The periostracum contains quinone-tanned protein or glycoprotein and varying amounts of polysaccharides; chitin and traces of lipids also may be present (Beedham, 1958, 1965; Hillman, 1961; Dunachie, 1963; Trueman, 1949). The brown colour of the periostracum in the majority of species undoubtedly results from the quinone-tanning. The green colour of the periostracum in *Mytilus viridis* probably results from a variation in the chemistry of the tanning reactions (Fox, 1966). Chemical analyses have been carried out on the whole periostracum of a variety of species, without consideration of the possible chemical variations in the different layers that Beedham and Owne (1965) have shown to occur in *Solemya parkinsoni*. In *S. agassizii* and *Viviparus georgiana*, also, the outer and inner layers of the periostracum differ in their chemical composition (Wilbur and Simkiss, 1968).

Small amounts of hydroxyproline and hydroxylysine have been found in the periostracum of some species (Degens *et al.* 1967). All analyses, except those from Opisthobranchia and certain marine Prosobranchia, show a high content of glycine which frequently accounts for more than half the residues. Acid amino acid residues tend to be few and there is a considerable variation in the amount of tyrosine present (Table 10.1); and dopa (3,4-dihydroxy-phenylalanine) which occurs in the metabolic pathway leading to tanning in the insect cuticle (see p. 219) has been reported from six species. Acid mucopolysaccharides have not been found in the periostracum, though considerable quantities of hexosamines may be present.

The underlying calcareous part of the shell in Monoplacophora and many bivalves is divisible into two layers, which differ in the form and arrangement of their crystals. In gastropods and cephalopods from three to seven layers may be distinguished (Bøggild, 1930; Haas, 1935; Bouillon, 1960). The calcareous substance is permeated by organic material called conchiolin.

Table 10.1 Amino acid composition of the periostracum of molluscs: Residues per 1000 total residues

	Bivalves		Gastropods			
No. of species analysed	*a* 7	*b* 8	Marine[a] opisthobranchs 4	Marine[b] prosobranchs 7	Fresh-water[b] prosobranchs 7	Terrestrial[b] pulmonates 3
Alanine	11–14	8–120	81–99	23–115	10–60	56–80
Glycine	391–671	400–546	68–87	71–398	460–628	250–632
Valine	16–55	17–42	66–72	22–67	12–105	50–103
Leucine	9–35	9–60	91–108	14–94	12–89	32–52
Isoleucine	6–26	13–40	49–66	13–52	7–35	33–46
Proline	28–54	24–38	32–43	46–92	16–43	15–38
Phenylalanine	24–80	10–70	13–46	12–99	9–91	17–35
Tyrosine	12–74	91–135	9–14	6–184	18–139	3–59
Serine	21–59	18–69	81–126	26–116	9–29	10–120
Threonine	8–35	8–14	37–70	16–64	5–24	5–50
Cystine (½)	11–14	4–37	1–5	6–46	0–39	—
Methionine	6–56	4–27	4–9	0–14	0–19	0–31
Arginine	13–55	13–90	23–45	14–105	7–26	4–38
Histidine	2–11	10–20	0–48	0–12	1–20	0–2
Lysine	14–29	3–24	20–39	7–136	7–25	7–19
Aspartic acid	13–72	26–60	131–158	90–164	16–60	8–109
Glutamic acid	10–38	10–38	119–142	30–146	14–69	9–101
Hydroxyproline	—	—	12–19	—	—	—
Hydroxylysine	0–12	—	—	—	—	—

[a] Degens *et al.*, 1967.
[b] Meenakshi, Hare and Wilbur: see Wilbur and Simkiss, 1968.

In all opisthobranch gastropods, nearly all pulmonates, and many prosobranchs, the calcium carbonate is in the form of aragonite though a little calcite also may be present, as in *Scurria*. The rudimentary shells of *Limacea* and *Arion* are calcite as are the shells of *Ianthina*, *Scalaria*, *Cerithium*, and *Melania*, while in *Patella*, *Haliotis*, and *Fusus* the outer layers are calcite and the inner aragonite (Prenant, 1927). Bouillon is of the opinion that archaic and primitive shells are aragonite and that calcite appears in later stages of evolution.

Among bivalves, the entire shell in ostreids, pectens, and *Inoceramus* is calcite, in *Unio* only aragonite is present. The embryonic shells of *Dreissenia* and *Anodonta* are calcite, the adult shells partly aragonite. The scaphopod shell-mineral is aragonite. In cephalopods the shells of *Nautilus*, *Sepia*, and *Spirula* contain aragonite; that of *Argonauta*, calcite (Prenant, 1927). Other inorganic substances besides calcium carbonate, also may be present (Clarke and Wheeler, 1922; Vinogradov, 1953). The concentrations of magnesium, barium, and strontium, and of sulphate ions, are thought by some to determine whether calcite or aragonite is present (Prenant, 1927).

In land and freshwater species, where less calcium is usually available, the shells are often thin and lightly calcified, though some freshwater species, e.g. *Unio*, may have massive calcified shells. Marine species, also, vary very considerably in the amount of calcification present.

The calcareous layer directly under the periostracum, whether it is calcite or aragonite, is frequently in the form of prismatic crystals arranged at right angles, or obliquely, to the surface of the shell. Each prism is surrounded by a conchiolin sheath formed of a fibrous network embedded in amorphous material. Organic matter also penetrates into the prisms, between the micro-crystals from which they are built up (Grégoire, 1967).

The inner layer, when of aragonite, is highly irridescent and called nacre. It consists of flat, polygonal crystals arranged in a regular manner in layers, like thin bricks in a wall, with conchiolin forming, as it were, the mortar between the bricks (Fig. 10.2). In the electron microscope the conchiolin is

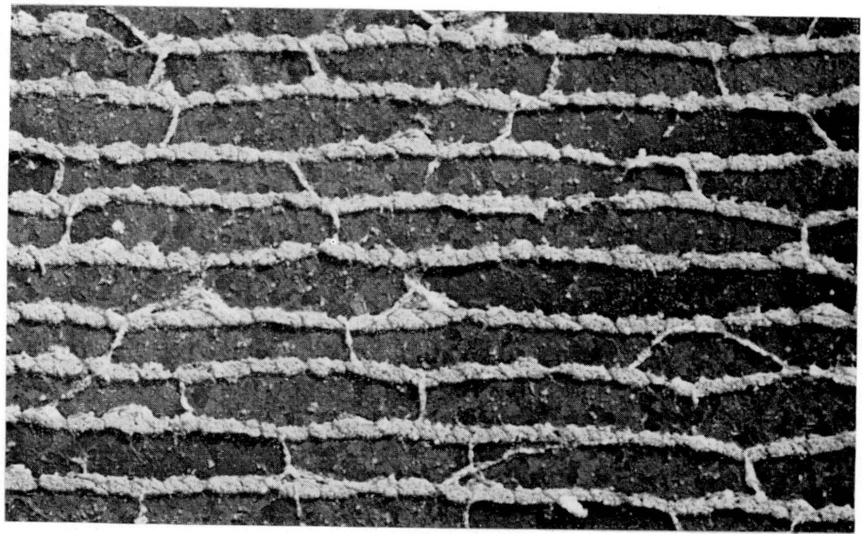

Fig. 10.2. Vertical section of nacreous region of decalcified *Pinctada mattensii* shell. Before decalcification, a single crystal occupied each chamber enclosed by interlamellar (horizontal white cords) and intercrystalline (vertical cords) matrix. The cords are composed of cabled bundles of granular fibrils 100–200 Å in diameter (Watabe, 1965). [×12 150]

seen as a membrane with perforations that give it a lace-like appearance. Differing membrane-patterns are found in the gastropods, bivalves, and cephalopods (Fig. 10.3(*a*) (*b*) (*c*)). Similar conchiolin patterns have been found in fossil shells from a few thousand to 450 million years old (Fig. 10.3(*d*)) and the fine structure of conchiolin is considered as useful in the identification of fossil and archaeological shell fragments (Grégoire, 1967).

Conchiolin from the nacre of shells of *Neopilina galathea* (Monoplacophora) forms rigid membranes of fibres embedded in granular material with occasional holes irregularly arranged (Fig. 10.3(*e*)). In *Spirula*, conchiolin from

(*a*)

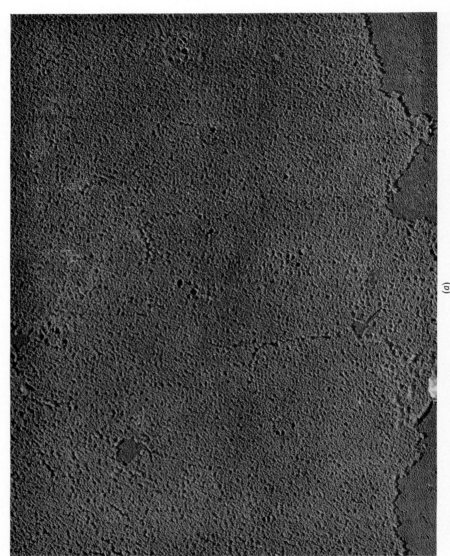

Fig. 10.3. Conchiolin from Mollusc shells. (*a*) *Pinna nigra* (bivalve) (Grégoire, 1960). [×32 500] (*b*) *Umbonium gigantum* (gastropod) (Grégoire, 1967). [×35 100] (*c*) *Nautilus pompilius* (Cephalopod) (Grégoire, 1962). [×40 000] (*d*) Unidentified nautiloid approximately 300 million years old (Grégoire, 1967). [×27 300] (*e*) *Neopilina galatheae* (Grégoire,

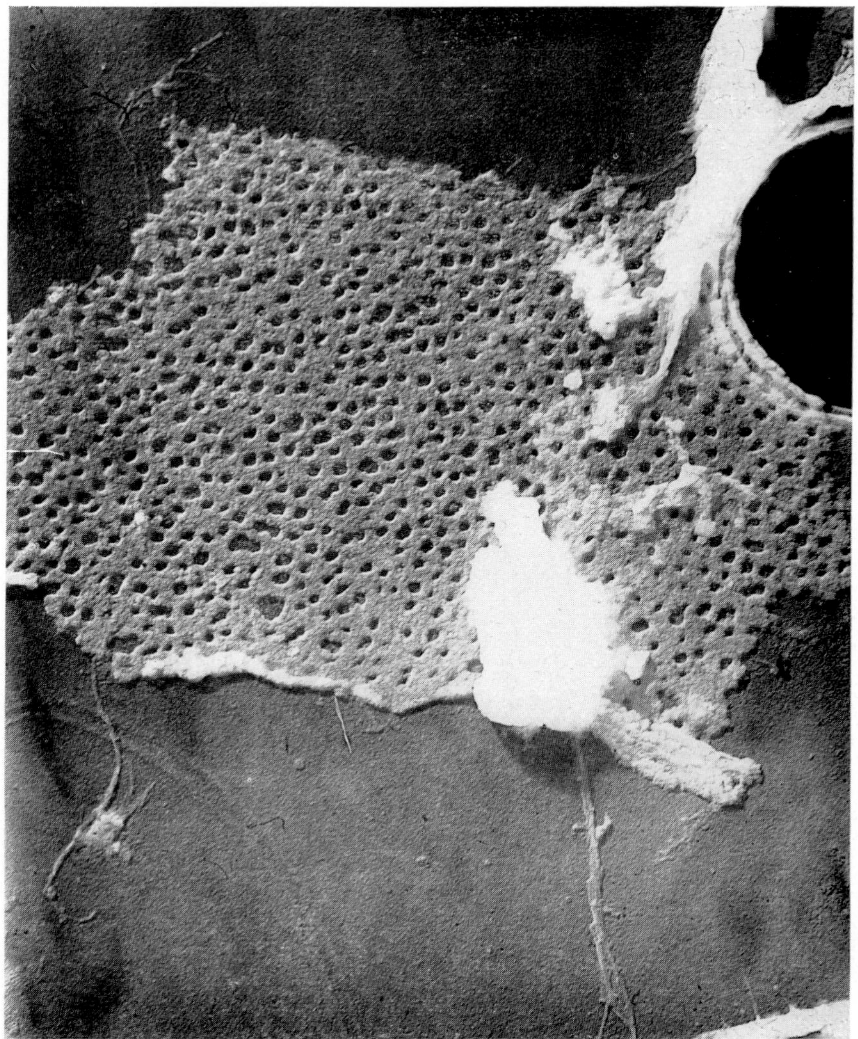

Fig. 10.3 (*b*)

the nacre, which only occurs in the septa, forms rigid membranes that maintain the domed shape of the septa after decalcification. The membranes are made up of thin, fibrous laminae, with the fibres all parallel in each lamina and at angles of 19–90° with the fibres in adjacent laminae. No perforations are present (Grégoire, 1967).

In Ostreidae, Pectinidae, and Anomiidae, the substance of the inner layer is called "calcitostracum" and contains layers of flat calcite crystals, overlapping like roof tiles (Fig. 10.4). The conchiolin is in the form of flat,

Fig. 10.3 (*c*)

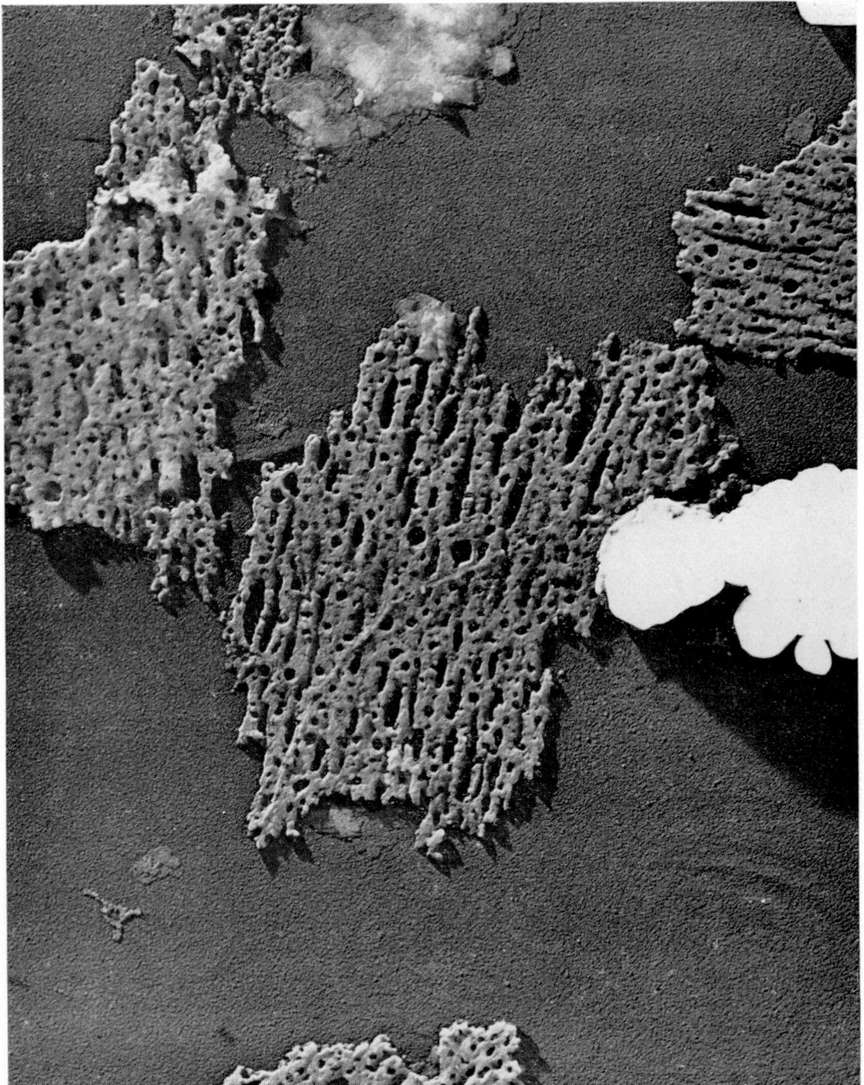

Fig. 10.3 (*d*)

transparent, fibrous lamellae which, in *Ostrea* and *Anomia*, may be perforated. It is less abundant than in nacre.

Marine bivalves, and marine and freshwater gastropods that have neither nacre nor calcitostracum in the inner layer have a porcellaneous inner layer in which needles or rectangular-shaped crystals, generally of aragonite,

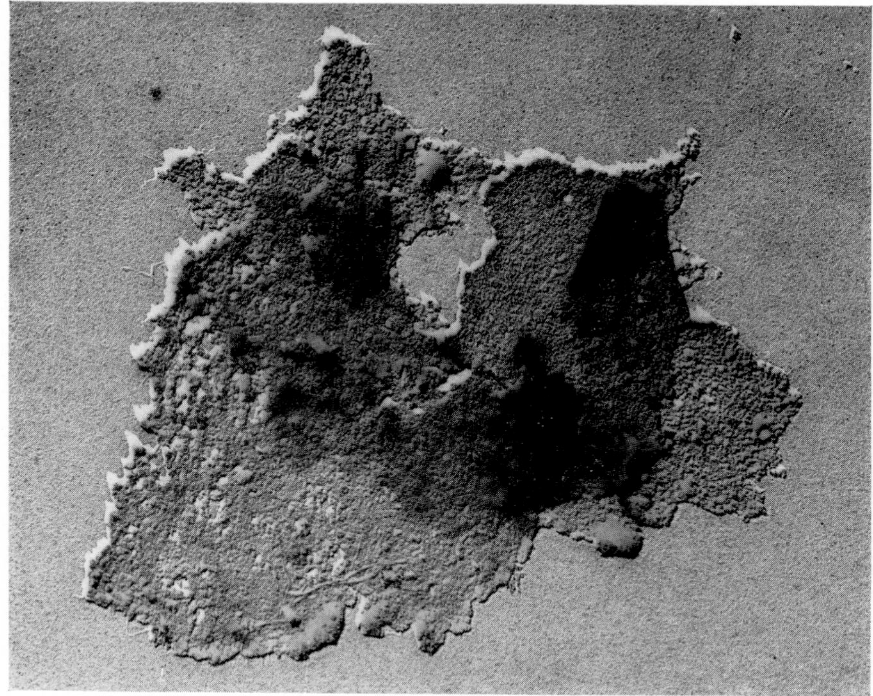

Fig. 10.3 (e)

have a complex arrangement and little is known about the conchiolin in these layers. In *Nautilus*, where the outer layer is porcellaneous, the material yields, on decalcification, a few, semirigid membranes made up of a network of fine fibres.

1. *Chemistry of shell*

Considerable variation exists in the ratio of conchiolin to calcium carbonate in molluscan shells, organic matter forming about 0·01 per cent of the shells of neogastropods and 5 per cent of the external shells of cephalopods. In the internal shells of *Loligo* and *Ommastrephes*, conchiolin forms 79 per cent and 90 per cent, respectively (Wilbur and Simkiss, 1968, for references).

Conchiolin is certainly not a homogeneous material. Travis *et al.* (1967) have found in it fibres with a periodic banding of 550 Å–655 Å, which they take to be collagen fibres. These banded fibres are frequently seen to change into fibres without periodic banding, and Travis *et al.* (1967) also found other fibres that appeared to be heavily coated with a second substance.

Many analyses of the amino acid content of the matrix of the whole shell have been carried out (Degens *et al.*, 1967; Hare and Abelson, 1965). Differences between families may be considerable, where similarities between

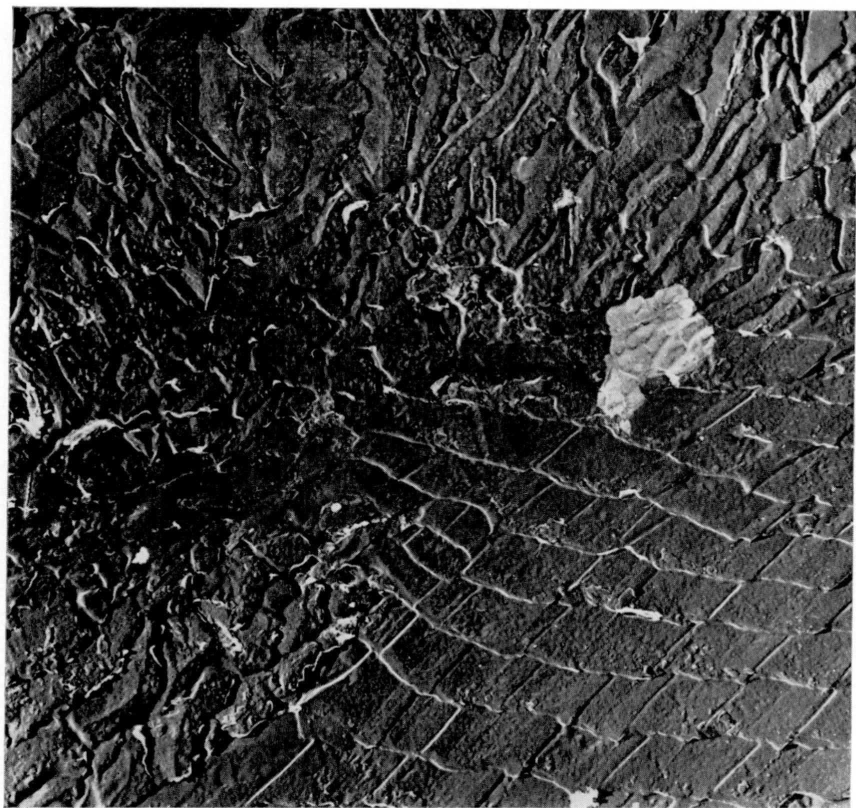

Fig. 10.4. *Ostrea edulis* (bivalve). Internal face of one of the valves with overlapping crystals in the calcitostracum layer. Positive carbon replica shadowed with palladium (Grégoire, 1967). [×6 500]

genera of a single family may be quite marked (Fig. 10.5). However, since it has been shown that the matrix of the separate layers of the shell differ in their amino acid composition, and that environmental factors that affect the shell affect the various layers differently, analyses of total matrix are of limited interest.

In spite of the fact that Travis *et al.* (1967) reported finding collagen fibres, though in small numbers, in the matrices of the shells they examined, neither they nor Hare and Abelson found hydroxyproline or hydroxylysine to be present. Degens *et al.* (1967) found hydroxyproline in the matrix of *Akera soluta* and *Oxynoe viridis* shells, and hydroxylysine in 26 of the 63 species they analysed. Piez (1961) found both hydroxyproline and hydroxylysine in the snail *Australorbis globratus*. Degens *et al.* (1967) also reported dopa from the matrix of eight species, presumably part of the system leading to the

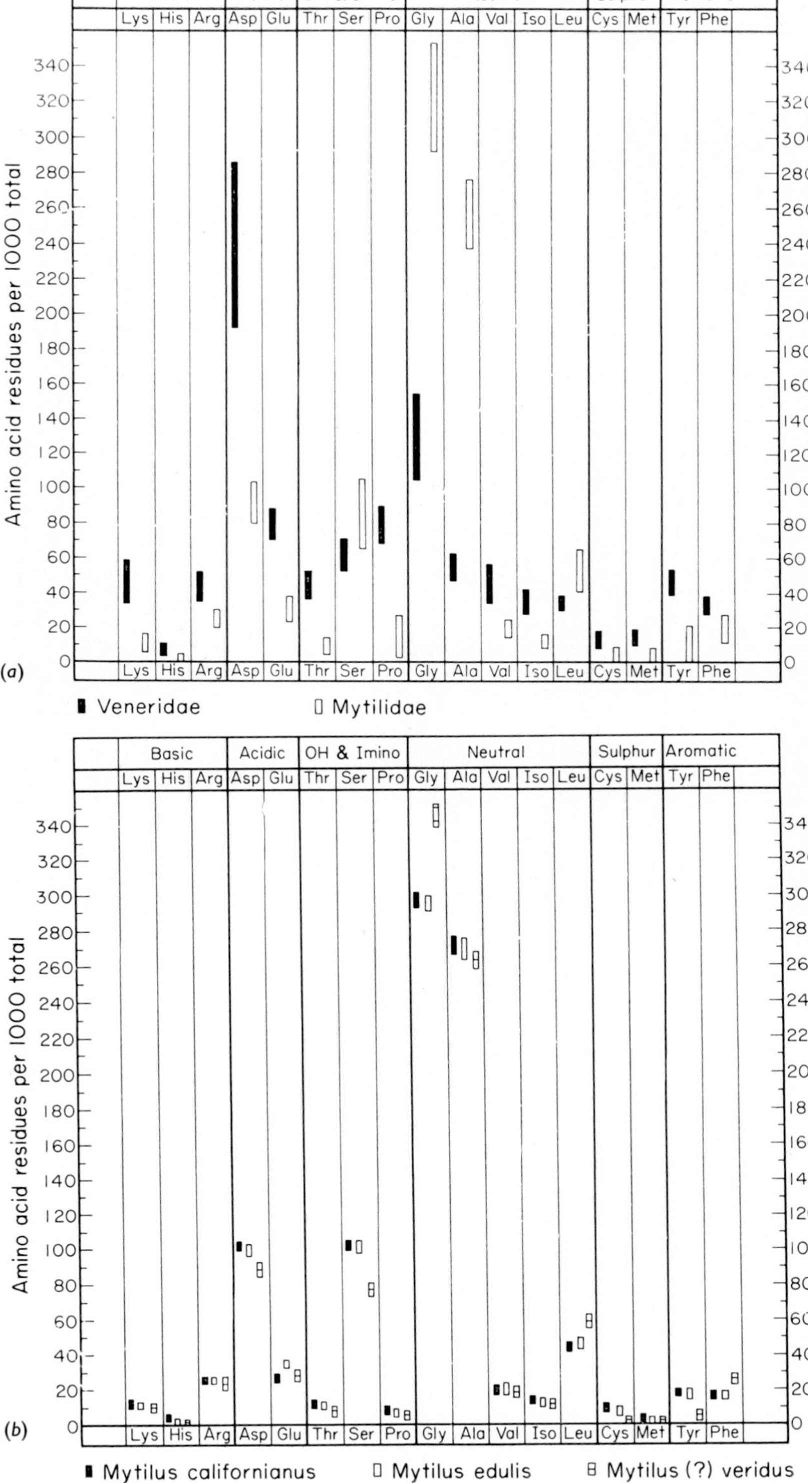

Fig. 10.5. Range of amino acid composition of organic matrix of mollusc shells: (*a*) in two bivalve families of widely differing shell structures and morphology, Veneridae (three genera) and Mytilidae (four genera); (*b*) in three species of *Mytilus*. (Hare and Abelson, 1965.)

Table 10.2 Amino acid composition of the neutral-soluble, insoluble, and whole matrix proteins from the nacreous and prismatic layers of the shell of *Mytilus edulis*. Residues per 1000 total residues. (Travis *et al.*, 1967.)

Fraction	Neutral soluble protein		Insoluble residue		Whole matrix	
Layer	Nacreous	Prismatic	Nacreous	Prismatic	Nacreous	Prismatic
Alanine	230	258	275	245	259	302
Glycine	261	290	293	261	282	329
Valine	28	17	24	27	27	24
Leucine	54	45	46	50	43	40
Isoleucine	21	16	15	21	19	14
Proline	19	11	10	18	44	31
Phenylalanine	21	16	18	20	12	5
Tyrosine	12	12	13	21	6	4
Serine	116	121	109	112	114	96
Threonine	21	13	13	21	24	14
Cystine (½)	6	7	5	7	1	0·2
Methionine	5	5	2	4	7	3
Arginine	33	29	26	30	25	22
Histidine	5	3	8	7	3	0·7
Lysine	26	16	20	29	18	12
Aspartic acid	99	108	93	95	99	94
Glutamic acid	41	35	31	43	114	96
Hydroxyproline	—	—	—	—	—	—
Hydroxylysine	—	—	—	—	—	—

tanning of the matrix which has been shown to occur in various species (Trueman, 1949; Beedham, 1958; Abolins-Krogis, 1963a, b).

Even analyses of the matrix of separate layers are not completely satisfactory as there is evidence (Grégoire *et al.*, 1955; Goffinet, 1965) that the different matrix layers contain more than one protein. Travis *et al.* (1967) obtained neutral-soluble and insoluble proteins from the different layers of *Mytilus* shell (Table 10.2). The proteins of the shell (Table 10.3, 10.4) have a high content of glycine and serine and generally of alanine, though in *Crassostrea*, *Mercenaria* and the prismatic layer of *Pinctada* shell the alanine content is low. Hare and Abelson (1965) found a general similarity in amino acid composition of the nacreous layer matrix between gastropods, bivalves, and cephalopods (Fig. 10.6), although the matrix of gastropods contains relatively more aspartic acid and less glycine and alanine than that of the other two classes; bivalve matrix is somewhat lower in arginine and glutamic acid and richer in glycine and alanine while that of cephalopodis is relatively low in leucine and high in phenylalanine.

Beedham (1954) showed histologically that the matrix of the calcite layer of *Anodonta cygnea* is more acid than that of the aragonite layer. Beedham and Owen (1965) found that in *Anodonta cygnea* and *Solemya parkinsoni* the matrix of the calcite layer is also more extensively tanned than that of the aragonite layer. Differences in amino acid composition between the calcite and aragonite layers have been found in a number of species (Roche *et al.*,

Table 10.3 Amino acid composition of the matrix of calcite and aragonite layers in shells of some species of bivalves. Residues per 1000 total residues

Species	*Mytilus*[a] *californianus*		*Mytilus*[b] *edulis*		*Pinctada*[c] *martensii*	
Layer	Prismatic (outer)	Nacreous	Prismatic	Nacreous	Prismatic	Nacreous
Crystal form	calcite	aragonite	calcite	aragonite	calcite	aragonite
Alanine	274·0	246·0	302	259	53·2	250·4
Glycine	298·0	269·0	329	282	313·9	320·0
Valine	24·6	29·4	24	27	88·0	17·5
Leucine	47·0	49·5	40	43	63·1	83·0
Isoleucine	13·5	17·1	14	19	54·8	13·1
Proline	11·4	14·7	31	44	11·6	4·4
Phenylalanine	15·7	18·7	5	12	29·9	23·3
Tyrosine	16·3	16·9	4	6	99·7	23·3
Serine	102·0	101·0	96	114	76·4	48·0
Threonine	13·4	19·2	14	24	26·6	8·7
Cystine (½)	11·4	16·5	0·2	1	38·2	13·1
Methionine	5·0	7·6	3	7	33·2	14·6
Arginine	26·2	30·9	22	25	18·3	8·7
Histidine	2·5	5·5	0·7	3	8·3	13·1
Lysine	16·7	26·7	12	18	23·3	27·7
Aspartic acid	98·6	102·0	94	99	36·5	106·3
Glutamic acid	24·4	28·6	96	114	24·9	24·8
Hydroxyproline	—	—	—	—	—	—
Hydroxylysine	—	—	—	—	—	—
Acidic residues (Asp + Glu amide)	65·0	63	—	—	—	—
Basic residues	45·0	63	—	—	—	—
Acid/Basic residues	1·4	1·0	—	—	—	—

[a] Hare, 1963.
[b] Travis *et al.*, 1967.
[c] Tanaka *et al.* (*see* Wilbur and Simkiss, 1968).

1951; Hare, 1963; Travis *et al.*, 1967). Although in *Mytilus californianus* the differences are small (Table 10.3) and mainly involve the ratio between acidic and basic amino acids, the calcite layer, being more acid than the aragonite layer, such differences might determine whether calcite or aragonite is deposited in the matrix. Unfortunately this ratio has not been determined for other species.

On the other hand, in all-calcite and all-aragonite shells differences between the amino acid composition of the different layers of the shell are as great as, or even greater than those between the layers in calcite/aragonite shells (Table 10.4). Nor is it possible to detect an amino acid pattern specific for calcite or for aragonite layers so there is no obvious relation between the chemistry of the matrix and the form of calcium carbonate deposited in it.

While the conchiolin from calcitic calcitostracum differs in organization from that of aragonitic nacre, Grégoire (1961a, b) could detect no visual difference in the conchiolin surrounding the prisms in calcitic and aragonitic

Table 10.4 The amino acid composition of the neutral insoluble matrix of the prismatic and inner layers of the calcite shell of *Crassostrea virginica* and the aragonite shell of *Mercenaria mercenaria* compared with that of the calcite/aragonite shell of *Mytilus edulis*. Residues per 1000 total residues. (Travis *et al.*, 1967.)

Species	*Mytilus*		*Crassostrea*		*Mercenaria*	
Layer	Nacreous	Prismatic	Calcitostracum	Prismatic	Nacreous	Prismatic
Alanine	275	245	41	80	53	78
Glycine	293	261	346	337	133	101
Valine	24	27	17	20	23	44
Leucine	46	50	18	38	36	47
Isoleucine	15	21	7	12	21	22
Proline	10	18	38	70	103	126
Phenylalanine	18	20	10	31	31	39
Tyrosine	13	21	34	46	24	46
Serine	109	112	133	126	110	71
Threonine	13	21	11	18	40	58
Cystine	5	7	35	7·3	—	—
Methionine	2	4	5	3·5	13	12
Arginine	26	30	19	27	44	44
Histidine	8	7	5	17	4	11
Lysine	20	29	45	15	54	73
Aspartic acid	93	95	180	122	239	146
Glutamic acid	31	43	58	33	69	84
Hydroxyproline	—	—	—	—	—	—
Hydroxylysine	—	—	—	—	—	—

prismatic layers. However Wilbur and Watabe (1963) obtained α-protein type X-ray photographs from the conchiolin of the calcitic shell of *Crassostrae virginica* and a β-protein type from the aragonite shells of *Elliptio complanatus* and *Mercenaria mercenaria*. Travis *et al.* (1967) found that all the conchiolin they tested, whether from aragonite or calcite shells, gave two unorientated reflections at 4·86 Å and 10·4 Å.

Pieces of decalcified conchiolin from the aragonite shells of *Elliptio complanatus* and *Atrina rigida* and from the nacre of *Pinctada martensii* inserted into sites of calcite deposition in *Crassostrea virginica* results in aragonite mixed with calcite being deposited in the inserted conchiolin in about 25 per cent of the experiments; in contrast, no aragonite was deposited on films of glass or plastic or in conchiolin from the prismatic layers of *Pinctada martensii* similarly inserted. Similar results were obtained with decalcified *Elliptio complanatus* conchiolin placed in solutions of calcium bicarbonate (Watabe and Wilbur, 1960; Wilbur and Watabe, 1963). These experiments suggest that the nature of the conchiolin may be of importance in determining the form in which the calcium carbonate is deposited.

Polysaccharides and acid mucopolysaccharides have been identified histochemically in the matrix of gastropods and bivalves (Trueman, 1949; Bevelander and Benzer, 1948; Beedham, 1958; Beedham and Owen, 1965; Watabe *et al.*, 1966), and galactosamines have been found in widely varying

increase in calcification and reduction in the relative amount of conchiolin in the younger shells. However, Armstrong and Tarlo (1966) call for caution in interpretation of estimates of amino acids in fossils. Not only may amino acids diffuse out of fossils but foreign amino acids may also diffuse into them from organic material in surrounding rocks. Change in salinity of the environment could also be correlated with a systematic change in the form of the shell. Salinity may also bring about changes in the composition of the periostracum in the gastropod, *Potomopyrgus* (Wilbur and Simkiss, 1968), and age and salinity both affect the composition of the periostracum of *Mytilus* (Hare, *see* Wilbur and Simkiss, 1968).

In species containing both calcite and aragonite in their shells Lowenstam (1954) found the proportion of aragonite higher in individuals living at higher temperatures. Hare (*see* Wilbur and Simkiss, 1968), in *Mytilus californianus*, found that the lower the environmental temperature at which the animal lived the lower the acidic amino acid content of the aragonite layer. He suggests that increasing the environmental temperature above 20°C would result in both layers becoming similar in acidic amino acid composition and that this might result in both layers having the same crystal form. In support of this suggestion he cites a species of *Mytilus*, living at 22°C, in which both layers are aragonite: from Hare's earlier results on *Mytilus californianus* one might have expected that at this high environmental temperature the acidic amino acid content of the layers would be high and both layers would therefore be calcitic.

Environmental effects and factors influencing the deposition of calcite or aragonite have been investigated by subjecting regenerating shells to varying conditions but here complications arise because, even under normal conditions, differences exist, both in the conchiolin and in the crystal form, between the regenerating and the surrounding shell. Wilbur and Watabe (1963) found that while conchiolin from a normal shell of *Crassostrea virginica* gave an α-protein type X-ray photograph, that from regenerating shell gave a β-protein type, and while the normal shell contains calcite, regenerating shell contains both calcite and aragonite. In the aragonitic *Elliptio complanatus* normal conchiolin gives a β-protein photograph, regenerating conchiolin gives both α- and β-type photographs, and contains aragonite and calcite, and also vaterite, the third form of calcium carbonate. Vaterite is known to occur also in the regenerating shell of *Helix* (Meyer, 1931; Meyer and Weineck, 1932), and calcite occurs in regenerating shells of the normally aragonitic *Murex fulvescens* (Muzii and Skinner, 1966).

2. *Shell formation*

Growth of the shell results from activities of the mantle. The mantle edge is three lobed. The periostracum is secreted by the inner face of the outer lobe, the prismatic layer by the outer face of the outer lobe, and the inner layer of the shell by the mantle surface underlying the shell (cf. the secretion of brachiopod shells).

Kapur and Gibson (1967, 1968a and b) have made, on the freshwater gastropod, *Helisoma duryi eudiscus*, an extensive study of the histochemistry of the mantle in relation to the secretion of shell-substances. The presence of dopa oxidase in the epidermal cells of the outer lobe of the mantle edge suggests that quinones formed from the oxidation of tyrosine or dopa are responsible for tanning the proteins of the periostracum and conchiolin. Melanin also is present in the tissues although it is absent from the epidermis secreting the periostracum, and it may act as a cation-exchanger in calcification (Whyte, 1958). Peroxidase, only present in the region of the mantle edge secreting the periostracum, may prevent dopa oxidase oxidizing tyrosine to melanin in this region. This may be one of the reasons why the periostracum remains uncalcified. The peroxidase may also contribute to the production of tanning quinones. A similar function for the peroxidase in the epithelium secreting the radula has been suggested by Ducros (1967). Kapur and Gibson have also examined the distribution of other substances and agents such as carbohydrates, alkaline phosphatases, ribonucleic acid, and vitamin A, and of calcium spherules, and discussed their possible function in shell formation. High concentrations of vitamin A are known to suppress keratinization of vertebrate epidermis (see p. 302) and Kapur and Gibson suggest that in *Helisoma* epidermis the vitamin A present may prevent the epidermis from keratinizing. Intracellular filaments have been observed in *Helix aspersa* and *Arion* epidermal cells, but the chemical nature of these filaments is not known. Intracellular, sulphur-containing protein structures are found in invertebrates (e.g. nematocysts, embryophores of cestodes, hooks of trematodes, epidermal filaments in echinoderms) but there is as yet no evidence that the epidermis of invertebrates is normally or potentially able to produce keratin, nor is there evidence that vitamin A is widely distributed in invertebrates, or of importance in their nutrition (Fisher and Kon, 1959). While, therefore, the suggestion as to the function of vitamin A in the epidermis of *Heliosoma* is of interest, there is as yet no evidence to support it.

There has been discussion whether individual cells secreting the shell are able to produce only one of the shell components, so that shell growth must result from cell multiplication within each of the secreting zones, or whether multiplication of cells in the trough between the middle and the outer lobes of the mantle pushes cells outward so that, in turn, they secrete periostracum, outer shell, and inner shell layers. Beedham (1965) found that in the repair of *Anodonta cygnea* shells the underlying cells did produce first periostracum, then outer and then inner shell layers, implying that each is capable of producing all types of shell material.

Electron micrographs of developing shells show small crystal nuclei, distributed at random, on the surface of conchiolin sheets. Although this suggests that conchiolin may play a part in initiating calcification, as does collagen in bone formation, it is not known which, if any, particular chemical groups are responsible for nucleation. Conchiolin has a higher ratio of acidic to basic amino acids than does the periostracum, and acid mucopolysaccharides are confined to conchiolin (Wada, 1964; Beedham and Owen, 1965). In the gastropod *Pila globosa* the conchiolin is mainly acid mucopolysaccharide with very little protein present (Meenakshi, 1963). These findings conform with the belief that acid groups play a part in calcification by binding Ca^{++}. Degens and Love (1965) suggest that the holes visible in some conchiolin sheets may be places where acid, alkali, or amide groups confront each other, due to the elimination of short lengths of neutral polypeptide chains and that these exposed groups may concentrate Ca^{++} or CO_3^{--}, leading to nucleation of the network. Abolins-Krogis (1963a, b, 1968) believes that calcification begins on organic crystals, containing mucopolysaccharide and quinone-tanned protein, that form in the matrix of mollusc shells.

Digby (1968) has shown, in *Mytilus edulis*, that at the mantle edge the outside of the periostracum is acid and the inside alkaline. This probably results from sea water being sucked through the periostracum by the mantle. The periostracum is quinone-tanned and acts as a semiconductor so that through the salt-flow a streaming potential is built up, the outside of the periostracum becoming electronegative and the inside electropositive. The alkalinity of the inner surface of the periostracum where exposed at the edge of the shell, brings about deposition of calcium from the calcium-rich organic colloids of the pallial fluids. Crystal orientation in the prismatic layer may be due to orientation of the matrix and to the sucking of water through the shell tissues at the edge of the shell. Formation of nacre, he suggests, is due to the bathing of the inner surfaces of the shell, not affected by the sucking action of the edge of the mantle, with extra-pallial fluids, the alkalinity of which is produced at the mantle rim. The flattened form of the crystals is attributable to their deposition in the matrix which has a laminar arrangement due to sliding movements of the animal within its shell. The way in which the crystals, once initiated, grow to form the shell has been described by Tsujii *et al.* (1958) and Watabe *et al.* (1958).

Deposition of calcium is not a continuous process, possibly being influenced by such factors as food intake, tides, and the alternation between day and night (Barker, 1964). The crystals are interrupted at intervals by thin films of conchiolin anchored to the inner surface of the surrounding conchiolin sheaths and forming so-called growth lines. Orton and Amirthalingen (1926) described in *Ostrea edulis* the formation, each autumn at the end of the growing season, of a thin, horny, brown membrane over the whole inner surface of the shell which next year becomes embedded in new deposits of calcium.

The pigments of mollusc shells are sharply divided into acid-soluble and acid-insoluble types. The acid-soluble types which include porphyrins, pyrroles, and a number of unidentified pigments of small molecular size, occur in archaeogastropods, opisthobranchs, and lower bivalves as well as in isolated groups, and in species throughout the marine genera. They probably represent excretory products. In the higher lamellibranchs and all pulmonates the acid-insoluble pigments are firmly attached to the conchiolin and may be by-products of quinone-tanning (Comfort, 1951).

The shell provides protection against mechanical damage, predators, and parasites, though many animals have acquired the ability to bore into mollusc shells. In terrestrial and intertidal forms the shell also provides protection against desiccation while in prosobranchs, and in *Amphibola*, among the pulmonates, protection is made more complete by the possession of a calcareous operculum on the foot which, when the foot is withdrawn into the shell, seals up the entrance (Fischer, 1940; Kessel, 1941; Hubendick, 1948). The pulmonate Clausiliidae have a moveable piece, the clausilium, attached to the columella of the shell by an elastic ligament which also serves to seal the shell mouth. Useful as shells are for protection they undoubtedly are a hindrance to locomotion; this may be why many bivalves and the prosobranch Vermilidae and Magilidae have become sedentary. *Pecten* has utilized rapid closing of the shell to develop a method of swimming and *Nautilus*, an active, pelagic form, has its shell made buoyant through gas secreted inside it. Further, many molluscs that have become pelagic have reduced or lost the shell completely. The loss of the shell can confer other advantages such as economy in calcium, in land-living forms, an ability to glide through narrow spaces for protection, or an ability to burrow actively after animal prey that can then be swallowed whole into a distensible body (Morton, 1967). However, in losing the shell they lose a source of carbonate that could be drawn upon in need for there is much evidence of resorption of shell material, where necessary, by shelled forms. Various cephalopods, e.g. *Nautilus* and *Sepia*, control their buoyancy by regulating the relative amount of liquid and gas in their shells (Denton and Gilpin Brown, 1961, 1966; Bidder, 1962). Trueman (1966) has shown that the rigidly walled space enclosed by the bivalve shell is an important element in the development of hydrostatic forces used in burrowing into sand and mud. In *Teredo* and *Pholas* it is the shell that is used for burrowing into wood and rock.

3. *Hinge ligament of bivalves*

In bivalves the two valves of the shell are joined by a dorsal ligament which is part of the structure of the shell (Trueman, 1942, 1949, 1950, 1951, 1953a, b; Beedham, 1958). It varies in shape, although it is primitively an elongated structure between the dorsal edges of the two valves. It is formed from modifications of the inner and outer layers of the shell and covered over by periostracum. Its elastic properties make the shell gape when the adductor muscles relax. In *Mytilus* the outer layer of the ligament is uncalcified and composed of quinone-tanned protein (Trueman, 1950). It also contains lipid (Brown, 1952). The inner layer, which also contains quinone-tanned protein (Trueman, 1950, 1951), is calcified. The calcification of the inner layer varies considerably, from *Anodonta* containing 40 per cent calcium carbonate by weight to *Pecten* containing only 0·6 per cent (Trueman, 1953a, b). The form of the calcium in the ligament may differ from that in the rest of the shell (Wada, 1961). Closing of the valves by the adductor muscles causes stretching of the outer layer and compression of the inner layer, relaxation results in the ligament regaining its resting shape and the lipid present may lubricate the movements of the protein molecules during these changes of shape.

Kelly and Rice (1967) found the hinge material of *Pecten irradians* and *Placopecten magellanicus* to contain about 80 per cent protein, 2·8 per cent inorganic material, and a little pentose and hexose, but no lipid. Neither stretched nor unstretched does it give an X-ray diffraction pattern and in this it resembles elastin. It differs from elastin in being digested by pepsin and in being insoluble in all the standard protein solvents. Although histologically the hinge appears to have inner and outer layers, chemically both layers are similar. The amino acid composition (Table 10.5) differs slightly in the two

Table 10.5 Amino acid composition of abductin from hinge ligaments of *Placopecten magellanicus* and *Pecten irradians* compared with resilin and elastin. Residues per 1000 total residues

	Whole ligament[a]		Ligament[a] cortex	Ligament[a] medulla		
	Pecten	*Placopecten*	*Placopecten*	*Placopecten*	Resilin[b]	Elastin[b]
Alanine	32·9	26·5	23·9	24·1	109	223
Glycine	629·9	620·3	622·5	621·9	379	332
Valine	6·0	3·5	2·5	2·7	28·0	141
Leucine	1·8	2·9	2·7	2·5	23·0	64
Isoleucine	7·5	4·0	3·7	4·0	16	27
Proline	12·6	7·4	7·5	7·9	76	109
Phenylalanine	92·6	51·3	52·1	52·1	25	35
Tyrosine	1·5	10·7	9·2	10·5	27·0	7
Serine	68·8	36·4	34·5	38·0	79	8
Threonine	11·3	7·4	6·7	6·5	30	8
Cystine	—	—	—	—	—	—
Methionine	90·4	117·5	124·0	120·5	0	—
Arginine	3·4	9·8	8·8	9·1	35·0	7
Histidine	6·0	0·3	0·3	0·3	9·0	1
Lysine	8·0	12·4	12·6	9·5	5·0	4
Aspartic acid	18·1	69·9	69·8	72·0	101·0	7
Glutamic acid	12·0	19·4	19·2	17·6	46	15
Hydroxyproline	—	—	—	—	—	1·1
Hydroxylysine	—	—	—	—	—	—

[a] Kelly and Rice, 1967.
[b] Andersen and Weis-Fogh, 1964.

species but in neither does it resemble that of elastin or resilin. Hinge material or *Abductin*, is distinguished by its high content of methionine, which Kelly and Rice suggest might give rise to methionine hydrophobic bends linking the protein chains into a three-dimensional network. Mechanically it behaves like a true rubber, with a low value of Young's Modulus of Elasticity 1·27 kg/mm^2 (Trueman, 1953a, b; Alexander, 1966; Kelly and Rice, 1967) compared with 0·064 kg/mm^2 for resilin and 0·61 kg/mm^2 for elastin (cf. Table 1.1).

Andersen (1967) thought he had found two phenolic tanning agents in the ligaments of *Mytilus edulis*, *Spisula solidissima*, and *Pecten maximus*. One of the compounds was 3,3′-methyl-bistyrosine, the other, though resembling it in several respects, was not fully identified. The interest in the identity of these particular phenolic substances was that they are closely related to dityrosine, the cross-linking agent in resilin, the elastic protein of Arthropods (see p. 221). Recently Andersen (1968a) found that the 3,3′-methyl-bistyrosine did not exist in the natural ligament but was produced during acid hydrolysis of the whole ligament. The inner layer of the ligament, in contrast to the outer, contains no tyrosine. When the inner layer alone is hydrolyzed in acid in the presence of added tyrosine, the 3,3′-methyl-bistyrosine is produced. The nature of the tanning agent therefore remains as yet unsolved.

4. *Cuticles*

The animals are firmly and permanently attached to the shell by muscles. Gastropods can extend head and foot from the shell and use the foot for locomotion. Bivalves have an extensible foot that may be used for ploughing and burrowing. These soft structures that are capable of changing their shape and surface area can gain protection by being withdrawn into the shell. They are often covered only by ciliated epidermal cells and the secretions of epidermal mucus cells, though sometimes a thin cuticle of unknown composition is present (Hyman, 1967). In *Neopilina* the cuticle is restricted to the lips round the mouth. Cuticle commonly covers over such forms as nudibranchs that lack shells.

The cuticle detected by the light microscope on the surface of the body and tentacles of *Helix aspersa* and *Arion* has been shown in electron micrographs to be a border of microvilli. They are arranged in a hexagonal pattern and are interconnected at their tips by fine fibrils; similar fibrils are also seen projecting down the sides of the microvilli. Beneath the microvilli on the optic and inferior tentacles there is a cytoplasmic layer of tangential and radial fibrils, 50–70 mm in diameter. This layer of fibrils is absent from the epidermis over the rest of the body (Lane, 1963). Lane suggests that one of the functions of the microvilli is to hold in place the mucus secreted on to the surface of the animal.

Whether or not cuticle occurs over the surface of the body it often lines the fore and hind gut and may also be present in the stomach. In *Loligo* the oesophageal and stomach cuticle contains protein and chitin (Bidder, 1950). While the chitin of the pen is in the β-form that of the oesophagus lining is in the α-form, and that of the stomach in the γ-form (Rudall, 1965). The pen is not much subjected to wear and tear, being embedded in the body, but the oesophagus is a flexible tube and the stomach undergoes considerable changes of shape. The different forms of chitin that occur in these situations may be related to their particular functions (Rudall, 1965). The cuticle of the stomach is one of the few chitin-containing structures that undergo extensive changes of shape.

B. The radula

The primitive molluscs rasped small particles of food from the ground and raked them into their mouth with an odontophore, a unique feature of molluscs, though it is absent in bivalves, which are almost all ciliary feeders The odontophore is composed of the radula, a membraneous ribbon carrying hard, tooth-like structures, formed in a ventral sac which opens into the buccal cavity. The radula is supported underneath by the odontophore cartilage which is moved by a complex of muscles that push the whole apparatus out of the mouth, draw the radula over the food and withdraw the structure, carrying the rasped-off food with it. In higher molluscs the radula has been adapted to various other methods of feeding such as rasping away flesh from dead or living animals, boring into soft bodies or eggs, or through shells so that the food within can be sucked out. The shape and arrangement of the radula teeth show adaptations to the type of food eaten. In *Loligo* the radula is too small and soft to be used for rasping off food but it functions as a tongue aiding the swallowing of food (Bidder, 1950). Some species have lost the radula; *Dendrodoris citrina* which lacks a radula, digests a hole in the test of the ascidians on which it feeds so that the proboscis can be inserted to suck out the contained animal. Many gastropods and cephalopods have, in addition, hardened "jaws" used for capturing and biting off pieces of their prey.

Chemically the radula contains protein, and often chitin and various inorganic salts (Sollas, 1907; Jones *et al.*, 1935; Runham, 1961). In Aplaco-

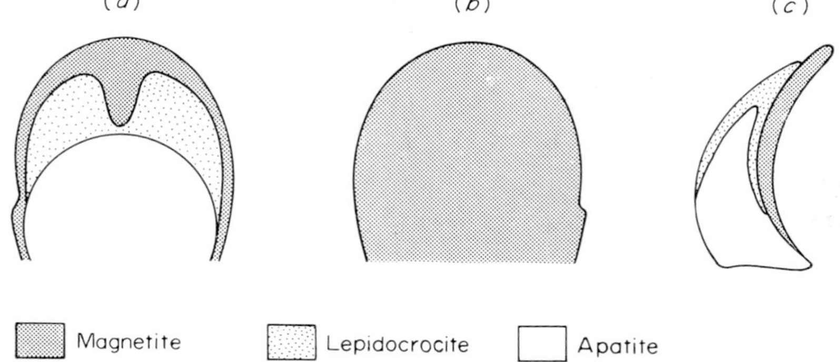

Fig. 10.8. Radula tooth of *Acanthopleura*. (*a*) anterior surface view; (*b*) posterior surface view; (*c*) median longitudinal section. (After Lowenstam, 1967.)

phora the teeth are not, as in other classes, mounted on a ribbon but embedded in cuticle. Nothing is known about them chemically, except that Hymna (1966) failed to find chitin in the radula of *Solenogaster*. Lowenstam (1967) found in the radula teeth of three species of *Chiton*, three different inorganic salts: lepidocrocite ($Fe_2O_3 \cdot H_2O$), an apatite (probably francolite), and magnetite ($Fe_2O_3 \cdot FeO$) each occupying a distinct region of the tooth (Fig. 10.8). Magnetite formerly was thought to be of volcanic origin but its occurrence in the teeth of chitons, and the large numbers of chitons present on some areas of the sea-bottom, suggests that part, at least, of the magnetite that occurs in marine deposits may be from their teeth.

In the early stages of the formation of the radula in gastropods it is possible to demonstrate in the teeth P.A.S.-positive (chitin), and argentaffin-positive material as well as protein. In older teeth these reactions are much reduced or absent (Runham, 1961, 1963a; Ducros, 1967). Gabe and Prenant (1957, 1958) suggest that these changes are due to mineralization, but in *Patella* and *Acanthochitona* (Runham, 1961) they precede mineralization.

Although Runham found that radula teeth are generally argentaffin-positive and that there is circumstantial evidence for quinone-tanning, he failed to detect a polyphenol oxidase, though he may have used an unsuitable substrate. The occurrence of considerable quantities of tryptophan in the radula suggests that this as well as, or in place of, tyrosine may be the source of the quinone (see Pryor, 1955, 1962; Brunet, 1967). Gabe and Prenant (1957) found sulphur in the radulae of cephalopods and a few gastropods.

If the chitin of the radula becomes covalently linked to the protein by aspartyl residues, possibly forming a *N*-acyl glucoseamine linkage, as is believed to occur in insect cuticle (Hackman, 1960), this would not prevent the chitin continuing to give a positive P.A.S. reaction. If, however, the chitin becomes attached by O-acyl linkages to aspartyl and/or glutamyl residues, such as those which Murphy and Gottschalk (1961) suggest link glucosamine to protein in bovine submaxillary gland mucus, this would eliminate the capacity of the chitin to give a positive P.A.S. reaction and would also account for the loss of positive dyeing with dyes for protein carboxyl groups (Runham, 1963a), and so account for the failure of mature radulae to give the same histochemical reactions as the young radula.

Ducros (1967) also failed to find polyphenol oxidase within the radula-producing epithelium though it was present in near-by tissues such as the buccal mass. She did, however, find a peroxidase in the epithelium of the sheath surrounding the radula, and basing her argument on the work of Booth and Saunders (1965), Jayle (1939), and Mason (1957) on the capacity of peroxidases to oxidize phenols to quinones, and also on that of Vovelle (1965) who found peroxidase in the glands secreting the quinone-tanned tube of the annelid, *Sabellaria alveolata*, she suggested that the peroxidase converts the tryptophan and tyrosine of the radula protein to quinones, which produce a form of auto-tanning in the protein. The enzyme also oxidizes phenolic material in the sheath epithelium, which contributes to the tanning of the radula protein and possibly forms linkages between the protein and the chitin, in this way accounting for the observed histochemical changes during the development of the radula.

The radula is being constantly worn away anteriorly and replaced by growth posteriorly. Runham (1963b, c) has described the growth and movement forward of the radula and the rate at which it is replaced in *Helix*.

C. Byssus of bivalves

Many bivalves develop a temporary post-larval structure, the byssus, and in many non-burrowing forms the byssus is preserved in the adult as a structure anchoring it to rocks or other surfaces. In the byssus of *Mytilus edulis* three different parts are distinguishable (Fig. 10.9(*a*) (*b*)): the root which lies embedded in the byssus gland at the top of the posterior face of the foot; the stem which is continuous with the root but lies outside the gland; and the threads, attached proximally by rings to the stem and ending distally in discs attaching the threads to the substratum.

The byssus gland is flask-shaped and divided by thin laminae secreting sheets of material from which the root is formed. There are two glands in the foot; one is the "white" gland, secreting into the bottom of the longitudinal groove on the posterior face of the foot material from which the core of the threads is formed. The "purple" gland (which is only purple after fixation in formalin) is most voluminous around the sucker-like depression at the distal end of the foot but also extends up each side of the posterior groove where it coats the secretion of the white gland. In the region of the sucker it discharges through ciliated ducts that drain the deeper regions of the gland (Fig. 10.9(*c*)).

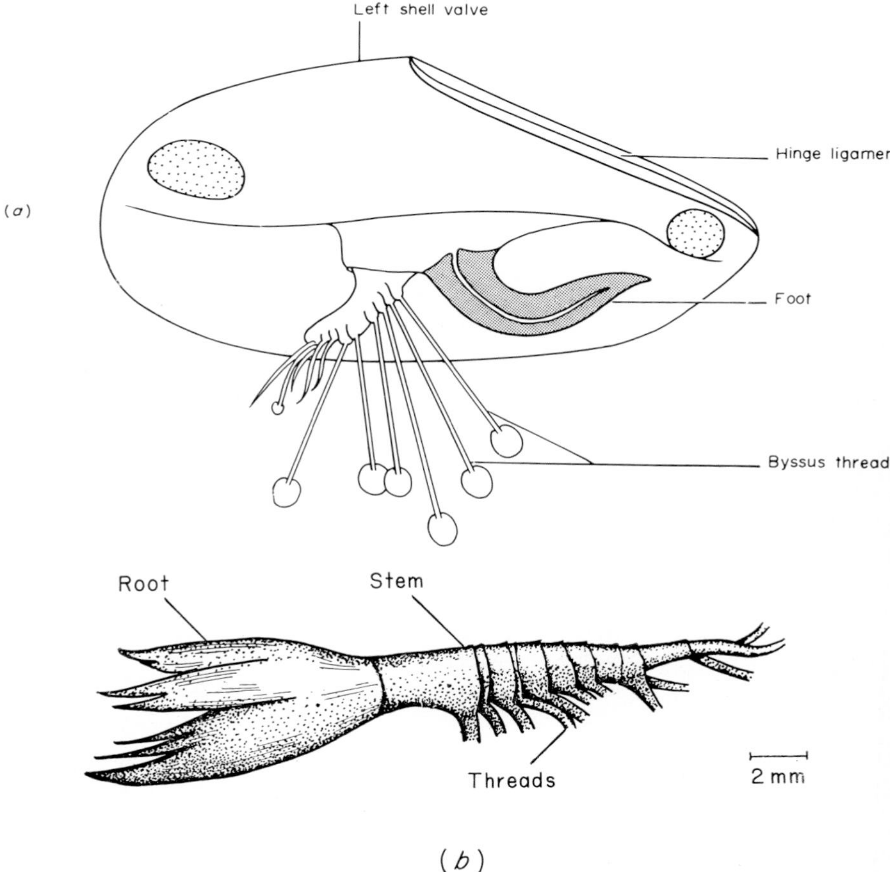

Fig. 10.9. (*a*) *Mytilus edulis* with right shell valve removed to show foot and byssus; (*b*) byssus; (*c*) diagram of longitudinal section of foot of *Mytulus edulis* to show arrangement of the glands. (After Brown, 1952.)

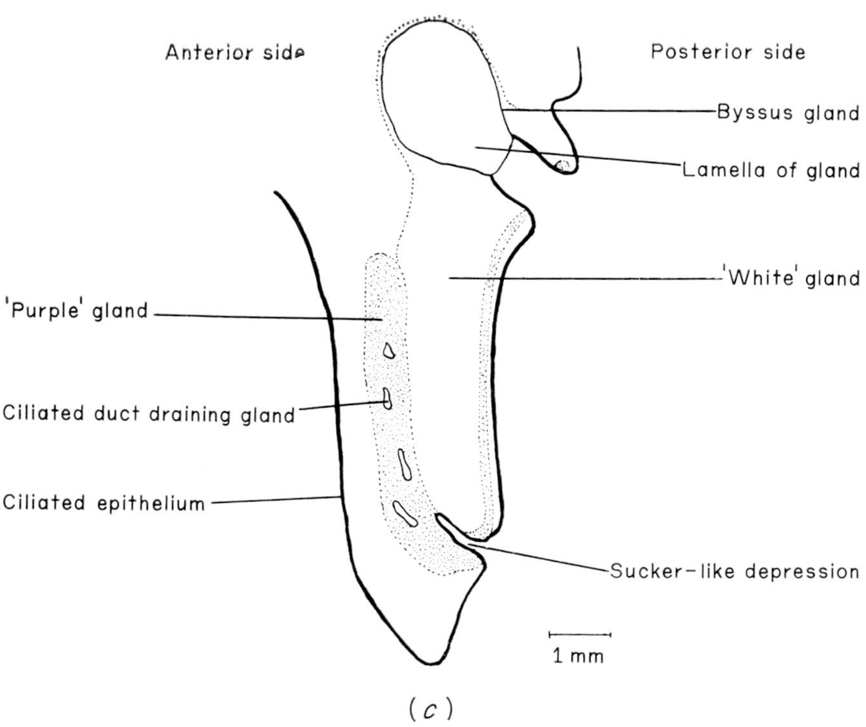

Fig. 10.9 (*continued*)

The structure of the byssus and the gland secreting it have been described also in *Barbatia* (Kusakabe and Kitamori, 1948) *Lasaea rubra*, *Turtonia minuta*, *Kellia suborbicularis*, *Montacuta ferruginosa*, and *M. substriata* (Oldfield, 1955, 1961) and in various species of *Anadara* (Kusakabe and Kitamori, 1948; Lim, 1965). Lim found in *Anadara* that three different glands are responsible for the secretion of the byssus and has considered previously published accounts of byssus secretion for evidence that in these cases also three different glands may be concerned.

There is evidence in *Mytilus edulis* and other species that the byssus is at least partly formed of a quinone-tanned protein (Pyefinch, 1945; Brown, 1952; Smyth, 1954). The tanning precursor is possibly a phenolic amino acid or protein since all attempts to extract a free phenolic material from the purple gland have been unsuccessful. A protein material with phenolic properties can be extracted from the gland with dilute hydrochloric acid and resembles in various ways the protein fraction extracted with acid from the vitellaria of *Fasciola hepatica* by Smyth and Clegg (1959; see also Clegg and Smyth, 1968). A polyphenol oxidase occurs in "purple" gland cells (Smyth, 1954).

The byssus threads are very elastic, most of the extension takes place in the upper, corrugated part of the thread which stretches between 300 and

400 per cent. All parts of the unstretched byssus are strongly birefringent, so that elasticity cannot depend upon a random network of linked protein chains as it does in elastin.

Stretched byssus threads of *Mytilus edulis* give an oriented collagen-type X-ray photograph (Champetier and Fauré-Fremiet, 1938a, Fitton-Jackson *et al.*, 1953; Randall *et al.*, 1952; Rudall, 1955). Randall *et al.* found no sign of a low angle, 640 Å period, and fibres of the material appear unbanded in the electron microscope. Rudall found evidence for the presence of a second protein, and Andersen (1968b) has calculated from the hydroxyproline content of the threads that they contain approximately 55 per cent collagen. Presumably the two proteins are the products of the two foot glands. The byssus stem gives quite a different X-ray photograph, with the meridional region showing up to 16 orders (the fifteenth is lacking) of an axial period of 43 Å suggesting a fully extended β-protein. The byssus threads of various species of *Pinna* do not give a collagen-type photograph but one closely

Table 10.6 Amino acid composition of collagen extracted from the byssus of *Mytilus edulis* with trichloroacetic acid. Residues per 1000 total residues. (Andersen, 1968a.)

Amino acid	
Alanine	84
Glycine	337
Valine	33
Leucine	22
Isoleucine	17
Proline	83
Phenylalanine	10
Tyrosine	8
Serine	42
Threonine	34
Cystine and cysteic acid	3
Methionine	2
Arginine	45
Histidine	11
Lysine	40
Aspartic acid	67
Glutamic acid	89
Hydroxyproline	71
Hydroxylysine	2

resembling that from the stem of the *Mytilus* byssus (Mercer, 1952; Fitton-Jackson *et al.*, 1953; Rudall, 1955). Andersen (1968) has analysed the collagen from the byssus of *Mytilus edulis* (Table 10.6) and Lucas *et al.* (1955) have given an incomplete analysis of *Pinna* threads.

Yonge (1962) suggests that the byssus and glands secreting it have evolved from the pedal glands of gastropods, primitively concerned with the secretion of mucus often used in locomotion.

D. Internal skeletal structures

In those aplacophorans in which the cuticle is thick it is probably sufficiently strong and inelastic to control the shape of the animal, because a thick dermis of connective tissue is only found in forms like *Neomenia carinala* which have a thin cuticle. In some species there are, in the body wall, well developed layers of circular, longitudinal, and diagonal muscles, as might be expected in animals that move or burrow, albeit sluggishly, by peristaltic waves of contraction. In others (that rely on cilia for locomotion) the musculature is reduced and in *Sandalomenia* there is only a loose network of muscle fibres. The space between the body wall and the digestive tract is occupied by strands and webs of connective tissue, the interstices being filled with blood.

Connective tissue is important in the construction of the chiton body. It is particularly abundant in the foot and mantle, supports the epithelium of the digestive tract and other viscera, and forms strands between the various organs. It contains collagen fibrils which are either single or in strands (Fretter, 1937; Gabe and Prenant, 1949). The fibres are associated with ground substance and various cells. Connective tissue also forms an inner supporting structure in the gills (Hyman, 1967).

In the remaining classes, collagen fibres surrounding and supporting tissues and organs have been demonstrated in many instances (Arvy, 1955; Rudall, 1955; Nesbit, 1961; Campion *et al.*, 1964; Plummer, 1966; Elder, 1966; Elder and Owen, 1967). Plummer has described, in various species of *Achatina*, fibroblasts or "pore cells" characterized by a cytoplasm having extensive and interconnecting vesicular systems, opening to the exterior either through wide and deep indentations of the cell surface or through elongated surface channels, the openings of which are subdivided by cytoplasmic tongues. These specialized fibroblasts secrete tropocollagen that forms the fibres of the nerve ganglion-capsule and the connective tissue of the collar. The surface openings of the pore cells show a parallel arrangement in surface view and this may be related to the parallel arrangement of the fibres into columns, particularly in the glandular tissue of the collar. The fibres show a periodic banding of 330 Å units.

Williams (1960) has analysed a gelatin obtained from *Helix* connective tissue collagen. Chemically (Table 10.7) it approximately resembles mammalian collagen, though it contains somewhat less alanine and lysine and more serine. With a total of 204/1000 imino acid residues, *Helix aspersa* collagen has a T_S of 52°C, though several other species with a lower imino acid content (*Melarapha unifasciata* 175/1000, *Nodilittorina pyrimidalis* 176/1000, *Helicella virgata* 199/1000) have a T_S between 56°C and 59°C (Rigby, 1967).

Elastic fibres have been found in molluscan connective tissue (Arvy, 1955; Elder, 1966; Elder and Owen, 1967), in the arteries of *Mya arenaria*, and the wall of the excretory organ of *Cryptochiton*. Doubtless elastic fibres are generally present in the walls of blood vessels and contractile organs.

Locally, connective tissue may become organized into compact structures closely resembling vertebrate cartilage (Person and Philpot, 1969). Such material is found underlying the radula as the odontophore cartilage already mentioned. Cartilage-like material is also found in the internal framework of gills and in the extensive internal skeleton supporting, protecting, and providing muscle attachments in the head, eyes, neck, funnel, mantle, diaphragm, fins, and gills of cephalopods. The odontophore cartilage of *Busycon* contains collagen fibrils with a banding period of 600–650 Å (Person and Philpot, 1963) and a non-aminated polyglucose sulphate (Lash and Whitehouse, 1960). *Loligo* and *Ommastrephes sloani pacificus* cartilage contains a

Table 10.7 Amino acid composition of collagen from *Helix aspersa*. Number of residues per 1000 total residues. (Williams, 1960.)

	Collagen
Alanine	72
Glycine	321
Valine	22
Leucine	24
Isoleucine	12
Proline	104
Phenylalanine	10
Tyrosine	9
Serine	61
Threonine	28
Cystine	—
Methionine	1
Arginine	51
Histidine	3
Lysine	8
Aspartic acid	67
Glutamic acid	99
Hydroxyproline	100
Hydroxylysine	8
Total imino content	204

highly sulphated chondroitin sulphate (Mathews *et al.*, 1962; Kawai *et al.*, 1966). Haliburton (1885) reported chitin fibres in the head cartilage of *Loligo*, and Stegeman (1963) obtained chitin from *Octopus* skeletal tissue. These molluscan cartilaginous structures never become calcified.

In some gastropods, calcium carbonate as spheruliths or spicules occur in the tissues (Graff, 1883; Odhner, 1952), particularly in nudibranchs and pleurobranchs. The dorsal papillae of *Rostanga* (Labbe, 1929) are supported by a circle of 6–9 upright spicules. In some Onchidiacea siliceous spicules are present (Labbe, 1933). Prenant (1924) describes the calcareous spicules as being formed intracellularly. These inorganic spicules, which do not form a continuous skeletal system seem well adapted to providing support and toughness in extrusible structures.

E. Egg shells

At the base of the gonoduct in *Solenogaster* there is the so-called shell gland but whether or not it secretes a covering to the eggs is not known. The eggs in *Epimenia* are laid as a pair of flattened egg bands held together by mucus.

The eggs of Polyplacophorans are fertilized externally and are generally held together loosely in masses or strings by mucoid material, presumably secreted by the slime sacs of the oviducts. The eggs are provided with two membranes, an inner, thin vitelline membrane and an outer thick shell or chorion, sometimes covered with projections. This shell is formed by a layer of flat, follicle cells that remain around the egg after it is shed. The chemical nature of the shell is not known.

The eggs of primitive gastropods and the majority of bivalves are fertilized externally and are covered only by a vitelline membrane, outside which there may be a layer of jelly. Such eggs restrict these species to a marine environment. The evolution of internal fertilization made possible oviparity, or the incubation of the eggs, in the comparative safety of the mantle cavity, or alternatively, the development of protective capsules secreted round the fertilized eggs by special glands of the genital duct. These in turn made possible the colonization of the more demanding marine environments, freshwater, and damp terrestrial habitats.

Fretter (1941) has described the structure of the capsule surrounding groups of eggs in several Prosobranchs. The *Buccinum* capsule has four layers, an inner layer of homogenous protein, a layer of longitudinally orientated protein fibres, a layer of circularly orientated fibres, and an outer layer of homogenous protein. A gap in the capsule wall through which the larvae eventually hatch, a common feature in hardened egg shells which are resistant to enzyme digestion, is filled with a mucus plug. The outermost layer of protein covers the whole structure including the plug. The shell gland, responsible for secreting the shell, contains two types of gland cell, one secreting mucus, the other a protein. The varying activity of the two types of gland cell, the direction of beating of the cilia in the gland, and the movement of the egg mass through the gland, all play a part in the formation of the different layers and the orientation of the protein fibres. Mucus plays a part in separating the protein into distinct fibres in the two middle layers. When first extruded, the capsule is soft and it is passed along a temporary groove in the foot to the ventral pedal gland which, by muscular activity, moulds it into its final shape and converts it into a tough, horny but elastic structure.

The structural element of the capsule is a very thin protein ribbon, in which Rudall (1968b) has demonstrated an extraordinary banded structure (Fig. 10.10(*a*) and (*b*)). The pattern is symmetrical, not polarized as in mammalian collagen, and it has a band period of about 960 Å. A similar periodic structure has been found in *Neptunae antigua*, (Buccinacea) but in *Urosalpinx cinerea* (Muricacea) the fibre-axis repeat-period is shorter than in *Buccinum* (Tamarin and Carricker, 1967). X-ray studies show that the material of these egg cases is an α-type protein that readily converts to the β-form on being stretched. Other α-type proteins besides this *Buccinum* material show periodic banding. It occurs naturally in mantid ootheca protein (Millard and Rudall, 1962), and possibly the banded protein from the shell-lining of the foraminiferan, *Haliphysema*, belongs in this class. Banding can also be produced in myosin by treatment with trypsin (Szent-Gyorgyi *et al.*, 1960;

oviposition in damp environments or in the soil, though the cytoplasm of the embryo is able to tolerate considerable loss of water (Morton, 1967).

Helix and cephalopods transfer sperm enclosed in a spermatophore, a narrow elongated tube which in *Octopus dofleini martini* is formed from a transparent and highly elastic material containing mainly protein, about 10 per cent orcinol-reactive carbohydrate and 10 per cent amino sugar (Mann *et al.*, 1966). When the spermatophore breaks open it is the sudden elastic recoil of the spermatophore wall that provides the force driving the sperm into the female.

Table 11.1 Amino acid composition of annelid cuticle and internal collagens: Residues per 1000 total residues

	Giant earthworms			Small earthworms								
Amino acid	*Rhinodrilus fafneri*[a] cuticle	*Digaster longmani* cuticle[b]	*Digaster longmani* body-wall[b]	*Lumbricus* sp. Cuticle[a]	*Lumbricus* sp. Cuticle[c]	*Lumbricus* sp. Cuticle[d]	*Lumbricus* sp. Body-wall[d]	*Lumbricus terrestris* Cuticle[e]	*Pheretima megascolidioides* Cuticle[b]	*Allolobophora* sp. Cuticle[b]	*Allolobophora* sp. Body-wall[b]	*Allolobophora* sp. Body-wall[f]
Alanine	89	92	54	99	103	97	432 (Ala + Gly)	100	92	102	59	
Glycine	255	272	358	304	324	333		334	334	324	352	189
Valine	36	29	17	25	17	20	22	20	21	21	21	
Leucine	38	40	38	32	29	24	34	30	36	32	35	
Isoleucine	24	27	15	19	15	13	7	16	24	19	14	
Proline	21	10	59	11	13	6·6	41	7·7	8	22	53	48
Phenylalanine	21	17	9	9·1	5·2	9	10	5·7	10	9	12	
Tyrosine	19	6	6	6·3	0	4	4	2·3	4	3	8	
Serine	89	98	59	85	105	121	86	83	86	85	52	56
Threonine	68	77	33	57	52	53	31	49	54	59	36	
Cystine	7	0	3	tr.	n.d.	0	0	0	4	1	4	
Methionine	4·4	1	2	1·9	0	0·5	2	1·0	2	tr.	3	
Arginine	27	26	50	25	21	n.d.	84	22	23	26	51	
Histidine	4·3	4	9	2	0	n.d.	3	1	4	2	9	
Lysine	28	17	15	20	15	n.d.	12	16	19	18	22	
Aspartic acid	76	72	60	68	56	53	51	62	65	63	65	
Glutamic acid	84	80	85	85	81	62	64	85	66	77	97	60
Hydroxyproline	99	116	115	153	165	155·9	133	165	148	137	95	
Hydroxylysine	n.d.	16	13	1·1	0	n.d.	n.d.	n.d.	tr.	0	12	
Total imino content	120	126	174	164	178	163	174	173	156	159	148	
T_D°C	n.d.	22	22	n.d.	n.d.	n.d.	n.d.	22	22	22	22	n.d.
T_S°C	n.d.	34	40	n.d.	45	n.d.	n.d.	n.d.	37	34	40	n.d.

n.d. not determined. [a] Maser and Rice, 1962. [b] Rigby, 1968. [c] Watson, 1958. [d] Fujimoto and Adams, 1964. [e] Josse and Harrington, 1964. [f] Rudall, 1968a.

Although quinone-tanning has never been reported in annelid cuticles, Gorbman *et al.* (1954) reported that the cuticle of *Amphitrite* takes up radio-iodine, and it would be worth while looking for quinone-tanning in this animal, and in related forms.

Rudall (1955) found chitin in the cuticle of the earthworm gizzard; but Gansen-Semal (1960) suggested that it was made of a non-fibrillar elastin. In sections containing both gizzard and the external cuticle the external cuticle always dyes like collagen while the gizzard lining dyes like elastin. The gizzard is highly muscular and its contractions, together with small pieces of grit taken up by the worm in its food, break up the food before it is passed back into the intestine. An elastic structure adhering closely to the epithelium of the gizzard would follow the changes of shape of the gizzard better perhaps than a collagenous membrane. However, Izard and Bronssy (1964) concluded from a histochemical investigation that the gizzard was lined by hardened acid mucopolysaccharide.

The majority of polychaetes and oligochaetes possess chaetae, bristle-like structures occurring in each body-segment and secreted by epidermal cells, they are absent in leeches. While in the oligochaetes they are simple, short structures, in polychaetes they may be much more complex and in some they are expanded to help in swimming. Chaetae are generally found to contain chitin, but Goodrich (1896) could not detect any in the chaetae of *Vermiculus* and *Enchytraeus* and Ebling (1945) found none in the opercular chaetae of *Sabellaria alveolata*. The chitin in *Aphrodite* chaetae is in the β-form (Lotmar and Picken, 1950). Brown (see Lotmar and Picken, 1950) found quinone-tanned protein in *Hermione* and Dennell (1949) found it also in earthworm chaetae. Though Brown failed to find sulphur in chaetae of *Hermione*, Bobin and Mazoné (1944) claimed that sulphur is present. The development of chaetae has been described by Bobin (1944) and Ebling (1945). Some polychaetes, e.g. *Nereis*, and various leeches have, in the buccal cavity, hardened structures which act as teeth and, in those of *Nereis*, Jeuniaux (1963) could find no chitin.

Earthworms secrete mucus which acts as a lubricant in locomotion, and what appears to be mucus cells are seen in the epidermis. How they discharge to the outside is not known but it is possibly through the microvilli (Krall, 1968). Some polychaetes build protective tubes by cementing together sand grains or other foreign particles. Serpulids build tubes of small granules of aragonite embedded in a mucus cement containing sulpho-mucopolysaccharide. Both the aragonite and the cement are secreted by glands of the peristomium. When the tube is first secreted it is plastic and the peristomial collar moulds it into shape before it sets hard (Robertson and Pantin, 1938; Hedley, 1956).

In *Sabellaria alveolata* Vovelle (1965) detected two proteins, one of which contained considerable quantities of aromatic residues; he also found an alcohol-soluble orthodiphenol and a polyphenol oxidase, all concerned in forming the cement of the tubes. He believes that the polyphenol oxidase not only oxidizes the orthodiphenol to a quinone capable of tanning the proteins present, but also oxidizes the aromatic residues in the protein itself to bring about a type of auto-tanning. He also detected a peroxidase which, he suggests assists in oxidizing the polyphenols. Cameron (1914) reported that the tubes of various polychaete worms often contain appreciable quantities of iodine and here again it seems that this may be correlated with the presence of quinone-tanned proteins.

The body wall of earthworms consists of the cuticle, the epidermis, a dermal layer of connective tissue, and layers of longitudinal and circular muscles. If a living earthworm is punctured coelomic fluid usually spurts out indicating that the body fluid is under pressure. If the animal is anaesthetized relaxing the body wall muscles, the body becomes limp indicating that the pressure is maintained by the muscles and not by the cuticle as in nematodes, and that the earthworm cuticle is not normally under tension. This allows the cuticle of annelids to be relatively thinner than that of nematodes. The cuticle must, however, be subjected to various stresses during movement and locomotion. Earthworms move and burrow by passing waves of local contraction backwards along the body. The helical arrangement of the cuticle fibres is well adapted to accommodate the changes of shape that accompany the contractions.

Earthworm cuticle certainly does not play the part in controlling the body shape or in locomotion that it does in nematodes. It cannot isolate the epidermis from the environment since the microvilli are in direct contact with the outside medium. The microvilli and not the cuticle must take the full force of the friction between the animal and the surfaces with which it is in contact during movement. The protective functions of the cuticle therefore appear limited, though it may prevent friction from pulling apart the epidermal cells as it might do if the epidermal cells were in direct contact with the substratum. As a tough and essentially non-elastic structure, it must define the limits of expansion and extension of the body, and offer resistance to osmotic swelling.

Polychaetes move and swim rather like nematodes, though by lateral undulations. In such a form as *Nereis*, the longitudinal muscle layer is broken up into four blocks, two ventral and two dorsal, and this allows the muscle blocks to be contracted individually. Circular muscles are still present but poorly developed and the layer is interrupted laterally where muscles originating from the circular layer pass out to the parapodia which play an important part in locomotion. The body wall of *Nereis* appears weaker than that of earthworms though whether the cuticle therefore plays a part in preventing lateral expansion of the body is not known. *Nephthys* completely lacks circular muscle and Clark and Clark (1960a, b) found the body was restrained from widening when the longitudinal muscles contracted by the combined influence of special semi-elastic ligaments, possibly aided by the septal muscles. They did not consider any possible influence of the cuticle.

Leeches, in spite of the fact that the coelomic cavity is much reduced and that the interior of the body is largely filled with connective tissue, show considerable change of shape, brought about by dorso-ventral as well as the longitudinal and circular muscles of the body wall. The solid construction of the body makes some leeches remarkably tough and almost unsquashable.

As the muscles beneath the epidermis contract and relax, the junction between the epidermis and the external cuticle must be considerably strained particularly as the cuticle, being non-elastic, has to be pulled into shape. The microvilli must help to anchor the epidermis to the cuticle and Coggeshall (1966) has described filaments that radiate into the cuticle from desmosomes at the end of short microvilli ending within the cuticle.

B. Internal skeletal structures

Besides the dermal layer of connective tissue, the various organs are supported, encased, and connected to the body wall by connective tissue. In leeches, as previously mentioned, the connective tissue has greatly increased and the coelomic spaces have become much reduced. In the earthworm dermis the connective tissue fibrils are arranged in an outer layer of approximately circular fibrils and an inner layer of longitudinal fibrils (Rudall, 1968a). The fibrils show periodic banding. In sections the periodicity varies from 280 to 600 Å (Ruska and Ruska, 1961; Coggeshall, 1966), though in fibrils isolated from fresh material the periodicity is 600–640 Å. Fresh fibrils stained with uranyl acetate show the banding broken up into three similar segments. If the definition is poor, the fibrils can be thought, mistakenly, to have a periodicity of 200–220 Å and something similar may account for the variable periodicity seen in sections (Rudall, 1968a). Elder (1966) in *Lipobranchius* found fibrils with approximately 500 Å periodicity and Bradbury and Meek (1958) reported a faint banding of 300 Å in fibrils from the leech. Earthworm fibrils give a well-defined collagen X-ray diffraction photograph (Rudall, 1955).

Cross sections of fibrils from earthworm dermis have a deeply staining periphery and an unstained or lightly staining core (Fig. 11.3) and a similar appearance is found in connective tissue fibrils from other parts of the earthworm body. Bradbury and Meek (1958) describe the fibres in leeches as composed of collagen around cytoplasmic processes originating from bi- or multipolar fibroblasts.

Collagens from the body wall of *Lumbricus* sp. (Fugimoto and Adams, 1964), *Digaster longmani*, and *Allolobophora* sp. (Rigby, 1968; Rudall, 1968a) differ in amino acid composition and in total imino acid content from that of their cuticle collagens (Table 11.1). Rudall found a very low glycine for *Allolobophora* collagen though the other analyses give a glycine content approximately similar to that of mammalian collagen. The internal collagens resemble the cuticle collagens in having a very low proline content, a high hydroxyproline content, and a low total imino acid content compared with mammalian collagen. In spite of differences in total imino acid content between *Digaster longmani* and *Allolobophora* body-wall collagens they both have a T_D of 22°C; the same T_D as that of their cuticle collagens which, as mentioned above, is about the same as the upper limit of their environmental temperature. The T_S of both collagens is 40°C, slightly above that of their cuticle collagens (40°C). Here again while T_D is related to environmental temperature neither T_D nor T_S are related to total imino acid content.

In Sabellidae, with a well developed crown of tentacles, the tentacles are

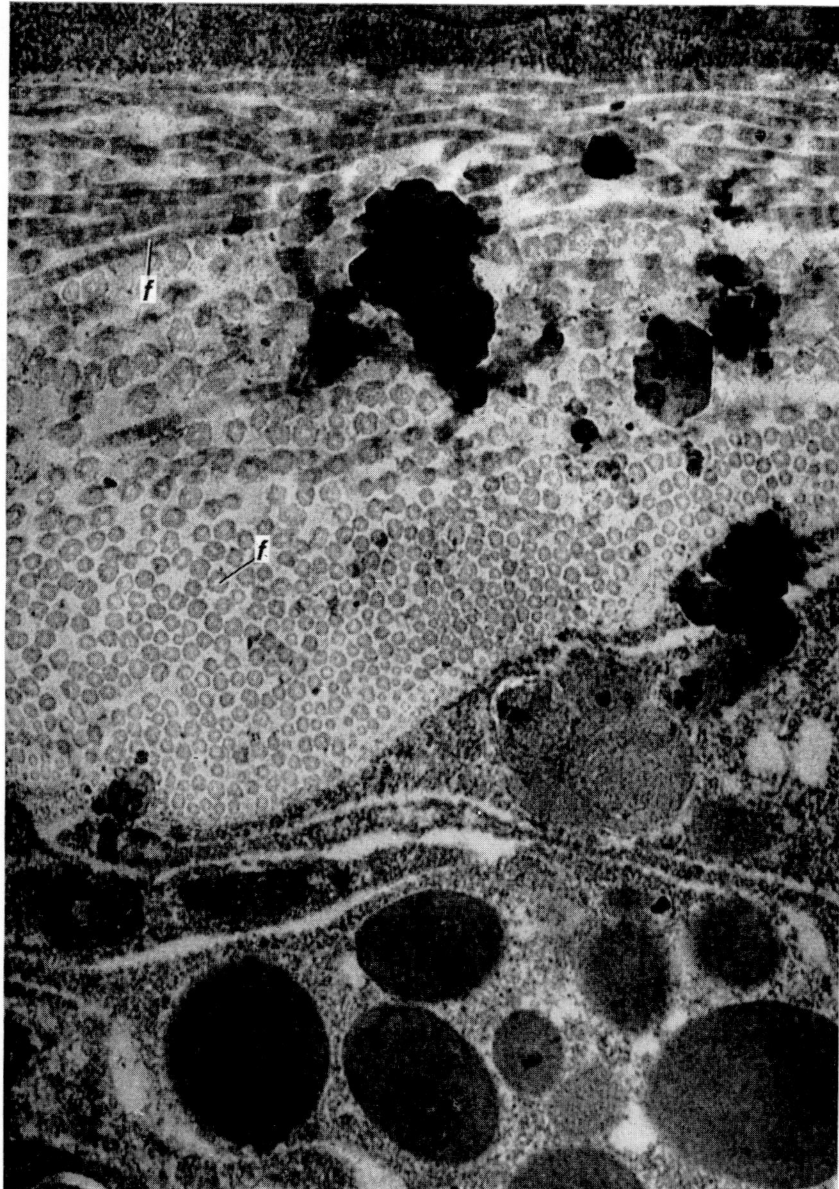

Fig. 11.3. Transverse section through connective tissue layers of earthworm body wall. Collagen fibrils mainly in cross section, where they are seen to have a densely stained periphery and a lightly stained core. *f*, collagen fibrils. (Rudall, 1968a.)

supported internally by cartilage-like tissue. In *Eudistyla* the tissue contains cells and collagen fibrils in a matrix composed of a highly sulphurated acid mucopolysaccharide (Person and Mathews, 1967).

Owen (1959) and Elder and Owen (1967) found elastic fibres in the earthworm, in the polychaete *Lipobranchius*, and in leeches. In the earthworm and *Lipobranchius* radial, branching fibres originate from a network of elastic fibres beneath the circular muscle layers and extend to a similar network beneath the epidermis, resembling the arrangement of elastic fibrils in some nemerteans (Owen, 1959). Elastic material also occurs in the walls of the blood system of *Lipobranchius* and is, doubtless, also present round the blood vessels of other annelids. Clark and Clark (1960a, b) have described also in *Nephthys*, the remarkable ligaments mentioned above, with bands of collagen-like fibrils alternating along the length of the ligament with bands of elastic material. Their main function appears to be to restrain the body wall and parapodia from deleterious dilations during changes in body shape. They probably also serve as shock-absorbers against high, transient, fluid pressures in the coelom which are thought to accompany the pounding of the proboscis against the sand when the animal is burrowing.

C. Egg shells

Little is known about the shells of annelid eggs. Many polychaetes protect their fertilized eggs with jelly coats or capsules of unknown composition. In oligochaetes and leeches the eggs are laid in cocoons produced by a specialized part of the ectoderm called the clitellum. Grove and Cowley (1926) have described the formation of the cocoon, which appears initially as a broad, elastic belt around the animal. Eventually the animal draws backwards out of this belt depositing fertilized eggs in it as it passes the openings of the oviducts. As soon as the animal has withdrawn completely the two ends of this cocoon close up to produce a lemon-shaped body which is at first white but later turns yellow and finally brown. Rudall (1955) found neither chitin nor collagen in the cocoon. Needham (1968) suspects that it is a protein tanned by a quinone derived from tryptophan.

In some leeches sperm is passed to the female in a spermatophore. In *Glossiphonia complanata* the spermatophore is formed by two glands developed round the ejaculatory channels and is composed of hardened protein (Damas, 1968).

II. SIPUNCULA

Sipunculids are marine, worm-like unsegmented, coelomate animals that may be as much as 440 mm in length, but are generally between 50–155 mm. The anterior part of the body can be invaginated into the hind part. They are sedentary forms living in burrows lined with secretions, but definite tubes are never formed.

A. External skeletal structures

The body is covered by cuticle that varies in thickness according to the region of the body. Since the cuticle dissolves completely in hot dilute NaOH, it cannot contain chitin. Structurally it is very similar to the cuticle of annelids (Moritz and Storch, 1970). Various spines and attachment organs are present which are formed by the epidermis over raised papillae of connective tissue.

B. Internal skeletal structures

The body wall is made up of a dermis of connective tissue under the epidermis and layers of circular and longitudinal muscles. In some, the muscle layers form continuous sheets, in others both circular and longitudinal muscles may be divided up into a series of bundles which, showing through the outer layers, give the animal a checkered appearance. The animal burrows by peristalsis, employing both sets of muscles. Elder and Owen (1967), by their special technique, found in *Golfingia* regularly arranged elastic fibres passing through the circular muscle layer.

III. ECHIURA

Echiura are worm-like coelomic animals covered by a thin cuticle, although the male of *Bonellia* has a ciliated epidermis. Beneath the epidermis is a dermis of fibres embedded in a jelly-like matrix and there are circular, oblique, and longitudinal muscle layers. Some live in tubes and Newby (1941) has described the secretion by *Urechis* of a mucus funnel, one end of which is attached to the mouth of the burrow, the other to the animal. Water is drawn through the funnel and then the funnel is eaten along with the food particles it has trapped; a somewhat similar mechanism for trapping food particles is found in *Chironomus* (Insecta) larvae (Walshe, 1947). Echiuroids possess a few spines rather resembling annelid chaetae and Tetry (1959) described them as containing chitin.

CHAPTER 12

Arthropoda

Arthropods are segmented animals with paired, segmental appendages which are often jointed and are used in feeding and locomotion. Arthropods make up more than three-quarters of all known species of animals and have managed to colonize virtually every possible habitat; on this basis they must be considered the most successful of animal types. Although the majority are less than 3 cm long, some extinct eurypterids were as much as 6 m.

A. External skeletal structures

The body of arthropods is completely covered by cuticle which also lines the buccal cavity, fore gut, distal parts of the genital ducts, and the tracheal system, where present. So far as its structure has been investigated, this cuticle is basically the same in all classes except possibly the diplopods and chilopods though this generalization is based on the investigation of an infinitesimally small fraction of the total number of species.

The bulk of the cuticle is made up of a laminated layer, sometimes called the procuticle, containing a protein—chitin complex and this layer can often be subdivided both visually and chemically into an exocuticle and an endocuticle. The exocuticle is covered by an epicuticle composed of lipoprotein and lacking chitin. Krishnan *et al.* (1955) claimed that the epicuticle of the scorpion, *Palamnaeus*, contained chitin but Kennaugh (1959) suggested they had mistaken a specialized outer layer of the exocuticle for the epicuticle. Although Langner (1937) and Cloudsley-Thompson (1950) have described an epicuticle as present in myriapods, Blower (1951) failed to detect an epicuticle in those diplopods and chilopods that he examined. The cuticle is secreted by the epidermis and in many species numerous fine canals, the pore canals, pass from the epidermal cells through the endocuticle and exocuticle to end beneath the epicuticle. The cuticle may also be penetrated by ducts from various glands in or below the epidermis which often secrete over the epicuticle a layer of wax and sometimes an outermost layer of cement material (Fig. 12.1).

In many species parts of the cuticle become considerably hardened. Part of the hardness and chemical resistance of the cuticle may be due to a regular arrangement of protein and chitin micelles (Kroon *et al.*, 1952). In addition, quinone-tanning of the protein has frequently been demonstrated in the

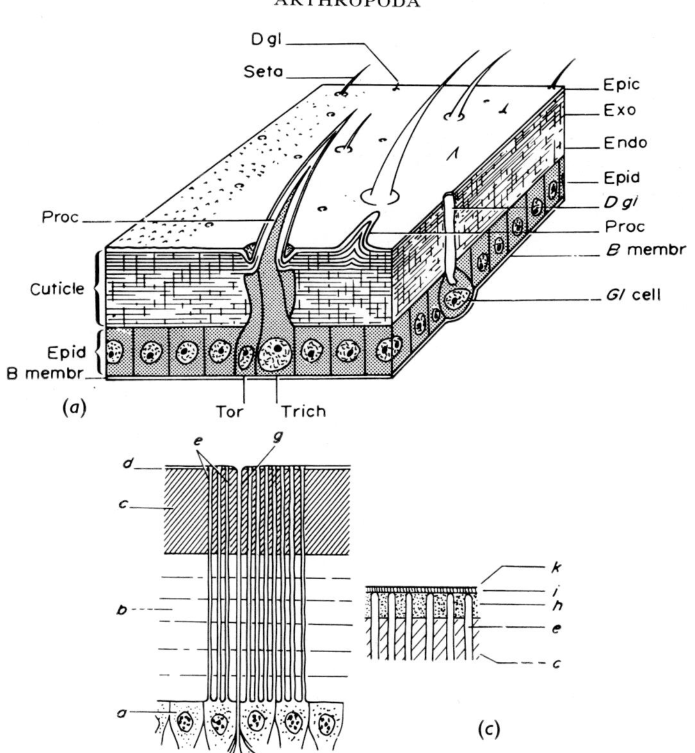

Fig. 12.1. (*a*) General organization of insect cuticle and underlying epidermis. B. Membr., basement membrane; D. Gl., gland duct; Endo, endocuticle; Epic., epicuticle; Exo, exocuticle; Proc., a type of immovable, non-cellular process; Seta., movable projection; Tor., cell secreting articular membrane of seta; Trich., Trichogen (seta-forming cell). (After Richards, 1951.) (*b*) Schematic section through typical insect cuticle. (*c*) Schematic section through the outermost layers. *a*, epidermal cells; *b*, laminated endocuticle of soft protein and chitin; *c*, exocuticle of sclerotin and chitin; *d*, epicuticle; *e*, pore canals; *f*, dermal gland; *g*, duct of dermal gland; *h*, cuticulin (outermost) layer of epicuticle; *i*, wax layer; *k*, cement layer. (From Ebling 1964), after Wigglesworth, (1948).

epicuticle and exocuticle, often giving them an amber to brown colour. Quinone-tanning is not found in the endocuticle. Inorganic salts, mainly calcium carbonate, occur in the exocuticle and outer layers of the endocuticle in many crustacea and diplopods. Disulphur bonding also is believed to be important in some groups.

Areas of hardening of the cuticle are generally related to the underlying body segmentation. Movement of the body is made possible by the cuticle between the hardened areas remaining soft and flexible, forming the arthrodial membranes. The degree of movement that can take place is determined by the nature of the arthrodial membrane and the shape of the opposing hardened cuticular parts.

Chitin in the arthropod cuticle is in the α-form and forms the fibrils that are seen in the laminae of the procuticle. The chitin is bound to the protein by covalent links, possibly through aspartyl or histidyl residues, or both (Hackman, 1960). It is believed that the chitin fibrils are first secreted, and orientated, and that subsequently protein diffusing into the cuticle becomes linked to them.

The structure of the lamellae forming the procuticle has been examined in crustaceans and insects. Initially, each lamella was interpreted as composed of fibrils arranged at right angles to the surface in the centre of the lamella but curving round to lie parallel to the surface at the inner and outer faces of the lamella (Dennell, 1960; Locke, 1964). However, in insects, Bouligand (1965) has reinterpreted the photographs from which the previous deductions were made. He suggests that each lamella is made up of a series of laminae within which all the fibrils are parallel to each other but the orientation of the fibrils changes slightly and progressively from one lamina to the next. Over the whole lamella the orientation in successive laminae changes by a total of 180°. The fan-like arrangement seen in sections results from the fibres of successive laminae being cut with regularly changing obliquity. Such fibril arrangement must make the lamellae equally strong in every direction parallel to the surface.

Hardening of the cuticle does not always produce complete rigidity, and movements of the body by contraction of localized muscles inserted onto the cuticle cause shearing stresses within the cuticle. The bonding between laminae is loose enough to permit slippage under such shearing forces. It also allows the cuticle to withstand compression from knocks and falls without breaking; the laminae, with their varying arrangement of fibrils, reduce the likelihood of any cracks that develop involving the whole thickness of the cuticle.

Under normal conditions the blood turgor pressure in the body of arthropods is low and muscles are arranged to hold the arthrodial membranes folded in beneath the hardened portions of the cuticle so that the body wall does not bulge outwards at these places. However, occasionally the blood turgor pressure rises, the arthrodial membranes become forced out tight, and movement becomes impossible (Needham, 1947). In some forms in which the cuticle remains unhardened and the limbs are jointed, e.g. most branchiopod crustacea, blood pressure plays a part in the movement of the limbs and it may also help in the operation of relatively hardened jointed limbs, e.g. in spiders (Parry and Brown, 1959).

The normally hardened, inelastic cuticle imposes problems in providing for growth and the cuticle is moulted from time to time to allow for increase in size. To minimize the loss from the animal of material used in making the cuticle, the unhardened parts of the old cuticle are generally reabsorbed before it is cast. Calcium carbonate deposited in the cuticle is also reabsorbed but it does not seem possible for the animal to reabsorb the quinone-tanned portions and these are lost, except possibly in those arthropods that are able to eat the cast cuticle, for moulting enzymes capable of digesting quinone-tanned protein have not yet been described.

The new cuticle is forming beneath the old cuticle before the moult takes place. The epicuticle is formed first and, although it is at first smooth, subsequently it becomes folded to allow for the increase of body volume that will take place after moulting through swallowing air or water (Bennet-Clark, 1963). The exocuticle, which also is partly formed before the moult, stretches to accommodate the increased body volume. Material is added to the endocuticle after moulting and Neville (1963a, b) has shown that, at least in the exopterygote insects, there are daily growth layers, a thick laminated layer being deposited during the night, and a thin, non-laminated layer with all the fibrils parallel, during the day. Most of the hardening of the new cuticle takes place after it has expanded to its full size. Moulting of the cuticle is under hormonal control.

1. *Onychophora*

If onychophorans are taken as primitive arthropods, then within this group one might expect to find an indication of the primitive condition of the cuticle. However, onychophorans are not marine, as primitive arthropods are believed to have been, but terrestrial, living in damp, decaying logs, and the cuticle may show adaptions to this specialized environment.

In *Peripatus* and *Peripatopsis* the cuticle is only 1–2 μm thick in animals about 5 cm in length and 0·5 cm in diameter. The epicuticle is composed of four layers. The outermost layer is a uniform lamina 10–15 mm thick which tends to lift away from its substratum. Below it is a thin, dark layer not strikingly osmophilic and not homogenous, for it may include fine striations or wisps of material which appear to connect the outer layer to the third layer, which is 3–5 nm thick and lies above a layer 0·1 μm thick which is osmophilic, stains with lead, and lacks resolvable texture. The remaining part of the cuticle is 0·2 μm thick. Evidence of quinone-tanning is found in the outer layers of the cuticle covering the spines, jaws, claws, and chemoreceptors, where hardness is required, and in the epicuticle covering the whole body. The whole cuticle lacks pore canals and is not penetrated by dermal gland ducts. The cuticle and underlying epidermis are much folded. The cast skin, flattened by floating on water, is about a third longer than the length of the resting animal. The folding of the cuticle allows for changes in shape in the animal, although the cuticle is inelastic.

Onychophorans probably represent the primitive arthropod condition in being essentially turgid tubes, their shape defined largely by their dermal connective tissue, which is well developed, and controlled by their circular and longitudinal body wall muscles working against the fluid in the haemocoel, with the cuticle defining the limits of length and diameter. In this onychophorans resemble annelids (except in the nature of the body-cavity). *Peripatus* undergoes extreme deformations of the body, being able to step slowly through a hole a ninth of the transverse sectional area of the resting animal (Manton, 1958). This enables it to retreat from predators through small cracks in decaying logs and may be a specialization rather than a primitive property. Although onychophorans are segmented animals, no trace of segmentation, except for the paired, segmentally arranged legs, is seen at the surface, furrowing of the cuticle probably having obscured it, and this again could be a specialization rather than a primitive character.

Living in damp, and sometimes even water-logged environments, onychophorans are better adapted to repel water than to withstand desiccation and they lose water rapidly in a dry atmosphere, although not as rapidly as an annelid under comparable conditions (Morrison, 1946). The hydrofuge properties probably depend on the outer layer of the epicuticle which, as in insects, may consist of orientated lipid associated with lipoprotein, although quinone-tanning alone may make it hydrofuge. The non-wetability of the cuticle which preserves a film of air around an animal submerged in water no doubt is also, in part, dependent on the micropapillae (usually ending in a sensory spine) that are formed at the surface of the cuticle (Morrison, 1946; Robson, 1964).

Onychophorans moult throughout life and little if any of the old cuticle is resorbed before moulting. The animal recovers the material by eating the old cuticle after it is shed.

2. *Crustacea*

Crustaceans are predominantly marine though some live in fresh water and a very few have become terrestrial. In branchiopods the cuticle is soft, pliable, and colourless and no quinone-tanning is demonstrable in it. It also lacks the chemical resistance usually found in arthropod cuticles. Krishnan (1958) found, in *Streptocephalus dichotomas*, that the protein from both the epicuticle and endocuticle resembles vertebrate collagen in chemical composition, e.g. containing hydroxyproline, although it lacks hydroxylysine, cystine, leucine, and valine; also it contains tryptophan, not usually found in collagen. The collagen-like component of the cuticle suggests a similarity with the cuticle of annelids.

In branchiopods the body cuticle is segmented but that of the limbs does not show any jointing. Blood turgor pressure is important in maintaining the shape of the body and in controlling movement of the limbs. In general, the blood pressure is opposed, not by circular and longitudinal muscles, but by the non-elastic cuticle, in this respect resembling nematodes. However, in the notostracan, *Triops*, the cuticle is thought to be too thin and delicate to preserve the body shape (Dennell, 1947).

In ostracods and copepods, the cuticle, although colourless, is firmer and the limbs jointed, but no details about the cuticle are known. The antennular stalks of pedunculate cirripedes are covered by a thick, leathery cuticle which is wrinkled to allow for the extension and contraction of the stalk brought about by its circular and longitudinal muscles; the contraction of these also drives blood up into the capitular region, to open the valves of the carapace and protrude the cirri. The carapace cuticle is strengthened by the deposition in it of very thick calcareous plates, and such calcareous plates may, as in *Scalpellum*, occur also in the stalk cuticle. Quinone-tanning has not as yet, been described in this class. Cirripedes, being sedentary, are unable to seek a refuge during moulting as do free-living Crustacea. A retention of the old cuticle outside the new cuticle, over the stalk and carapace, provides protection during this vulnerable period. Over the rest of the body the old cuticle is shed in the normal manner (Thomas, 1944). Parasitic copepods and cirripedes are described as having very thin cuticles but no details of structure are available.

Only in Malacostraca have the hardened parts of the cuticle been studied in any great detail (Dennell, 1960). In decapods the epicuticle is covered by a layer of lipid. The procuticle is divisible into three layers. The outermost, the *pigment layer*, which is comparable to the exocuticle, is amber coloured, quinone-tanned, and calcified; granules of melanin-like pigment and carotenoids also may be present. The middle layer, the *calcified layer*, lacks quinone-tanning. The innermost layer is both untanned and uncalcified. All the layers except the epicuticle are penetrated by pore canals, by which ducts of the tegmental glands pass through the cuticle to open at the surface. As in insects, these glands may play a part in quinone-tanning.

The surface view of the decapod and other crustacean cuticles often shows a polygonal pattern which in vertical sections shows up as vertical lines demarcating the secretory territories of individual epidermal cells. In some cases the individual cellular prisms of secretion are separated from each other by amber coloured material resembling the quinone-tanned material in the pigment layer.

The chemistry of the crustacean tanning process has been investigated in the fiddler crab, *Uca pugilator*, by Summers (1967, 1968). He found free tyrosine in the epidermis, a phenol oxidase capable of converting the tyrosine to dopa, a decarboxylase that converts the dopa to dopamine, and an acetyl-CoA-acetyltransferase that converts the dopamine to *N*-acetyl dopamine; this, after oxidation, serves as the tanning agent, and is the same agent as that found in the cuticles of some insects.

The calcium carbonate is mainly in the form of calcite, although a little amorphous material also may be present, and vaterite very occasionally has been detected. In decapods the inorganic material may form from 60 per cent to 80 per cent of the dry weight of the external skeleton. It does not necessarily make the cuticle harder than quinone-tanning alone could make it, for Bailey (1954) has shown that the completely uncalcified mouthparts of insects are sufficiently hard to scratch calcite; however, calcite is the material mainly employed to harden crustacean exoskeletons. It reduces the deformability of the cuticle but makes it brittle and adds to its weight. The cuticle of actively swimming forms is much less heavily calcified than that of slow-moving, reptantian forms. The weight of the calcified cuticle unsupported by water may have been one of the factors preventing effective colonization of dry land by malacostraceans. Crustaceans having no calcification have a higher ratio of protein to chitin in the cuticle than those with calcification and, in general, there is an inverse relationship between the amount of calcification and the amount of protein present (Travis, 1960). Calcification in relatively large crustaceans may represent a method of achieving a sufficiently hardened cuticle without having to use large quantities of protein.

The calcite occurs in plate-like crystals, or crystal aggregates, of extreme thinness arranged parallel to the surface of the cuticle, the crystal shape presumably being related to the laminated nature of the cuticle in which they are formed. The crystals form a mosaic, the units of which have irregular outlines. As the crystals are believed to form more or less simultaneously—in *Gammarus* calcification of the cuticle occurs in considerably less than 24 hours (Reid, 1943)—each crystal can develop only to the extent permitted by its neighbours. In each single crystal plate the optical axis is normal to the plane of extension but in different plates the optical axes are orientated at random in the plane of the surface, so that between crossed nicols the structure shows a mosaic of light and dark areas. Surprisingly, the mosaics show constant specific patterns (Dudich, 1931) showing that the arrangement of centres of calcification, the speed of crystallization, and the direction of growth of each crystal must be genetically controlled. In *sphenocycles*, consisting of wedge-shaped, plate-like crystals radiating from a centre, the centre is usually occupied by a bristle or the orifice of a gland (see Picken, 1960).

Ingrowths from the external cuticles, particularly in the larger decapods, give rise to an endoskeletal system, the *endophragmal* skeleton, which not only provides for a more efficient anchorage and arrangement of the muscles moving the appendages but also, through a series of girder-like structures, gives added strength and rigidity to the floor of the cephalothorax.

In the joint between the merus and ischium in the legs of *Astacus fluviatilis* is found an elastic protein, *resilin*, which will be described below (p. 220). Functionally, it replaces the extensor muscles of the leg (Andersen and Weis-Fogh, 1964).

3. *Diplopoda and chilopoda*

The remaining arthropod classes are predominantly terrestrial and their cuticles show various adaptations to terrestrial life. The cuticle of diplopods and chilopods probably lacks an epicuticle. Although there is a layer of wax over the surface, the cuticle remains relatively permeable to water and these forms are restricted to damp environments, generally living in litter on forest floors, or in soil. These environments are likely to become water-logged. Julids and geophilomorph centipedes are less susceptible to desiccation than other forms and their cuticles are more hydrofuge, giving them a better chance of survival in soils that dry out or become water-logged. All forms are particularly vulnerable to dessication during moulting, and *Polymicron* and other nematophora build silken chambers in which they moult, while polydesmids build earthen moulting chambers (Blower, 1955).

Quinone-tanning is the typical method of hardening the cuticle of terrestrial arthropods and it occurs in the cuticle of diplopods and chilopods (Blower, 1951). In the majority of diplopods calcification also is present and probably gives the cuticle the necessary rigidity for the development of the strong muscular forces necessary for these animals to bulldoze their way into the soil. In *Polyxenus*, a non-burrowing diplopod, calcification is lacking (Manton, 1953, 1961; Barrington, 1967). The geophilomorph centipedes burrow like earthworms, producing a zone of broadened body segments that

other components by hydrogen bonds, and 3 per cent by electrovalent or double covalent bonds or both, the remaining 56 per cent is bound covalently to chitin (Hackman and Goldbert, 1958). Each of these components is itself heterogeneous.

Amino acid analyses of untanned cuticle show that no cystine, cysteine, or methionine are present. Hydroxyproline occurs in extracts of cuticle, suggesting the presence of collagen amongst the cuticular proteins (Hackman and Goldberg, 1958). It is not possible to analyse the tanned protein since degradation leads to its complete destruction.

The precursors of the tanning agent pass out through the cuticle, via the pore canals, to the epicuticle. Here, enzymatic reactions produce the tanning agent which then diffuses inwards tanning the proteins of the epicuticle and exocuticle. The tanning agent that converts the soft maggot cuticle into the hard pupal case, and hardens the adult cuticle of blow flies has been found by Karlson and Sekeris (1962) to be *N*-acetyl-1-3,4-dihydroxyphenyl-B-ethylamine (acetyldopamine) derived from tyrosine by the action of a series of enzymes; these convert it first to dopa, then to dopamine, and finally to acetyl dopamine which after oxidation serves as the tanning agent.

Since Karlson and Sekeris also found *N*-acetyldopamine in the cuticle of *Schistocerca*, and probably in the beetle *Tenebrio*, they suggested that it was the universal insect cuticle-tanning agent, but in the adult cockroach 3,4-dihydroxybenzoic acid (protocatechuic acid) is the tanning agent (Brunet, 1967). It is not known whether one molecule of the tanning agent directly links together two protein molecules or whether the tanning molecules form a polymer which becomes attached only here and there to the proteins.

Quinones are coloured, and quinone-tanned cuticles are usually coloured brown, but undoubted cases of pale, hardened cuticles exist. Mason (1955) suggests that the development of brown colour depends upon the presence of excess quinone. The initial reaction of the quinone with protein reduces it to a colourless phenolic substance; only when there is excess of quinone will the phenol then be reoxidized to a quinonoid protein, which is coloured.

Normally tyrosine must be supplied in the diet. There are, however, some insects that can synthesize tyrosine, and other essential amino acids, from glucose (Brunet, 1963). This capacity depends upon the presence of certain bacteria in special cells of the fat bodies (Henry, 1962).

Calcium carbonate is occasionally deposited in the cuticle of insects. It occurs on the surface of the cuticle of some *Pericoma* when they live in water rich in calcium. The larvae of *Sargus* and other stratiomyids deposit calcium carbonate in the pupal case in a series of wart-like structures formed in shallow pits. The calcium is probably obtained from the Malpighian tubules (Wigglesworth, 1965). This calcium carbonate, which may form as much as 75 per cent, by weight, of the cuticle helps to protect the pupa, particularly those that float on the surface of the water, from predators such as *Dytiscus*, although it is unable to protect against invasion by parasitic insects (Myall, 1895). Calcium also occurs in the puparium of *Rhagoletis cerasi* and *Acidia heraclei*.

The epicuticle, though extremely thin, is generally divisible into two layers. It is not possible, as yet, to separate and analyse the individual layers so that knowledge about them depends on histochemical techniques. The *cuticulin layer*, which is the first layer to be secreted in the formation of the new cuticles, is approximately 100 Å thick, about the thickness of a cell membrane, and may originate as a lipid double layer (Locke, 1964). It contains lipid and protein, probably cross linked by tanning, into an extremely inelastic substance (Bennet-Clark, 1962). Below the cuticulin layer, in most species, is a much thicker, homogeneous, dense layer of tanned protein. The pore canals generally stop beneath this layer.

Ducts from the wax glands pass through the cuticle to secrete over it a layer of wax and beyond this wax layer a cement layer is secreted in many insects. The wax and cement layers are often considered as layers of the epicuticle but they differ from the epicuticle proper in being secreted by special dermal glands and not by the general epidermis.

In the wax layer the wax molecules nearest the cuticulin layer are orientated into a polar monolayer, though the outer molecules are more randomly arranged. In some species the wax in the outermost layer forms a bloom of thin sheets or plates providing the maximal surface for volume of wax. The surface of the insect has, as a result, a high angle of contact for water, which breaks up into spheres, covered with wax fragments, when it comes in contact with the insect and these then run off the cuticle. This wax layer thus reduces the tendency of the insect to become trapped in a film of water (Locke, 1964). Because the wax can be extracted from the cuticle, a certain amount is known about its chemistry (Gilmour, 1961) but only in a few species, and it is not known to what extent it varies from species to species.

The arrangement of the wax layer controls the movement of water through the cuticle and by reducing water loss through the cuticle has been largely responsible for the success of insects in colonizing dry land. In some aquatic larvae, e.g. *Dytiscus*, the wax layer is missing. Fatty acids, particularly caprylic or caproic acids in the wax layer of *Bombyx* and *Chilea* larvae, are important in protecting them from fungal attack (Koidsumi, 1957).

Cement forms a layer of variable thickness over the wax layer. It is thickest over exposed areas and thin or lacking in protected areas such as the dorsal abdominal tergites under the wings of *Rhodnius*. Shellac of commerce is probably derived from the cement layer of *Laccifer lacca* (Beament, 1955) and is composed of a sugar, laccose, and lipids. In the lepidopteran *Diataraxia oleracea* quinone-tanned proteins also are present in the cement. The mixture forms a varnish and it functions as a varnish whatever its exact composition in the various species. It protects the underlying wax layer. Sometimes it forms not a continuous layer but a network which cannot give much protection but may form a reservoir of lipids which can leak out to seal over broken surfaces in the wax layer so preventing loss of water from the body (Locke, 1964).

(*i*). *Resilin.* In certain parts of the cuticle there is formed a hyaline, isotropic, colourless, rubber-like protein, *resilin.* Resilin has great deformability and perfect elastic recovery and it occurs in places where its rapid elastic recoil can supplement or replace muscular contractions. Since it has only recently been discovered (Weis-Fogh, 1960), the full extent of its occurrence in insects and other arthropods and in other groups of animals is not known. It has been found in the main wing-hinges and ligaments of locusts, dragon-flies, and cockroaches, and in minor wing-hinge ligaments of all insects investigated. In grasshoppers it forms the clypeolabral spring that holds the labrum pressed against the mandibles both in feeding and at rest. In the abdominal springs of beetles it replaces muscles in producing inspiration movements (Andersen and Weis-Fogh, 1964). Miller (1960) found it in a spiracular valve of locusts, Edwards (1964) in the spitting mechanism of certain predaceous bugs, and Thurm (1963) at the base of sensory hairs in bees.

Resilin occurs either associated with normal cuticular materials or as pads of the pure material. The elastic, transparent, hyaline pegs or prealar arms on the mesosternum of the desert locust are formed from concentric lamellae of chitin, less than 0·2 μm thick, sandwiched between layers of resilin 1–3 μm thick. In flight the pre-alar arm is subjected to shearing stresses and the layers of resilin permit the structure to shear in response to these stresses, the chitin sheets controlling the extent of the deformation. In the elastic tendon of the hind wing of a dragon-fly, formed by a hollow ingrowth of cuticle so that a canal passes right through it, the tough, inextensible ends of the tendon are formed from normal cuticle. In the region of change from inextensible to rubber-like tendon, resilin first appears as irregular masses within the cuticle and then becomes organized into layers interleaved with chitin lamellae, the normal protein component of the cuticle being absent. In the elastic portion there is only resilin, covered internally by a buckled layer of epicuticle and on the outside by a vaguely differentiated layer between the resilin and the epidermis. This tendon is subjected to tensile stresses and the interleaving of chitin and resilin gives a firm attachment of the resilin to the rest of the tendon. The wing-hinge of grasshoppers has a complex arrangement of pure resilin and resilin laminated with chitin, the arrangement doubtless related to the stresses that develop during flight.

Although resilin somewhat resembles vertebrate elastin in its physical properties (Bailey and Weis-Fogh, 1961), it is quite distinct from it in its amino acid content (Table 10.5). It appears to be a network of protein molecules in which the primary chains show little or no tendency to form secondary structures. The chains are linked together by di- and tri-tyrosine units (Andersen, 1964). These are probably formed, by the pairing of free radicals arising from the enzymatic oxidation of tyrosine residues that are part of the resilin molecule (Andersen, 1966), so representing a form of auto-tanning.

5. *Chelicerata*

Although not all workers (Krishnan, 1953, 1954; Sewell, 1955, and Kennaugh, 1959, on Scorpions; Cloudsley-Thompson, 1950, and Sewell, 1955, on spiders; Lees, 1947, on ticks and B. M. Jones, 1954, on mites) have reported the presence of an epicuticle, it is now generally believed to be present in this class of arthropods. There is an outer wax layer, and in the spider *Tegenaria* Sewell (1955) found indications also of a layer of cement. Quinone-tanning has also been found in these cuticles. There have been various reports that appreciable quantities of sulphur are present in the cuticle of scorpions and *Limulus*. Lafon (1943) found 2·85 per cent sulphur in that of *Limulus* and suggested that it contributed to the hardness of this cuticle by forming disulphur bonds, although Sewell (1955) failed to confirm the presence of sulphur in acid hydrolysates of *Limulus* cuticle. In the cuticle of some scorpions sulphur can be demonstrated by colour reactions (Pryor, 1962). In *Pandinus imperator* and *Scorpiops hardwickii* Kennaugh found a hyaline layer of exocuticle between the epicuticle and the quinone tanned exocuticle and concluded that the sulphur was confined to this layer. Shrivastava (1957) found no sulphur in the cuticles of *Palamnaeus bengalensis* and *Buthustamulus gangetianus* so that sulphur is not universally present in scorpions, and disulphur bonds are probably not very important in stabilizing the scorpion cuticle. Sewell (1955) found no sulphur in the cuticle of the spider *Tegenaria*.

Nothing is known about the cuticle in the remaining arthropod groups, Pycnogonida, Pentastomida, and Tardigrada.

6. *Conclusions*

It is sometimes said that the success of the arthropods in general, and of insects in particular, has been due to the possession of a chitin-containing cuticle (Rudall, 1968a). It is generally believed that arthropods evolved from annelid-like ancestors and judging from modern annelids and arthropods this evolution has been accompanied by a change from a cuticle consisting of collagen fibrils embedded in a matrix of, as yet, unknown composition to a cuticle containing chitin fibrils embedded in a protein matrix.

The properties of the arthropod cuticle have certainly contributed to the success of the phylum, having allowed the evolution of limbs and appendages capable of complex movements and of a waterproof for the body. It is, however, difficult to see that it is the presence of chitin that has made these developments possible. Physiological relations between the arthropod and its environment depend largely on the properties of the epicuticle and this contains no chitin. Hardening of the cuticle depends upon quinone-tanning that takes place in the epicuticle and exocuticle. Although chitin occurs in the exocuticle, it is not necessary for tanning to take place. Chitin is most concentrated in the untanned layers of the endocuticle. Being fibrous, it may provide, as it were, a fabric backing to the outer hardened layers which, unsupported in this manner, might tend to be brittle. Alternatively, it may represent no more than a padding material, used to reduce the requirements for protein in the cuticle. In crustaceans, the protein content of the cuticle has been further reduced by substituting calcium carbonate.

The absence of chitin from the cuticle of annelids and its presence is the cuticle of arthropods poses problems for the possible evolution of arthropods from annelids. However, some annelids at least can produce quinone-tanned proteins mixed with chitin in their chaetae, although here the chitin is in the β-form and not the α-form as in arthropods. This does suggest that annelids have the capacity to produce, if only locally, the components of arthropod cuticle. The occurrence of hydroxyproline in the cuticle protein of branchiopods, and amongst the amino acids extracted from some insects, suggest that collagen has not been entirely eliminated from the cuticle of arthropods.

Although the cuticle of arthropods is generally penetrated by pore canals which, certainly in the early stages of formation of the cuticle, contain cytoplasmic processes from the epidermal cells and, although ducts from the dermal glands also pass through the cuticle, these cytoplasmic processes and ducts are concerned only with the secretion of the cuticle. They appear different from the microvilli that penetrate the cuticle of annelids to protrude at the surface of the cuticle.

It is generally believed that the mechanical properties of the arthropod exoskeleton have limited the animals in this phylum to a small size. Thompson (1942) points out that every hollow structure grows weaker as it grows larger and concludes that "a hollow shell is admirable for small animals but Nature does not and cannot make use of it for large".

While the majority of living arthropods are less than 5 cm long, the Japanese crab, *Macrocheira kaempfera*, is nearly 4 m across the outstretched claws, although the body remains comparatively small. The American lobster sometimes attains a length of nearly 1 m and a body weight of over 13 kg. Both these arthropods are marine. The largest living land arthropods are the Goliath beetles which are up to 15 cm long and some butterflies which may have a wing span of about 20 cm. If only living arthropods are considered they must therefore be considered as small compared with cephalopods on one hand and vertebrates on the other. However, fossil eurypterids sometimes were as much as 270 cm in length and while these were aquatic, with their bodies supported by water, some species of *Arthropleura* (Rolfe and Ingram, 1967) which were terrestrial and probably related to myriapods must have been at least 150–180 cm long. These are dimensions that would place these forms amongst the larger marine and terrestrial animals. There is as yet no information on how these large arthropods overcame the apparent weakness of the exoskeleton at these dimensions, whether by a great development of endoskeletal struts and braces, or by increasing the relative thickness of the cuticle, but at least these animals demonstrate that an exoskeleton can be adapted to support large bodies.

When these large arthropods first appeared in the fossil record there were virtually no animals that could prey upon either their larval or their adult stage. Their disappearance coincides with a great increase in jawed fishes, that could both devour planktonic eurypterid larvae and attack the bottom living adults, and with the appearance of amphibians and reptiles that could have attacked *Arthropleura*. Under these circumstances it was probably of advantage to arthropods to be small and so possibly more agile in escaping from vertebrate predators, or better able to obtain hiding places from them. A small size would generally also represent a shorter time to reach sexual maturity, and therefore a greater likelihood of reproducing before being eaten, and greater ability to replace rapidly individuals lost to predators. But, whatever the factors that have operated to keep arthropods small, it has not been the mechanical limitations of the exocuticle.

Haldane (1930) suggested that insects have been limited in size by the inability of tracheae to supply an adequate amount of oxygen to tissues lying more than about 7 mm from the surface of the body. This would be true if oxygen was transported to the tissues only by diffusion but in the majority of insects this is aided by pumping movements of the abdomen and there is "no reason to suppose that the respiratory mechanisms available to insects could not meet the requirements of species larger than any known, either amongst living or fossil forms" (Miller, 1964).

B. Silks

All the components of the arthropod cuticle, both larval and adult, are secreted by the epidermal cells. However, the epidermis also is able to secrete other substances. A number of millipedes, insects, and arachnids produce, from glands that are ingrowths of the epidermis, fine, strong, insoluble threads of varying composition. These are used for a variety of purposes: to wrap round themselves to form cocoons in which they pupate, to line nests, to cover eggs, or to make net-like structures that are used for a variety of other purposes.

1. *Insects*

The best known of these threads is the silk of commerce, produced by the silkworm, the larva of the moth, *Bombyx mori*, to form a pupation cocoon. Numerous other lepidopterans also produce silk, some of which is of commercial value, e.g. those of the Tussah moths of the genus *Antheraea*, and *Philosmaia cynthea ricini*. The very fine silk of *Lasiocampa otus* was dyed purple and woven with gold thread for the use of Roman patricians; it's thinness and transparency gave rise to adverse comment amongst some Latin authors (Lucas *et al.*, 1958).

Silkworms live on the leaves of mulberry. Both *Morus nigra* and *M. alba* leaves are readily eaten, but a higher quality silk is said to result from feeding silkworms on the latter species. *Philosamia cynthia* fed on *Ailanthus glandulosa* produces a darker, less acceptable silk than when it is fed on the caster oil plant, *Ricinus communis*. The nature of the effect of the food on the silk is not known (Lucas *et al.*, 1958).

Silk of lepidopterans consists of a core of the protein, fibroin, coated with a second protein, sericin. It is secreted by a pair of labral glands in the head. In *Bombyx mori* each gland has three sections, a posterior section forming a long, thin much coiled tube about 30 cm long, closed at one end and leading into a shorter, thicker middle or reservoir section; this connects with the exterior through a long, very thin tube which, just before it reaches the spinneret through which the silk is extruded, joins with the tube from the other gland so that one opening serves the pair of glands. At the junction of the two tubes is another pair of glands, the glands of Filippi, of unknown function (Fig. 12.5).

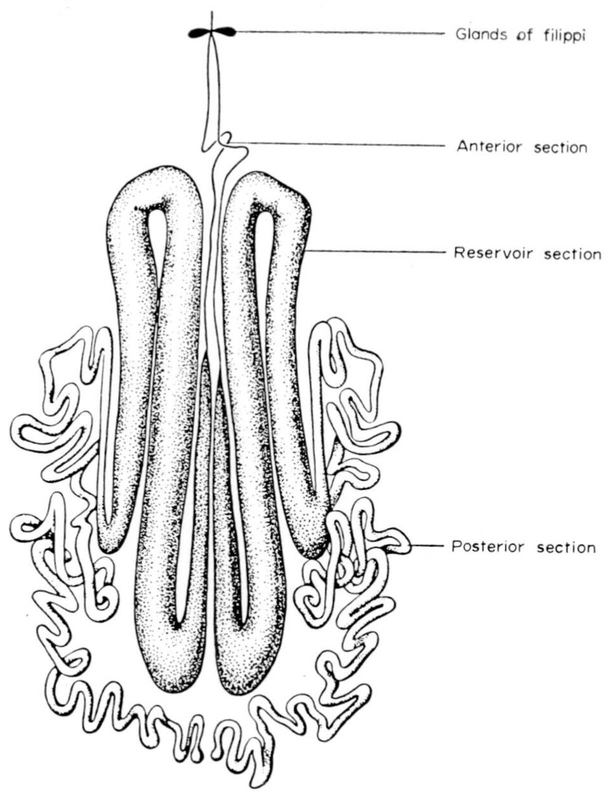

Fig. 12.5. Silk glands of the larva of *Bombyx mori* (after Lucas *et al.*, 1957).

Fibroin is produced in the posterior portion of the gland and passed into the reservoir where it is stored as a very viscous, aqueous solution. The reservoir not only stores the fibroin but also secretes the sericin, again as a viscous, aqueous solution. This surrounds the fibroin mass, but the two materials remain distinct. The silk is then extruded down the long, thin tube and through the spinnerets, the fibroin from each gland remaining distinct to form a two strand core while the sericin from the two glands fuses to form a common coating round the core.

When forming a thread, the insect exudes a droplet of silk solution onto a leaf or twig and then, pulling back its head, draws out a fine thread from the spinneret. The process of extrusion converts the fibroin rapidly and irreversibly into an insoluble, chemically resistant fibre. The process is spontaneous for, if the contents of the silk gland are diluted and allowed to stand, masses of fine fibrils separate out (Mercer, 1951).

Fibroin from a variety of Lepidoptera has now been analysed (Table 12.1). They are all distinguished by their large amounts of the simplest amino acid residues, glycine, alanine, and serine. These three residues rarely account for less than half the residues and may account for over 90 per cent. Of the larger residues, there is usually more tyrosine than phenylalanine. The total of acidic groups is always higher than that of the basic groups. The proline content is usually low and no hydroxyproline has been found. Until recently cystine and methionine were thought to be absent but now there is evidence for small amounts of cystine in the fibroin of *Bombyx mori* (Earland and Raven, 1961; Lucas, 1966). Although it only forms 0·23 per cent of the residues, it may be important in contributing disulphur bonds to the structure.

X-ray studies show that within the gland the fibroin molecule is in the coiled α-state but in all extruded lepidopteran silks it is in the fully extended β-form, though only that part of the fibroin in which there is a concentration of small residues is so organized. The larger residues mainly occupy less ordered regions. In the various lepidopteran silks that have been examined the repeat-period along the fibre axis is always 6·95 Å and the unit dimension at right angles to the fibre axis which depends on the hydrogen bonding between the fibrils is always 9·44 Å. The spacing between the sheets, the side-chain spacing, depends upon the length of the side chains and so shows some variation. Warwicker (1960) has divided lepidopteran fibroins into four groups on the basis of these X-ray variations and these groups correlate satisfactorily with what is known of the chemical composition of the different fibroins (Table 12.1).

Group I contains the fibroins of the Bombycidae where, in the crystalline part of the material, every other residue along each chain is glycine, the alternate residues being mainly alanine, with an occasional serine. All the glycine residues, therefore, lie on one side of each sheet; there is evidence that they face the glycine-residue side of the adjacent sheet and not the alanine side, which reciprocally face, and dovetail with, the alanine residues in the next sheet (Fig. 12.6(*a*)). The resulting structure is obviously very compact, richly bonded, and therefore very strong.

When the two glycine residues are opposite each other there is a minimum distance between the two chains and the X-ray diffraction photograph gives a side-chain spacing (repeat unit of two sheets) of 9·3 Å, the smallest found in the fibroins. Species of Westermanniinae (*Bena prasinana*), although not closely related systematically to the Bombycidae, have a fibroin with approximately equal numbers of glycine and (alanine + serine) residues (Table 12.1) and an X-ray picture with a 9·3 Å side-chain spacing, which places it also in Group I.

Group II, contains the *Anaphe* fibroins, with more than half their residues alanine and the remaining residues mainly glycine; the side-chain spacing is 10 Å and this also occurs in *Clania* fibroins, which have one third of the residues alanaine and one third glycine. In these fibroins alternating alanine residues may face glycine in the adjacent sheet (Fig. 12.6b).

The *Saturniidae* (*Antheraea assamensis* and *Cricula andrei*) fibroins, of Group III, also have a high alanine content but much less glycine and much more serine than *Anaphe* fibroins. Here the side-chain spacing is 10·6 Å which is

Table 12.1 Amino acid composition of β-type Arthropod silks. Residues per 1000 total residues

	GROUP I			GROUP II		GROUP III						
	Bombyx[a] *mori* L	*Bombyx*[b] *meridionalis* L	*Bena*[b] *prasinana* L	*Anaphe*[e] *moloneyi* L	*Clania*[b] *sp.* L	*Antheraea*[e] *assamensis* L	*Criccula*[b] *andrei* L	*Pachypasa*[b] *otus* L	*Braura*[b] *truncata* L	*Galleria*[b] *mellonella* L	*Chrysopa*[b] *flava*	*Nephila madagascariensis*[b] Drag-line A
Alanine	293	298	248	530	338	422	341	271	293	249	212	321
Glycine	445	449	426	424	370	291	275	348	347	318	246	406
Valine	22	21	20	21	17	5	10	8	5	52	—	9
Leucine	5	15	13	2	9	7	18	22	20	67	—	29
Isoleucine	7		8	1	13	3				35	—	—
Proline	3	—	3	2	10	4	—	—	—	47	—	—
Phenylalanine	6	10	3	1	2	4	35	5	9	2	—	6
Tyrosine	52	49	59	1	32	44	37	30	37	3	8	32
Serine	121	114	136	3	104	105	154	92	86	166	427	43
Threonine	9	10	15	2	17	3	7	10	10	18	31	6
Cystine ($\frac{1}{2}$)	2	—	—	—	—	—	—	—	—	—	—	—
Methionine	1	—	—	—	—	—	—	—	—	—	—	—
Arginine	5	5	11	1	11	26	27	82	80	13	5	24
Histidine	2	—	1	2	3	10	—	—	—	1	—	—
Lysine	3	2	4	1	19	2	2	3	4	2	—	—
Aspartic acid	13	15	21	5	29	50	94	113	98	21	60	9
Glutamic acid	10	11	31	3	27	12		12	11	5	8	116
Hydroxyproline	—	—	—	—	—	—	—	—	—	—	—	—
Hydroxylysine	—	—	—	—	—	—	—	—	—	—	—	—
Tryptophan	2	—	—	—	—	11	—	2	—	—	3	—
Cysteic acid												

Table 12.1 (continued)

	GROUP VI			GROUP IV				GROUP V			
	Pachylota[e] *audouinii* H	*Digelansinus*[e] *diversipes* H	*Arge*[e] *ustulata* H	*Thaumelo-poea*[b] *pityocampa* L	*Arctia*[b] *caja* L	*Lasiocampa*[b] *quercus* L	*Araneus*[f] *diadematus* dragline A	*Araneus*[f] *diadematus* cocoon A	*Nephila*[b] *senegalensis* cocoon A	*Araneus*[g] *diadematus* web radii A	*Araneus diadematus*[h] Swathing band A
Alanine	362	382	409	138	226	225	226	239	285	262	214
Glycine	21	22	21	239	216	187	376	79	118	291	298
Valine	21	19	25	29	39	54	12	54	23	13	43
Leucine	29	41	28	43	25	29	14	53	64	21	26
Isoleucine	9	19	12	39	22	19	6	14	25	17	38
Proline	—	—	17	—	33	—	106	11	—	143	93
Phenylalanine	4	—	—	14	13	11	4	32	37	1	—
Tyrosine	12	21	10	43	36	54	46	13	20	25	21
Serine	121	91	82	220	112	171	63	301	224	66	117
Threonine	13	5	25	29	36	32	12	33	43	12	17
Cystine ($\frac{1}{2}$)	—	—	1	—	—	—	6	0	—	—	—
Methionine	—	—	—	—	—	—	5	0	—	—	—
Arginine	4	5	8	32	41	49	9	10	19	12	23
Histidine	12	6	6	—	14	20	0	0	1	—	—
Lysine	13	5	12	4	25	24	9	13	11	13	16
Aspartic acid	16	22	13	117	104	59	14	62	32	17	17
Glutamic acid	361	363	331	48	57	67	94	85	97	100	78
Hydroxyproline	—	—	—	—	—	—	—	—	—	—	—
Hydroxylysine	—	—	—	—	—	—	—	—	—	—	—
Tryptophan	—	—	—	4	—	—	—	—	—	—	—
Cysteic acid										4	

[a] Lucas *et al.*, 1958. [b] Lucas *et al.*, 1960. [d] Lucas *et al.*, 1957. [e] Lucas and Rundall, 1968. [f] Lucas, 1969. [g] Fischer and Brander, 1960. [h] Peakall, 1964. L: Lepidoptera. H: Hymenoptera. A: Araneida

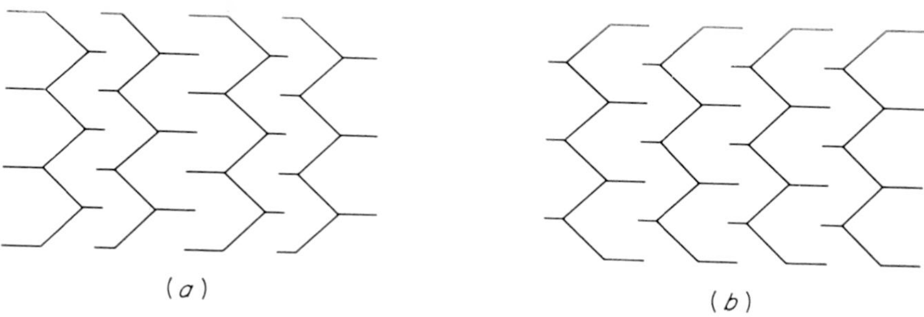

Fig. 12.6. (*a*) Diagram to show intersheet separation in silk of *Bombyx mori*: alternating glycine-glycine and alanine-alanine layers of side chains. (*b*) Intersheet separation in silks with regular repetition of mixed glycine-alanine layers of side chains. (After Warwicker, 1960.)

about the same as that in synthetic polyalanine, suggesting that in these fibroins alanine residues lie on both sides of the sheet and determine the side-chain spacing. Also in this group are placed fibroins from various species of Lasiocampidae (*Pachypasa otus* and *Braura truncata*) and Galleriidae (*Galleria mellonella*) with a chemical composition somewhat similar to that of Saturniidae.

In fibroins with considerable quantities of serine, glutamic acid, and aspartic acid the sheets naturally lie further apart; these constitute the fibroins of Group IV. In this group is the fibroin from *Thaumetopoea pityocampa* which, although closely related systematically to *Anaphe*, differs considerably from it in the chemistry of its fibroin. The side-chain spacing is 15·0 Å. Also in this group is the *Lasiocampa quercus* fibroin, which contains considerably more serine and large residues than other lasiocampid fibroins that have been analysed and which belong in Group III (Lucas and Rudall, 1968). Silks from other orders of insects and from spiders can, on the basis of their X-ray pictures and chemical composition, be fitted into these four classes or into two additional classes of silks with spacings different from those found in Lepidoptera.

The great strength of silk, both when wet and dry, results from its high crystallinity and strong hydrogen bonding between the chains. Its extensibility and elasticity depend largely upon the amorphous regions of the fibrils. The load/extension properties of the different types of silk correlate well with the amount of short side-chain residues present. In *Anaphe* silk, with 90 per cent of the total protein accounted for by glycine and alanine, the stress/strain diagrams are straight lines up to breaking point which occurs at 12·5 per cent extension. Saturnid tussah silks, with many more amino acids with bulky side chains, have a sigmoid stress/strain curve with a pronounced yield point and this may be related to the considerable buckling of the chains that must exist in the amorphous regions, the chains in these regions being held together by weak links that rupture at a certain low stress, allowing the chains to unfold and extend till complete rupture occurs at 35 per cent extension. *Bombyx* silks are intermediate in their behaviour, with a 24 per

Table 12.2 Amino acid composition of silk sericins. Residues per 1000 total residues.

	Bombyx[a] *mori*	*Antheraea*[b] *mylitta*	*Philosamia*[b] *cynthia*
Alanine	43	58	31
Glycine	147	145	279
Valine	36	52	12
Leucine	14	25	19
Isoleucine	7	13	9
Proline	7	8	3
Phenylalanine	3	20	7
Tyrosine	26	68	65
Serine	373	123	107
Threonine	87	111	39
Cystine (½)	5	—	—
Methionine	—	—	—
Arginine	36	17	53
Histidine	12	21	19
Lysine	24	165	197
Aspartic acid	148	102	101
Glutamic acid	34	69	58
Hydroxyproline	—	—	—
Hydroxylysine	—	—	—
Tryptophan	—	3	—

[a] Lucas (from Lucas and Rudall, 1968).
[b] Dhavalikar, 1962.

cent extension before breaking (Lucas *et al.*, 1958). Unfortunately there is as yet no knowledge of the correlation between the properties of the silks and the stresses they are subjected to in life.

Sericin from *Bombyx mori* is characterized by a high content of serine which accounts for about a third of the residues (Table 12.2). Aspartic acid and glycine are the next most abundant residues. Sericin is remarkable for its high content of hydrophilic groups; almost half the residues carry hydroxyl groups and about a quarter carry ionizable acidic or basic side-chains. There are far more acidic than basic groups (Lucas and Rudall, 1968). When extruded, the sericin remains moist and gummy for a little time. When it dries it forms an amorphous film round the fibroin fibres and also cements together the criss-cross of filaments making the cocoon and forming a protective envelope round the larva.

Analyses of Saturniid sericins have been made by Dhavalikar (1962) but they are not entirely satisfactory (see Lucas and Rudall, 1968). He records for these sericins a high content of lysine although in other respects they resemble *Bombyx* sericin in having high contents of serine, threonine, glycine, dicarboxylic acids, and the basic amino acids.

When silks are naturally coloured, the colouring lies in the sericin layer. In some species development of colour can be prevented if cocoon-formation takes place under dry conditions. The sericin of the robin moth (*Samia cecropia*) is very dark brown and Brunet (1967) has found that it is tanned by 3-hydroxyanthranilic acid, derived from tryptophan. In the gland it exists as the O-β-glucoside which is subsequently converted to the free acid and then oxidized to bring about tanning. Brunet suggest that tryptophan may be used to reduce the requirements for tyrosine, used in the tanning of the pupal case and the cuticle of the adult. It would be of interest to know if *Samia* lacks the bacteria in the fat bodies that enable some insects to manufacture essential amino acids, including tyrosine.

The imago of *Bombyx* escapes from its pale cocoon by dissolving away the sericin with an enzyme, cocoonase, formed by epidermal cells of the maxillary galeae and deposited, as small crystals, on the mouth parts. It is brought into action by a buffer solution secreted by the anterior ends of the silk glands. The sericin dissolved, the imago is able to push aside the threads of the cocoon and escape (Kafatos *et al.*, 1967). Cocoonase has no effect on the tanned sericin of the robin moth and other Saturniidae, and in these species the imago escapes through special holes left in the cocoon.

Shimizu (1958) reported finding, between the sericin and fibroin, in *Bombyx mori* and other lepidopteran silks, a fibril composed of cellulose, and his findings have been confirmed by Buonocore (1958). Whatever the significance of this cellulose fibril, the use of cellulose rather than chitin in this situation may be an economy measure and may be compared with the use of

Table 12.3 Amino acid composition of insect silks. Residues per 1000 total residues

	Four species[a] of Vespidae and Apidae	*Nematus*[b] *ribesii*	*Phymatocera*[b] *aterrima*	*Chrysopa*[c] *flava*
X-ray photograph type	α-type	Collagen type	Polyglycine II	Cross β Group III
Alanine	161–367	138	15	212
Glycine	46–65	275	662	246
Valine	19–65	38	20	—
Leucine	47–95	22	34	—
Isoleucine	10–51	20	22	—
Proline	13–31	9·7	8	—
Phenylalanine	1–22	4	1	—
Tyrosine	1–25	25	57	8
Serine	87–227	110	34	427
Threonine	28–53	35	3	31
Cystine	some	—	—	—
Methionine	—	—	—	—
Arginine	1–53	40	29	5
Histidine	2–13	5	11	—
Lysine	23–44	17	21	—
Aspartic acid	82–127	58	76	60
Glutamic acid	83–140	98	7	8
Hydroxyproline	—	—	—	—
Hydroxylysine	—	18	—	—
Trytophan	—	—	—	3

[a] Lucas, 1967.
[b] Lucas and Rudall, 1968.
[c] Lucas *et al.*, 1957.

tryptophan rather than tyrosine as the tanning agent precursor in the robin moth.

Hymenopterans produce silk from labial glands which they use for lining nests or forming cocoons. Amongst the more primitive hymenopterans, the Argidae produce silks with 70–80 per cent of the residues either alanine or glutamic acid (Table 12.3) and there is evidence that the crystalline areas consist of alternating alanine and glutamic acid residues. No sericin layer is present in hymenopteran silks. In *Pachylota audouinii* and *Digelansinus diversipes* the molecules are in the extended β-form. The side-chain spacing, because of the long glutamic acid residues, is 13·8 Å and for these silks Lucas and Rudall (1968) propose a new Group VI which fills the gap between Warwicker's Group III and Group IV. In the silk of *Arge ustulala* there is an α-helix structure superimposed on the extended β-type picture. Whether, on stretching, this α-helix structure changes to β-type structure, similar to that already present in the fibre, is not known.

Silk from the symphytan, Nematinae, the serigenous glands of which have been described by Kenchington (1969), gives a collagen type X-ray photograph, though it differs in detail from that of vertebrate collagen. In *Nematus ribesii* its chemical composition approximates to that of vertebrate collagen, one third of the residues being glycine and one tenth proline, but no hydroxyproline is present (Table 12.3). In the nearly related Solomon's Seal sawfly, *Phymatocera aterrima*, although it is possible to draw out threads from the contents of the silk glands, it has not been possible to get the larva to form cocoons in the laboratory, so fibre structure cannot be investigated. Two thirds of the material consists of glycine residues and it gives an X-ray photograph similar to that of polyglycine. When the larvae fail to pupate the contents of the silk glands become hard and dark suggesting some tanning process, though there is no evidence for the presence of a separate, low-molecular weight tanning agent.

Aculeate hymenopteran silk has a high content of alanine, serine and (aspartic + glutamic acid) but much less glycine than lepidopteran fibroins (Table 12.3). The extruded thread gives a special α-helix-type X-ray photograph. As yet the molecular arrangement to fit the photograph has not been worked out completely but Atkins (1967) suggests that a 4-chain rope unit seems the most satisfactory structure. The α-form readily converts to the extended β-form on stretching. Flower and Kentchington (1967) have described the silk glands and the state of the stored silk in *Apis*, *Bombus*, and *Vespa*. In *Apis* the stored silk molecules aggregate into spindle-shaped tactoids showing strong positive birefringence and a regular axial cross striation (Fig. 12.7(*a*) and (*b*)). In *Bombus*, although a few, non-striated tactoids are produced, the bulk of the material is in "fibrous bars" composed of fibres a little over 1 μm long stacked side by side at right angles to the long axis of the bars. In *Vespa* all the material is in the form of fibrous bars. Silk from many species develops a distinct brown colour some time after it is exuded and therefore it is likely that some tanning process takes place.

Silks from the non-aculeate Apocrita, gives a β-type X-ray photograph that would place them in Warwicker's Group IV silks, but there has as yet

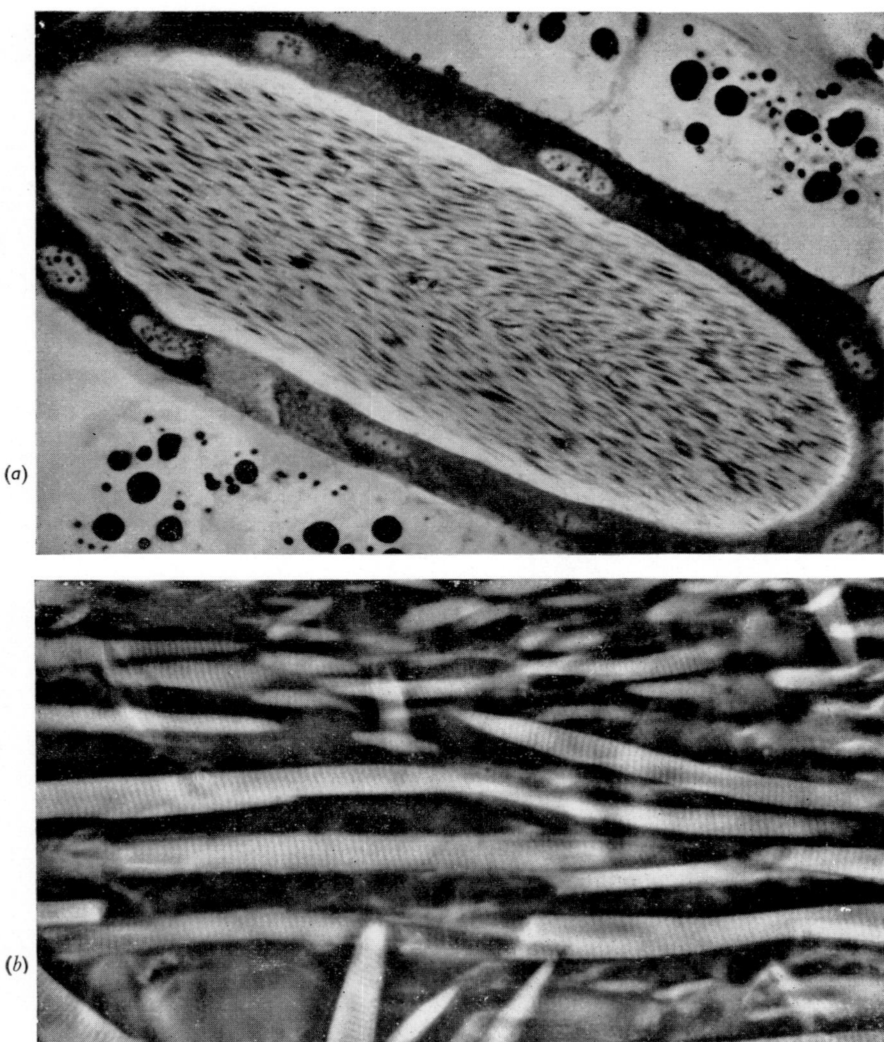

Fig. 12.7. (*a*) Tactoids in the lumen of the silk gland of *Apis mellifica* (× 300). (*b*) Tactoids of silk glands of *Apis mellifica*, highly magnified (phase contrast × 1500). (Lucas and Rudall, 1968.)

been no amino acid analysis to indicate whether they are similar chemically to the lepidopteran silks in this group.

Hymenoptera, therefore, show, even within the few species that have been examined, a considerable variation in the chemistry of their silks, but nothing is known of their mechanical properties, nor of the stresses they are subjected to in life. It is not possible, therefore, to decide whether these variations are adaptations to special needs, or a variety of ways of producing mechanically similar threads.

Larvae of the carabid beetle *Lebia scapularis*, and various neuropterans, produce silk from modified Malpighian tubules. There have been arguments as to whether Malpighian tubules are epidermal or endodermal in origin, since they develop from the region where ectoderm meets endoderm. Silk production by Malpighian tubules is evidence for their ectodermal derivation in insects. The female neuropteran *Chrysopa* deposits a drop of silk on a leaf and draws out from it a thread to the end of which she attaches an egg. X-ray photography shows that the molecules of silk, which are in the β-configuration, instead of lying parallel to the length of the thread, are folded at right angles to it, the folds being 25 Å long, an arrangement of molecules somewhat resembling that found in thermally contracted myosin and keratin. This folded structure is called the cross β-structure and is also found in the silk of the beetle *Hydrophilus*, secreted by a pair of papillae on the sternum, and used to form a floating container for the eggs. It occurs also in the hanging threads of the New Zealand glow worm, *Arachnocampa luminosa*, secreted by labial glands, either to trap insects as food or to form a cocoon. On stretching, the chains are pulled out and give a β-type diagram similar to that of Group III fibroins, (Table 12.3) although the chemical composition is rather different from that of other members of the group (Lucas, *et al.*, 1957) serine forming over two fifths of the residues.

The chysomelid beetle, *Donacia*, produces a cocoon of chitin fibres orientated at 45° and 90° to the long axis of the cocoon (Picken *et al.*, 1947). Praying mantis larvae wriggle out of their ootheca and become suspended by two fine threads, which contain a chitin–protein complex. These threads are not formed by extrusion but are hollow tubes secreted at the surface of a column of cells that grows out from each embryonic cercus and becomes twisted into a helix (Ketchington, 1969). The silks produced by the tarsal glands of Embioptera and Empidae have not been studied and neither have millipede silks.

2. *Spiders*

Spiders are the other large group that produce silk. All spiders can produce silk, although in some the capacity is much reduced. Some spiders use it only for protecting the eggs, others use it for many purposes. It is secreted by abdominal glands and five different types have been distinguished. (1) Ampullaceal glands of which there may be up to twelve, produce the drag-line which many spiders lay down as they move about and which holds them if they fall. The ampullaceal glands also secrete the frame and radii of orb webs. (2) Pyriform glands make the short threads that at intervals attach the drag-line to the substratum. (3) Some spiders have a cribellum, a perforated plate through which numerous small aggregate glands discharge sticky drops on to the threads of the web. (4) Female spiders have numerous aciniform glands that secrete the silk used to make cocoons for the eggs. (5) There are also cylindrical glands which produce the wadding round the egg masses inside the cocoons (Savory, 1964.)

Spider silks are much finer than insect silks and although they have been made into fabrics they have no commerical value except for making cross-lines in the eye-pieces of optical instruments. The drag-line of *Araneus diadematus* is, weight for weight, as strong as steel and much stronger than *Bombyx* silk. The cocoon silk, which is not subjected to the stresses that the drag-line may have to withstand, has less than a third of its strength. These silks are also remarkable in their extensibility; the drag-line extends 31 per cent, and the cocoon silk 46 per cent, before breaking. Chemically the two silks differ, the drag-line having a high content of glycine, alanine, and proline, while in the cocoon silk serine and alanine are most abundant and the proline content is quite low (Table 12.3). Apart from the absence of hydroxyproline, the amino acid composition of drag-line silk approaches that of collagen but X-ray photography classifies it as a Group IV fibroin (*Nephila madagascariensis* drag-line silk belongs in Group III) while the cocoon silk, with a wider side-chain spacing, has been placed in Group V.

The cocoon thread is 6-ply and much crinkled, resembling the high bulk yarns from man-made fibres that have been developed in recent years for making very elastic knit-wear. A cocoon around a larva made from such a thread will contain air trapped amongst the fibres and provide thermal insulation besides providing a bar to predators (Lucas, 1964). The cocoons of lepidopterans undoubtedly provide the same sort of protection.

C. Connective tissue

A well-developed, collagen-containing dermis beneath the epidermis of *Peripatus* has been described by Rudall (1955); its great development is undoubtedly correlated with the extensibility of the cuticle. The collagen fibres in *Peripatopsis* have a band period of 650 Å (Robson, 1964).

In Crustacea there are well developed, connective tissue membranes round the various organs, although Rudall (1955) found evidence for only a slight development of collagen in the subcuticular tissues of *Homarus* and the stalk of the barnacle, *Lepas*.

Insects also have sheets of connective tissue around their organs but the sheets are very thin (Ashhurst, 1964; Harper *et al.*, 1967). In the hypopharyngeal cavity of various apterygote insects (Francoise, 1968) and the dictyopteran, *Blabera craniifer* (Moulins, 1968), there is, stretching across the cavity, a broad band of collagen fibres which functionally replaces the endoskeletal tentorium found in other insects. In the anterior wall of the ejaculatory tube of *Locusta migratoria* the connective tissue assumes a cartilage-like consistency (Martoja and Bassot, 1965). The occurrence of cartilage-like connective tissue in *Limulus* was first described by Gegenbaur (1858). It occurs in the gill books and, as a hard, flat shelf supported by tendinous strands, in the anterior third of the cephalothorax dorsal to the oesophagus. The outer layers are composed of dense bundles of collagen fibres, orientated longitudinally, amongst which are elongated cells. The central portion has a structure similar to mammalian fibrous cartilage. Cartilage-like material has also been reported from scorpions, spiders, and mites (Person and Philpot, 1969).

Branchiopods possess a tendinous plate, the endosternite, ventral to the anterior part of the alimentary canal, which does not appear to belong to the endoskeleton formed by ingrowths of the cuticle; in *Cypridia lavis* it dyes quite differently from the endoskeleton and the rest of the cuticle but its chemical nature is not known (Calman, 1909).

In *Astacus* Schneider (1902) found three orders of fibre-forming Leydig cells concerned in the formation of the connective tissue of the blood vessels and blood spaces. These three orders display a range from cells with intracellular fibres to those producing extracellular collagen fibres. Whitear (1962, 1965) found there were two types of connective tissue associated with the chordotonal organs in the legs of the shore crab, *Carcinus maenas;* one was composed of fibrous collagen with a band-period of about 500 Å, while the other was granular but non-fibrous. Although this non-fibrous material may be only the ground substance of the connective tissue, its concentration as a band round a core of collagen fibres suggests that it is in some way specialized. Special "rosette cells" located within the amorphous material may be responsible for its secretion but as no fibroblasts were found amongst the collagen fibres the rosette cells may first secrete the collagen fibres and then move outwards to secrete the amorphous material. The histochemistry of the amorphous material has not yet been described.

In *Rhodnius* Wigglesworth (1956) described amoeboid blood cells that apply themselves to the various organs and contribute to the formation of the connective tissue sheaths, but Ashhurst (1964, 1965) believes that, in *Galleria* and *Schistocerca*, the fibroblasts associated with the connective tissue round the nervous system originate from the outer ganglion cells. In *Galleria* the fibroblasts contain fibrils which sometimes have a periodic banding of 150–200 Å and a similar condition has been described in *Blaps gibba* by Baccetti (1961). These fibril-containing fibroblasts recall the fibril-containing Leydig cells in *Astacus*. Marked gaps are found in the plasma membranes of the fibroblasts, opposite the accumulations of intracellular fibrils and Ashhurst has evidence that the fibrils are extruded in these regions to become extracellular. This is not a unique method of secreting collagen fibrils (see Ashhurst, 1964). The secreted collagen fibrils which remain small in diameter are partly obscured by masses of ground substance containing both acid and neutral mucopolysaccharides. Hyaluronidase, while it reduces the mucopolysaccharide content of the ground substance, still leaves a sheath round the collagen fibrils that tends to obscure any fibre-banding. In *Schistocerca* there are no intracellular fibrils in the fibroblasts and only very limited gaps in the plasma membrane. The collagen fibres have a well developed banding with a period of 550–600 Å and the ground substance contains neutral but no acid mucopolysaccharides. Mucopolysaccharides, particularly acid mucopolysaccharides, are important in stabilizing collagen. When collagen deposition takes place in the presence of acid mucopolysaccharides large numbers of thin fibres are formed but when acid mucopolysaccharide is absent deposition takes place partly on previously formed fibres so that individual fibres become thicker (Wood, 1960). The absence of acid mucopolysaccharide from the ground substance of *Schistocerca* connective tissue probably accounts for the collagen fibres being thicker in this insect (Ashhurst, 1964).

Banding has been observed in collagen from other insects and generally lies within the range of that found in mammalian collagen (Rudall, 1968a; Smith and Treherne, 1963; Harper *et al.*, 1967), although both Pipa and Woolever (1965) in *Galleria* and Baccetti (1961) in *Blaps gibba* report periods of 150–200 Å. Collagen fibres from *Limulus* have an axial period of 500 Å (Dumont *et al.*, 1964), shorter than that in mammals.

Table 12.4 Amino acid composition of cockroach collagens. Residues per 1000 total residues. (Harper *et al.*, 1967.)

	Corpus cardiacum and c. allatum	Carcass
Alanine	81·2	104·7
Glycine	120·7	235·2
Valine	58·5	44·0
Leucine	72·3	35·3
Isoleucine	42·6	25·3
Proline	63·8	88·1
Phenylalanine	34·0	27·6
Tyrosine	24·8	31·7
Serine	50·3	62·5
Threonine	44·6	40·1
Cystine (½)	5·0	7·9
Methionine	10·5	3·8
Arginine	45·6	34·5
Histidine	20·3	18·2
Lysine	59·6	20·7
Aspartic acid	107·4	66·4
Glutamic acid	130·0	86·5
Hydroxyproline	8·9	41·2
Hydroxylysine	7·8	26·4
Cysteic acid	3·9	—

Harper *et al.* (1967) have analysed a protein extracted by autoclaving from the corpus cardiacum and corpus allatum of the cockroach. Since this protein is degraded by collagenase they have good reason to assume that it is a collagen. It is remarkable for a very low content of glycine (120·7/1000 residues), and low total imino residues (73/1000) and also for containing a considerable quantity of tyrosine and cystine (Table 12.4). They also extracted with trichloroacetic acid a protein, from the whole carcass of the cockroach, which also is degraded by collagenase. This has a higher glycine 235/1000 residues) and total imino acid (129/1000 residues) content and also contains cystine and tyrosine. Harper *et al.* do not suppose that their materials represent purified proteins.

The cartilage of *Limulus* contains 10–15 per cent collagen (Person and Philpot, 1969) and a highly sulphated chondroitin sulphate (1·5 moles SO_4

per mole of hexosamine) very similar to that from the head cartilage of *Loligo*. It was also reported as containing chitin (Haliburton, 1885). Presuming that this observation is correct it is one of the few cases that have been reported of mesodermally produced chitin.

Rudall (1958, 1968) says that, of all animal groups, insects show the least development of collagen but this is rather a subjective assessment since only in one insect (cockroach, Harper *et al.*, 1967) and in the mouse (Harkness, 1961) has any attempt been made to measure the *quantity* of collagen present. Rudall suggests that the small quantity of collagen present in insects may be due, not to the functional replacement of the collagen of the dermis by the hardened cuticle, but to chitin production hindering in some way the production of collagen by both epidermal and mesodermal cells.

If the product of epidermal cells alone is considered, then in certain instances this suggestion appears justified. Nematodes and annelids have collagen-containing cuticles in which no chitin is present, while the cuticles of arthropods contain chitin and protein which is generally held not be be collagen. Although the protein in insect cuticle gives an extended β-type X-ray photograph, hydroxyproline has been found amongst the amino acids extracted from insect cuticle, indicating that collagen may not be entirely absent from the cuticle. In branchiopod crustaceans the protein component of the cuticle appears chemically closely related to collagen. Further, the Nematinae (Insecta) produce, from localized epidermal glands, a silk that gives a collagen-type X-ray photograph although it does not contain hydroxyproline). In brachiopods the protein from the chitin-containing periostracum and shell matrix of the inarticulates contains hydroxyproline and therefore probably collagen, while that from the chitin-lacking periostracum and shell matrix of articulates contains no hydroxyproline. The cuticle-covered stalk of the inarticulate, *Lingula*, contains chitin and the X-ray photograph suggests that collagen also may be present (Rudall, 1955). The chitin here is in the γ-form, and not in the α-form as in arthropod cuticles. Annelids produce, from epidermal pockets, chitin-containing chaetae and this does not prevent them from producing abundant collagen in the cuticle, though certainly the chitin in the chaetae is in the β-form. The above evidence does not support the suggestion that chitin inhibits the production of collagen by epidermal cells. Rudall, however, has suggested that it is only α-type chitin that inhibits collagen-production, although even α-chitin does not suppress the production of limited quantities of collagen by the arthropod epidermis.

Evidence of chitin-supressing collagen-production by mesodermal cells is even less convincing. Certainly insects, and arthropods generally, lack a collagen-containing dermis, although in *Peripatus* where the cuticle, which contains α-chitin, is soft and extensible, there is a well-developed dermis. Nematodes and other aschelminthes that have firm, non-chitin-containing cuticles also have no dermis. The evidence, therefore, suggests that the dermis is absent where it has been replaced functionally by a firm cuticle of any kind. However, in the barnacle *Lepas*, with α-chitin in the cuticle of the stalk, a dermis is lacking while in the brachiopod *Lingula*, with γ-chitin in the cuticle, a dermis is present. In *Lingula* the stalk cuticle, is described as extensible, however. The mollusc, *Loligo*, produces abundant collagen in the connective tissue though it has α-chitin in the oesophageal cuticle and beaks, β-chitin in the pen and γ-chitin in the stomach lining.

While it is agreed that β- and γ-chitin do not suppress collagen-production Rudall suggests that at least they may have an influence on its morphology and chemistry, since he believes that all collagen from chitin-producing animals is atypical when compared with mammalian collagen, the formation of which is uncomplicated by a capacity to produce chitin. However, all invertebrate collagens depart more or less from mammalian collagen, sometimes in their periodic banding, or in even lacking any banding, and particularly in chemical composition, whether or not the animal also produces chitin. Rudall suggests that the peculiar cored collagen fibres that have been observed in the earthworm may result from the production of β-chitin by these animals, but similar collagen fibres occur in the leech which does not produce chitin in any form. Until more is known about the significance of variations in invertebrate collagens it is difficult to establish which, if any, are due to the presence of chitin.

Collagen, produced internally, supports the various organs of the body. Harkness (1968) has shown that, in vertebrates, the larger the organ the greater, proportionally, is its content of collagen; logically this same relationship should also exist in invertebrates, so that the amount of collagen in and around an organ is a function of its size. Vertebrates, with many of their organs suspended from the dorsal wall, require strong connective tissue mesenteries to hold them in position. Invertebrates, with their organs mainly attached to, and resting on, the ventral body wall do not need such extensive mesentries to hold the organs in position. Further Miller (1964) has suggested that the tracheal system of insects (and presumably of other tracheate arthropods as well) may serve as a connective tissue by binding together organs in the body cavity. In arthropods ingrowths of the cuticle largely replace the collagenous tendons found in vertebrates for the attachment of muscles. Therefore, one would not expect insects to have a great development of collagen, because of their small size, their body architecture, and their hard cuticle. In the cockroach, *Leucophea madesae*, Harper *et al.* (1967) estimated that less than 0·1 per cent of the carcass, minus appendages, was made up of collagen. Compared with a 5 per cent collagen content in the body of a mouse (Harkness, 1961), this is certainly low but as Harkness estimated that half the collagen was present in the dermis and, if the dermis is replaced functionally in insects by the hardened cuticle, then the quantity for comparison may reasonably be reduced to 2·5 per cent in the mouse. A cockroach weighs approximately 1 g while a mouse weighs about 25 g, so that the mouse is 25 times heavier than the cockroach and it also has 25 times as much collagen per unit weight (or 50 times as much if the mouse's dermal collagen is included). The final test of Rudall's theory must await the determination of the collagen content of more animals and how it varies both in relation to body size and organization, in both chitin-producing and non-chitin-producing animals.

D. Egg cases and shells

The eggs of terrestrial arthropods are sometimes contained in a protective ootheca. The eggs of lithobiomorphs and millipedes are covered with a sticky secretion which hardens on exposure and to which soil particles adhere (Blower, 1955).

It was through investigating the formation of the protein ootheca of the cockroach, *Blatta orientalis*, that Pryor (1940a, b) obtained evidence for the hardening of the ootheca by quinone-tanning of its protein, and from this went on to demonstrate quinone-tanning in the cuticle of insects. The ootheca, when first formed, is a soft, white structure but later becomes hard and brown, resembling in appearance the cuticle of the parent, although it does not contain chitin. Pryor believed that the left colleterial gland secreted the untanned protein and the right gland the tanning agent, protocatechuic acid (Pryor *et al.*, 1946). There was also evidence for a polyphenol oxidase responsible for converting the protocatechuic (3:4 hydroxybenzoic) acid to the *o*-quinone but the site of its production was not determined. Brunet and Kent (1955) however found that the left gland contains, besides the protein to be tanned, the tanning agent in the form of the β-glucoside of protocatechuic acid, and the polyphenol oxidase as well. The right gland contains only a β-glucosidase. When the secretions of the two glands are mixed the β-glucosidase converts the glucoside to free protocatechuic acid which is then oxidized by the polyphenol oxidase, and tans the protein. The protocatechuic glucoside is derived from tyrosine (Brunet, 1963). The chemistry of the protein that becomes tanned is not known, but in its tanned condition it gives an extended β-type X-ray photograph somewhat resembling β-keratin (Pryor, 1940a). Oothecae of *Periplaneta* are tanned in the same manner as those of *Blatta*.

The ootheca of *Mantis* differs considerably from that of *Blatta* both in morphology and chemistry. It is formed from ribbons of material giving an α-helical-type X-ray photograph. The glands responsible for secreting this protein have been described by Kenchington and Flower (1969). It has not been examined for tanning (Rudall, 1968a, b).

Oothecae of the beetle, *Aspidomorpha* (Adkins *et al.*, 1966), are formed from tactoid bodies secreted by the epithelium of the ovariole pedicels and these tactoids maintain their identity in the oothecal wall. Each tactoid is made up of cylindrical rods about 75 Å in diameter separated from its neighbours by matrix material. The internal structure of the tactoids is probably helical. Glycine, alanine, and serine are all abundant, accounting for over 40 per cent of the residues, but there is also a very high content of proline (Table 12.5).

Oothecae of cockroaches and mantids contain crystals of calcium oxalate and calcium citrate respectively and the cockroach secretes the oxalate from the left colleterial gland while mantids secrete the citrate from a special gland beneath the seventh ventral sternite. *Aspidomorpha* oothecae also contain small crystals but they have not been identified. The function of these simple salts is not known. The significance of the variations in the

Table 12.5 Amino acid composition of *Aspidomorpha* ootheca protein. Residues per 1000 total residues.

Alanine	194
Glycine	127
Valine	41
Leucine	71
Isoleucine	22
Proline	174
Phenylalanine	13
Tyrosine	7
Serine	103
Threonine	37
Cystine	—
Methionine	—
Arginine	7
Histidine	19
Lysine	24
Aspartic acid	74
Glutamic acid	41
Hydroxyproline	—
Hydroxylysine	—

chemistry of oothecae also is not known. Oothecae protect the eggs during all stages of development, and probably are particularly important in protecting eggs when they are deposited in soil and in preventing the sharp particles of the soil from damaging the soft cuticles of newly hatched larvae.

Whether or not the eggs of arthropods are protected by an ootheca or by some form of cocoon, the individual eggs are surrounded by protective membranes, although only in crustaceans and insects have these membranes been described in any detail.

In the fairy shrimp, *Chirocephalus bundyi*, the egg reaches the ovisac where it is fertilized without having acquired any membranes. It then becomes covered with secretions from the shell glands, which produce a shell with indications of quinone-tanning. The inner layer of the egg shell of *Chirocephalus* and *Artemia salina* is spongy (Linder, 1960; Morris and Afzelius, 1967). These are freshwater Crustacea living in bodies of water liable to dry up. The egg shell is not water proof but the egg within the shell can survive complete desiccation, to develop when the egg becomes wet again. Morris and Afzelius suggest that the spongy layer in *Artemia* is a flotation layer as these eggs are always found floating on the top of the water. By the time the embryo has developed to the blastula stage, a thin membrane, or blastoderm-cuticle, is secreted by the embryo below the egg shell.

In decaped crustaceans Yonge (1935) describes the egg in the ovary as having no membranes. In the oviduct it becomes surrounded by a chitin-containing membrane secreted by glands of the oviduct which are believed to be of epidermal origin. An outer covering is not formed until the eggs have passed out of the oviduct and become covered by a secretion from tegumental glands, or "cement glands"; this also cements them to the pleopods. However, Cheung (1966) has evidence that in *Carcinus maenas*, *Homarus vulgaris*,

Nephrops norvegieus, and *Astacus pallipes* all the membranes are formed, not by the oviduct and the cement glands, but by the egg itself. The outer membrane is three-layered, the outermost of these is formed from the vitelline membrane, the middle layer from solidification of liquid extruded by the egg, and the inner layer also is formed by the egg. The formation of this triple-layered membrane is probably initiated by fertilization and may represent a highly specialized fertilization membrane. Both the outer and inner layers contain neutral mucopolysaccharides but there is little evidence of protein.

The outer layer of the membrane becomes locally drawn out—how, it is not known—to form the stalk by which the egg is attached to the pleopods. Nor is the origin of the cement which actually sticks the eggs to the pleopods known, but Cheung suggest it may be derived from the material of the middle layer of the membrane while it is still fluid, being squeezed through holes in the outer layer. The "cement gland" secretions may be responsible for providing a suitable medium for fertilization to take place externally in the incubation chamber, or it may be concerned in hardening the outer membranes. Brachyura (crabs), with internal fertilization, lack cement glands. During embryonic development one or two inner, chitin-containing membranes are formed round the embryo.

The eggs of geophilomorph centipedes are protected by a shell of tanned protein (Blower, 1955). In insects, as soon as the egg has developed its complement of yolk, it forms about itself the vitelline membrane, and outside this the chorion is formed by the follicular cells. In locust eggs the chorion is divisible into endochorion, which forms a meshwork round the egg, and the exochorion. After the egg has passed into the common oviduct, an extra-chorion is formed around it (Harley, 1961). In *Rhodnius* it is possible to distinguish seven different layers in the chorion. From outside inwards the exochorion is formed of two distinct layers, an outer resistant and an inner soft layer of lipoprotein. Below this is a soft layer of endochorion, then a thin amber layer of lipidized, tanned protein, an outer layer of small polyphenol granules, separated by a resistant layer of tanned protein from an inner layer of larger polyphenol granules. These outer and inner granule layers are probably formed from small quantities of heavily tanned protein. The occurrence of quinone-tanning in structural proteins derived from the follicle cells shows that quinone-tanning is not restricted to products of epidermal cells. Beneath the chorion is a wax layer (Beament, 1946). These eggs are not protected by an ootheca.

The vitelline membrane in *Drosophila* and other Diptera is added to by the follicle cells to produce a membrane of some thickness. It contains proteins which may have disulphide bonds, lipids, and both neutral and acid mucopolysaccharides. The chorion also contains proteins, lipid, and polysaccharide but these differ from those in the vitelline membrane and no disulphur bonds are present (King and Koch, 1963).

Wax layers often occur on egg shells and are probably important, as usual, in controlling water loss from the egg. In *Rhodnius*, where one occurs between the vitelline membrane and the chorion, it is believed to be secreted through the vitelline membrane by the oocyte. In *Culex* an oily layer exists between the exo- and endochorion, in *Lucilia* between the chorion and the chorionic-vitelline membrane.

Many endoparasitic hymenopterans have a very thin chorion and this is capable of stretching to a considerable extent with the growth of the embryo, which obtains its nourishment not from yolk stored in the egg but from the host tissues. Eggs laid in protected positions, like those of *Tenebrio*, also have thin chorions. Shells are absent from ovoviviparous eggs.

The insect chorion is often sculptured or provided with a stalk, or with prolongations. Since the chorion is formed before the egg is fertilized, one or more pores or micropyles are left in it through which the sperm can reach the egg. Inorganic salts are often deposited in the chorion (Moscona, 1950).

Developing embryos of insects also secrete beneath the vitelline membrane a cuticle, the *serosal cuticle*, which in locusts appears very similar to that of the adult cuticle, with proteins tanned by acetyldopamine (Furneaux and MacFarlane, 1965). Development of the egg is often accompanied by absorption of water and an increase in volume of the embryo; this ruptures the chorion so that the embryo is then protected only by the serosal cuticle. When the insect hatches the serosal cuticle is discarded.

The developing embryo needs air and, while in some species there is no special arrangement to get air to the embryo, many insects have special structures that aid the respiration of the embryo. *Rhodnius* has a series of pseudomicropyles which admit air to the embryo. The inner layer of the chorion of many Diptera is spongy, the spaces being filled with air. In *Blatella*, within the ootheca the egg shells have spongy caps that communicate by a tube with a spongy chamber in the top of the ootheca.

CHAPTER 13

Chaetognatha and Pogonophora

I. CHAETOGNATHA

Chaetognatha, or arrowhead worms, of which *Sagitta* is the best-known example, occur only in the sea, mainly in the plankton. They are small, generally about 40 mm long as adults, with transparent, torpedo-shaped bodies expanded laterally into a pair of horizontal fins and a tail fin. These fins contain no muscles; they are aids to flotation and not organs of locomotion. The body is stiff and turgid and capable of little bending.

A. External skeletal structures

The body is covered by a thin cuticle, thickened on the head to form lateral and ventral plates supporting the teeth and spines, and serving for the attachment of head muscles. It also forms a thick lining to the vestibule leading to the mouth.

The head is often armed anteriorly with two or four short rows of small teeth. The number of teeth in each row varies with the species and within the species, the number increasing with age up to sexual maturity, after which loss of teeth may occur (Hyman, 1959).

The sides of the posterior part of the head have an arc or chevron-shaped row of hard, curved spines, the number of which again varies with the species and within the species. Each spine has a broad base to which muscles are attached, a curved shaft, generally wedge-shaped in cross section, and a tip, set in a socket at the end of the shaft (Fig. 13.1). Both teeth and spines are used to catch and hold prey.

Cuticle, teeth, and spines are all secreted by the underlying epidermis. Although Schmidt (1940) failed to detect it, Hyman (1958) found chitin in the grasping spines of *Sagitta elegans* but failed to find it in the remaining cuticular structures. Nothing else is known of the chemistry of these structures, although the brown colour of the teeth and spines suggest possible quinone-tanning.

B. Internal skeletal structures

The dermis must play an important part in the support of the body. As the body-wall contains only longitudinal muscles the dermis, or the cuticle, or a

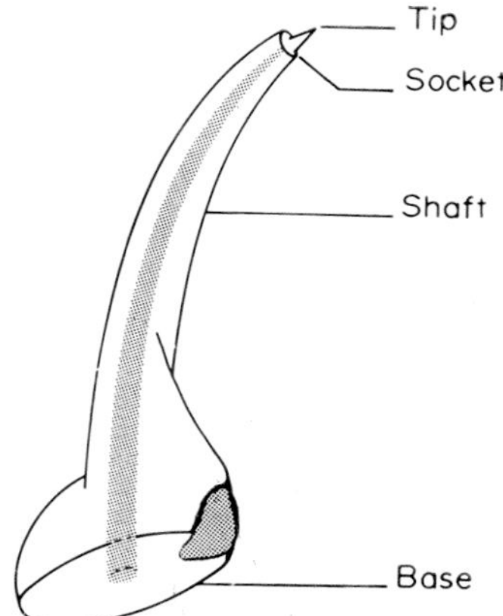

Fig. 13.1. Chaetognath spine (after Burfield, 1927).

combination of both, must define and maintain the body-shape and oppose the turgor pressure of the body, and the contraction of the longitudinal muscles. In the head-region it serves for the attachment of the head muscles and it is thickened locally to form capsules supporting the eyes. In the neck, and sometimes in the adjacent trunk region, it forms skeletal plates for the attachment of muscles that move the head. The rays that support the horizontal and tail fins appear also to be formed from thickenings of the dermis. Chemically and ultra-structurally the basement membrane has not been examined.

C. Egg membranes

Sagitta eggs are fertilized internally and are shed, covered with a layer of jelly, to float in the plankton; *Spadella* eggs are covered, as they escape, by secretions of the cement glands around the vagina. This secretion forms an adhesive coat and stalk to the egg, which becomes fastened to seaweed. In *Krohnitta* the eggs become fastened to the back of the parent. Chemically, nothing is known about these various egg coverings.

II. POGONOPHORA

Pogonophora are solitary, tube-living, marine animals, coelomate but without digestive tract. They are worm-like with one to many anterior fringed tentacles. While pogonophors may be from 10–35 cm in length, their diameter is generally less than 1 mm, although the largest species may be 2·5 mm thick.

A. External skeletal structures

The whole of the body is covered by cuticle, even where there is a ciliated epithelium, as in the dorsal ciliated band of the metasoma. The thickness of the cuticle varies considerably from one part of the body to another and Ivanov (1963) describes it as possessing great extensibility and elasticity.

The cuticle over the tentacles and pinnules in *Nereilinum* and *Oligobranchia* is 0·2–0·4 mm thick and has a distinctly fibrous structure, the component fibrils being about 30 Å wide and of indeterminate length. Nearest to the epidermis the fibres are loosely packed and orientated roughly parallel to the apical surface of the cell, although randomly orientated in this plane. Towards the outer surface, however, the fibres appear more closely packed but still lack any definite orientation. In *Siphonobranchia* the fibrils are arranged in layers in which the fibrils, although loosely packed, are roughly parallel to each other and at approximately 90° to the fibrils in adjacent

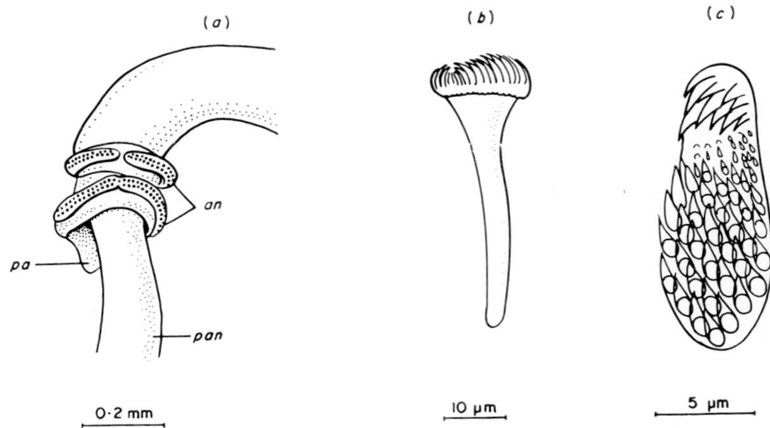

Fig. 13.2. (*a*) The girdles of *Siboglinum caulleryi*. (*b*) Toothed platelet and stalk of *Polybranchia annulata*. (*c*) Toothed platelet in surface view. *an*, girdles; *pa*, median papilla behind the girdles; *pan*, postannular region of the metasoma. (Ivanov, 1963.)

layers. In *Nereilinum* and *Oligobranchia* the outer surface is highly irregular and pitted, but in *Siphonobranchia* there is distinct, densely staining layer at the surface of the cuticle. Beyond the surface of the cuticle is a layer, believed to be of mucus. Passing through the cuticle and mucus layer and protruding at the surface are microvilli from the epidermal cells (Gupa *et al.*, 1966, 1969), as in annelids. Since pogonophora are without mouth or gut, they must absorb food through the body surface and the microvilli may in some way be concerned with this. The fine structure of the cuticle over the rest of the body has not been examined in detail.

The anterior part of the metasome has longitudinal rows of papillae that are probably concerned in secreting either the tube or adhesive material. However, in some species the papillae are topped by hardened plates of cuticle.

In the forepart of the mesosoma the cuticle is thickened, producing a pair of keels or crests running obliquely forward to form the so-called "bridles" which probably smear the papillary secretions over the inner surface of the tube (Webb, 1965). These bridles are usually coloured brown or black. Just in front of the posterior third of the metasome there are generally two bands or girdles bearing two to several rows of small oval platelets, their surface covered with tiny teeth. Each platelet shows externally only its head of teeth projecting through the cuticle (Fig. 13.2). The teeth are arranged in rows in two groups, a small anterior group of short teeth projecting backwards and larger posterior group of longer teeth projecting forwards. Each platelet, which is produced within a single epidermal cell, is drawn out into a long stalk lying in the epidermis and still attached by tonofibrils to the formative cell. It is not connected to any muscle fibres. These teeth, almost certainly, serve to anchor the animal in its tube.

Southward and Southward (1966) found the outer mucus layer probably contained a sulphated acid mucopolysaccharide, while the cuticle itself was

Table 13.1 Amino acid composition of the protein from the tube of *Siboglinum*. Residues per 1000 total residues. (Foucart *et al.*, 1965.)

Alanine	57
Glycine	128
Valine	53
Leucine	47
Isoleucine	38
Proline	63
Phenylalanine	34
Tyrosine	56
Serine	87
Threonine	50
Cystine	42
Methionine	12
Arginine	86
Histidine	23
Lysine	39
Aspartic acid	113
Glutamic acid	68
Hydroxyproline	—
Hydroxylysine	—

composed of neutral mucopolysaccharide, perhaps with collagen and some phospholipid.

The tubes in which the pogonophores live are arranged upright in the bottom ooze; they fit closely to, and are considerably longer than, the contained animal. The tube, which may be smooth or composed of a series of rings, differs chemically from the cuticle. Brunet and Carlisle (1958) found it contained chitin which Blackwell *et al.* (1965) showed to be highly crystalline and in the β-form. It is associated with a very stable protein (Brunet and Carlisle, 1958). Chitin forms 33 per cent of the dry weight of the tube of *Siboglinum* and protein 47 per cent, the protein (Table 13.1) being distinguished by a relatively high content of glycine and aspartic acid, and some cystine (Foucart *et al.*, 1965), so that its stability could be due to disulphide bonds.

B. Internal skeletal structures

The dermis is always well developed in pogonophores (Ivanov, 1963). It becomes locally attached to connective tissue membranes surrounding the various organs, forming mesenteries holding them in place. There are well-developed circular and longitudinal muscles in the body wall. The elasticity of the cuticle enables it to accommodate the changes in shape of the animal brought about by contractions of the body wall muscles.

C. Egg membranes

The eggs are laid, unfertilized, into the anterior part of the tube. After fertilization they become surrounded by a fertilization membrane. They are protected during development by the parent's tube and no further membranes form round the egg.

CHAPTER 14

Echinodermata

Echinoderms are marine, coelomate animals, which though bilaterally symmetrical in the larval stages, become radially symmetrical as adults. Almost all of them achieve a size that makes them easily visible members of the fauna.

A. External skeletal structures

Cuticles are of irregular occurrence in this phylum. Some species of crinoids have a cuticle, other do not. Holothurians, asteroids, echinoids, and some ophiuroids are generally described as possessing a thin cuticle (Hyman, 1959). In some asteroids the cuticle is divisible into an outer, homogeneous layer and an inner layer of platelets corresponding to the underlying epidermal cells (Smith, 1937). In asteroids and echinoids, in which the epidermis is ciliated, the cilia pass through the cuticle. When the cuticle is stripped off the epidermis of asteroids it becomes crinkled as if, *in situ*, it is under tension. There is nothing known about the physics and chemistry of echinoderm cuticles. It is unlikely that in these animals the cuticle plays any part in supporting the body, which has a well developed mesodermal skeleton but presumably it gives some protection to the epidermis.

B. Internal skeletal structures

Asteroids, echinoids, and holothurians have a nervous plexus underneath the epidermis and the epidermal cells have basal extensions penetrating into the nerve plexus. Crinoids and ophiuroids have the nerve plexus restricted to the tube feet. Fibres extend through the epidermal cells, at right angles to the body surface, from the cuticle to the layer of connective tissue bounding the nerve plexus internally. These fibrils, which are probably a form of tono-fibril, presumably give support to the extended epidermal cells which, in turn, provide a scaffolding for the nerve plexus. In sections of material that have shrunken in the fixative, the fibres are wavy or coiled, suggesting that normally they are under sufficient tension to hold them out straight. Brown (1950c) found disulphur bonds in the fibres.

Within the Echinoderma it is the dermis and its derivatives that provides support for the body. In all classes it is a well developed tissue containing collagen fibres (Marks *et al.*, 1949; Randall *et al.*, 1952) and ossicles of calcium carbonate. In holothurians the ossicles may be overlapping plates completely covering the animal, or small, detached spicules of a wide variety of shapes. In asteroids the ossicles are generally packed close together on the lower surface but on the upper surface have spaces between them. Crinoids are sedentary forms, fixed to the substratum by a flexible stalk. The body consists of a calyx, containing the gut and reproductive organs, and arms, which collect food and pass it to the mouth along ciliated grooves. Stem, calyx, and arms are supported by ossicles. Articulations between the ossicles of the arms and the stalk facilitate muscular movement of these parts. The free-living ophiuroids have a very small, ossicle-covered central disc and long, often spiny, arms in which certain ossicles have become centrally located to form a longitudinal series of "vertebrae", each hinged to its neighbour, so that horizontal movement between the vertebrae is possible. Surrounding the "vertebrae" are four longitudinal series of plates. In echinoids the ossicles are joined together by sutures to form a rigid test, or corona. Associated with the endoskeleton in echinoids, asteroids, and ophiuroids are various types of spines. In asteroids and echinoids there are also highly specialized, pincer-like structures, the pedicellariae, which help to prevent the settling of other organisms on the surface of these animals. Both spines and pedicellariae are supported by a calcareous endoskeleton.

The bodies of echinoderms exhibit a pentamerous symmetry and this is shown in the arrangement of the ossicles, particularly those of the apical disc which are the first formed and play an important part in protecting the larva during metamorphosis. Whatever the evolutionary pressures that brought about pentamerous symmetry, Nichols (1966, 1967) sees, in the arrangement of the five basal plates around the central plate in the apical disc, the minimum number of units that can be arranged in a circle without producing inherent weaknesses in the structure. Assuming that the sutures between the plates are regions of weakness, a circle of three plates would produce long sutures approaching straight lines. Even numbers of plates in the ring would result in any radial suture always being opposite a radial suture on the other side of the ring. With five plates, radial sutures lie opposite the middle of plates on the other side of the ring so that lines of weakness do not extend across the apical disc (Fig. 14.1).

Moss and Meehan (1967) do not believe that the sutures are regions of inherent weakness in the test. When the test of *Arbacia punctulata* is decalcified the sutures are seen to form a completely inter-connected lattice. Moss and Meehan suggest that the sutures act as "stress-breakers" for externally applied loads, reducing the effect of the load by transmitting the force through the whole test.

Although the skeleton is basically composed of calcite it has also a high content of magnesium, mostly in the form $MgCO_3$ in solid solution with the calcium carbonate (Chave, 1952). Clark (1911) first noticed, in crinoids a correlation between the magnesium content and the temperature of the water in which they lived. This was also found in other echinoderm classes by Clarke and Wheeler (1922), Vinogradov (1953), and Chave (1952). The warmer the sea, the higher the content of magnesium, although there are

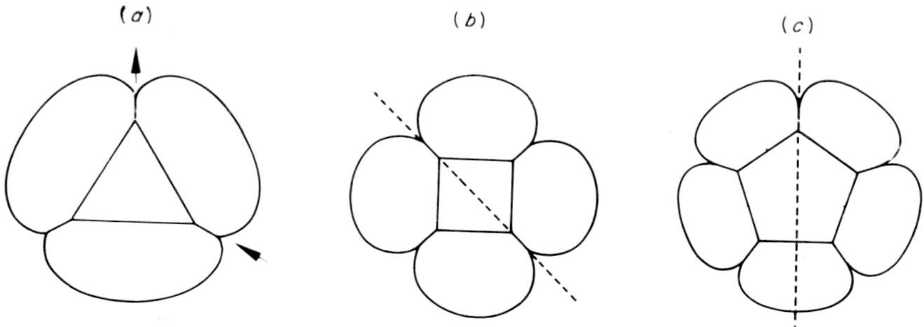

Fig. 14.1. Diagrams to show a suggested reason for the establishment of five plates in the first ring of the apical disc of an echinoderm. (*a*) Three plates, line of weakness shown by arrows. (*b*) Four plates-opposing sutures of apical ring in line across apical plate. (*c*) Five plates (actual situation). No suture of apical ring in line with any other. (After Nichols, 1966.)

species variations at any given temperature and variations in the content of different parts of the skeleton in the same species.

In micro-detail the ossicles have a spongy texture in which the spaces inter-communicate and these spaces may account for more than 50 per cent of the volume of the ossicle. Nichols (1966) suggests that the spongy texture produces a light, strong skeleton while conserving calcium. The open pores at the surface also provide spaces into which connective tissue can penetrate to effect firm junctions between skeleton and muscles.

The ossicles and spines behave, in polarized light, as single crystals, just as the spicules in calcareous sponges (p. 87). Garrido and Bianco (1947) and Nissen (1963) have produced X-ray evidence to suggest that they are, in fact, not single crystals but aggregates of micro crystals with their crystallographic axes all parallel. Currey and Nichols (1967), from studies of the exposed surfaces of broken echinoid spines, concluded that the spine was definitely a single crystal. Towe (1967), using a similar technique, found that while the inner portions of the ossicles have the morphology of a single crystal, the outer parts are formed by polycrystalline aggregates with a common preferred orientation that subsequently become converted, by some process involving fusion and recrystallization, to a single crystal structure.

The orientation of the crystallographic axes of the different ossicles varies in different species. In *Strongylocentrotus purpuratus* the *c*-axes of the coronal plates are everywhere at right angles to the surface of the ossicle and the surface of the animal. In *Eucidaris thouarsii* the *c*-axes are nearly tangential to the ossicle surface, and parallel to the ossicle columns. There is no direct relationship between the orientation of the *c*-axes in the apical system and those in the coronal ossicles adjacent to them, in spite of the fact that new coronial ossicles are always formed in contact with the apical system. Within the apical system the *c*-axes show a bilateral symmetry and not a pentamerous radial symmetry (Raup, 1965).

Millott (1954) has shown that light can pass through the test of echinoid and excite underlying light-sensitive parts of the nervous system. The amount of light that can be transmitted through the test of a sea urchin will be partly determined by the difference in refractive index between the calcite and the organic material filling the spaces in the plates. The refractive index of the organic material is 1·58, that of the calcite 1·48–1·65, depending on the crystallographic orientation in relation to the incident light. When the refractive index of the calcite is almost the same as that of the organic material, the light can pass through the ossicle; this happens when the *c*-axis is tangential to the surface of the plate. Maximum difference occurs when the *c*-axis is at right angles to the surface of the plate and then internal reflection and diffraction considerably reduce the amount of light that can pass through the plate. Sea urchins that have the *c*-axes tangential to the surface are photo-negative, always attempting to hide from incident light or through the "covering reaction," collecting, with their tube feet and spines, opaque material such as shell-fragments and seaweed with which they cover themselves to provide protection from the light.

Urchins with the *c*-axes at right-angles to the plate surface are insensitive to incident light. The perpendicular *c*-axes might reasonably be expected to be an adaptation to life in shallow water in a sunlit environment while the tangential axes would occur in forms living in deeper water. However, there is no correlation between environment and the orientation of the *c*-axes. The perpendicular orientation appears, from fossil evidence, to have been the primitive condition. The tangential orientation cannot represent an adaptation to let light through the test, as forms with this orientation go to great lengths to protect themselves from the light. Species with perpendicular *c*-axes have more highly curved tests and fewer plates in the test than forms with tangential *c*-axes and so the crystallographic orientation of the plates may control the amount of curvature that the plates can develop. Whatever the advantage of the tangential *c*-axes apparently it was sufficient to outweigh the disadvantage of increased light transmission by the test (Raup, 1960, 1962, 1965).

In the development of the larval skeleton of echinoids and ophiuroids and the adult skeleton of all classes, the ossicles make their initial appearance as tiny granules of calcite which develop into triradiate spicules. Woodland (1907) and Bevelander and Nakahara (1960) regard the initial spicule as being intracellular but Okazaki (1960) describes it as forming in an organic matrix secreted by mesenchyme cells. Woodland describes the subsequent division of the spicule-containing cell to form a syncytium, in which the spicule completes its development. Bevelander and Nakahara believe that subsequent growth of the spicule into an ossicle takes place extracellularly. Wolpert and Gustafson (1961) have described the larval skeleton of sea urchins as developing within a sheath of mesenchyme cells which they observed moving over the surface of the spicules but never moving away from them. They further describe the part the mesenchyme cells play in orientating the spicule, which initially is randomly orientated, and in controlling the direction of growth of its various branches.

In the region of the mouth of asteroids and echinoids there are various structures which help in the collection of food. These are believed to have evolved from specialization of ossicles lying round the mouth. Except for the teeth, all the complex calcareous elements of "Aristotle's lantern" in echinoids develop from spicules, as do the plates of the test. The teeth differ from the rest of the lantern both in their development and in their construction. Each tooth originates in the larva as a pair of round calcite particles which, through addition of calcium carbonate on one side only, broaden out into triangular lamellae which become curved round to unite with each other to form a cone-shaped body, the apex pointing downwards to the future mouth. Further cones develop behind the initial cone building up a tooth composed of a series of nesting cones fused together, the tip of which becomes greatly hardened (Devanesen, 1922). Growth of the teeth continues throughout life to make good the wearing away through use.

Whether or not the ossicle is an intracellular production, there is no real evidence to support the belief of Travis *et al.* (1967) that echinoderms calcification takes place in association with collagen fibres, as in vertebrate bone. Foucart (1966a, b), by careful decalcification, has shown that collagen fibres are restricted to the sutures holding the ossicles together. The organic material left after decalcification of the ossicles is a slightly granular material with no trace of fibrils and with a chemical composition completely different from that of collagen, hydroxyproline and hydroxylysine being absent, and the main components being glutamic and aspartic acid (Table 14.1). Moss

Table 14.1 Amino acid composition of the matrix of *Paracentrotus* shell plates. Moles per 1000 moles. (Foucart, 1966b.)

Alanine	74
Glycine	104
Valine	66
Leucine	79
Isoleucine	52
Proline	55
Phenylalanine	50
Tyrosine	29
Serine	86
Threonine	71
Cystine	—
Methionine	5
Arginine	64
Histidine	24
Lysine	tr.
Aspartic acid	125
Glutamic acid	117
Hydroxyproline	—
Hydroxylysine	—

and Meehan (1967) also failed to find collagen fibres in the ossicle matrix of *Arbacia*. Currey and Nichols (1967) concluded there was no organic matrix within the calcite crystals of echinoid spines. Towe (1967) could not be certain whether calcification took place in or on an organic matrix. If there is no matrix within the calcite crystal, the material analysed by Foucart may represent the material filling the pores of the ossicles.

The dermis contains, besides the ossicles, various amounts of collagen fibres and ground substance. In holothurians there is a loose meshwork of fibres surrounding the spicules, while in the inner part of the dermis the fibres are closer together forming a tougher layer. The body-wall also contains circular and longitudinal muscles. The ground substance of the dermis of crinoids is described as gelatinous (Hyman, 1955) and fibres are abundant. Here the ossicles form a rigid wall and the body-wall lacks regular layers of circular and longitudinal muscles. Asteroids also have a thick layer of fibrous dermis but poorly developed circular and longitudinal muscles. In echinoids the dermis is almost wholly occupied by the ossicles, and body-wall musculature is absent, except in Echinothuriidae in which the ossicles are separated, or overlap, and the test is flexible. In Ophiuroidea, again, the ground substance is described as gelatinous (Hyman, 1955),and no muscles are present in the body-wall.

Echinoderm collagen fibres have an axial repeat period of 625–670 Å (Marks *et al.*, 1949; Tarvis *et al.*, 1967) and a typical amino acid composition, (Table 14.2) although the hydroxyproline and total imino acid content is

Table 14.2 Amino acid composition of echinoderm collagens. Residues per 1000 total residues.

	Paracentrotus lividus: Ligament[a] *	*Thyone:* bodywall[b]	*Holothuria forskali:* cuvierian tubules[b]	*Strongylocentrotus droebachienis*[c]	*Lytechinus variegatus atlanticus*[c]
Alanine	83·3	71	114	72	92
Glycine	310·6	306	308	339	300
Valine	24·5	30	21	15	20
Leucine	33·0	22	28	25	23
Isoleucine	17·7	13	13	8	11
Proline	86·9	109	81	90	95
Phenylalanine	9·5	8·9	11	5	5
Tyrosine	4·5	7·9	11	6	2
Serine	63·3	54	55	88	52
Threonine	40·7	39	51	33	32
Cystine	—	2·5	—	—	—
Methionine	12·4	2·5	6·9	23	12
Arginine	52·8	54	46	57	52
Histidine	—	2·8	4·8	3	2
Lysine	14·4	7·5	12	4	12
Aspartic acid	63·5	62	70	56	57
Glutamic acid	111·0	110	77	97	127
Hydroxyproline	67·1	60	54	72·5	100
Hydroxylysine	4·4	11	4·7	6	7
Proline + Hydroxyproline	154	169	135	162·5	195

* Moles per 1000 moles.
[a] Foucart, 1966b.
[b] Gross, 1963.
[c] Travis *et al.*, 1967.

below that of mammalian collagen. In spite of the low amino acid content, the T_S of *Holothuria forskali* collagen is 55–59°C (Watson and Silvester, 1959), similar to that of hog collagen, indicating that some factor in addition to imino acid residues contribute to its stability.

C. Egg shells

In echinoderms fertilization is almost always external and the eggs are never covered by more than a layer of jelly and a fertilization membrane, although in crinoids the fertilization membrane sometimes develops spines. In a few species development takes place in a brood chamber and the ciliated larval stage is suppressed, but in the majority of species the egg develops rapidly into a free-swimming, ciliated larva and protective egg shells are not required.

PART THREE

Chordata

Vertebrates, which possess backbones and, at least as embryos, a notochord, together with those animals which, although they lack a backbone, possess a notochord at some stage of development are classed together in the phylum Chordata. Members of extant subphyla have a hollow, dorsal nerve-tube in at least part of the body during at least part of their life. They all lack a cuticle, although some secrete a surface layer of mucus and sometimes they form tubes or tunics in which they live.

CHAPTER 15

Hemichordata, Graptolitoidea, Urochordata, and Cephalochordata

I. HEMICHORDATA

Hemichordates, which are all marine, are small animals. The Enteropneusta, or acorn worms, vary in length from 2–3 cm up to 50 cm, although one species reaches a length of 1·8 m. The Pterobranchia are smaller still: Cephalodiscida and Rhabdopleurida do not exceed 2 mm.

A. External skeletal structures

The epithelium of hemichordates is ciliated and is also richly supplied with various sorts of gland cells. Acorn worms mostly live in burrows in sand, and these may become lined with mucus which in some cases becomes hardened to form definite tubes. If the animal is removed from its burrow and placed on a hard surface it covers itself with a sheath of mucus, most of which is secreted by the epithelium of the proboscis and passed back over the body by the cilia. If the lining is removed from the burrow it appears yellowish, smooth, and shining on the inside but on the outside brown, fibrous, and with splits that look as if the surface has been broken by being stretched (Knight-Jones, 1953). The brown colour of the burrow lining, together with the fact that the surface secretions of *Dolichoglossus* absorb radioactive iodine from seawater (Gorbman *et al.*, 1954), suggest that there may be quinone-tanning of proteins in the tube lining.

Among Pterobranchia, the Atubarea are sedentary and naked, living curled round hydroids. Both the Cephalodiscida and Rhabdopleurida are colonial and both produce permanent tubes or coenoecia in which they live. In both the tube is not secreted by the general epidermis but by the cephalic shield which is homologous with the proboscis in the Enteropneusta. The arrangement of the tubes of the colonies shows considerable variation and has been described by Hyman (1959). In *Cephalodiscus* sp. the tubes are often ornamented with projections which may be supported by spines of hardened material. In some species hardened fibres form a meshwork in which the animals are housed. The tubes are built up of interdigitating half rings, generally of a drab or yellowish-brown colour, although some are reported as being orange, red, or brown. The colour is lost from material preserved in alcohol.

In *Rhabdopleura* the horizontal tube of the colony is formed from overlapping half rings but the upright tubes that house the zooids are formed from a series of rings that overlap each other (Fig. 15.1).

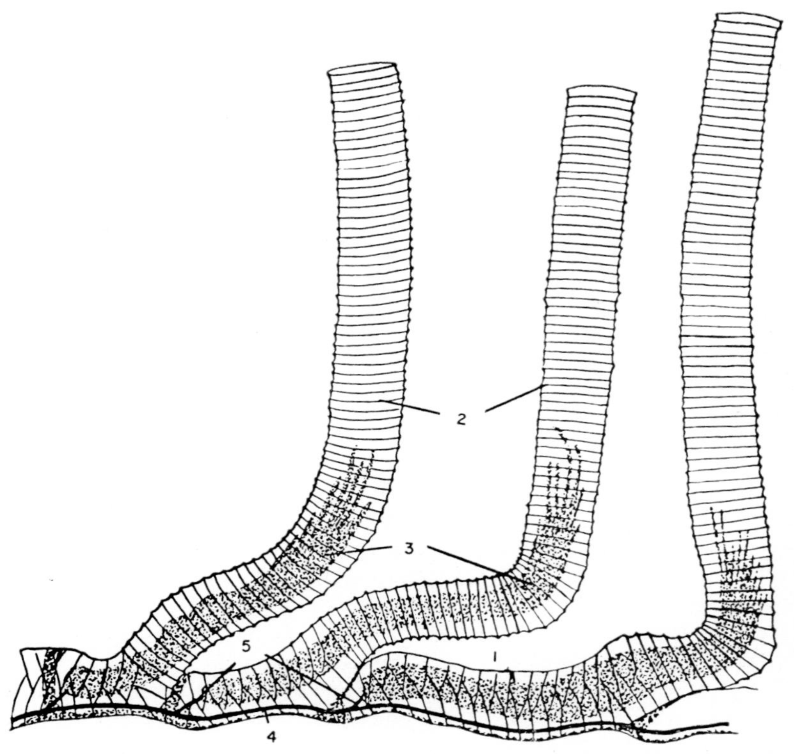

Fig. 15.1. Creeping tube and three erect tubes of *Rhabdopleura*. 1, creeping tube; 2, erect tubes; 3, retracted zooid; 4, black stolon; 5, branch of black stolon attached to base of zooid stalk. (After Schepotieff, 1907.)

No chitin is present in the tube (Rudall, 1955; Foucart *et al.*, 1965). In *Cephalodiscus* 19 per cent of the dry weight of the tube is a protein distinguished by its high glycine content (Table 15.1) but the main bulk of the organic material of the tube has yet to be identified (Foucart *et al.*, 1965).

Cephalodiscus is free in its tube and can move out of it, although its powers of movement are very limited. In *Rhabdopleura* colonies the individuals are connected by the "black stolon", a tube of hard, black material enclosing a core of living cells. The black stolon produces branches at the ends of which the zooids develop. The chemistry of the "black stolon" sheath unfortunately has not been investigated.

Table 15.1 Amino acid composition of the protein from *Cephalodiscus inaequatus* coenoecium and *Pristiographus gotlandicus*, Monograptidae sp. and *Climacograptus typicalis* (Graptolitoideit). Moles per 1000 moles.

	Cephalo-discus inaequatus[a]	*Pristio-graptus*[b]	*Mono-graptidae*[b]	*Climaco-graptus*[b]
Alanine	66	63	95	89
Glycine	350	201	208	234
Valine	47	53	41*	tr.
Leucine	31	78	61	2*
Isoleucine	32	40	36	23
Proline	95	39*	60*	—
Phenylalanine	21	29	13	—
Tyrosine	7	12*	11*	—
Serine	42	110	106	228
Threonine	38	49	49	43
Cystine	—	—	—	—
Methionine	—	—	—	—
Arginine	66	40	22	tr.
Histidine	—	22	11*	37*
Lysine	41	35	44	94
Aspartic acid	82	90	86	100
Glutamic acid	83	139	153	128
Hydroxyproline	—	—	—	—

* Approximate determination only.
[a] Foucart *et al.*, 1965a.
[b] Foucart *et al.*, 1965b.

B. Connective tissue

The maintenance of shape in these animals must rest largely on the connective tissue beneath the epidermis. In Enteropheusta the dermis is very well developed and in the region connecting the proboscis to the collar it is specialized to produce the proboscis skeleton, having a median plate and two posterior extensions or horns; this skeleton contains typical collagen (Rudall, 1955). In the gill region the septa and tongue-bars also are supported by a collagenous skeleton formed from thickenings of the dermis of the pharyngeal epithelium. The dermis is also the main structure opposing contraction of the longitudinal muscles of the trunk since circular muscles are lacking in this region. In the proboscis, which is used for burrowing, both circular and longitudinal muscles are present enabling the proboscis, with the assistance of its cilia, to push into the sand and then contract, pulling the rest of the body in after it.

In *Cephalodiscus* the dermis, although generally well developed, does not give rise to specialized skeletal structures in the gill bars or in the dorsal shield, though it is markedly thickened in the dorsal wall of the tentacles. *Rhabdopleura* also lacks any specialized internal skeletal structures and the musculature of the zooid also is poorly developed. The stalk of *Cephalodiscus*, however, is muscular and is used to move the zooid in and out of the tube.

There is considerable doubt whether the diverticulum from the buccal cavity of hemichordates, which Bateson (1885) called the notochord, is in fact homologous with the notochord in the remaining chordates. It is a hollow structure lined with epithelium similar to that of the buccal cavity and as such can have little skeletal function. It is completely different from the solid chord of cells that forms the notochord of other groups where it forms, at the stage when it is present, an important part of the skeletal system.

The eggs which are fertilized externally are only provided with a thin outer membrane which probably represents a fertilization membrane.

II. GRAPTOLITOIDEA

The systematic position of these extinct, colonial animals is undecided. Certain aspects of their organization suggest a relationship with the Rhabdopleurida. The periderm in graptolites has two layers, the inner of which consists of interdigitating, hardened half rings, recalling the structure of the tube of *Rhabdopleura*. But graptolites have, in addition, an outer layer of quite different nature, which is continuous and laminated. Although it is impossible to know with certainty how the periderm was secreted, there are reasons for believing that it was not secreted from a specialized region as in *Rhabdopleura* but from the whole of the ectoderm. Similarities have also been seen between the virgula of graptolites and the "black stolon" of *Rhabdopleura*. Foucart and Jeuniaux (1966) have analysed organic material obtained from three fossil graptolites and found no indication of chitin being present. The amino acid composition resembles slightly that of the coenoecia of *Cephalodiscus*, although its glycine content is lower (Table 15.1). Although Kozlowski (1947) also favoured a relationship between graptolites and the Pterobranchia, Hyman (1959) would reject graptolites as hemichordates and place them, as in the past, with the coelenterates.

III. UROCHORDATA

Urochordata or tunicates are either sedentary or free-living, solitary or colonial, marine animals varying from about 1 mm to 14 cm in length.

A. External skeletal structures

Urochordates are always surrounded by a test or tunic secreted by epidermal cells. These tests have a superficial resemblance to tubes secreted by various animals, but in detail they have features not found in other secreted encasements.

The tunic varies considerably in thickness and texture and may be leathery, cartilaginous, gelatinous, or viscous. The surface may be smooth and clean, or bits of shell and sand may adhere to it. In *Bostrichobranchus molguloides* the surface is covered by a mass of loose fibres that makes it look like cotton wool. *Chelyosoma* has horny plates surrounding the inhalant and exhalant openings and in Pyuridae the tunic is prolonged into spines. In some species the basal part of the tunic forms root-like extensions that anchor the animal into or on to the substratum.

In pelagic forms the tunic is thin. In Larvacea the test completely but loosely encloses the animal and the inhalant opening is protected by a filtering device to prevent large particles entering the alimentary canal. In *Oikopleura* the filter is formed by a lattice of fibres.

At the microscopic level the tunic, which is often laminated, is seen to be composed of a ground substance in which fibres may or may not be visible. In laminated tests, the fibres in one lamina run parallel to the long axis of the body while those in the next run at right angles to this (Schulze, 1863). There is a differentiated outer layer which is generally very thin but in a few species, such as *Cynthia microcosmos* and *Boltenia echinata*, this outer layer is much thicker and it is generally thick over spines, where they occur (Saint-Hilaire, 1931; Godeaux, 1964). In *Pyura stolonifer* it appears to be absent (Endean, 1955).

The tunic of Ascidians is generally penetrated by tubular outgrowths of the epidermis and mesenchyme forming blood containing vessels that ramify through the tunic. Cells of various types pass from the blood, through the walls of the vessels, into the tunic. Although, generally, it is thought that the tunic is secreted by the general body epidermis, in Didemnidae, Millar (1951) found the tunic secreted only by the epidermis of the terminal dilations of the tunic blood vessels and both Das (1936) and Endean (1961) found that cells lying in the tunic contributed material to it. Inorganic spicules also may be present in the tunic. In *Herdmania* Das suggests the calcareous spicules, some of which bridge the junction between the test and the epidermis, may serve to anchor the tunic to the animal.

Hecht (1918) describes the tunic of *Ascidia* as being continually worn away at the surface and replaced from below. He also found that damage resulting from removal of bits of tunic could be made good and Brien (1930) and Das (1936) found the same for *Cavelina* and *Herdmania*. Das found sense cells and nerve cells in the tunic. Stimuli applied to the tunic result in response from the animal. The test is the only part of the animal in direct contact with the outside so that the presence of nervous elements is not surprising. Das draws a comparison between the tunic with its blood supply, wandering cells, fibres, and nerves and vertebrate connective tissue.

Chemically, the test is composed of up to 95 per cent water (Saint-Hilaire, 1930; Endean, 1961; Godeaux, 1964). It has been known for over 100 years that the test of ascidians contains cellulose similar to that in the cell wall of plants. Cellulose also occurs in the tunic of Thaliacea (Salensky, 1892) but not in the tunic of Larvacea (Brien, 1948).

How closely the tunicate cellulose resembles plant cellulose is a matter for argument. It resists solution by acids that readily dissolve plant cellulose (Berthelot, 1858) but this may be due to the highly crystalline state in which the cellulose occurs (Marrinan and Mann, 1956—see Picken, 1960), or to protection by the associated protein. In the large size of its crystals it resembles cellulose from bacteria and *Valonia*. Its unit structure and infra-red spectrum resemble cellulose from higher plants, rather than that from bacteria and *Valonia*. Its molecular weight is higher than that from either higher plants or bacteria.

It was formerly believed that ascidians were unique in the animal kingdom in being able to produce cellulose. However, it has now been found in the fertilization membranes of acanthocephalids, in lepidopteran silks, and in mammalian connective tissue (Hall *et al.*, 1958). Here it occurs in small amounts as a helical configuration round protein fibres which, Hall *et al.* believe, result from the degradation of collagen fibres to form chemically more elastin-like structures. This is very similar to the association between cellulose and protein that Hall and Saxl found in *Ascidiella aspersa*.

The occurrence of cellulose in ascidians and mammals has been put forward as an indication of a relationship between Urochordata and Vertebrata. It was not, however, until 1958, in spite of many years of intensive investigation of mammalian connective tissue, that cellulose was detected. No invertebrate connective tissue has been subjected to the same scrutiny. It would be unwise, as yet, to conclude that cellulose is not a general component of connective tissue in all animals.

Hall and Saxl (1961) found also collagen and elastin in the tunic of *Ascidiella aspersa*, together with a third protein resembling vertebrate elastin in containing 50–60 per cent proline and glycine, only 2 per cent hydroxyproline, and a high concentration of valine. It differs from elastin and resembles collagen, however, in having relatively high concentrations of basic (*c.* 11 per cent) and acidic (9 per cent) amino acids. This third protein is closely associated with the cellulose.

In *Dendrodoa grossularia* Barrington and Thorpe also have evidence for the presence of a polyphenol oxidase capable of oxidizing dopa, and wandering, granular cells containing polyphenol. They suggest, therefore, that the leathery nature of the tunic in this species is due to quinone-tanning of the contained protein, particularly in the outer layer of the tunic which is formed from compacted fibres. The brown colour of the outer layer of *Phallusia mammillata* (Endean, 1961) suggests that quinone-tanning may also be present in this species. The high concentration of iodine found at the surface of *Ciona intestinalis* (Barrington and Barron, 1960) again implies, in this species, that quinone-tanning is present.

Where the matrix of the tunic has been examined, it has been found to contain acid mucopolysaccharides (Endean, 1961; Barrington and Thorpe, 1968).

B. Internal skeletal structures

Connective tissue layers underlie the epidermis, but they show no specialized development. Hertwig (1873) and Salensky (1892) both reported finding cellulose in the connective tissues. Hertwig found it internal to the body-wall musculature and in the connective tissue of the gut of *Cynthia mytiligera* and Salensky found it in *Pyrosoma*. Both longitudinal and circular muscles are present and the sedentary forms are able to contract down, squirting out the contained water, and deforming the tunic material. In Thaliacea and Larvacea the muscles are better developed and the very thin tunics offer no resistance to their contractions.

Although a notochord is lacking in adult urochordates, it is present in the long tail of larval forms. It forms a flexible support to the tail which is the swimming organ. At metamorphosis the tail with its notochord is lost. The notochord is described as composed of a core of vacuolated cells surrounded by a connective tissue sheath. The fine structure of the notochord in this group has not yet been examined but in *Amphioxus* there is reason to believe that the vacuolated appearance of the cell is a fixation artifact so that the same may also be true here.

C. Egg membranes

Eggs of tunicates are generally surrounded, before fertilization, by a chorion secreted by follicle cells. According to Leonardi (1965) the chorion contains protein and polysaccharide. Ascidians are hermaphrodite and the chorion may be responsible for the block to self-fertilization, sperm failing to penetrate the chorion of eggs produced by the same individual (Morgan, 1942). In many species the eggs develop protected in the tunic. Where eggs are shed into the water, the follicle cells, which remain attached to the chorion, may become large and vacuolated and serve to float the egg. The embryonic stages of ascidians may be passed through very rapidly, *Ciona* taking only twenty-four hours to pass from fertilized egg to free-living larva, so that protective membranes are hardly required.

IV. CEPHALOCHORDATA

The cephalochordate, *Amphioxus*, has a strong resemblance to the vertebrates and is, in appearance, very different from hemichordates and adult urochordates. It is about 5 cm long, with the body thin and compressed from side to side. Both ends of the animal are expanded into thin, vertical fins, joined by a shallow fin running along the middle line of the back, and the fin also extends a short distance forward from the hind end on the ventral surface.

The epidermis is only one cell thick and in this resembles invertebrates. The epidermal cells are richly provided with microvilli. In the epidermal cytoplasm there are fibrils which, though they are not abundant enough to give histochemical reactions, appear like fibrils in vertebrate keratin-containing cells. Homogeneous bodies, rather like keratohyaline granules in vertebrate keratin-containing cells, are sometimes seen amongst the fibres. The epidermis is covered by a thin layer of material containing mucoid substances together with proteins containing —SH and —S—S-groups (Olsson, 1961).

A. Notochord

The notochord in *Amphioxus* is a rod-like structure running through the body above the gut and below the nerve cord. It develops from endodermal cells in the roof of the archenteron, which suggest that it is an endodermal structure but the mesoderm which gives rise to endoskeletal structures also arises from endoderm of the archenteron, on either side of the notochord, so that distinctions between the derivations of mesodermal skeletal structures and notochord are of doubtful significance. In *Amphioxus* the notochord is the principal skeletal structure, and not only supports the body but also prevents the contraction of the myotomes from causing a shortening of the body. Instead, contraction of these muscles, first on one side then on the other, brings about a typical piscine undulatory movement that enables the animal to swim.

When first formed, the notochord is a rod of compact cells but these then grow and interlace with each other (Willey, 1894) until each cell comes to occupy the whole width of the notochord, forming vertically arranged, disc-shaped cells, each bounded by an irregular cell membrane. Within the cytoplasm are embedded long, electron-dense fibrils, orientated transversely, but they do not reach to the marginal zone on either side of the cells. At the dorsal and ventral surfaces of the plates is a row of cells, Muller's cells, which contain no fibrils but have a system of intracellular canals. On the dorsal surface the Muller's cells, at intervals, send paired processes down between the plate cells. The "*elastica interna*", a more or less homogeneous membrane surrounds the plate cells and Muller's cells and fills up the irregularities at the surface of the plate cells. It probably represents a basement membrane. Outside the *elastica interna* is the notochord sheath of inner, circular and outer longitudinal fibres (Edkin and Westfall, 1962). That these fibres are collagen is concluded from their periodic banding of 600 Å. Edkin and Westfall could find no trace of the vacuoles which have frequently been described in notochord cells and concluded that the spaces described as vacuoles resulted from the cells shrinking away from each other during fixation.

B. Dermis

The epidermis rests on a two layered basement membrane from which fine processes pass down into the underlying dermis or cutis. The dermis is composed of lamellae, the number of which varies in different parts of the body, from 16 to 40. Each successive lamella, in which the collagen fibres are all parallel, shows a change of 90° in the orientation of the fibres compared with the lamella above (Olsson, 1961). Of the general connective tissue the dominant component is a finely granular ground substance which is somewhat gelatinous in texture and is penetrated by fibres from the dermis, but lacks any cellular components. Locally it is organized to form a definite skeleton supporting the oral tentacles and gill bars, and a series of "fin ray boxes", small blocks of specialized connective tissue that support the median fin. The connective tissue fibrils have a periodic structure of 610 Å and are assumed to be collagen.

The eggs are fertilized externally and a vitelline membrane is secreted by the egg itself. There are no further membranes.

CHAPTER 16

Vertebrata: External Skeletal Structures

The epidermis that forms the covering of vertebrates is never a single layer of cells as it is in invertebrates, but is always stratified. Its thickness depends not only on the class of animal but also on the species and on the site within the species from which the material is taken. A variety of cells occur in vertebrate epidermis. In the fishes the cells more closely resemble the cells of invertebrate epidermis, in amphibia and higher vertebrates the majority of cells produce keratin. Outgrowths of the epidermis in birds give rise to feathers and in mammals to hairs, both of which contain keratin, and these outgrowths considerably influence the properties of the surface of these animals.

The term keratin describes not a single protein but a class of proteins which have certain features in common. These proteins have been defined by Mercer *et al.* (1964) as "proteins produced in epidermal cells and usually retained within the cell. They are insoluble in the usual protein solvents due to the presence of numerous disulphur bonds between the protein chains. Keratins may be predominantly filamentous, predominantly amorphous or a mixture of filamentous and amorphous." Mercer (1961) has reviewed the whole field of keratin physics, chemistry, and development within the cell. Production of keratin generally leads to the ultimate death of the cell and these dead, keratin-containing cells form at the surface of the animal a tough and sometimes horny layer. It would be wrong, however, to think that the physical properties of this outer layer depend only on the physical properties of the keratin. Since keratin is contained within cells and does not form a continuous system through the whole structure, the cell membranes and the forces holding the cells together also are of importance in determining the properties of the epidermis. The mechanical properties of these structures will not, therefore, be considered until their formation and cellular organization has been described.

Indication that keratin has been formed in a cell rests upon the dead appearance of the cell, the nucleus and cell organelles having disappeared or being represented only by traces of cytoplasmic débris. It is possible to test, by histochemical means, for the presence of disulphur groups within the cellular contents. Often the disulphur bonds are associated with fibrils that give a distinct X-ray picture of either α- or β-type fibrous protein. But not all the fibrils within the cell are necessarily keratin, just as keratin within the cell is not necessarily fibrous. Keratin forms the basis of the diet in the clothes moth, in *Anthrenus* and other dermestid beetles, and in Mallophaga. In the midgut of these forms, there is a strong reducing agent which at the high pH of these guts, breaks the disulphur bonds and the protein is then digested by keratinase; however, this enzyme-digestion has not yet been used for histochemical identification of keratin.

In all vertebrates, the epidermis can be divided into three regions. There is the basal, germinal layer of columnar cells that divide to supply cells to the layers above, which are therefore gradually pushed outwards. Above is an intermediate layer in which cell differentiation takes place. The outer layer is generally composed of greatly flattened cells which have achieved their final, differentiated form and later, having served for a while to protect the surface of the animal, are finally lost from the surface, either in small groups or as a continuous layer, in a process of moulting, to be replaced by cells below, so that the surface is kept in good repair.

Blood vessels do not occur in the epidermis—a reasonable precaution against loss of blood from lightly abraded surfaces. This means that the epidermis must obtain all its requirements for nutrition and respiration by diffusion from the blood vessels in the dermis or, in the case of oxygen, by diffusion inwards from the surface.

A. Agnatha

The epidermis of cyclostomes is made up of four or five layers of cells, most of which differentiate into mucus cells. In the lamprey the outermost layer of cells, besides containing mucus, may also be ciliated. There are besides two other types of cell, club cells which are in the inner layers of the epidermis, and granular cells which are amongst the outer cells. Pfeiffer and Pletcher (1964) believe club cells are developmental stages of granular cells. These, they suggest, secrete the substance that makes the lamprey so distasteful to potential predators.

In *Myxine* small and large mucus cells, thread cells, and sensory cells differentiate from the epidermal basal layer. The basal cells have, round the periphery of the cell, what light microscopists have described as a "capsule" separating ectoplasm from endoplasm. Electron microscopy has shown this capsule to be formed from a concentration of fine filaments, some of which establish contact with desmosomes on the cell membranes. Although Rudall (1947) obtained an α-type protein X-ray picture from the epidermis of *Myxine*, there is no evidence for the filaments containing disulphur bonds. While keratin fibrils may have evolved from such fibrils, the presence of fibrils in epidermal cells does not necessarily indicate that keratin is present.

The small mucus cells make up the bulk of the cells at the epidermal surface. They discharge through a series of short canals that open between microvilli (Blackstad, 1963). Possibly a similar arrangement accounts for what Pfeiffer and Plecher (1964) describe as a striped cuticle on the surface of the lamprey ammocoete epidermis. From the evidence of dyeing reactions, the mucus in the large mucus cells is different from that in the small mucus cells. The large mucus cells, when they have grown to maximal size, reach the surface of the epidermis and are expelled. The thread cells contain a spirally coiled thread, which is released into the water to become entangled in the mucus.

Both *Myxine* and *Polistrotrema* have, in addition, special multicellular slime glands; in *Polistrotrema* there are two per segment, in the ventro-lateral position. Each gland is a flask-shaped invagination of epidermal cells surrounded by dermis, and muscles whose contractions expel the secretions of the gland. Within each gland are many thousands of cells, each containing a coiled thread (Newby, 1946). The whole thread-cell is expelled into the water where it liberates its thread, which is about 1·3 μm broad and several centimeters long, Threads from different cells become entangled to form a fibrous network which absorbs water, forming a slimy mass. The threads are protein but chemically, and from X-ray diffraction photographs, this protein differs from any other known protein; Ferry (1941) proposes the name *mitin* (μίτος = a thread) for it. In *Myxine*, mucus cells also occur in the slime gland (Blackstad, 1963). Their contents differ in dyeing reactions from those of the large and small mucus cells in the general epidermis. No specific function has been found for the great amount of mucus that is produced.

Lampreys and hag-fish have a circular mouth, surrounded by horny teeth, and a rasping tongue which is also covered with similar structures. The teeth are formed by the epidermis and replaced from underneath when worn away. Although no keratin is found in the general epidermis, it has been found to occur in these horny teeth (Barrnett and Sognnaes, 1955; 1960; Dawson, 1963).

B. Elasmobranchii

No full description of the epidermis of elasmobranchs has yet been given. The body surface in these fish is effectively covered by closely apposed denticles. Although the outermost layer of the denticle is formed by the epidermis, it develops mainly from the dermis and will be considered below (p. 312) with other dermal scales. The denticles protrude above the general surface of the epidermis which is thin and supplied with mucus cells. The epidermis also gives rise to the supposedly poison-producing cells at the base of the sting in sting-rays, and in some deep sea forms to light-producing organs.

The denticles provide a tough surface offering little attraction to the larger ectoparasites, to the attacks of which other fish appear more prone. Denticles not only provide a limited protection from predators, but may themselves be used offensively for raking the bodies of animals that fall victims of elasmobranch attacks. *Cephaloscyllium uter* gulps water to swell the body and erect the denticles to anchor it in crevices in the rocks. Modified denticles cover sensory pits or together with the gills form a food-filtering apparatus. In the males of most species they also form copulatory spines. Around the mouth they form teeth and they also contribute the teeth to the saw-like rostrum of the saw-fish. The function of some of the specialized spines and projections is not known (Applegate, 1967). The presence of denticles at the surface of elasmobranchs may have reduced the requirement for elasmobranchs to evolve a thick and complex epidermis.

C. Actinopterygii and Sarcopterygii

Amongst the bony fishes, only the epidermis of teleosts and of lung-fish has been examined in any detail. Henrikson and Matoltsy (1968a) claim that teleost epidermis differs from that of the higher vertebrates in that even cells above the basal layer undergo cell division.

Teleost epidermis is largely made up of cells containing filaments round the periphery of the cell. The cells in the outer layer are considerably flattened, although not so much as in the outer layers of higher vertebrates. In the guppy extensive microvilli or surface pleats are often formed at the outer cell surface. These microvilli are covered by a finely fibrillar substance or "fuzz" (Henrikson and Matoltsy, 1968a). Rudall (1947) obtained α-protein type X-ray diffraction diagrams from teleost fish skins but Burgess (1956) could detect no sulphydryl groups in the epidermis of any teleost. Keratin is produced locally in the epidermal nuptual tubercles of *Carassius auratus* (Gaupner and Fischer, 1933) and Harms (1929) claims to have found it in the epidermis at the base of the ventral fins of the mud-skipper, *Periophthalmus*. Mucus cells are always present, together with a variety of other cells, the function of which is not certain (Mullinger, 1964; Welling *et al.*, 1967; Henrikson and Matoltsy, 1968a).

Mucus, which is such a regular product of all fish epidermis, plays an important part in the wellbeing of these animals. Not only does it serve to reduce friction between the body and the various surfaces with which the body is in contact, but it also helps to keep the body surface clean (grooming as carried out by birds and mammals being impossible in fish). Small quantities of nitrogen waste may be excreted in the mucus. In some fish living in muddy water, the mucus precipitates mud and maintains clear water around the gills. Some chemical pollutants of rivers unfortunately form insoluble compounds with the mucus; these, deposited on the body, particularly over the gills, lead to suffocation and death of the fish (Van Oosten, 1957). Sticklebacks use mucus to line their nests, where it hardens like mortar. Fighting fish, *Betta*, and paradise fish, *Macropodus*, attach their eggs to rafts of bubbles trapped in mucus. The young of *Symphysodon discus* feed on mucus produced over large areas of the body of both parents (Hildemann, 1959).

Goldfish do better in water previously inhabited by other goldfish than in clear water, and the secreted mucus is believed to contribute some growth substance. Amongst species of carp, this effect is greater in mirror and leathery carp than in the scaled species, in which mucus glands are not so well developed (Van Oosten, 1957).

In eels special slime-producing cells are packed with coiled threads (Henrikson and Matoltsy, 1968b) that recall the thread cells of *Myxine*. This slime is important in controlling the water relations of the animal. If, in the freshwater phase, the slime is scraped away, more water enters the body osmotically than the kidneys can secrete, so that the animal becomes waterlogged and dies (Baldwin, 1937). It also serves to protect the body, both mechanically and from water loss, during its travels over dry land. In the mud-skipper, also, mucus serves the same purpose when the fish is on land; in both eels and mud-skippers it probably maintains the skin in a fit condition for respiratory exchanges when they are out of water.

Catfish and goldfish have "club cells" filled with fine fibrils. Henrikson and Matoltsy (1968c) suggest that they produce the warning substance which wounded fish are believed to secrete into the water, causing similar and related species in the neighbourhood to swim rapidly away and retreat into a hiding place. These species also contain, in their epidermis, cells remarkable for their high content of smooth-surfaced endoplasmic reticulum, and resembling cells in the gills of *Fundulus* that Philpott (1963) and Philpott and Copeland (1965) believe to be responsible for the transport of electrolytes through the skin (Henrikson and Matoltsy, 1968c).

The lung-fish, *Protopterus*, produces large quantities of mucus which it uses to line burrows in the mud, in which it survives periods of drought. Prior to aestivation, the epidermis is five to six layers thick and not only special goblet cells, but also all the cells of the epidermis above the basal layer are concerned in the production of mucus (Kitzan and Sweeny, 1968). The goblet cells become very large and some extend through the whole depth of the epidermis. At the end of aestivation the goblet cells are small and are separated from the surface by a layer of small epidermal cells (Smith and Coates, 1936). Kitzan and Sweeny found filaments concentrated round the periphery of the cells in the middle region of the epidermis. They did not test for the presence of keratin. Such tests would have been of great interest as it is from ancestors of the lung-fish that amphibia, which are the first vertebrates to produce keratin in the general epidermis, are believed to have evolved.

Mucus plays an important part in the protection of the surface in these aquatic vertebrates and is used for many other purposes as well, but it cannot perform these functions in terrestrial animals because the properties of mucus are dependent upon its high content of water; in land-living forms this high water content could not be maintained. Terrestrial vertebrates, therefore, needed to evolve some material functionally to replace the mucus, and the material that they evolved was keratin.

D. Amphibia

The majority of amphibians start life as aquatic larvae. In both anurans and urodeles the larval epidermis is a layer only two cells deep. By the time the larva of *Rana pipiens* is 8 mm long, tonofibrils associated with desmosomes make their appearance in the basal layer of epidermal cells, concentrated round the periphery of the cell. By the time it is 24 mm long, the outer layer of cells have short, stumpy microvilli and beneath the microvilli numerous secretory granules are embedded in a fibrillar condensation (Leeson and Threadgold, 1961).

In the epidermis of larval newts, five different cell types have been described. Both the basal and outer cells contain tonofilaments. The outer cells are occasionally ciliated and have mucus vesicles in the distal portion of the cell. Surrounded by outer, and basal cells and nowhere reaching the free surface, are large, clear cells containing a network of tonofilaments, the Leydig cells, of unknown function. Mucus goblet cells are confined to the anterior head region. Amongst the outer cells are a few with a high content of rough endoplasmic reticulum and densely staining, membrane-bound granules, the function and fate of which are not known (Kelly, 1966). Although tonofilaments, in places aggregated into fibrils, are a constant component of the larval epidermis, keratin is thought not to be present. However, the horny teeth of tadpoles, each formed from a single epidermal cell shaped like a cap, are said to contain keratin. These teeth are shed from time to time and replaced from beneath, in this resembling the teeth of Agnatha. Horny structures may also occur at the digit tips of various larval forms, as in mountain-brook species of hynobiids and in ambystomiids and plethodontids. In *Onychodactylus* larvae sharp claws are developed.

The epidermis of adult, land-living *Rana pipiens* is six to seven layers of cells thick. Cells of the basal layer contain tonofilaments with a moderate content of sulphydryl groups. In the mid-part of the epidermis all the cells are concerned with the production, storage, and secretion of mucus, the mucus being liberated into the intercellular spaces where it forms small globules. Cells of the outer layer are dead and almost entirely filled with fibrils, nuclear fragments, and other cell debris; globules of mucus sometimes occur within the fibril network. The fibrils give an intense positive reaction for disulphur groups (Spearman, 1968) and this, together with the α-protein type X-ray photograph obtained from amphibian epidermis by Rudall (1947) indicates the presence of a fibrous keratin. No keratohyalin granules, which occur in keratin-forming cells in mammals, have been seen in amphibian material. At the stage in development at which keratohyalin granules occur in mammals, amphibian epidermal cells develop globules of mucus (Parakkal and Matoltsy, 1964). The development of keratin in the general epidermis is closely correlated with the development of terrestrial life. *Necturus*, which remains aquatic in adult life, develops keratin only in the epidermis of the hands and feet; *Siredon pisiformis* develops it also at the end of the snout (Dawson, 1920).

In many parts of the body, especially in anurans, epidermal cells proliferate locally to form hard spikes or warts. Generally they form over underlying dermal papillae. Similar to these, but of temporary occurrence, are nuptual excrescences such as those often developed in male anurans on the thumb and the under surface of the fingers and toes, and used in holding the females while mating.

The tips of the fingers and toes in many anurans, but less frequently in urodeles, have horny sheaths forming claws or nails. In *Pelobates* a metatarsal tubercle develops a thick, horny cover to form a digging spur or spade, with which it burrows in the sand.

Although they have not been described in electron microscope studies, light microscope studies of amphibian skin always show epidermal, multi-cellular glands. The more generally distributed glands secrete mucus while glands restricted to certain parts of the body, for instance the parotid glands of anurans, may produce poisons, some of considerable potency. In Colombia, poison from *Dendrobates tincterius* is used to tip arrows with which the natives shoot monkeys (Gadow, 1901). Many poisonous species exhibit warning colouring, the pigment cells being located in the dermis, but granules of pigment may also occur in the epidermis.

Associated with the development of an outer layer of dead, keratin-containing cells, is a habit of periodically casting off or moulting the layer of dead cells. The first moult generally occurs towards the end of metamorphosis, preparatory to life on dry land. As long as the animal is growing rapidly, moults occur frequently. Full grown urodeles do not moult often, mostly only when they take to the water in the breeding season. *Necturus*, although its epidermis lacks keratin, has been observed to moult from time to time (Dawson, 1920), particularly if the animal has been much disturbed or the skin is allowed to become dry. Anura moult more frequently, at least every few months. After moulting, the new skin is at first quite wet and soft, but it soon becomes dry and hard. The cast skins of amphibians are rarely found, as usually they are immediately eaten (Gadow, 1901).

Moulting is under the control of the thyroid gland. If this is removed, in the majority of amphibians the old skin is not cast, although the basal layer of the epidermis continues to divide and produce keratinized cells, resulting in a greatly thickened epidermis. In some anurans (e.g. *Bufo*), sloughing of the skin continues after thyroidectomy and here it is believed that adreno-cortical hormones control moulting (Jørgensen and Larsen, 1960).

The layer of dead cells containing keratin, at the surface of terrestrial vertebrates, is said to water-proof them against water-loss and the capacity to produce keratin is believed to have played an important part in the evolution of life on dry land. However, not only in aquatic forms like *Necturus* but also in fully terrestrial forms like *Bufo*, water can pass readily through the skin in either direction, and under dry terrestrial conditions the major route for water-loss is through the skin, water evaporating from it at a rate comparable with evaporation from a free-water surface (see Deyrup, 1964, for references). Cohen (1952) and others found a lower permeability in skins of terrestrial amphibians, although Thorson (1955) failed to find, in the five species with which he worked, that terrestrial forms had less permeable skins, and similar results were found by Littleford *et al*. (1947). Amphibians can, however, tolerate a loss of water between 7 and 60 per cent of the body weight, and it has been shown by Thorson (1955) and Ray (1958) that those amphibians which have become most completely adapted to terrestrial life can survive dehydration longer, and tolerate larger losses of water, than can aquatic or semi-aquatic forms. Desiccated amphibians placed in contact with water rehydrate very rapidly. Water is taken in only through the skin, for drinking of water does not occur (Deyrap, 1964). The epidermis is also an important respiratory surface.

E. Reptilia

Reptiles are more fully adapted to life on land than are amphibians, and these adaptations are assumed to involve the structure and physiological properties of the epidermis. This is dryer and harder than it is in amphibians.

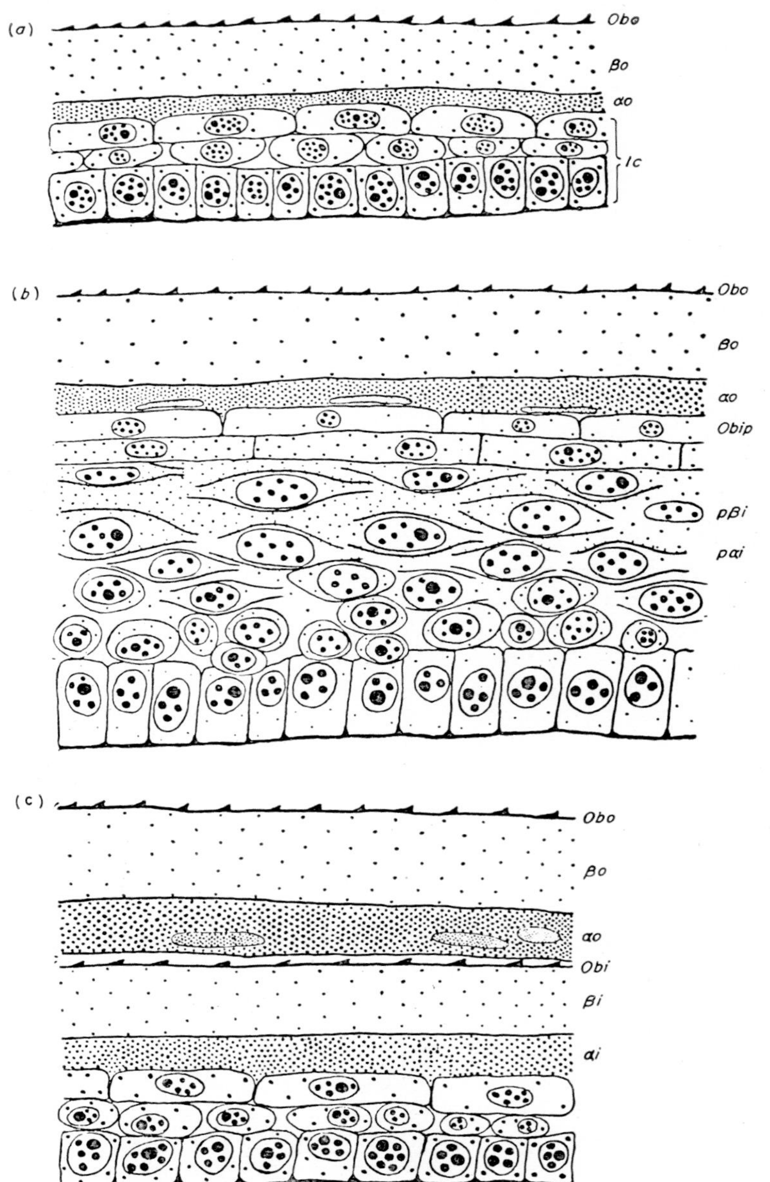

Fig. 16.1. (*a*) Diagrammatic drawing of the epidermis of a snake immediately after moulting. (*b*) Diagrammatic drawing of the epidermis of a snake at the beginning of moulting. Living cells beneath the scale divide to produce the cells from which the new scale will be formed. (*c*) The new scale formed beneath the old scale which has not yet been moulted. *αi*, α-keratin layer of new scale; *αo*, α-keratin layer of old scale; *βi*, β-keratin layer of new scale of new scale; *βo*, β-keratin layer of old scale; *clo*, layer of clear cells that forms between the old and new scale; *lc*, living cells of epidermis; *obi*, Oberhautchen layer of new scale. obip, presumptive Oberhautchen layer; *obo*, Oberhautchen layer of old scale; *pαi*, presumptive α-keratin layer; *pβi*, presumptive β-keratin layer. From Maderson (1965).

In most species it is organized into hard areas or scales separated by more flexible regions. In most snakes the scales have an overlapping arrangement. In some lizards they are reduced to small tubercles and in Amphisbaenidae they have practically disappeared. Below the epidermal scales the dermis shows corresponding thickenings and it is these that give the familiar pattern to crocodile leather in the preparation of which the epidermis is removed.

Crocodile epidermal scales contain a mixture of keratin in the α- and the β-form (Rudall, 1947) and disulphide groups are evenly distributed through the scales. Crocodiles do not undergo periodic moultings, but the scales continually flake away at the surface and are added to from below. Snake and lizard scales have a well-marked outer layer containing β-keratin and an inner layer containing α-keratin (Rudall, 1947) and the two layers are separated from each other by an intermediate layer of greatly flattened cells, the *mesos* layer devoid of fibrils but containing bodies of unknown chemical composition that appear dense in the electron microscope (Maderson, 1965a, b; Roth and Jones, 1967). The outer layer is bounded by the "Oberhautchen" layer of cells in which the outer cell walls have been formed into small corrugations or spines (Fig. 16.1(*a*)). The presence of β-keratin in the scales must make them virtually inextensible and probably contributes to their hardness. There are relatively more disulphur groups in the outer than in the inner layers of the scales. Each scale is joined to adjacent scales by regions of much thinner epidermis in which only the Oberhautchen and the inner α-keratin-containing layers are present. This connecting membrane is often corrugated and the underlying dermis reproduces these corrugations. This makes the skin flexible and allows for local body expansion such as is required to accommodate the periodic intake, for instance, by some snakes of relatively enormous volumes of food. These flexible regions between hardened areas of the surface resemble the arthrodial membranes of arthropods. Snakes and lizards moult periodically, snakes usually in one piece as a complete cast-skin, lizards generally in pieces.

Although the outer layer remains unaltered during the intermoult period, the inner layer is added to from the underlying layer of living cells. Prior to moulting, a new outer and inner layer forms below the old hardened layers and at moulting separation occurs between the new Oberhautchen layer and a layer of clear cells beneath the old inner layer of the scales. The projections of the Oberhautchen layer fit into pockets formed by the cell membranes of the clear cells. At moulting, the old hard layer becomes detached along the boundary between the clear cells and the new Oberhautchen layer (Fig. 16.1(*b*) and (*c*)).

Lizards are able to climb up smooth, perpendicular surfaces by means of digital lamellae which adhere, possibly by suction, to surfaces against which they are pressed. These lamellae are formed from specializations of the Oberhautchen layer and are moulted with the rest of the old epidermis. The whole process of growth and moulting of these structures has been described by Ernst and Ruibal (1966).

In tortoises the epidermis covering the dermal shell is not shed. Layers of keratin-containing cells, each rather broader than the previous layer to allow for the growth of the body, are continually added to the lower surface of the epidermal scales to build up the characteristic protruding scales with sloping, laminated sides, that cover over the dermal shell and form the tortoise-shell of commerce.

Spearman (1960) says that keratohyalin granules are not found in reptiles, although various workers have reported finding them (see Maderson, 1966 for references). Maderson did not find them in the snake, *Elaphe*, or in *Gecko*, nor did Ernst and Ruibal in *Anolis*, although in the boa constrictor, Roth (see Maderson, 1966) found that they make a very fleeting appearance.

Another feature of reptilian epidermis is the lack of epidermal glands. They are completely lacking in lizards and snakes. Crocodiles have two pairs of multicellular glands, both producing musk, one pair in the neck region, the other at the cloaca. In crocodiles, each scale of the sides, belly, and tail has a small central pit in which the epidermal cells remain unhardened. At the bottom of the pit is a nerve-ending and a sensory corpuscle.

The epidermis of reptiles is much less permeable to water than is that of amphibians. This is indicated by the fact that in a dry atmosphere the body temperature of amphibians falls, due to evaporation of water from the body surface, while the temperature of reptiles remains steady. Pettus (1958), using isolated pieces of epidermis, found negligible water loss through them. However, Bentley and Schmidt-Neilson (1966) found that cutaneous evaporation was the major source of water-loss in all the reptiles they examined. There was, nevertheless, a clear correlation between the loss of water from a particular species at a given temperature and the dryness of its normal environment. At 23°C total evaporation from the desert lizard, *Sauromalus obesus*, was only 5 per cent of that lost from the crocodile, *Caimen seterops*.

Ditmars (1933) and Bentley and Blumer (1962) observed the capacity of some reptiles to absorb water through the skin, which mops it up like blotting paper. It is possible to watch water travelling through the skin in *Moloch horridus*, when placed in contact with water. When it reaches the mouth the animal sucks it in and swallows it. Movement through the skin seems to be by capillarity along fine channels in the horny layer, but it is not possible, from the published photograph, to determine in which region of the epidermis these channels occur.

F. Aves

The epidermis of birds is characterized by its capacity to produce feathers, a capacity it shares with no other class of animals. The epidermis under the feathers is generally thin and more delicate than that in other vertebrates, though in areas not covered by feathers, it is thick; it may be scaled like reptile skin over the unfeathered regions of the legs and it is hardened where it forms the beak.

The general epidermis, and the epidermis between the toes of web-footed forms, and that from bare regions of the body such as the neck of guinea-fowl, contains α-keratin. In scales, claws, and beaks it is in the β-form and this may account for its hardness; softer parts of the beak may give an α-type picture.

Keratohyalin granules are said not to occur in bird epidermis (Spearman, 1966), although Fitton-Jackson (see Fell, 1964) has described, in the chick embryo, cytoplasmic granules that participate in making the hard keratin of the beak and metatarsal scales. They do not occur in the epidermis of the back and scalp.

1. *Feathers*

Feathers are produced in follicles arranged in definite tracts separated by areas of naked skin. The follicles make their appearance as small papilale (Fig. 16.2) early in development, in the common fowl towards the end of

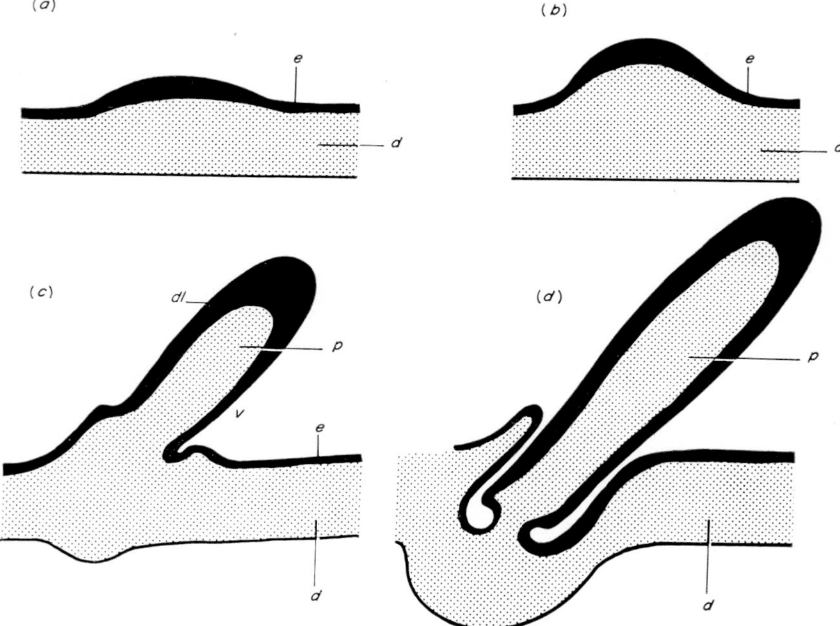

Fig. 16.2. Diagrams of longitudinal sections of a feather follicle in different stages of development. *e*, epidermis; *d*, dermis; *p*, feather pulp; *dl*, dorsal; *v*, ventral.

the first week of incubation, later the follicle sinks down into the skin. By the twelfth day the various tracts are well defined. With minor exceptions, new follicles are not added after hatching.

During development a bird generally passes through a succession of three plumage types. When first hatched they are generally covered by down feathers. These are simple, fluffy feathers suited to provide thermal insulation for the nestlings. Each consists of a number of fine filaments, or barbs beset with barbules, and attached to a basal calamus embedded in the follicle (Fig. 16.3(*c*)).

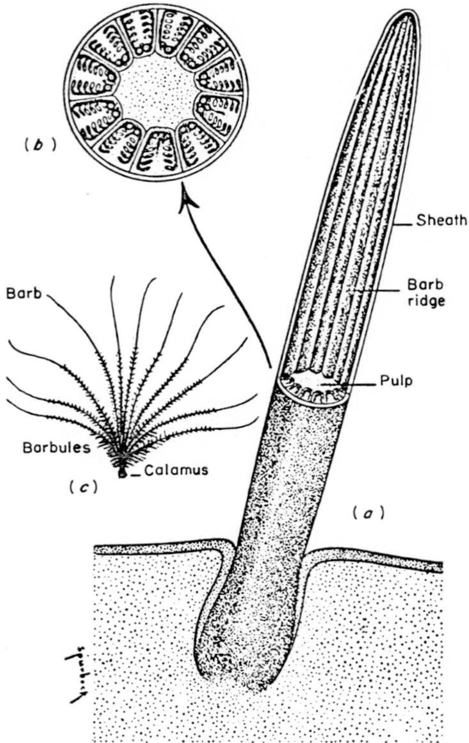

Fig. 16.3. Diagrams showing structure of down feather. (*a*) Feather, near time of hatching, sunken beneath surface of skin in tubular follicle lined with epidermis. Dermal papilla now located permanently at base of follicle. Epidermal walls of feather cylinder are divided into a series of longitudinal ridges—primordia of barbs and barbules—surrounding central pulp, and protected by external sheath. (*b*) Transverse section showing eleven barb ridges, central pulp, and external sheath. (*c*) Completed down feather of newly hatched chick. Sheath split, barbs and barbules released. Note absence of shaft, and circular arrangement of barbs around short calamus. (Rawles, 1960.)

These down feathers are soon displaced by feathers of the juvenile plumage that form beneath them in follicles. Juvenile feathers are more complex than down feathers and show regional variations in structure. Some down-like feathers occur in the juvenile plumage together with contour feathers which provide the body with a smooth surface, and on the wings provide the large wing surface area necessary for flight. In the Australian bush turkey replacement of the down feathers by the juvenile plumage takes place before hatching so that the freshly hatched bird is able to fly. The juvenile plumage is gradually replaced by the adult plumage which differs from it in colour and some structural details.

In contour feathers there develops, beyond the calamus, an elongated rachis with barbs on either side forming the vane (Fig. 16.4(*a*)). The barbs

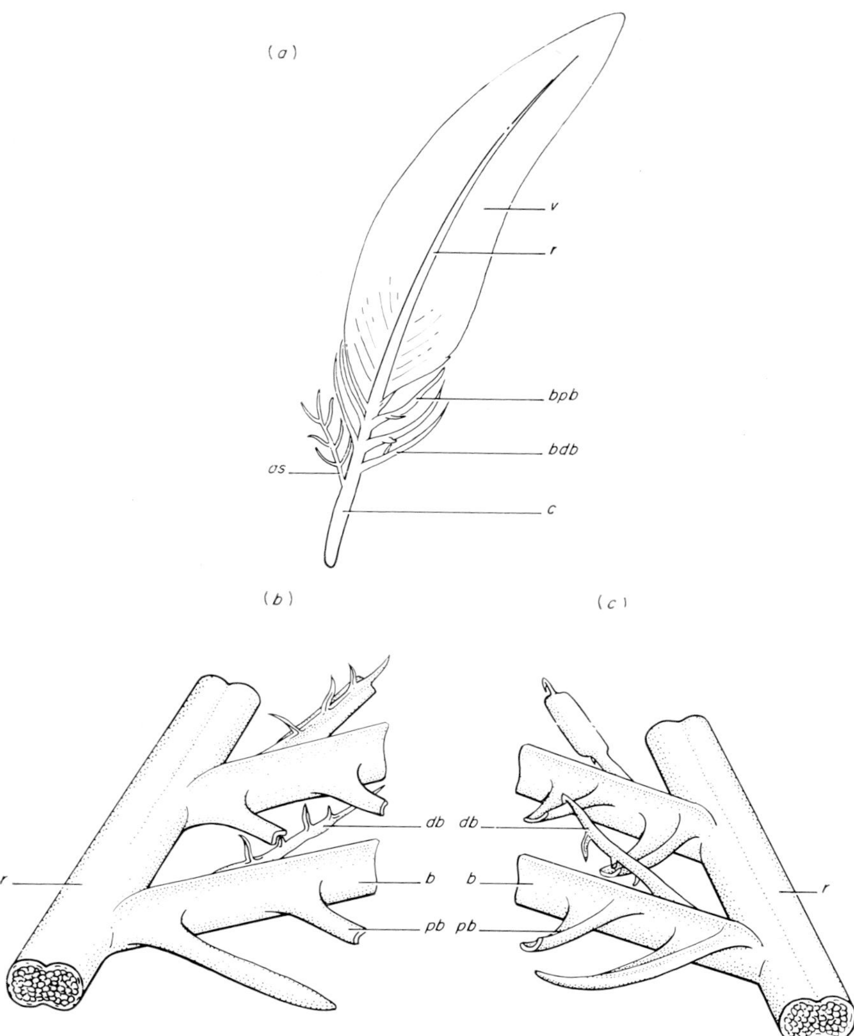

Fig. 16.4. (*a*) Diagrammatic representation of the principal parts of a generalized feather, seen from the ventral side, to show the essentials of their relations to one another. Three barbules have been shown much exaggerated in size; two of them link the two most proximal barbs. *as*, aftershaft; *bdb*, barb showing distal barbule; *bpb*, barb showing proximal barbule; *c*, calamus of quill; *r*, rachis; *v*, one side of the vane. (*b*) Enlarged diagrammatic view from the ventral side of a length of rachis in a generalized feather, with two barbs attached to it. Three of the proximal barbules have been cut off short so as to show their curled structure. The upper, distal barbule has a flattened and expanded dorsal surface. For clarity, the barbules have been drawn much more widely spaced than they are in life. *b*, barb; *db*, distal barbule; *pb*, proximal barbule; *r*, rachis. (*c*) A piece of feather similar to that shown in Figure 16.4(b), viewed from the dorsal side. *b*, barb; *db*, distal barbule; *pb*, proximal barbule; *r*, rachis. ('Espinasse, 1964.)

are held together by rows of barbules almost at right angles to the barbs and carrying, on their anterior surfaces, hooks that hook onto ridges on the posterior surfaces of the adjacent barbules (Fig. 16.4(*b*) and (*c*)). These connections can be broken but stroking the feathers, as in preening, serves to re-establish them. This interlocking device occurs only on the exposed region of the vane. Where the base of the vane is overlaid by other feathers, the barbules do not interlock and the feather is fluffy.

Contour feathers may develop on the ventral surface where the calamus joins the rachis, an after feather. In some birds, such as the emu, the after feather may equal the main feather in size and shape, although in others it is much smaller, or it may be absent as in the flight and tail feathers of the domestic fowl.

The distribution of plumules, the downy feathers of juvenile and adult plumage, is very variable. Usually concealed below the contour feathers, in water birds they form a thick undercover beneath these feathers, and by trapping air in this region provide thermal insulation and probably prevent water reaching the skin surface. They are lacking in Ratitae. In some species, for instance herons and bitterns, the tips of the plumules disintegrate to a very fine powder of unknown function (Rawles, 1960). Some filoplumes, thread-like feathers with a long thin rachis and only a few barbs at the far end, occur in all birds except the Ratitae. All types of feathers may be modified for ornament or display.

The primordium of the down feather first appears as a slightly raised, local aggregation of dermal cells covered by a somewhat thickened layer of epidermis. This papilla at first grows outwards but unequal growth on the two sides of the feather primordium causes the apex to be directed backwards and ultimately to lie almost parallel to the surface of the skin. By continued growth round the base of the primordium it sinks below the surface of the skin enclosed in a follicle. Growth continues at the base of the follicle to produce an elongated cone of fibril-containing cells which comes to project beyond the surface-opening of the follicle. It is from this cone of cells that the feather develops. The dermal papilla sends a core of pulp, containing blood vessels, into the developing feather, to supply it with food and oxygen.

The outermost layer of cells of the developing down feather differentiates as a protective sheath, and the innermost layer, the cylinder cell layer, also differentiates, separating the feather from the pulp cavity. The cells between the sheath and the cylinder cell layer develop into the feather proper. Longitudinal ridges appear in this middle layer of cells the feather barbs develop between the ridges while the barbules develop at right angles to the barbs up the sides of the ridges (Fig. 16.3(*a*) and (*b*)). In the completed feather the barbs do not reach to the base of the cone which remains as an intact smooth cylinder within the follicle, forming the calamus ('Espinasse, 1939). As the feather becomes fully differentiated, disulphide groups are detectable in the cells and the structure becomes hardened. At the completion of growth, the intercellular material of the pulp becomes resorbed but pulp cells and blood cells remain enclosed by the cylinder cell layer and become cut off from the dermal papilla projecting into the calamus by the formation of a pulp cap. When the bird hatches, the feathers dry, and the sheath ruptures liberating the barbs; the strand of pulp material enclosed in the cylinder cell layer is broken away (Watterson, 1942.)

The initial development of the contour feathers that replace the down feathers in the follicle is very similar, with a cone of epidermal cells differentiating into an outer sheath, a layer from which the feather develops, and an inner layer of cells separating the feather from the pulp. However, the rachis differentiates as a large "dorsal" longitudinal ridge, while the barbs with their associated barbules form on either side of this, as ridges sloping upwards, round the cylinder, to meet on the opposite "ventral" side (Fig. 16.5). As

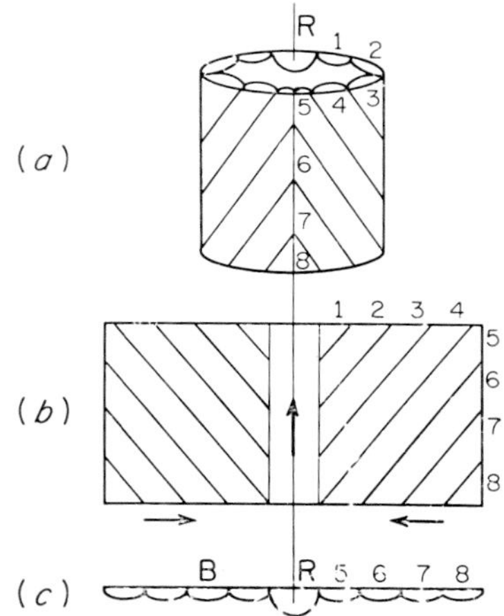

Fig. 16.5. Illustrating rachis and barb formation in the contour feather follicle. (*a*) A short length of the feather cylinder; (*b*) the cylinder opened out; (*c*) a section cut across (*b*). R rachis; 1–8 barbs. (Mercer, 1961.)

the feather hardens and the pulp is withdrawn it leaves behind little caps of inner membrane material which break away when the feather opens out along the ventral meridian, but the last one remains to form a cap over the pulp in the calamus (Hosker, 1936).

Feathers are made up of three morphologically distinct cell-types. On the outside is the cuticle, a single layer of cells, solid and flattened. Beneath is the cortex of spindle-shaped cells arranged with their long axes parallel to the long axis of the feather component, although in the calamus and the rachis the outermost layer of cells are orientated at right angles (Astbury and Bell, 1939; Earland *et al.*, 1962). This layer of transverse cells protect the feathers, particularly the flight feathers which are subjected to a certain amount of bending, from splitting longitudinally. The medulla, at the centre, is a loose tissue of largish cells containing a lower concentration of disulphide groups than the cortex. Within the cortex the keratin appears as microfibrils about 30 Å in diameter embedded in a densely staining matrix (Filshie and Rogers, 1962).

Although the cytology of keratin production in the various parts of the feather and its inner and outer sheaths shows no marked difference, the keratin in both the sheaths is in the α-form while that in the feather itself is in a condition closely related to the extended β-form, although from its spacings it appears not to be fully extended (Rudall, 1947).

The colours of bird plumage result either from pigment deposited in the feathers or from differential diffractions and reflections caused by the structure of the feather and the optical properties of the keratin. The colours are used in camouflage, for recognition, and extensively in the development of special plumage in sexual display. The development and use of colour in the life of birds is associated with the development of good colour vision in these animals, which also have good visual acuity and sensitivity to movement correlated with their speed of movement in flight.

Feathers produce a durable, but lightweight, water-repellant covering. They also provide highly efficient thermal insulation enclosing an air jacket round the body and aiding the body in maintaining the high body temperature that allows the rapid functioning of the muscles necessary for flight. This must also be aided by the reduction, through the feather coverage, of evaporation of water from the body-surface with its consequent cooling effect. The evolution of feathers must have preceded, and then made possible, the evolution of flight. The larger feathers can be moved by muscles at their base, providing for the control of heat loss, the adjustment of their position during flight, and their manipulation during sexual display. Nerve endings round the follicles allow them to be used as sensory organs.

2. *Moulting*

Moulting of feathers, which is under hormonal control, not only allows for the replacement of down feathers, by first the juvenile and then the adult plumage, but is continued throughout life. It occurs at least once a year, generally after the breeding season. In some there may be one or more partial moults in addition, as in the development of special plumage in some males during the breeding season. Moulting is generally a gradual and orderly process with a relationship maintained between feather loss and gain, so that birds are rarely deprived of their power of flight or the protection and thermal insulation that the feathers provide. In some water birds, however, all the large flight feathers are shed together, leaving them for a short time, relatively helpless. Bills, claws, and the general body epidermis are continually shed in small flakes, although scales on the legs are shed periodically as in snakes. Some periodic changes in bird plumage, as in the indigo bunting and the purple finch, result not from moulting but from loss, through wear, of the tips of the feathers, exposing areas not previously seen (Rawles, 1960).

3. *Gizzard lining*

The gizzard of birds is lined by a thick, tough membrane composed of cystine-containing protein, and mucoid material (Broussy, 1932) which protects the cell-surfaces from wear and tear. The protein is accepted as a keratin, although the lining is not intracellular but is secreted by the underlying cells. Droplets pass out between short, stubby microvilli on the luminal surface of the cells and break up into small granules which then appear to unfold into numerous short, fine filaments. These associate to form first, thick tapering rodlets, and then the horny membrane (Mercer, 1961). The components are too irregularly orientated for the membrane to produce a good X-ray photograph but the spacings of the three halos produced suggest that the protein is in the β-form (Longley, 1950).

4. *Uropygeal glands*

The uropygeal gland, and a few glandular cells said to be present near the ear passages of some gallinaceous birds, are the only epidermis glands in this class. Chemically, the product of the uropygial gland varies from species to species. It contains lipid material, protein, and inorganic salts. Since it is usually large in aquatic birds, its secretions are said to waterproof the feathers, although there is no evidence to support this (Rawles, 1960). It is equally well developed in some land birds, for instance in the oil bird, *Steatornis*. In some it produces an offensive odour, in others a secretion odorous during the breeding season (Rawles, 1960). It is absent from bustards, some pigeons, and parrots, ostriches, cassowaries, rheas, etc., although present in the embryos of these birds. In others, its condition suggests that it is non-functional. Although its secretion is spread over the feathers during preening, birds without these glands are able to maintain their plumage in good condition (Rawles, 1960).

Hou (1928, 1930) suggested that under the action of the sun, the secretions spread on the feathers, produced vitamin D which is absorbed by the mouth during preening. However, Knowles *et al.* (1935) showed that the gland was not essential for calcium metabolism in birds. Friedman (1935) suggested that it confers, on those birds that possess it, a relative independence of environmental sources of vitamin D.

G. Mammalia

The epidermis of man, in that it differs considerably from that of species covered with thick hair, such as sheep and rodents, is undoubtedly a special adaptation to the exceptional fineness of human hair which leaves the body virtually naked. The general epidermis has to provide services that in hairy mammals are provided by the hair, i.e. protection against abrasion and cuts, and chemical, thermal and radiation damage. The most striking way in which it differs from hairy mammalian epidermis is in total thickness. There is, in all mammals, an inverse relationship between the thickness of the hair coat and the thickness of the epidermis. In man, there are six to ten layers of living cells, in rabbit, mouse, and dog, only two to four layers. In man there are about fifteen layers of horny cells and they can be stripped off as a thin, transparent, but tough, membrane. In cattle and rabbits the horny layer cannot be stripped off, and in guinea pigs and rats it comes off as a gossamer-like membrane, easily broken up. The horny layer of the nearly hairless new-born rat is far thicker than that of the hairy adult rat (Kligman, 1964).

In hairy mammals the fur is a considerable barrier to loss of water from the body. Sweat glands, which in man help to regulate temperature by the evaporation of their watery secretions, are absent under fur and are restricted in many hairy forms to the pads of the feet. On the other hand, the horny layer of human epidermis is remarkably impermeable to water; water-loss through intact, abdominal epidermis, *in vitro*, at room temperature, and 5 per cent relative humidity, is in the region of 0·10–0·2 mg/cm^2 per h, which rises to 18 mg/cm^2 per h when the horny layer is removed. The horny layer of guinea pigs, because it is much thinner, allows water to pass through it at about three times the rate of the human abdominal horny layer (Kligman, 1965). A large increase in the rate of diffusion of water through the epidermis takes place after the application of fat solvents, indicating that natural lipids are an important component of the barrier to water loss, although keratin also contributes the barrier (Matoltsy *et al.*, 1968).

The horny layer is very hygroscopic, however, the capacity to absorb and hold water resting in the presence of hygroscopic, water-extractable substances in the horny layer (Szakall, 1957). These may be mucopolysaccharides in the intercellular cement (Blank, 1953; Blank and Shappirio, 1955). Water absorbed in the horny layer is of great importance in maintaining its plasticity and Blank (1952) maintains that water is the only known plasticizer of the horny layer. The water is obtained mainly by diffusion from the dermis and from sweat on the surface. Dried horny layer, although strong, has very little elasticity, only stretching about 5 per cent before it breaks. Wet horny layer, although reduced in strength, can be stretched to over twice its original length without breaking. The capacity of the horny layer to stretch allows for swellings of the body in infections, inflammations, and pregnancy. The underlying dermis is unable to stretch as much as the epidermis, so that during pregnancy, while the epidermis can stretch to accommodate the increasing size of the uterus, splits occur in the dermis. The actual surface structure of the epidermis under the microscope shows considerable sculpturing into a variety of patterns of fine folds, although their condition under tension has not been studied. Regions about joints are often crossed by definite permanent creases, or flexure lines, where bending occurs. They develop in relation to the underlying musculature and involve not only the epidermis but also the dermis.

The horny layer on the palms of the hands and soles of the feet is much thicker than that on the rest of the body. Contrary to expectation, this thick layer is not as resistant to abrasion or to chemicals as the general horny layer of the body, nor is it as resistant to water-loss. Its function appears to be the provision of a cushion rather than a resistant layer in these areas. Callus material, somewhat similar to the horny layer of the hands and feet, may develop in other parts of the body in response to constant friction, although the pads of the hands and feet are already developed in the embryo.

Scales resembling those of reptiles are occasionally found in mammals, in the tails of many marsupials, insectivores, and rodents, and over the body of armadillos. Scale-production on the tail of the mouse differs from that in lizards for, in the mouse, keratin production is a continuous process, the scales being gradually worn away and replaced.

While some amphibians have thimble-like caps of horny material at the ends of the digits, the fingers and toes of all vertebrates above the amphibia are capped by specially hardened keratin-containing structures which serve various purposes. These generally take the form of nails, claws, or hoofs. Claws are structurally similar wherever they are found, and consist of two unequally developed, scale-like structures which meet over the end of the digit. The dorsal structure is tougher and harder than the ventral and is strongly arched which gives it added strength. Where the two meet, a sharp projecting edge is formed (Fig. 16·6). The nails of primates are formed from

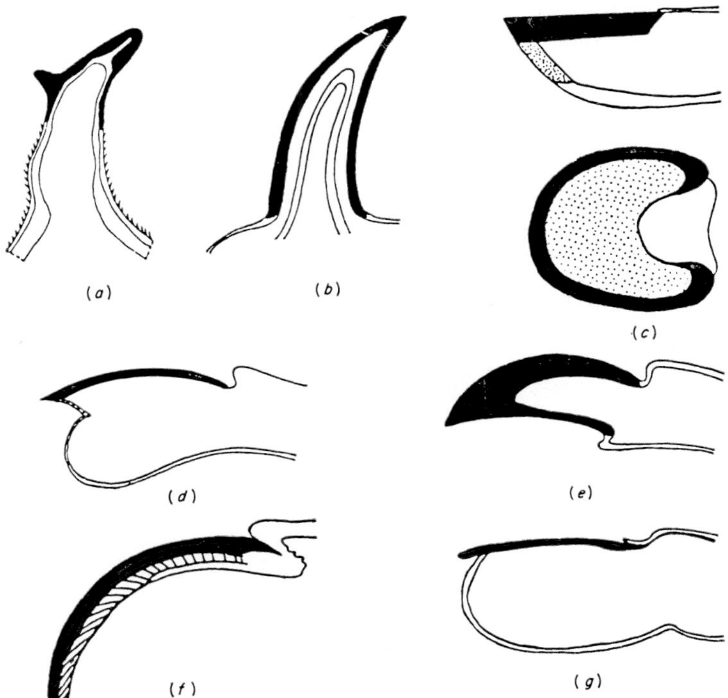

Fig. 16.6. Horns, nails, and claws of mammals. Hard keratin shown black, softer keratin stippled. (*a*) and (*b*) are two kinds of horns: (*a*) is the pronghorn consisting of a "thimble" of true horn capping a "bony horn" covered by hairy skin. The cap is shed annually. (*b*) is the true horn covered entirely with horn keratin which is not shed. (*c*) The principal parts of the hoof of a horse. This consists of an outer covering of hard keratin (the unguis) and an inner subunguis of softer keratin. This combination of a harder and softer keratin recurs in claws. (*d*), (*e*), and (*f*)—claws of a carnivore, a bird, and a rat. The wearing away of the softer subunguis helps to maintain the sharp cutting edge of the carnivore claw (Le Gros Clark, 1936). (*g*) The human nail in which only the hard keratin layer remains. (Mercer, 1961).

claws, by the loss of the ventral component and the flattening of the dorsal. In hoofs the ventral portion, although soft, takes the main weight of the body but it is encased in a sheath of the dorsal, harder structure (Mercer, 1961).

Horns of cattle, sheep, and goats are permanent bony structures covered by a sheath of horny cells which, as they are worn away, are replaced by the proliferation of an underlying germinal layer. Antlers of deer which develop in the male, and are annually shed, are not horns. They are bony structures covered by a layer of hairy skin, the velvet, which gets rubbed away, exposing the bone. The prong-horn of the prongbuck, *Antilocapra*, is a bifurcating structure half way between a horn and an antler. It has a bony core which is not shed, and this has a sheath of horny cells which is shed periodically (Mercer, 1961).

The fine structure of the epidermis between the hairs in mammals, particularly in man and sheep, has often been described, but differing cytological and electron-microscope techniques have shown up differing details and resulted in differing interpretations. Rudall (1968a) has most usefully summarized these various points of view.

Intracellular filaments, generally attached to desmosomes, make their appearance in the basal layer and undergo changes in numbers, appearance, state of aggregation, and affinities for dyes, as the cells move up into the various layers of the epidermis. In the prickle-layer the cells are not as closely packed as in the basal layer below, and are in contact with each other only in the region of the desmosomes, giving the cells a spiny or prickly appearance. Within these cells membrane-bound granules, which were first described by Odland (1960) and which often show internal laminations, make their appearance. Cells of the next layer, the granular layer, contain also keratohyalin granules. These are difficult to define other than as bodies that appear dense in the electron microscope. Jarret *et al.* (1959) describe them as fluorescing orange after treatment with titian yellow. Keratohyalin granules are generally closely associated with the intracellular fibrils. In the granular layer the membrane-bound bodies make their way to the periphery of the cells and then pass out into the intercellular spaces, after which the cell membranes become thicker, presumably through the addition to them of material from these bodies. Similar membrane-bound bodies have been seen in the epidermis of turtles and chicks (Matoltsy and Parakkal, 1965).

In the outer layer of the epidermis, within the now flattened cells, the keratohyalin granules disappear and the fibrillar material becomes converted to a system of fibrils mainly orientated parallel to the surface of the epidermis and embedded in a matrix. At the centre of each cell is a space into which the cytoplasm and nucleus are concentrated before they are broken down and resorbed (Spearman, 1966). In reptile and bird flexible epidermis the keratin completely fills the cells. The restriction of the keratin to the periphery of mammalian epidermal cells provides a more flexible surface layer associated with the increased movements and agility of mammals. Each flattened cell is cemented to its neighbours by a dense, granular cement, possibly derived from the membrane-bound granules. This cement finally breaks down, liberating groups of cells from their firm attachment to the rest of the epidermis, from which they then become detached and lost (Kligman, 1964). Detachment of the cells at the surface is an active process continuing even when the skin is covered over and protected from abrasion. Bullough (1962) has evidence that division of cells in the basal layer, and loss of cells at the surface, are in equilibrium, and the whole process under hormonal control and related to food intake.

1. *Hair*

Hair is the characteristic product of mammalian epidermis, although it is believed to have occurred also in the fossil aquatic reptile, *Ramphorhynchus* (Boili, *see* Matoltsy, 1962b). Completely hairless skins do not occur in any normal mammals, except in one or two cetaceans. There is, however, considerable variation in the amount and texture of the hair developed in different species.

Generally two types of hair can be distinguished. Long, thick, over-hairs which are the first to develop embryologically and softer, shorter hairs that form the undercoat. Hair is most important in providing thermal insulation and for this the undercoat is mainly responsible. The capacity of many mammals to erect the hairs by means of muscles at their base provides a way of controlling heat-loss. There is a correlation between normal environmental temperature and the thickness of the fur, tropical mammals tending to have thinner coats than arctic types. There is also some correlation between thickness of coat and body-size; the smaller the animal, the greater the relative loss of heat, and the greater the thickness of coat required to provide effective insulation. Very small animals, however, could not carry the weight and length of coat that would be necessary to provide efficient insulation in very cold weather. This is one of the reasons why small mammals in high latitudes have to hibernate, and why they do not occur at all in the arctic and antarctic. The smallest mammal in Spitzbergen is a fox. Scholander *et al.* (1950) found that, in mammals from the size of foxes upwards, there was no correlation between the insulation achieved and size; they had all about the same thermal insulation per unit area of surface. The medullation (see p. 289) of rodent hairs results in a lighter, more bulky hair and would therefore give greater thermal insulation than the same weight of unmedullated hair.

Many large mammals, such as elephant and hippopotamus, with thick, rather rigid skin, have a poor development of hair. Marine mammals also generally have little hair which in aquatic animals would be useless for thermal insulation. The northern fur seal which spends most of the summer on land has a better development of fur than the common seal which spends more of its life in water. It is perhaps surprising, therefore, that it is in an aquatic fossil reptile that indications of premammalian hair development are found, although perhaps, like the fur seal, *Ramphorhynchus* spent part of its life out of water.

Hairs develop from tubular follicles embedded in the skin. With the single exception of the eyelashes, hairs do not grow vertically out of the skin, but generally slope backwards so that they lie flat on the surface. They are

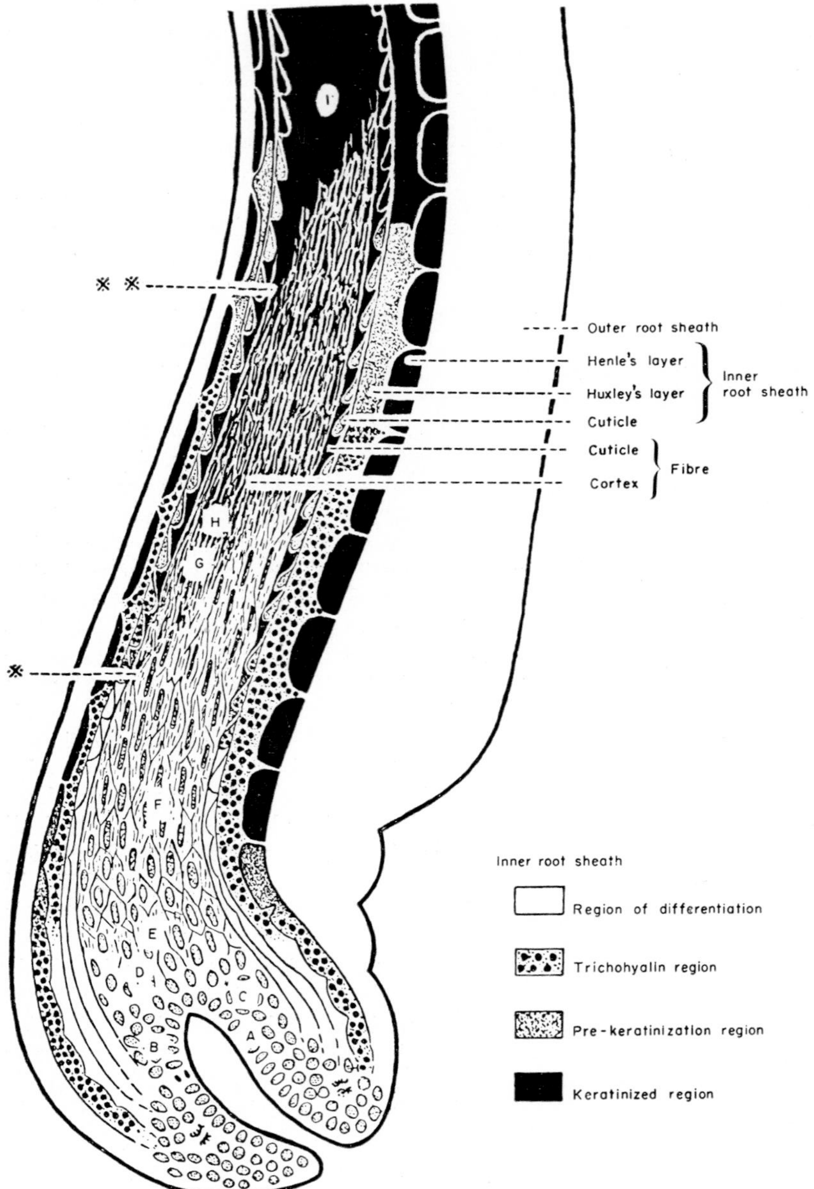

Fig. 16.7. Diagrammatic longitudinal section through the proximal portion of a non-medullated follicle indicating differentiation and keratinization (Auber, 1956).

arranged in definite tracts which are constant for a given species. Primitively, these tracts ran from head to tail giving a simple streamlining related to the normal movements of the body in a forward direction, although this simple arrangement shows modifications, usually related to habits of posture, movements, and the direction in which animals comb or scratch the fur (Le Gros Clark, 1945).

Hairs first appear, in the human embryo, in the region of the eyebrows, lips, and chin, at the end of the second month. Development over the rest of the body does not occur till the fourth month. Each arises as a solid down-growth into the dermis of a narrow column of epidermal cells. A small papilla of dermal cells, serving to provide nutriment, pushes up into the base of the down-growth. The epidermal cells immediately above this papilla form a bulb, and it is from this that the hair develops. The hair forms as a column of cells, tapering at the top, which penetrates through the solid column of cells above it and passes out beyond the surface. That part of the hair that lies inside the follicle is called the hair root. The wall of the follicle continuous with the surface epidermis, is called the outer root sheath and is distinguished by its content of glycogen. At its base it is continuous with the differentiating cells of the hair shaft. Six different concentric layers of cells can be distinguished in the developing hair papilla. From outside inwards there are Henle's layer, Huxley's layer, and the sheath cuticle layer, which together form the inner root sheath. This is a structure which supports and protects the developing hair as it pushes upwards through the follicle. After having served its purpose the inner root sheath disintegrates. The three layers of cells inside the inner root sheath form the hair proper (Fig. 16.7).

Opinions differ about events leading to the hardening of the inner root sheath. Dense bodies called trichohyalin granules appear in the cells and Birkbeck and Mercer (1957) describe these granules as becoming transformed into the filaments, about 100 Å in diameter, with which these cells become filled. This description receives strong support from the photographs published by Rogers (1964a), which show the apparent development of filaments from trichohyalin granules. Unlike the filaments in mammalian epidermis, the inner root sheath filaments do not appear to be embedded in a matrix but coalesce to form the rigid structure of the sheath, which gives an α-protein-type X-ray photograph.

In the hair the outer layer of cells forms the cuticle, this surrounds the cortex and in the centre of some hairs, such as those of rodents but not of man, there is a medulla. The cells that give rise to the cuticle contain very few filaments. Granules appear in the cytoplasm and move to the cell-periphery where they collect, mainly on the outer wall of the cell, forming a continuous layer while the rest of the cytoplasmic contents collect towards the inner part of the cell. The final hardened material is isotropic, without visible fibres, and X-ray photographs show that it is amorphous. In the fully formed hair the cuticle, which varies in thickness, consists of thin over-lapping cells. At the surface the overlapping portions are free and project away from the body forming ratchet-like scales that give the hair a rough feel when it is stroked towards the skin. They hinder dirt from working up the hair towards the root, and also prevent the hairs matting. In spun wool the hairs may lie in either direction so that the scales of one hair may interlock with those of adjacent hairs. It is relative movement between these interlocking hairs, generally resulting from rubbing, that causes matting and shrinkage in wool. The cuticle, as distinct from other parts of hair, has only a limited extensibility and when a hair is stretched by over 50 per cent the cuticle cells part company. When, as in some fur, a very thick cuticle is present, the extensibility of the whole hair is modified.

As the cortex differentiates, the cells become long and spindle-shaped with their long axes parallel to the length of the hair. The cells develop abundant filaments which later become aggregated into fibrils and orientated parallel to the length of the hair. Subsequently, a matrix appears between the fibrils. The relative amount of fibre to matrix varies in different regions of the hair. The matrix first contains a concentration of —SH groups, later oxidized to —S—S— groups when the hair hardens (see Rogers, 1964a, for review). The cells become separated by the deposition of a dense cement but there are no cytological indications of the source of this intercellular material. The cortex gives an α-protein type X-ray photograph. The arrangement of the cells gives the hair its fibrous character and a tendency sometimes to fray into fibrils, although this is held in check by the cuticular layer. In the innermost layer of the cortex, surrounding the medulla in the whiskers of lions and tigers, the cells are orientated at right angles to the long axis. This prevents these relatively thick, stiff hairs from splitting longitudinally when they are bent (Earland *et al.*, 1962).

In the development of a medulla, changes occur that resemble those leading to the hardening of the inner root sheath. Granules, similar to trichohyalin, develop and appear to change into fibrous material which condenses against the cell walls, but as not enough material is produced to fill the cell, the centre remains empty and finally becomes filled with air. Intercellular gaps appear which also become air-filled so that the medulla assumes a rather open, girder-like structure which is light but stiff (Auber, 1956). The X-ray photograph indicates that the molecules are in the cross β-type arrangement, somewhat like that which has been found in *Chrysopa* silk (see p. 234), although traces of the α-pattern also are present (Rudall, 1947; Rogers, 1964a).

After the hair has completed its growth and hardening, and on the human scalp growth may continue for up to six years, the middle region of the hair root becomes constricted by an ingrowth of cells from the outer root sheath. This severs the connection between the hair and the germinal matrix over the dermal papilla. Above the constriction the base of the hair expands and hardens into a club-shape and the whole hair is gradually pushed upwards in the follicle. When the hair falls out, it is replaced by a new hair developed from the germinal matrix remaining at the base of the follicle.

While feathers are thought to be evolved from the scales of reptiles, since they also contain β-keratin, hairs are believed to have evolved in the soft, interscale α-keratin-containing region. In rodents hairs occur on the tail between the scales, usually in groups of three, a central, thick, long hair, flanked on either side by a softer, smaller hair. Over the general body surface follicles develop in small groups often, e.g. in pig, of three follicles, suggesting that intervening scales were once present (Mercer, 1961; Spearman, 1964).

2. *Epidermal glands*

Mammalian epidermis, in contrast to that of birds and reptiles, is amply provided with multicellular glands. Opening into each hair follicle towards the surface of the skin is one or more sebaceous glands. These secrete sebum, composed of fatty matter and cell debris, which covers the hair shaft, lubricating and conditioning it, and spreads out over the epidermal surface. Since it has bactericidal properties, it helps to keep the skin free of infection. Sunlight on the fatty material may convert sterols to vitamin D, which is absorbed by the blood, or in some mammals by licking and cleaning the fur. Sebaceous glands are lacking from the soles of the feet and the palms of the hands, which also lack hairs. Such a secretion in these regions would cause slipping between the hands and feet and the substratum.

Sweat glands secrete a very dilute fluid, containing less than 2 per cent solids, mainly sodium chloride. While they may function as accessory excretory organs, they play an important part in regulating the temperature of the body and, by aiding the epidermis to remain damp, help to keep it pliable. Sweat glands are particularly abundant on the soles and palms where their secretion helps these regions to grip the substratum.

Mammary glands, which characterize the whole group, are morphologically similar to sebaceous glands but secrete milk instead. Other glands, in the ears, are modified to produce wax which traps dirt and dust, preventing it from reaching the tympanic membrane. Certain glands, resembling sweat glands, although larger and more elaborate, occur associated with the hair follicles in the pubic region of man. Similar glands occur also in the axilla of humans, and around the anus in some mammals. In the lower mammals they function as scent glands but in man their function is unknown.

Passing up amongst the epidermal cells, sometimes connecting with hair follicles and glands, are numerous nerve endings, which may end without special terminal formations or they may form sensory corpuscles or end on special sense cells. These various end-organs make the skin sensitive to pressure, heat, cold, and chemical irritation.

3. *Moulting*

Besides the gradual replacement of the horny layers of the epidermis of mammals, regular moulting and replacement of hairs also occurs. What pressures led to the evolution of moulting are uncertain. Although dead cells have been observed in the epidermis of fish, and these must be eliminated and replaced, moulting of the outer layers of the epidermis of fish has rarely been described. Applegate (1950) describes the sloughing of the skin in lampreys after spawning, and moulting has been described also in the horse-fish, *Agriopus* (Norman, 1963). Henrikson and Matoltsy (1968a) describe membranes lying beyond the outer layer of cells in teleost epidermis and these they believe are formed from cast, dead cells. Thyroid hormone, which will induce moulting in some vertebrates, leads to a considerable thickening of the epidermis in goldfish, although no sloughing takes place (Lagler *et al.*, 1962).

Amphibia, the first class with a complete covering of keratin-containing cells are the first to moult regularly. Moulting appears therefore to be connected with the evolution of keratin in the epidermis. Although Gadow (1901) says that moulting occurs more frequently in the growing stages of amphibia than in the adult, the fact that the horny layer is elastic, the basal layer is capable of cell division, and that moulting is continued into adult life, imply that it is not primarily for the accommodation of growth, as it is in arthropods and aschelminthes. An efficient system of replacement of the worn and damaged horny layer in vertebrates, and particularly in those that scrape a considerable part of their ventral surface on the ground during locomotion, would on the face of it be more economically provided for by gradual localized replacement in response to wear and tear, rather than by periodic total replacement. That moulting has evolved in this manner suggests that it fulfils other functions in addition to the provision for growth and the making good of wear and tear, although what these functions were initially is not apparent.

Moulting does provide, in amphibia and reptiles, a mechanism for periodically cleaning away from the surface, dirt, bacteria, and ectoparasites in land-living forms which, because of their anatomy, are not as well able to groom their surface as are birds and mammals. In birds and mammals moulting has come to serve additional purposes. In birds it provides for seasonal sexual display, and in birds and mammals for seasonal adaptation in colour. Cott (1940) claims that the white colour assumed by some birds and mammals each winter in snowy regions serves to camouflage these animals but Hadwen (see Seyereid, 1945) claims that the white colour is more important in preventing radiation of heat from the body. In mammals moulting of hairs is important in providing regulation of thermal insulation; a thick winter coat can be replaced by a thinner coat for summer. Moulting may also be related to the sexual cycle, some species of rabbits moult in spring before the breeding season, some in autumn at the end, and some moult at both seasons. Moulting at the beginning of the breeding season may provide fur to line the nest, the female pulling out the fur that has been loosened by the moult, and it may also serve to denude the nipple area (Ebling, 1965).

Except in young mammals, growth and replacement of hair is not synchronous over the body. The moult usually starts in one region and progresses, as a wave, over the body as in rats where it starts on the belly and then moves to the flanks and then to the back. In other mammals, such as the guinea pig, the hairs are replaced by a mosaic of activity, in which each follicle is independent of its neighbours.

Vertebrates in general, and mammals in particular, demonstrate the very wide variety of functions that can be performed by a cellular epidermis and its outgrowths. Some of these functions could never be performed by an extracellular cuticle, however complex. There are no functions performed by a cuticle which a cellular epidermis cannot be modified to perform.

CHAPTER 17

General Properties of Keratin-Containing Structures

Hairs and feathers are strong and resistant to abrasion and the horny layer of the epidermis and hairs are extensible and elastic, although strength, extensibility, and elasticity depend upon the water content and the temperature of the material. Dry wool fibres usually break at about 20 per cent extension. When the fibres are stretched in cold water extension rarely exceeds 70 per cent, but in steam, an extension of 100 per cent may be obtained. When fully extended α-keratin is converted to the β-form.

The physical and chemical properties of the horny layer and other epidermal products do not, as mentioned before, depend only on the properties of keratin. Since the keratin is intracellular, the tensile strength of the material must be largely determined by the material cementing the cells together, as this is the only component to form a continuous network through the system.

A. Intercellular cement

In most tissues, other than keratinized ones, the gap between cells is about 150 Å wide but in hair cortex it may reach 400 Å. By transmitted light the material between the cells may appear banded. Intercellular cement and cell membranes together form about 10 per cent of the total weight of hair. This material is able to withstand strong reducing agents and hydrogen bond breakers which attack and dissolve keratin, implying that hydrogen bonds and disulphide bonds are not responsible for the strength of this material. Mercer (1961) suggests that some sort of quinone-tanning may be responsible for the great chemical resistance of the cement and this suggestion receives some support from the high phenylalanine (6·7 per cent) and tyrosine content (11 per cent) found in it by Dedeurwaeder *et al.* (1964). The cement also contains 27 per cent glycine, 1·7 per cent tryptophan and a little cystine. There may be some biological advantage in surrounding keratin by a material that resists dissolution by precisely those reagents most injurious to keratin (Mercer, 1961). The cement must have a tensile strength at least equal to the strength of the whole structure and elasticity adequate to the elasticity in the whole.

B. Cell membranes

Matoltsy and Matoltsy (1966) analysed the thickened cell membranes from human foot epidermis (Table 17.1). Adhesion between the intercellular

Table 17.1 Amino acid composition of the membrane protein from human stratum corneum. Grammes per 1000 g protein. (Matoltsy and Matoltsy, 1966.)

Alanine	38·2
Glycine	86·7
Valine	50·1
Leucine	62·5
Isoleucine	35·7
Proline	128·9
Phenylalanine	31·4
Tyrosine	16·2
Serine	64
Threonine	36
Cystine ($\frac{1}{2}$)	75
Methionine	—
Arginine	78·8
Histidine	36·6
Lysine	99·3
Aspartic acid	63·1
Glutamic acid	199·3

cement and the membranes, and between the membrane and the keratin fibrils, probably localized at the desmosomes, must all contribute to the mechanical properties of the whole structure. The area of contact between the cell membrane and the materials on either side of it is increased by folds or crests in the membrane that interlock with folds in the membranes of adjacent cells. In some instances the interlocking cell membranes virtually resemble press-studs fastening the cells together.

C. Keratin

Rudall (1952) extracted from the prekeratin layers of the thick but soft epidermis of the cow's nose an α-type, low-sulphur component which he called *epidermin.* He also found a small quantity of non-fibrous high-sulphur material. A different low-sulphur α-type protein, *tonofibrin,* has been extracted from human epidermis (Roe, 1956; Flesch, 1958). Matoltsy (1964, 1965) has obtained from the cow's nose, by treatment with urea, a protein he calls *pure-prekeratin* which he holds to be analogous to tropocollagen. It is monodispersed by ultracentrifuge and electrophoresis, with a molecular weight of 640 000, a particle-length of 1050 Å and a calculated diameter of 37 Å, and this is in agreement with the diameter of the finest filaments seen in epidermis by Rhodin and Reith (1962).

Solutions of proteins obtained from sheep wool can be fractionated into two groups, one of which contains less sulphur than the parent structure (Table 17.2), is fibrous, and yields an α-type X-ray photograph and is believed to originate from the fibrillar component of keratin. The other,

Table 17.2 Amino acid composition of the low-sulphur fibrous protein and the high-sulphur non-fibrous protein from Merino wool. Amino acid N as percentage total N. (Corfield *et al.*, 1958, prepared from oxidized wool.)

	Low-sulphur fraction	High-sulphur fraction	Whole wool
Alanine	4·83	2·58	3·99
Glycine	5·16	4·97	6·09
Valine	3·98	4·15	3·97
Leucine	7·30	2·55	5·78
Isoleucine	2.49	2·14	2.34
Proline	2·69	9·85	4·89
Phenylalanine	1·94	1·15	1·90
Tyrosine	2·44	1·41	2·37
Serine	6·70	9·70	7·50
Threonine	3.45	7·46	4·66
Cysteic acid	3·72	14·5	6·50
Methionine	—	—	—
Arginine	20·8	19·0	19·2
Histidine	1·24	1·57	1·68
Lysine	4·60	1·03	4·02
Aspartic acid	6·25	1·79	4·56
Glutamic acid	10·9	5·87	8·50

which contains more sulphur and is amorphous, yielding only indistinct diffraction halos, is believed to originate from the matrix (see Crewther *et al.*, 1965). Starch gel electrophoresis has shown that neither high nor low sulphur proteins are homogeneous but contain a number of closely related proteins. The high-sulphur fraction is relatively richer in cystine, serine, threonine, and proline and poorer in glutamic acid, aspartic acid, alanine, lysine, leucine, and tyrosine than the low-sulphur fraction. In the high-sulphur protein aspartic and glutamic acids are present largely as their amides and there is therefore an excess of basic groups (Gillespie *et al.*, 1960) and the protein in its native state (before the conversion of cystine to acid groups by reduction and alkylation) is strongly basic. The low-sulphur protein is acidic (Crewther and Dowling, 1960); this could result in strong binding between the two fractions and the disulphur bonds also may be mainly concerned in linking the two fractions together. In feathers it has not been possible to separate the dissolved keratin into high- and low-sulphur fractions, although fibres and matrix are visible in the cortex. No comparable studies have been done on material from amphibians and reptiles.

Material extracted by similar methods from the unhardened hair roots again yields high- and low-sulphur fractions (Table 17.3) differing little from those of the fully formed wool, although the cystine content of both is lower (Rogers, 1964a). This is consistent with the belief that cystine is added to the system, possibly from the inner root sheath, during the hardening process. Injection of radioactive cystine is followed rapidly by the appearance of radioactivity in the keratin-producing region of the hair but not in the

Table 17.3 Amino acid composition of prekeratin protein fractions from wool roots. Micromoles per gramme.

	Fraction 1[a] (Low-sulphur protein)	Fraction 2[a] (High-sulphur protein)	Fully[b] keratinized wool
Alanine	56	40	41
Glycine	44	51	68
Valine	47	46	42
Leucine	77	55	57
Isoleucine	31	27	23
Proline	26	55	62
Phenylalanine	19	16	20
Tyrosine	25	19	35
Serine	53	74	85
Threonine	37	52	54
Cystine ($\frac{1}{2}$)	28[c]	148[d]	92
Methionine	8	3	3·5
Arginine	56	44	59
Histidine	8	7	6
Lysine	42	24	19
Aspartic acid	80	51	50
Glutamic acid	133	92	100
Hydroxyproline	—	—	
Hydroxylysine	—	—	

[a] Rogers, 1964b.
[b] Simmonds, 1955.
[c] As S-carboxymethyl cysteine.
[d] As S-carboxymethyl cysteine + cystine.

base of the hair root, suggesting that this is not the route by which it enters the hair (Ryder, 1958; Bern *et al.*, 1955). On the other hand, labelled carbon and phosphate compounds can be seen to pass into the hair via the hair base (Mercer, 1961).

Chemical analyses of whole structures, since they contain keratin fibrils, matrix, and cement, do not contribute directly to the understanding of the chemistry of keratin but they are of use in indicating variations in composition between regions of a structure and between different structures, and also differences between breeds and species (Table 17.4). There are consistent differences between avian and mammalian keratins. Such analyses also provide information about such things as the effects of nutrition and environment on the various structures. Thus Makar (see Crewther *et al.*, 1965) found that adding sodium sulphate to the diet of sheep increased the cystine, tyrosine, and phenylalanine content of wool, and reduced the content of glutamic acid, aspartic acid, valine, leucine, and isoleucine, resulting in wool that was longer, stronger, and of greater diameter than that of controls on an unsupplemented diet.

Apatite (calcium phosphate) has been reported from various epidermal products. It is found in the calamus of goose feathers, in the body hairs but not in those of the tail of the platypus, in the horn of the Indian rhinoceros,

Table 17.4 Amino acid composition of various keratin-containing structures. Grammes per 1000 g dry keratin.

	Sheep[a] wool	Human[a] hair	Cattle[a] horn	Porcupine[a] quill	Chicken[a] feather	Turkey[b] barbs	Turkey[b] calamus	Turkey[b] medulla	Turkey[b] rachis	Goose[b] barbs
Alanine	34–44	28	25	—	54	401	712	58·9	76·6	41·0
Glycine	52–65	41–42	96	57	72	725	96	89·0	101·4	83·8
Valine	50–59	55–59	53–55	—	83–88	860	84·3	85·9	86·5	73·4
Leucine	76–81	64–83	76–83	—	74–80	726	88·5	80·7	93·7	76·8
Isoleucine	31–45	47–48	43–48	—	53–60	498	39·4	39·6	39·0	45·8
Proline	58–81	43–96	82	—	88–100	1050	109·8	108·7	109·7	100·5
Phenylalanine	34–40	24–36	32–40	36	47–53	496	57·6	55·9	57·5	40·4
Tyrosine	40–64	22–30	37–56	33	20–22	232	39·7	38·1	29·1	44·6
Serine	72–95	74–106	68	61–62	102–140	129	150·9	123·7	140·9	125·3
Threonine	66–67	70–85	61	39–54	44–48	468	47·3	43·5	45·1	49·4
Cystine	110–137	166–180	105–157	80–95	68–82	868	82·9	81·0	84·8	107·5
Methionine	5–7	7–10	5–22	8	4–5	36	3·4	4·4	3·9	2·5
Arginine	92–106	89–108	68–107	76–80	65–75	644	66·9	65·7	61·8	60·4
Histidine	7–11	6–12	6–10	6	3–7	39	5·9	7·8	3·4	4·4
Lysine	28–33	19–31	24–36	26	10–17	123	9·8	13·2	8·8	13·0
Aspartic acid	64–73	39–77	77–79	87	58–75	655	70·9	70·1	74·1	74·7
Glutamic acid	131–160	136–142	138	176	90–97	908	87·4	86·0	88·4	89·9
Hydroxyproline	—	—	—	—	—	—	—	—	—	—
Hydroxylysine	2	0	—	—	—	—	—	—	—	—
Tryptophan	18–21	4–13	7–14	9	7	—	—	—	—	—

[a] Ward and Lundgren, 1954.
[b] Schroeder and Kay, 1955.

and in lion whiskers (Blakey *et al.*, 1963). Pautard (1961, 1962) has reported it from various structures, including the baleen of rorqual and sei whales. Blakey *et al.* also found it in the finger nails of 16 per cent of the patients in a geriatric unit. Apatite appears generally to be associated with stiff, dense

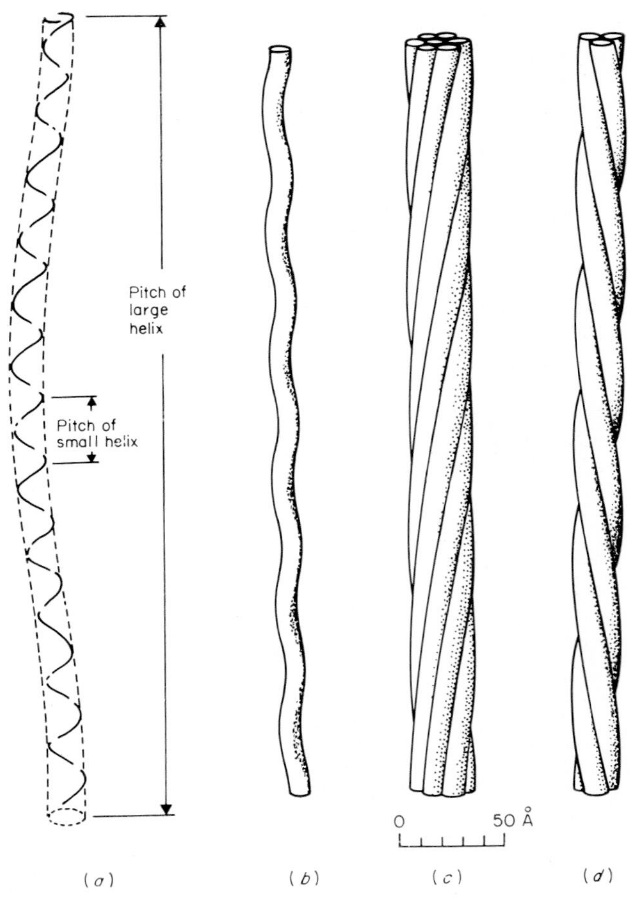

Fig. 17.1. (*a*) and (*b*) illustrating the coiling of an α-helix into a super helix. (*c*) and (*d*), two possible combinations of superhelical structures to give compound helices (coiled coils). (*c*) The seven strand cable. (*d*) Three strand coiled coil. (Pauling and Corey, 1953.)

masses of keratin and may contribute to the stiffness of these structures. Baleen of rorqual whales has a somewhat bone-like consistency.

It has not yet been possible to determine the molecular construction of keratin. The α-helical fibrous molecule of keratin has 3·6 amino acid residues per turn, each residue occupying 1·5 Å of the length of the helix, which has a pitch of 5·4 Å. Each complete repeat unit is composed of 18 residues, makes five turns and so measures 27 Å long. As yet there is little evidence on the actual amino acid sequence along the chain. The interpretations of the small-angle, long-spacing X-ray photograph that indicates the arrangement of the helices has proved very difficult. Crick (1952) and Pauling and Corey (1953) suggested the helices themselves might be twisted into super helices or coiled coils (Fig. 17.1). Crick suggested three such chains coiled together. Pauling and Corey suggested six chains coiled round a central chain. Swanbeck (1964) has suggested a more complex, multistrand cable composed of five concentric rings of chains and containing, in all, some 55–57 chains. One of the difficulties with all these arrangements, as pointed out by Huggins (1967), is that while the single α-helix can be stretched to produce a β-type extended chain structure, if α-helices are twisted together in coiled coils, stretching becomes virtually impossible without breaking the chains, particularly if inter-chain linkages have been established.

Taking the length per residue for an α-helix arrangement as 1·48 Å, then taking the molecular weight of 640 000 (5534 residues) and the molecular length, 1050 Å, of Matoltsy's pure prekeratin, this would give approximately 710 residues in a chain 1050 Å long and provide for nearly eight chains. That there might be seven different chains in a filament, is suggested by the finding of seven different end groups in the keratin molecule (Crewther *et al.*, 1965; Crounse, 1965; Rudall, 1968b).

Electron microscopy of ultra-thin sections of hair cortex by Filshie and Rogers (1961) suggests that each filament contains a ring of nine protofilaments of 20 Å diameter and possibly two more protofilaments are present in the centre of the filament, each protofilament being a coiled coil of three α-helices. Johnson and Sikorski (1962), however, think that the apparent protofilaments are electron-optical illusions. Electron microscopic evidence in favour of the existence of the protofilaments was provided also by Dodd (*see* Crewther *et al.*, 1965) who found filaments 20 Å in diameter in partially disintegrated wool.

Amorphous keratin is made up not only of the usual matrix surrounding the filaments but also of portions of fibrils in which the α-helices are disorganized. These amorphous regions may play an important part, as in silk fibroins, in determining the elasticity of the fibres and their reactions to various swelling agents. From the scatter observed in X-ray photographs, from the penetration of small molecules into the material, and from birefringence studies, it is suggested that more than 70 per cent of the material is disorganized. Probably, as in silk, the crystalline regions are composed of sequences of amino acids that can be packed close together because they have short side chains, the amino acids with long side chains being concentrated in the amorphous regions. The low-sulphur, fibrous component of keratin contains a considerable proportion of long side chain amino aids (Table 17.2) so that the crystalline regions are not likely to be extensive. Indeed, surprisingly, there is a higher proportion of long side chain amino acids in the low-sulphur than in the high-sulphur component (Corfield *et al.*, 1958). How far any of these ideas about the structure of mammalian keratin are applicable to keratins in other vertebrates is not known.

D. Trichohyalin

Analysis of the lower parts of the inner root sheath and of the medulla, because they are filled with trichohyalin granules, are believed to indicate the nature of this material. The material from both these sources (Table 17.5) is similar in amino acid content (Rogers, 1964a and b) and differs considerably from that of keratin from sheep wool (Table 17.4). Both have only a trace of cystine, and for this reason neither Rogers (1964a) nor Roth and Clark (1964) consider that this material should be called a keratin. In addition to containing a high proportion of glutamic acid residues, they both are distinguished by their high content of the rare amino acid, citrulline (G. E. Rogers, 1958, 1959). In the medulla of porcupine quill as many as one in every five residues may be citrulline. G. E. Rogers (1958) suggests that the citrulline may increase the number of hydrogen-bonding possibilities, and may therefore be an important stabilizing component.

Table 17.5 Amino acid composition of the medulla protein from porcupine quill and the protein from inner root sheath cells of wool roots. Micromoles per gramme. (Rogers 1964b.)

	Medulla protein	Inner sheath protein
Alanine	40	44
Glycine	30	46
Valine	32	31
Leucine	70	70
Isoleucine	13	23
Proline	16	24
Phenylalanine }	32	21
Tyrosine }		17
Serine	27	45
Threonine	14	25
Cystine	tr.	tr.
Methionine	4	9
Arginine	18	25
Histidine	10	9
Lysine	74	70
Aspartic acid	47	54
Glutamic acid	270	155
Citrulline	75[a]	28[a]

[a] Not corrected for losses during hydrolysis.

E. Keratohyalin

Since it is difficult to characterize keratohyalin either histologically, histochemically, or by electron microscopy, it is difficult to be certain that the various granules observed in the epidermis of different vertebrates are, in fact, similar structures. While keratohyalin has been reported occasionally from reptile and bird epidermis, Spearman (1964, 1966) believes that they occur only in mammalian soft epidermis tissue.

Keratohyalin granules have not yet been chemically analysed, although Matoltsy (1962b) managed to isolate them in small quantities. The granules may have a protein stroma (Mercer, 1958) and also contain phospholipids (Jarrett and Spearman, 1964). Barrnett and Sognnaes (1955) found neither —SH nor —S—S— groups in them. Enzymes and other substances may be adsorbed onto their surfaces. They are often surrounded by ribonucleoprotein (Spearman, 1966).

What part keratohyalin granules play, if any, in the process of keratinization is in doubt. Since they are always closely associated with keratin fibrils, Brody (1959, 1960) believes they may give rise to material that impregnates the fibrils and forms the interfibrillar matrix. Swanbeck and Thyresson (1965) describe them as the source of a hydrophobic material, possibly lipid, which coats the fibrils and is responsible for producing the barrier to the diffusion of water through the epidermis. Though Roth did find keratohyalin granules in the reptile material he worked with, Ernst and Ruibal (1966) failed to observe them, but did find lipids associated with the fibrils, and the difference between reptile and mammalian material may be only that in some reptiles the lipid coating to the fibrils does not pass through a keratohyalin stage. Even in mammals the appearance and definition of the granule varies. They are well developed as discrete bodies in guinea pig epidermis (Brody, 1959), but in human epidermis they are less well defined (Brody, 1960). In the gibbon, they do not occur at all in the general epidermis of the body, only in the epidermis of the soles and palm (Parakkal, 1962). They are also poorly developed in the skin of many marine mammals.

Spearman does not believe that keratohyalin granules play any part in keratinization. As they appear when the nucleus and cytoplasmic organelles are beginning to disintegrate, he suggests that they represent temporary aggregations of debris from these components which later are metabolized to provide energy for keratinization. Spearman, however, believes that there is a fundamental difference between keratin in mammal epidermis and that in the epidermis of lower vertebrates. Amphibian horny layer, reptile and bird scales, and the tail scales of mammals all fluoresce blue with congo red, while mammalian epidermis fluoresces red, although the hard skin of elephants and hippos, and that of many marine mammals fluoresces blue. Reptile interscale epidermis and bird body epidermis, while mostly fluorescing blue, do show a little red fluorescence, as if in these regions some mammalian-type keratin is present in spite of the absence of keratohyalin granules. He brings forward evidence that keratohyalin granules first evolved in cells around the mouth of hair follicles formed behind the scales. In the tail of the opossum, *Chironectus*, they are restricted to this position. In rodent tails, the granules spread beyond the mouth of the follicle to the interscale epidermis. Normally, there are no keratyohyalin granules in the cells that give rise to the scales of rodent tails, but in certain mutant strains of mice, keratohyalin granules do occur in these cells, and keratohyalin granules are found in the cells that form the scales of the armadillo, possibly representing an incipient change from a scaly to a soft skin. The epidermis of whales and other marine mammals often contains large amounts of phospholipids which may prevent the epidermis becoming waterlogged. In the human pathological condition, parakeratosis, the epidermis morphologically and histochemically resembles that of marine mammals and no keratohyalin granules are formed. Much lipid material is present and the epidermis fluoresces blue.

F. Vitamin A and keratinization

It has been known since the work of Mori (1922) and Wolbach and Howe (1925) that vitamin A deficiency in mammals causes certain excretory epithelia to keratinize. It also produces, in man, dryness and wrinkling of the skin accompanied by hyperkeratosis. Hypervitaminosis often results in inflammation of the skin and a reduction in the growth of hair. Fell and Mellanby (1953) found that, when the skin of a 7-day chick was cultured in the presence of excess vitamin A, not only was keratin production completely inhibited but the epidermal cells went over to the production of mucus, formed microvilli at their surface, and occasionally became ciliated, a condition resembling that of the epidermis of many invertebrates. Similar changes have since been observed in mammalian tissues in response to excess vitamin A, although the extent to which they respond varies with the site from which the explant is taken and the age of the material. In some cases, cells result that contain both keratin and mucus. Laurence and Bern (1960) obtained the same effect *in vivo* by the local application of vitamin A to the hamster's cheek pouch. All this implies that the epidermis of chicks and mammals is able to produce both keratin and mucus but which it produces depends upon the concentration of vitamin A in the environment. It is believed that vitman A in some way prevents the uptake and utilization of cysteine (Pelc and Fell, 1960); vitamin A being an antioxidant possibly could prevent cysteine from being oxidized to cystine.

Other factors also control keratinization. The epidermis of a human embryo does not become keratinized until it is 14–16 weeks old; up until that time the cells are microvillous and probably secrete the mucopolysaccharide which coats the microvilli (Hoyes, 1967). In the mammalian vagina an alternation between keratin production and mucus production occurs normally in the oestrus cycle. Neither of these conditions can be controlled directly by vitamin A. In addition, the chorionic ectoderm of the chick, if it is cultured in air, becomes extensively keratinized and this can be prevented by raising the concentration of CO_2 in the culture vessel (Moscona, 1961). Vitamin A is readily destroyed by oxygen and this may be responsible for the keratinization here.

Little is known about the requirements for vitamin A in invertebrates. It does not appear to be a required food factor in any of these phyla. It is sometimes found in the tissues of invertebrates, e.g. in two nematodes, a chaetognath, and an annelid, and possibly in a starfish. It may be present in these animals only because it occurs in their food. In cephalopods and many crustaceans, it occurs in the eyes where it is probably, as in vertebrates, a precursor of retinene (Wald, 1963) and it is stored in the hepatopancreas of these forms. Carotene, from which some vertebrates are able to manufacture vitamin A, is much more widely distributed in invertebrates, but what part it plays, if any, in their physiology, is not known (Fischer and Kon, 1959). It appears unlikely, therefore, that the activities of the invertebrate epidermis which often produces mucus, but which only rarely produces proteins which are in any way comparable to vertebrate keratin, are related to vitamin A concentration.

Vitamin A is an essential vitamin, at least for birds and mammals. It is necessary for the correct functioning of a variety of physiological processes, including the well-being of the epidermis. As well as in the eyes, it is found in the blood and other tissues, and when intake is more than sufficient to provide for requirements, the surplus is stored in the liver. The concentration of vitamin A in the liver shows considerable variation according to diet and does not reflect the vitamin A requirements of an animal.

Vitamin A occurs in the liver of the hag-fish and reaches a very high concentration in the livers of some bony fish; in halibut, it may reach a concentration of 10 000 i.u./g of liver. Fish live on a diet that is relatively rich in vitamin A or in other carotenes. Hag-fish feed partly on detritus on the sea bottom, which Fox and his fellow workers have shown may contain substantial amounts of carotenes (Moore, 1957). Other fish live on planktonic crustacea which contain vitamin A or on other fish that have acquired their vitamin A from crustacea. One may be justified, therefore, in suggesting a correlation between the rich vitamin A diet of fish and the production of mucus by the epidermis. Unfortunately, the effect of vitamin A deficiency in fish does not appear to have been studied. The capacity of limited areas of some fish epidermis to produce keratin, which has already been mentioned, suggests that perhaps like mammalian epidermis it has the capacity to produce both keratin and mucus but that over the general body surface this capacity is held in check by a high concentration of vitamin A in the tissues. A reduction in the vitamin A concentration might unmask this capacity.

While larval amphibians are aquatic, eating more or less the same food as fish, their epidermis produces mucus. When they become terrestrial they eat mainly insects which have never been shown to contain vitamin A, although phytophagus insects may contain carotenoids believed to be derived from the plants they eat. Vitamin A is not a required vitamin for amphibia. Since it occurs in amphibian eyes and in very low concentrations in the liver; frog liver contains only about 30 i.u./g (Moore, 1967) they can probably synthesise it from carotines. Although nothing is known about the response of amphibian epidermis to varying amounts of vitamin A, it is tempting to correlate the appearance of keratin in the epidermis of terrestrial amphibians with a reduction in the vitamin A content of the diet; also this may have been helped by the exposure of the epidermis to air, as in the case of the chorionic membrane of the chick.

Nothing appears to be known about the vitamin A requirements of reptiles. While some, like amphibians, live mainly on insects, others are phytophagus or eat small mammals which contain vitamin A; in the python and the giant monitor, both of which are carnivorous, the vitamin A content of the liver may be between 3000 and 4000 i.u./g. Since these animals are relatively heavily keratinized the effective concentration of vitamin A in the epidermis may be kept low by this ability to transfer the surplus to the liver. In mammals the vitamin A content of the liver ranges from about 10 i.u./g in the guinea pig to 13 000–18 000 i.u./g in the polar bear, which is a fish-eater.

CHAPTER 18

Vertebrata: Internal Skeletal Structures

Connective tissue forms the complex endoskeleton of vertebrates. It can be divided into the soft connective tissues and the specialized connective tissues, cartilage, and bone, which form the more rigid elements of the supporting skeleton.

A. General connective tissue

The soft connective tissue is formed basically from collagen fibres embedded in a mucopolysaccharide ground substance, although it also contains elastic fibres, the cells responsible for secreting the components of the connective tissue, blood vessels, and nerves. It is found around and within the various organs of the body, supporting them and attaching them by sheets of connective tissue or mesenteries to the skeleton, so that they are held in place in the body. No connective tissue is found, however, within the nervous system, except where it is associated with the blood supply.

The texture and arrangement of the fibres in the connective tissue varies according to its specific function at any given site, and the various types of connective tissue have been classified by Congdon (1937). Around cylindrical organs such as blood vessels, the gut, and muscles, the collagen fibres tend to be arranged in two spirals at right angles to each other. They are probably important in setting a limit to the amount these organs can expand (Harkness, 1968). Elastic fibres are particularly abundant in the connective tissue of the lungs and blood vessels which are continually changing their shape.

The rupture of the follicle in ovulation, the expansion of the uterus during pregnancy, rupture of the foetal membranes, and changes in the birth canal that make birth possible, all involve changes in the connective tissues, particularly the collagen fibres of these organs. After parturition the whole reproductive system has to return rapidly to normal and this involves the resorption of the additional connective tissue that developed to support the enlarged reproductive system. This subject has been extensively studied by Harkness (1964).

The collagen content of connective tissue never seems to rise above about 30 per cent of the dry weight of the tissue. In most sites where increased support is required it is met by increased amounts of connective tissue, not by increasing the collagen content above this percentage. In some organs, however, such as the liver, increased size does lead to an increase in the percentage of collagen present, the collagen content of the liver of rat is 0·05 per cent, that of elephant 4·0 per cent, although this again may represent varying amounts of a standard type of connective tissue (Harkness, 1968).

1. *The dermis*

Under the epidermis of vertebrates, as in many invertebrates, there is a specialized development of soft connective tissue, the dermis. Together with the epidermis this makes up the skin in this phylum of animals. While the epidermis largely controls the physiological relations of the animal with its environment, the dermis is mainly responsible for the mechanical properties of the skin. It does not play the part the invertebrate body-wall often plays in controlling the shape of the animal. A skinned vertebrate maintains its shape by virtue of its muscles and the skeleton to which they are attached.

In cyclostomes the dermis resembles that of Amphioxus, being formed from layers of collagen fibres, the fibres in each layer being parallel, and at right angles to the fibres in the layers above and below, and making an angle of 45° with the long axis of the animal. Within these layers fibroblasts are very rare (Porter, 1967: see Fitton Jackson, 1968) nor are there any blood vessels present. In elasmobranchs, teleosts, and amphibian larvae the dermis has a structure similar to that of cyclostomes, although the further the animal departs in shape from a cylinder the further is the departure from an orthogonal arrangement of the fibres (Garrault, 1937; Weiss and Ferris, 1954; Rosin, 1946). In these dermal structures the collagen fibrils have well-defined periodicity and in any one ply the periodic bands are aligned in register in adjacent fibrils.

The laminated dermis in fish and larval amphibians is well adapted to the shearing stress that must develop in the body wall through the mode of swimming by flexing the body. In some ways it resembles the cuticle of nematodes that swim in a somewhat similar manner. The cuticle of earthworms, while resembling the dermis of fish in its fibrillar arrangement, is more concerned in accommodating changes in overall length of the body which in nematodes and fish, only occur locally, on one side at a time. In limbed vertebrates, where body-flexing does not play such an important part in locomotion, the more random arrangement of the fibres makes the dermis relatively strong and resistant to shearing in all directions.

Although various theories have been advanced to account for the orthogonal arrangement of the collagen fibrils in fish and larval amphibian dermis (see Picken, 1960, and Fitton Jackson, 1968, for reviews) no satisfactory theory has as yet been put forward. The larval amphibian dermis grows by the addition of new layers directly beneath the epidermis. During the formation and orientation of the fibrils in the dermis they are separated from the epidermis only by the lamina densa. Porter (1956) has suggested that the basal epidermal cells are largely responsible for the organization, although Weiss and Ferris (1954) point out that the cells of the basal layer of the epidermis are continually dividing and moving outwards and this would seem to preclude such an influence. Weiss (1957) and Weiss and Ferris (1954) suggest that the arrangement may be due to the self-positioning of the collagen fibrils relative to one another. Since the average axial period and the average width of the collagen fibrils when fully formed are approximately equal, i.e. 600 Å, and since the periods are aligned in register in adjacent fibrils, Weiss and Ferris suggest that the first layer of fibrils to be formed is laid down in a plane defined by the epidermal surface, with the fibrils 600 Å apart; the periodic structures, being in register, form a squared grid with sides 600 Å long. Subsequent layers of fibrils would be laid down on this grid, each layer being separated from the preceding layer by 600 Å and with the periods in register. The added fibrils would be parallel with the fibrils in the initial grid until the thickness of the ply had been built up. Weiss and Ferris (1954) suggest that at a certain thickness a condition of equilibrium or "saturation" is reached which prevents the addition of more fibrils with the same orientation. For the addition of further fibrils the only permissible direction might be that of greatest deviation, i.e. at right angles to the preceding fibrils, but with the periodic square structures still aligned with those in the layer underneath.

This theory does not attempt to account for the orientation of the initial grid. Picken *et al.* (1947), in nematode and earthworm cuticle, and G. Chapman (1953a), in the mesogloea of *Calliactis*, have considered the effect that tension, compression, and shearing would have on fibrils being formed around a structure which is in essence a cylinder (to which the bodies of many fish and larval amphibia more or less approximate) and concluded that the fibres would adopt a crossed fibrillar, although not necessarily a laminated, arrangement. While Weiss and Ferris reject mechanical stress as the only factor orientating the dermal fibrils because of its failure to account for the formation of laminations, they do suggest that it may be responsible for the orientation of the initial grid layer. As an alternative Weiss and Ferris (1954) suggest that the collagen fibrils may take up positions in a previously orientated ground substance but how the ground substance becomes organized is not suggested.

Probably there are various forces that will produce an orthogonal arrangement of collagen fibrils, for whether or not Weiss and Ferris' theory explains the ordered deposition of collagen fibrils in larval amphibian dermis, it cannot account for the arrangement of the fibrils in nematode and annelid cuticle because here the fibrils have no periodic structure. Rudall (1968a) suggests that the regularly arranged cilia-like processes that project into the cuticle of annelids are responsible for the orientation of the collagen fibrils in this cuticle but in nematodes that lack similar processes this cannot be so, and in pogonophorans, where processes somewhat similar to those in annelids are present, the fibrils in the cuticle are randomly arranged, although here the fibrils may not be collagen.

When the fibres of the dermis have a regular arrangement this gives the tissue a considerable degree of transparency. In larval amphibia the dermis is transparent and the animal is coloured by pigment lying beneath the dermis. In adult amphibia the dermis is opaque and the pigment lies above the dermis. In the cornea of the vertebrate eye, which is formed from connective tissue beneath a layer of epidermis, the collagen fibres also have an orthogonal arrangement (Jakus, 1956).

In adult amphibia and in the higher vertebrates the fibres of the dermis are arranged in bundles which run a characteristically wavy course and are more or less randomly arranged, with fibres running at right angles to the surface as well as parallel, so that the dermis loses its laminated structure. However, traces of an orthogonal arrangement of the fibres can still be found in local areas in reptiles and birds, in the nail-bed of man, and in the thick dermis of the hippopotamus (Niizima, 1960), and also in the dermis of the whale. In so far as an orthogonal arrangement of the dermal fibres appears to be primitive, possibly associated with providing a firm, strong, ply structure under the soft epidermis in fish and larval amphibians, it might be expected to occur in embryos of higher vertebrates. In these embryos the dermis, however, generally remains very thin, but in the neck of a fifteen-day-old chick it reaches a thickness of more than 1 μm and contains orthogonally arranged fibres (Niizima *et al.*, 1954). Orthogonal fibres also occur in the dermis of embryo sheep.

The organization of the fibres in the dermis of higher vertebrates cannot be entirely random for consistent variations in the strength and extensibility of the skin from various parts of the body are found. Cattle hide has up to twice the strength in the direction along the axis of the body that it has at right angles to this, and variations have also been found in man and hippopotamus (see Harkness, 1968, for references). The back and side, but not the belly, skin of hippos is thick and relatively inextensible, and the dermis in these regions is a solid slab of collagenous fibres. Hippos frequently attack each other with their teeth and the rigid nature of the skin makes it difficult to buckle into folds which could then be bitten through (Harkness, 1968).

It is the arrangement of the wavy collagen fibres into a deformable network that makes it generally possible for the skin to be stretched, sometimes as much as 30 per cent in any one direction; this allows for the distortion of the skin in movement and its expansion due to areas of inflammation, increase in subcutaneous fat, or the increased volume of the uterus in pregnancy. Expansion beyond the basic extensibility of the dermis must be met by some breakdown and rearrangement of the original network and by addition of new dermal tissue. However, addition of new material to the dermis may not be sufficiently rapid; in rapidly growing tumours and in pregnancy splits occur in the dermis; these become filled with fibrous material and are visible on the surface as a series of white streaks (striae).

The elastic fibres in connective tissue are weaker than the collagen fibres and do not contribute to the strength of the skin. Normally they are under slight tension as is shown by the gaping of skin wounds, which results from the contraction of severed elastic fibres. Their main function is probably to assist in the restoration of the resting state of the dermis by pulling the meshwork back into shape after it has been deformed. What special function is served by the small number of cellulose fibres found in mammalian dermis by Hall *et al.* (1960) is not known. Under certain pathological conditions they become more abundant.

The mucopolysaccharide ground substance of connective tissue may be responsible for lubricating slight movements between the laminae in fish and larval amphibian dermis under shear stress, and also the movement of collagen fibres within the connective tissue meshwork, though little is really known about the mechanical properties of this mucopolysaccharide material. There is evidence in higher vertebrates that special matrix material cements the collagen fibres into bundles and also forms sheaths around the bundles (see Harkness, 1968 for references). This may prevent the fibres from being damaged by compression stresses that develop when the bundles are flexed and may be responsible for the high tolerance shown by vertebrate connective tissue, and leather (which is made from dermis), to repeated flexing. Alteration of the shape of the collagen network will involve movement of the ground substance and introduce a viscous element into the system contributing to the mechanical properties of the skin.

Elkan (1968) has found, between the inner and outer layers of the dermis in many terrestrial anurans, a special layer of mucopolysaccharides which he believes provides some protection against desiccation in these amphibia. This mucopolysaccharide is almost entirely absent in species that spend their life in water and it does not occur in urodeles.

The dermis as well as the rest of the connective tissue is generally believed to be derived from mesenchyme but there is convincing evidence (Hay, 1964) that, in the amphibian larval stage, the dermis is largely formed by secretions from the overlying epidermis, although mesodermal cells lying against the inner boundary secrete into it proline-rich protein which could contribute to the collagen. Such evidence would account for the growth of the larval amphibian dermis by the addition of layers immediately beneath the epidermis, and for the rarity of fibroblasts in this type of dermis. It also lends support to Porter's suggestion that it is the epidermis which is responsible for the initial organization of the dermis. At amphibian metamorphosis, however, fibroblasts from the mesenchyme pass through the lamellae and accumulate beneath the epidermis; their secretions of fibrils and ground substance displace the lamellae inwards and these are subsequently at least partly broken down by collagenolytic enzymes from the epidermal cells, and by hyaluronidase released by the mesenchyme. The products of this breakdown may be used to form the new dermis which forms through the activity of the subepithelial fibroblasts (Eisen and Gross, 1965; Usuka and Gross, 1965).

Whether or not the epidermis is responsible for organizing the dermis of fish and larval amphibia, the dermis plays an important part in organizing the epidermis in birds and mammals. It has been shown to induce feather formation in the overlying epidermis of birds (Sengel, 1958; Gomot, 1959). McLoughlin (1961a, b) found that epidermis removed from a five-day-old chick was unable to produce keratin and survived only a few days, but if it was recombined with limb bud mesoderm it grew well and produced keratin; only mesoderm from a limb bud would produce this effect, and similar results have been obtained by Wessells (1962) and by Dodson (1963) working with the twelve-day-old chick. The presence of living dermal cells is not essential; dermis killed by repeated freezing and thawing induced keratinization but trypsin-digestion of this killed dermis destroys its induction effect. Dodson also found normal epidermal growth on the surface of collagen gels, suggesting that part of the importance of dermis is simply in providing a firm substratum for the epidermis to rest on.

B. Hardened skeletal systems

Collagen-containing connective tissue also forms the evolutionary and developmental basis for the highly specialized skeletal tissues of vertebrates, cartilage and bone. These are used in two anatomical and morphogenetic components, the dermal and the endochondral skeletons. The first is developed best in fishes but to some extent persists in higher vertebrates. The second is the skeleton proper—skull, backbone, limb-bones, and limb-girdles—all lying deeper in the body.

1. *The dermal skeleton*

Being mesodermal this is an endoskeleton and in contrast to the exoskeleton of so many invertebrates. However, since it is formed in the superficial layers of the dermis, and the overlying layers tissues often wear away, it is usually more or less exposed at the surface. Consequently it is frequently analogous to the invertebrate exoskeleton.

Even the earliest known vertebrates, the Heterostraci, remains of which occur in Ordovician deposits, had in the dermis massive bony plates containing calcium phosphate completely encasing the animal. Such a dermal armour, while it might represent a sudden evolution, could be the result of a gradual development of dermal ossification of which no early record has yet been found.

Romer (1933, 1954) believes that vertebrates originated in fresh water and that the dermal armour evolved as a protection against the strongly jawed freshwater euryptids amongst which they lived. It is now generally believed that vertebrates evolved in the sea and that Heterostraci only later, after they had acquired their dermal armour, invaded fresh water and associated with eurypterids (Denison, 1956; Robertson, 1957; Tarlo, 1962). Smith (1939, 1953) suggests that the dermal armour waterproofed primitive freshwater vertebrates against excessive intake of water by osmosis, but the marine origin of the vertebrates discounts this suggestion. Berrill (1955), again believing in the freshwater origin of the group, suggested that the dermal armour represented a deposition of excess phosphate obtained from the fresh water, but there is no reason to believe that there has ever been a significant difference in phosphate content between fresh water and the sea. However, he did, also suggest that subsequently, the early vertebrates came to use some of their excess phosphate in energy-transfer just as Pautard (1959, 1961a, 1962) has suggested that certain protozoa use internal calcium phosphate deposits. Tarlo (1964), as well as Pautard, believes that the dermal plates of primitive vertebrates were primarily phosphate stores. Tarlo supports his suggestion by pointing out that the phosphate content of sea water tends to be a limiting factor in the development of marine life, and that it varies with the seasons, so that any animal that could accumulate a store of phosphate would be at an advantage. In mammals the mineral components of bone are in physicochemical equilibrium with their ions in the body fluids, and calcium and phosphate are withdrawn from the bone when the concentration of these for any reason falls below normal. Fish also are able to resorb bone-salts to control the ionic content of their blood but, as they normally obtain calcium by absorbing it from the respiratory current of water through the mucous membrane of the mouth, pharynx, and gills, there is little need for them to take it from the bone (Moss, 1962).

There is evidence, also that the jawless Heterostraci were inactive, bottom-living forms, sucking up detritus from the sea floor. This detritus and the water immediately above it are relatively rich in phosphates derived from dead and decaying organisms that sink down to the bottom of the sea. It is from here that the surface waters, depleted of phosphates by the spring growth of plankton, are replenished when the autumn and winter storms bring this phosphate-rich water to the surface again. The food of heterostracians, therefore, was probably rich in phosphates so that phosphate-stores would be of little importance to these bottom-living fish. Their food would have been rich, possibly too rich, also in calcium, from the disintegration of calcified invertebrates. May it not have been that biochemical conditions evolved in the connective tissues that led to the deposition in them of calcium and phosphate, with which the blood was saturated, and the removal of which from the blood could easily be made good from the food? Subsequent evolution then worked on these deposits, controlling and moulding them into usefulness. Whether or not the heavy dermal armour of the Heterostraci gave them advantages over unarmoured ancestors, or placed restrictions on their possible ways of life, subsequent vertebrate evolution in fact has been largely concerned with reductions in the amount of calcified skeleton present (Romer, 1942), not only in land vertebrates, where the weight of the skeleton was a major consideration, but also in aquatic vertebrates where it was not. If a phosphate store was of importance one would not expect evolution to have led to its complete elimination in the modern jawless fish, the lampreys and hag-fish. Furthermore, if phosphate stores are advantageous it is surprising that so few invertebrates have developed them, particularly when there is evidence (Bevelander and Benzer, 1948) that the mineral of mollusc skeletons is first laid down as phosphate and later exchanged for carbonate (see also Ørvig, 1968, for a discussion of this problem).

The earliest jawed fish, the placoderms, were also heavily armoured, although later forms have the dermal armour reduced. At a time when jaws and predatory habits were evolving amongst fishes and there was an increasing chance of being bitten by their associates, so that a dermal armour might appear to be an advantage, it actually tends to be lost. This may be correlated with the evolution of a more efficient swimming mechanism which would provide for escape from predators by flight. In lampreys and hag-fish the loss of dermal armour has not been accompanied by the development of active swimming. Chondrichthyes and Osteichthyes have the dermal armour reduced to a series of dermal scales, or even scales may be absent.

Each plate (Fig. 18.1) of the primitive dermal armour has a basal layer of laminated calcified tissue, a middle spongy layer of calcified deposits round vascular spaces, and an outer layer of hard calcified tissue which is often

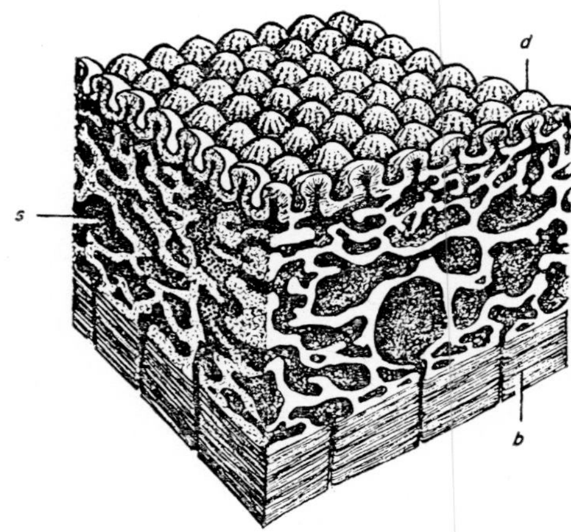

Fig. 18.1. Block diagram of a piece of dermal plate from a heterostracan. *b*, basal lamellar layer; *d*, dentine; *s*, spongy layer. (After Halstead, 1969.)

formed into protruberances or denticles; the material of this layer therefore has been called dentine. Although the morphology of this tissue is different from bone in modern vertebrates, the middle layer approaches fairly closely to that of true bone; Øevig (1967) has discussed in considerable detail the relationship between primitive and more recently evolved calcified tissues and Moss (1964) also has discussed the phylogeny of vertebrate calcified tissues: the subject will not be considered here.

In Selachia, the scales or denticles represent only the upper or dentine layer of the primitive dermal plate; to the outer surface of the dentine has been applied a layer of very hard calcified tissue, enamel, derived not from the dermis but from the epidermis. In Crossopterygii (Sarcopterygii) also, dentine and enamel layers are present, as well as the deeper layers; similar scales are found in the surviving *Latimeria cholumae* and in fossil lung-fish, although living lung-fish have a different type of scale. Fossil Chondrostei have a "palaeoniscoid" scale in which the dentine layer is much reduced and a thick, outer layer is formed from "ganoin", which differs from enamel in various ways. In fossil Holostei the dentine layer has completely disappeared, leaving only the bony layer and enamel. In the related living garpike a similar scale is found but in the bowfin, *Amia*, only the bony plate survives. In living teleosts and the more advanced Holostei the scales are reduced to thin bony plates, although Moss (1964) claims to have found in them evidence of dentine elements. In teleosts the scales are generally arranged in a single layer but overlapping, the anterior portion of the scale being in a dermal pocket, the posterior portion projecting up into, and sometimes through, the epidermis which may become rubbed away and expose the scales. They differ from the denticles of selachians not only in size, shape, and structure but also in not normally being lost and replaced. They grow continually, the rate of growth varying with the season, so that growth rings are formed from which the age of the fish can be calculated. If, however, for some reason the scales are lost they can be regenerated.

The actual size of the scale shows considerable variation. In *Megalops* they are more than two inches across, and in *Barbus* each scale may be the size of a man's palm. In *Thunnus* they are very small; in the common eel, *Anguilla* they are microscopic, while they are completely absent from *Gasterosteus*, the three-spined stickleback. In forms with reduced scales, or no scales, the dermis is often leathery. In puffers and globe-fish the scales are replaced by small moveable spines which stand erect when the body is inflated.

Some bony fishes have re-evolved a heavy dermal armour. The South American cat-fish, *Hoplosternum*, has the body completely encased by a double row of broad, overlapping bony shields. The mailed cat-fishes (Loricariidae) have the sides protected by overlapping plates, although the chest and abdomen are naked or covered by much smaller plates. Sometimes the side-plates have denticles similar to those of Selachians. Pipe-fish and sea-horses have the body almost entirely encased by rings of thin bony material.

Dermal ossifications are not confined to fish. The primitive labyrinthodont amphibian, *Cacops*, had dermal plates on the back, above the neutral spines, and living Apoda have small scales buried in folds of skin. Dermal bony plates form the bulk of the carapace and plastron of the tortoise, and bony scales are also found in *Cricotus*, various Microsauri, and crocodiles, and special structures are found on the head of Theromorpha. Even in mammals, such as the armadillo, dermal bony plates are formed, but these various dermal plates in higher vertebrates do not represent a survival of primitive structures; they are new developments.

The dermal skeleton, even when it formed a continuous covering over the animal, in fact, resulted from the fusion of a number of separate plates. Presumably the gradual growth of these plates allowed for the general growth of the body, complete coverage not being obtained till the animal reached its full size. No doubt the plates, even when in contact with each other, were able to grow round the edges, like the scales of bony fish, for there is no evidence of moulting of the dermal skeleton to allow for growth in either fossil or modern fish; in this the vertebrate dermal skeleton differs from the typical exoskeleton of invertebrates. The loss and replacement of denticles in the elasmobranch dermal skeleton might perhaps be considered as the relic of a primitive moulting process.

2. *Teeth*

One modification of fish dermal scales survives right through the whole range of vertebrates, namely teeth. Scales round the mouth became modified as teeth in various jawed fish. This relationship between scales and teeth is still clearly seen in elasmobranchs. It is to the scales of elasmobranchs that the teeth of higher vertebrates most closely approximate, not only in their structure, but also in the fact that they may be lost and replaced from below. Essentially, teeth consist of a cap of enamel over a cone of dentine which may rest on a bony base, or in some forms possibly the bony part of the original scale may be represented only by the cement that attaches the tooth to the jaw. At the centre of the tooth is a core of "pulp" containing blood vessels and nerves.

The actual form of the teeth is related to the type of food eaten so that completely unrelated animals eating the same food have very similar teeth. The walrus, various reptiles, rays, and the Port Jackson shark, *Heterodontus*, all of which feed on molluscs, have rounded or flattened tooth plates for crushing the shells of these animals. Simple, sharp-pointed teeth for catching fish are found, not only in predaceous fish, but also in some fish-eating reptiles and mammals. In all their complexity, even the dentitions of unrelated carnivorous mammals are similar (Halstead, 1969).

Some animals, like turtles and modern birds, lack teeth, although they were present in the earliest birds. In both turtles and birds the teeth are functionally replaced by a horny epidermal formation, the beak, which is supported by the jaw bone. Tusks, particularly that of the narwhal, are the most striking of the many specializations of teeth.

3. *Endoskeleton*

The function of the endoskeleton is to support the body, while joints in the skeleton make movement possible. It is also important in protecting the more vulnerable organs such as the brain which is covered by the skull, and the heart, lungs, liver, stomach, and kidneys which are covered over by the ribs.

In cyclostomes, the most primitive living vertebrates, the skeleton is formed of stiff cartilage (p. 320) and no bony material (p. 320) is present. Modern elasmobranchs also have a cartilaginous skeleton, although calcium salts are deposited in the cartilage of the backbone and jaws (Ridewood, 1921; Von Klement, 1938; Urist, 1961) and Moss and Freilich (1963) consider that this represents true ossification. In the development of higher vertebrates a cartiliage skeleton is first formed but later it is largely broken down and replaced by bone. It has therefore been held that the primitive vertebrate skeleton was cartilaginous and that the bony skeleton was a later evolution. However, in the Heterostraci, the earliest vertebrates known, and from which modern cyclostomes evolved, parts of a bony skeleton have been found beneath the dermal armour (Romer, 1945); elasmobranchs also evolved from bony ancestors, so that the cartilaginous condition in modern cyclostomes and elasmobranchs in now held to be due to reduction of ossification and is not a primitive condition. A similar tendency to reduction in ossification can be found amongst some modern bony fish. *Acipenser* and *Polyodon* (Chondrostei) both have largely cartilaginous skeletons and in modern crossopterygii also bone has been largely replaced by cartilage.

While modern cartilaginous fishes therefore, do not, indicate a primitive condition this does not necessarily imply that bone is the more primitive skeletal material.

Ultimately, vertebrates must have evolved from an invertebrate type, although the nearest invertebrate ancestor is by no means certain. Skeletal material, very similar to cartilage, both morphologically and chemically, occurs in many phyla of invertebrates (Person and Philpott, 1969). On the other hand, while mesodermally produced calcareous deposits also occur in invertebrates, none really approach vertebrate bone in structure. From this one might argue that cartilage was more likely to be the primitive material and that bone was evolved later.

Romer (1942) describes cartilage as an ideal embryonic skeletal material. Bone grows by deposition of material at the surface so that surface attachments of tendons and muscles have to be continuously re-established. Cartilage, he says, grows by internal expansion and the surface attachments of muscles and tendons remain undisturbed. Le Gros Clark (1945), however, describes young cartilage as growing mainly by surface accretion through the progressive differentiation of mesenchyme cells from the perichondium (the connective tissue sheath that surrounds cartilage structures) into chondroblasts which are then responsible for secreting the components of the cartilage, although some subsequent interstitial growth also takes place.

After the embryonic cartilage has been replaced by bone, growth by no means comes to an end. Cartilage does persist at the ends of the long bones and is progressively replaced by bone as growth continues, but growth of the skeleton does not only consist in increase in length of the bones; it also involves increase in diameter and the continual remodelling of the bones to proportion them to the increasing size of the animal. Therefore, bone can not only grow, but also is able to maintain the necessary muscular attachments under conditions that are much more demanding on the bone—muscle junction, due to the activity of the young animal, than are those in the relatively quiescent embryo.

Furthermore, the bone of the dermal skeleton, where it occurs, and some parts of the endoskeletal bone in higher vertebrates, develop directly in connective tissue membranes without a cartilaginous forerunner. The membrane bones of the endoskeleton, which include mainly the bones of the skull vault and face, and the clavicle, are believed to have been derived in fact, from bones of the dermal skeleton that sank into the body and became associated with the endoskeleton. Therefore, cartilage is not an absolute prerequisite for the development of bone. It would be as reasonable to speculate that cartilage evolved as a modification in the development of a primitive bony skeleton as vice versa.

Elasmobranchs, some of which are much larger than any bony fish, demonstrate that for aquatic animals cartilage is a highly satisfactory material. Therefore, with the advantages of cartilage as a skeletal material in the growing animal and its suitability for forming the skeleton in aquatic animals of considerable size, it is difficult to see, if it was the most primitive skeletal material, what evolutionary pressures led to its general replacement by bone. While none of the above facts prove that bone was the primitive material, they at least suggest that it could well have been.

However, cartilage would not have been suitable for the skeletons of terrestrial animals since it would not have been stiff enough to support their weight. Ossification, already present in bony fish, and particularly in the sarcopterygii, made possible the evolution of a skeleton adapted to support the body when early amphibia became terrestrial. In amphibians evolution has since reduced the heavy, clumsy skeleton of the early forms to a proportionately much lighter structure in modern species. A reduction in weight of the skeleton when feasible, would have considerable advantages for a terrestrial animal. Moreover, obtaining an adequate supply of calcium to form a heavy skeleton may have posed a problem for amphibia no longer immersed in calcium-containing water.

Amphibia are the first vertebrate class to possess a parathyroid gland, which stimulates resorption of calcium from bone to maintain its level in the blood. This level cannot now with certainty be maintained by immediately available outside sources. While bone may not have been important as a store of salts in fishes, it does become so in land vertebrates.

The bones of birds are particularly light, most of them are hollow and some of the bone spaces may connect with air-sacs which extend from the lungs into many parts of the body. The lightness of the bird skeleton is an obvious advantage in flight, and the hollowness of the bone, while reducing its weight, does not necessarily reduce its strength (see below).

Thompson (1942) has discussed the mechanical principles that are exemplified in the construction of the skeleton of tetrapod vertebrates, and points out the similarity between the body supported by the fore and hind limbs and a two-armed cantilever bridge with its load distributed over two piers. In such a structure the beam of the bridge between the supports is prevented from sagging under its own weight by a system of struts and ties attached to the beam (Fig. 18.2), the size and strength of the struts and ties depending on the weight that they have to support. In tetrapods the bodies of the vertebrae form, as it were, the beam of the bridge while the vertebral spines and their system of interspinous ligaments provide the equivalent of the struts and ties preventing the vertebral column from sagging under the weight of the body it supports. While a bridge is rigid, the tetrapod skeleton

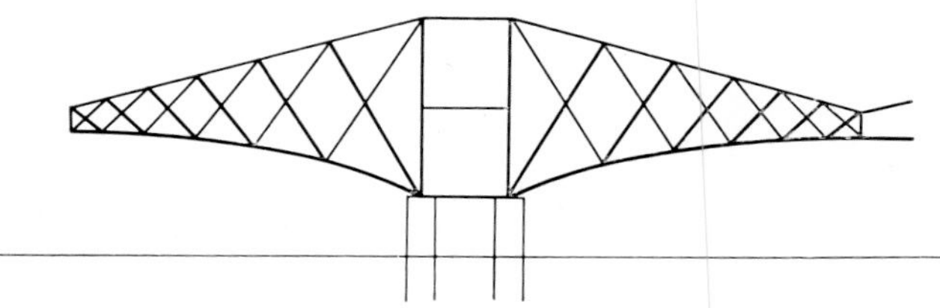

Fig. 18.2. A two-armed cantilever of the Forth Bridge. Thick lines, compression members (bones of vertebral column); thin lines, tension members (ligaments). (After Thompson, 1942.)

is jointed and flexible but manages to remain in equilibrium under considerable modifications of its curvature, such as occur when a cat arches, or flattens its back. The weight of the body is not necessarily evenly distributed between the fore and hind limbs. In both horse and ox the fore part of the animal is much bulkier than its hind quarters. The anterior two-armed cantilever is formed from the head and neck on one side and the trunk on the other, and this weight is so balanced that the structure transmits little of its weight to the hind legs, the amount that is transmitted varies with the position of the head. The hind legs, with their smaller load, are mainly used for propulsion. To support this concentration of weight at the front end of the animal, the dorsal spines of the vertebrae in this region are very elongated compared with those in the hind region, where they have less weight to support (Fig. 18.3). This arrangement of vertebral spines is even more

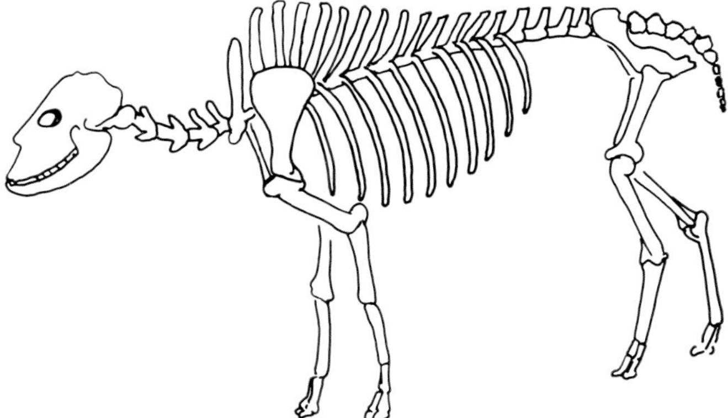

Fig. 18.3. Diagram of skeleton of an American bison showing the very long vertebral spines in the region of the forelegs. Drawn from photograph in Thompson (1942).

marked in *Titanotherium*, a large extinct animal somewhat related to the horse but which had in addition heavy, horn-like processes developed on the head. In dinosaurs and such as *Stegosaurus*, with their small heads and massive tails, the hind end of the body was much heavier than the front and here the vertebral spines were much longer in the hind region of the body (Fig. 18.4), for the same reason this condition is found also in kangaroos. In other animals, like the mouse, in which the distribution of weight between the front and back part of the body is more even, there is no marked variation in length of the vertebral spines along the vertebral columns. In aquatic vertebrates the weight of the body is largely supported by the water and again there is little variation in the length of the spines.

Limb bones too, have been moulded by evolution into structures highly adapted mechanically for the work they have to do. The moment of inertia of a structure, and consequently its resistance to bending, is increased when the matter of the structure is concentrated far from its central axis, so that

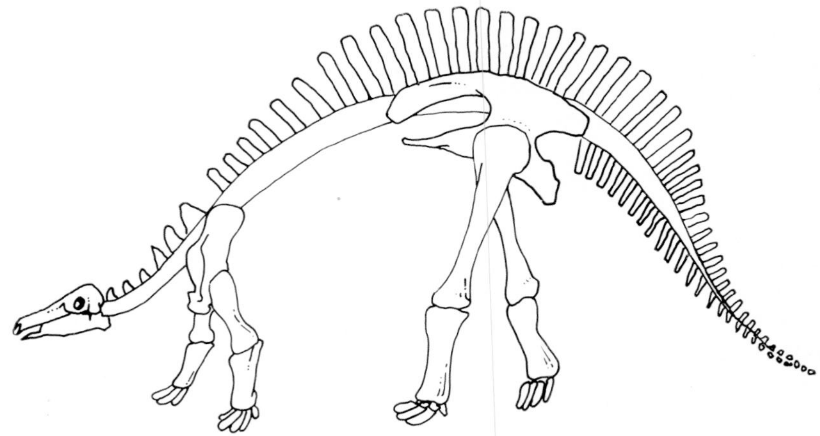

Fig. 18.4. Diagram of skeleton of *Stegosaurus* showing longer vertebral spines in the hind than in the fore part of the body (redrawn from Thompson, 1942).

tubes have a greater resistance to bending than solid rods of the same cross-sectional area, and very much more than a rod of the same total weight. The tubular nature of the long bones makes for lightness and economy in the use of bone material without reduction in the strength of the bone.

If a long bone is subjected to a bending force the maximum strain is taken by the middle of the shaft; correlated with this the wall of long bones is thickest in the middle of their length and gradually thins out towards the extremities. Not only the gross morphology but the organization of the calcareous deposits, particularly in the extremities, are also related to the stresses to which the bone is subjected (Thompson, 1942; Murray, 1936).

The vertebrate endoskeleton would appear to be a much more efficient method of supporting the body than is found in invertebrates, and the generally larger size of vertebrates compared with invertebrates is believed to have been achieved through the mechanical advantages of the endoskeleton. However, in simple aquatic animals the requirements for any type of skeleton are limited, and jelly-fish, supported and held together only by a soft, collagen-containing mesogloea, may weigh as much as half a ton, their body being supported mainly by water and not by a hard skeleton. However, on land where the body has to be supported against gravity the vertebral endoskeleton has an obvious advantage over the soft exoskeletons of such animals as oligochaetes which could not, in large animals, prevent the body shape being determined mainly by gravity. On the other hand, the mechanical advantages of an endoskeleton over a hardened exoskeleton such as that possessed by arthropods is by no means so certain. Chetverikov (1918), Kennedy (1927), and Thompson (1942) have all shown that, weight for weight, the hard exoskeleton is superior mechanically to the endoskeleton. In the exoskeleton the material is further removed from the central axis relatively, than it is in vertebrate long bones. Only just over half the material of a human femur would be required to provide the limb with an exoskeleton of the same strength and if all the femur material were used, the exoskeleton would be seven times stronger (Currey, 1967). Therefore, it has not been a complete mechanical superiority of the endoskeleton that has enabled vertebrates generally to be far larger than invertebrates, nor the unqualified mechanical disadvantage of the exoskeleton that restricts the size of arthropods, for some extinct arthropods, both aquatic and terrestrial, achieved a size comparable to that of the larger vertebrates.

Currey (1967) has, however, pointed out that the calculations of Chetverikov and others, showing the mechanical superiority of exoskeletons over endoskeletons, have been based only on consideration of static loading of the skeleton. If dynamic or impact loading is considered, Currey has shown that the advantage of the exoskeleton is considerably reduced, because under these conditions the stiffer the material subjected to the impact the greater the stress developed. In addition, the flesh around the endoskeleton serves to absorb some of the force of impact, adding further to the advantage of the endoskeleton. In large, active, fast-moving land vertebrates this resistance to impact loading may be of considerable significance, although the endoskeletons of the earliest, slow-moving, aquatic vertebrates cannot have been subjected to greater impacts than the exoskeletons of their associated invertebrates.

The endoskeleton also has apparent advantages over hardened exoskeletons from the point of view of growth. The endoskeleton is able to grow continuously until full size is attained, and there is no interruption during development of the connections between the skeleton and the nerves and muscles on which its movements depend. The moulting of the exoskeleton which is necessary to permit growth, not only involves the arthropod in periods when its body is relatively unprotected, but also requires the periodic severing and re-establishment of connections between the exoskeleton and the nerves and muscles connected to it; this may have reduced the possibilities of the evolution of complex and adaptable movement-patterns.

Whatever factors operated to keep invertebrates generally small, a lack of strength in hard exoskeletons has not been one of them. Since vertebrates evolved from invertebrates, they must have been initially subjected to the same size-limiting influences as invertebrates and the earliest vertebrates were only a few inches to a foot or so long. But some characteristic or combination of characteristics of vertebrate organization promoted the development of larger body size. Possibly this was the presence of a notochord and vertebral column which allowed the evolution of more efficient, and therefore more rapid, locomotion than is found in invertebrates. This in turn allowed them to escape from slow moving predators to live longer and grow larger. With increased size the absolute quantity of nervous tissue in the brain also increased, permitting the development of complex and variable patterns of behaviour which the smaller invertebrates, with their small brains, could not develop. Consequently, when vertebrates and invertebrates came into competition, the vertebrates were able to dominate the slower moving, more reflex-controlled invertebrates, which perhaps under these circumstances tended to become even smaller in order to occupy the environmental niches that the larger vertebrates could not utilize.

The one group of invertebrates, the cephalopods, that has a range of size comparable to that of vertebrates also shows striking similarities to them. Except for *Nautilus*, the extant cephalopods have their shell sunken into the body, although *Octopus* has lost the shell completely, and they have a fairly extensive cartilaginous endoskeleton. They are, except for *Octopus*, actively swimming, pelagic forms with large brains and complex patterns of behaviour that demand serious comparison with those of vertebrates. The earliest cephalopod remains are found in Cambrian deposits and, by the time vertebrates were gaining ascendency, the cephalopods had doubtless evolved sufficiently in a manner analogous to vertebrates to hold their own against these newcomers.

(*i*). *Histology of the endoskeleton*

(*a*) *Cartilage*. Hyaline cartilage, which forms the skeleton of embryos, has the cells or chondroblasts that secrete the cartilage fairly evenly scattered through the matrix, although somewhat concentrated at the outer surfaces. They are generally in discrete groups of two to four cells which have resulted from the division of a single initial chondroblast. The matrix consists of translucent material and the whole tissue is both resistant to compression and has considerable elasticity. The arrangement of the fine collagen fibres embedded in the matrix is generally related to the tensional requirements of the structure for it is the collagen fibres that enable the cartilage to resist tension. If the mass of cartilage is large it is penetrated by blood vessels. Cartilage very similar to hyaline cartilage, covers the articular surfaces of bones in diarthrodial joints such as those at the elbow or knee. These articular cartilages may represent the remains of the cartilage from which the whole bone has been developed, but they are very important in providing bearing surfaces with a low coefficient of friction and this in turn probably depends upon the mucopolysaccharides of the cartilage. In fibrocartilage the collagen content is considerably increased and the cartilage cells are sparse. Such cartilage is found, in the adult, forming symphysial joints between bones in which movement is very restricted, and it also forms part of certain tendons. Elastic cartilage, containing elastic fibres which give it increased resilience, forms the cartilage of the outer ear and certain of the laryngeal cartilages.

(*b*) *Bone*. In bone, calcium salts are deposited in the collagen-containing matrix and these may form as much as 60 per cent of the total weight of bone. A transverse section of a bone such as the femur of man shows a central cavity filled with a yellow, fatty tissue—the yellow marrow. The walls are formed by compact bone and the extremities contain a sponge-work of bony trabeculae, the spaces between which are filled with yellow marrow mixed with a different type—the red marrow. Except for the articular surfaces, the whole bone is covered by a connective tissue sheath, the periosteum, comparable with the perichondrium. In mammals the bone structure is penetrated by fine canals, the Haversian canals, carrying blood-vessels and nerves, and abundant osteocytes are present in them. The calcareous material is deposited in thin lamellae. Immediately beneath the periosteum the lamellae are parallel to the surface of the bone but deeper in they are arranged in concentric circles round, the Haversian canals, and each canal with its associated lamellae comprise "Haversian system". The canals branch and anastomose with each other to some extent and extend both to the marrow cavity and to the surface of the bone. In the spaces between adjacent Haversian systems there are more irregular, interstitial lamellae (Fig. 18.5). Between the lamellae are small cavities or lacunae from which

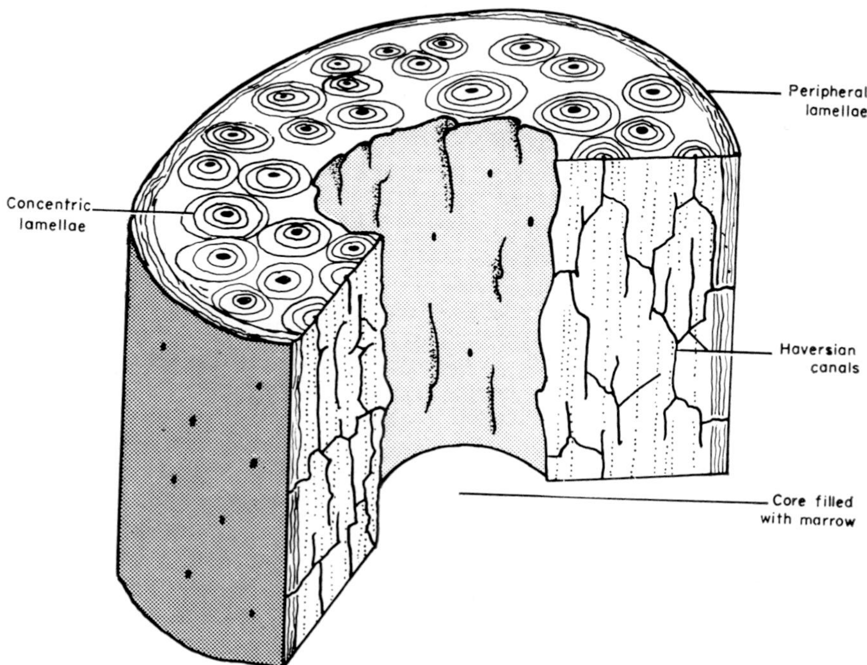

Fig. 18.5. Diagram illustrating the essential structures of bone. The Haversian systems are represented entirely out of proportional scale in order to indicate schematically their arrangement in a long bone. (After le Gros Clark, 1945).

extend very fine canaliculi, connecting up adjacent lacunae with each other and also extending to the marrow cavity, but the lacunae of adjacent Haversian systems are not connected by these structures. Each lacuna contains an osteocyte derived from an osteoblast originally responsible for secreting the organic material of the bone. From the osteocyte fine protoplasmic processes extend along the canaliculi.

(*c*) *Dentine.* The dentine layer of the tooth (Fig. 18.6), which is a bony tissue (p. 322), is penetrated by minute tubules which pass from the pulp cavity to the periphery of the layer. These tubules often branch, particularly in the outer layers of dentine. Apatite crystals and collagen fibrils fill the spaces between the tubules. No cells lie in the dentine, but odontoblasts lying on the outer surface of the pulp cavity send processes into the dentine along the tubules. No nerves or blood vessels pass into the dentine and the sensitivity of the dentine to stimuli may rest on the transmission of stimuli by the odontoblast processes to the nerves in the pulp cavity.

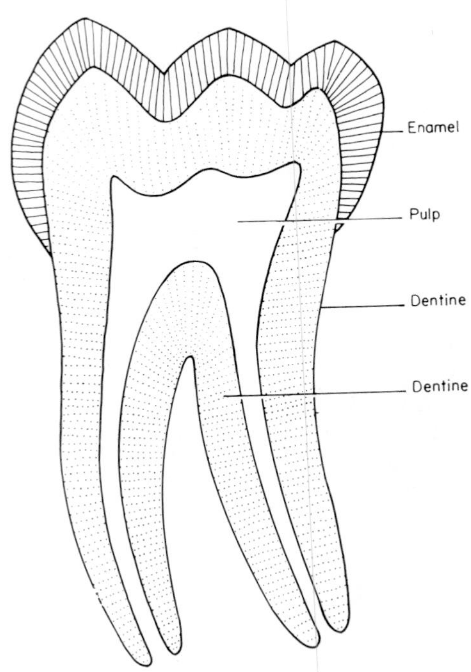

Fig. 18.6. Diagram of a sagittal section of an adult human permanent molar (after Schour, 1932).

(*d*) *Enamel.* In enamel large, prismatic, apatite crystals, standing upright on the surface of the dentine, run outwards through the whole thickness of the enamel layer to the surface. Each prism is separated from its neighbour by a thin film of organic matter. No cells lie in the enamel, which is secreted by the enamel organ formed in the epidermis above the tooth. Initially the tooth is covered by two organic membranes, very resistant to acids and alkalis but these are soon worn away after the tooth has erupted.

C. Chemistry of connective tissue

Chemically there is little variation in collagens from different vertebrates (Table 18.1), and they all have a periodic structure of 600–700 Å. Probably the most significant variation is in the percentage of imino acid groups present and with this varies the shrinkage temperature, T_S and the denaturation temperature T_D: The variation in T_S and T_D is not primarily related to the evolutionary status of the animal but to the normal maximal temperature to which the collagen is likely to be subjected (Fig. 5.6(*a*)). In cold-blooded animals this is determined by the temperature at which the animal normally lives, and in warm-blooded animals by the temperature of the blood. Unfortunately T_S and T_D have not yet been recorded for a really wide range of collagens but T_D is always very near that of the environment of the collagen. Deep-sea fish like cod and dogfish that live in cold water, the temperature of which rarely rises about 14°C, have a T_D of 16°C. Fish like shark, pike and carp that live in the warmer surface water, generally at a temperature between 24 and 28°C have a T_D between 27 and 29°C. No analyses or determinations of T_S or T_D have yet been made for the collagens from various species of *Tilapia* which live in alkaline lakes in Africa, at temperatures which in some lakes approach 40°C (Coe, 1967). The T_D has not been recorded for collagen from the lung-fish, *Neoceratodus*, that lives in rivers in the hot, arid interior of Australia, but the T_S is 63°C which is comparable to that of mammalian collagens. The African clawed toad, *Xenopus laevis*, although it inhabits the same type of environment as the lung-fish and although lung-fish and amphibians are believed to have a common ancestor, has a T_S of only 59°C.

Reptile collagen chemically resembles mammalian collagen rather than amphibian collagen and has a T_S of 59–62°C (Leach, 1957). The only bird collagen to have been examined is that from the leg tendon of the chicken. Although the body temperature of birds is higher (42°C) than that of mammals, the chemical composition of the avian collagen lies within the limits of composition found in mammalian collagens and it has a similar T_S. Unfortunately its T_D does not appear to have been recorded. The T_D of mammalian collagen lies between 38 and 39°C (Bailey, 1968). In birds one assumes it must be slightly higher. The toughness (in the culinary sense) of vertebrate flesh is said to depend largely on the properties of its contained collagen. The tenderness of fish flesh compared with mammalian flesh is taken to exemplify this. Although T_S for avian collagen is approximately the same as mammalian collagen, culinary practice stipulates a higher minimal temperature for cooking poultry (90°C) than for cooking mammalian flesh (60°C).

The biological significance of the correlation between T_D and the environmental temperature is not known but it is difficult to believe that this is purely coincidental, particularly when T_D for various invertebrate collagens is also taken into account. It has recently been suggested that the T_D of an animal's collagen might underlie the mechanism of thermal regulation in warm-blooded animals (Mason, 1966). A mechanism based on some temperature-controlled reaction might explain how the body maintains the same sensitivity to temperature changes throughout life.

On the basis of chemical analysis (Table 18.2) and X-ray diffraction photography collagen from bone appears essentially similar to that from soft

Table 18.1 Amino acid composition of vertebrat

	Dogfish (Lewis and Piez, 1964)	Shark: skin (Eastoe, 1957)	Sturgeon: swim-bladder (Eastoe, 1957)	Pike: skin (Piez and Gross, 1960)	Carp: swim-bladder (Piez and Gross, 1960)	Cod: skin (Piez and Gross, 1960)	Lungfish: skin (Eastoe, 1957)	*Xenopus*: skin (Leach, 1957)
Alanine	110	119	118·9	114	126	107	128	98·0
Glycine	339	333	337	328	325	345	311	301
Valine	27·8	21·5	18·0	18	18	19	21·3	21·9
Leucine	25·5	23·9	17·7	20	21	23	25·2	28·8
Isoleucine	16·7	19·4	11·4	9·2	10	11	12·2	14·0
Proline	99	113·4	102·2	129	116	102	126·0	109·7
Phenylalanine	12·4	13·9	14·1	14	14	13	15·3	19·3
Tyrosine	2·7	1·4	2·4	1·8	2·0	3·5	1·1	6·1
Serine	59	44·5	50·5	41	37	69	43·7	66·3
Threonine	24	25·8	29·2	25	29	25	26·1	26·4
Cystine (½)		0	—	<1	<1	<1	—	—
Methionine	15·7	10·0	8·8	12	13	13	4·0	8·7
Arginine	53	50·3	52·4	45	53	51	51·0	49·2
Histidine	11·7	7·4	4·8	7·4	3·8	7·5	5·1	6·5
Lysine	25·7	24·3	21·8	22	26	25	24·2	29·1
Aspartic acid	44	42·6	47·5	54	47	52	48·6	54·9
Glutamic acid	69	65·8	70·5	81	71	75	78·9	77·9
Hydroxyproline	60·4	78·5	82·0	70	81	53	73·1	77·5
Hydroxylysine	6·3	4·7	10·7	7·9	7·4	6·0	5·3	4·3
Total imino acids	159	192·9	184·2	199	197	155	199	186·5
T_S	40°C	53°C	50°C	45–57°C	49–58°C	37–45°C	63°C	54°C
T_D	16°C	29°C	—	27°C	29–32°C	12–17°C	—	—

connective tissues, although it differs from these collagens in being virtually insoluble in neutral salt solutions and in solutions of organic acids. This almost certainly results from increased non-covalent intermolecular linkages in the bone collagen. These may be due in part to a progressive loss of water from the maturing bone matrix. With loss of water from between the collagen fibrils the distance between adjacent molecules decreases and the strength of the intermolecular bonds increases. While very small amounts of organic phosphorus occur in soft tissue collagen, mainly in the α_2 chains, bone collagen contains slightly more organic phosphorus than soft tissue collagens, again mainly concentrated in the α_2 chain, which may have three to ten times more than α_1 chains. Collagen from dentine contains as much as 56 atoms of P per mole of collagen, although its distribution between the α chains has not yet been worked out (Glimcher and Krane, 1968).

1. *Elastoidin*

Fins of elasmobranchs contain fibrous rods of "elastoidin" that differs in various ways from collagen. Similar rods are found also in the fins and tail of larval teleost fishes. These are replaced in the adult by bony structures, although small rods of elastoidin persist at the end of the bony rays (Garrault, collagens. Residues per 1000 total residues.

Crocodile: skin (Leach, 1957)	Python: skin (Leach, 1957)	Chicken: tendon (Leach, 1957)	Man: skin (Fleischmajer and Fishman, 1965)	Rat: tendon (Eastoe and Leach, 1958)	Wallaby: tendon collagen (Eastoe and Leach, 1958)	Whale: skin gelatin (Eastoe and Leach, 1958)	
114·0	125	114·6	114·5	99·3	112·5	110·5	Alanine
324	315	331	324·4	351·0	320	326	Glycine
15·4	20·2	19·8	24·5	22·5	23·2	20·6	Valine
20·1	25·7	23·8	24·8	22·2	26·3	24·8	Leucine
11·4	11·7	10·9	10·4	13·2	8·9	11·0	Isoleucine
127·9	119·4	129·5	125·1	123·0	119·1	128·2	Proline
17·7	14·2	14·3	12·6	14·3	16·0	13·0	Phenylalanine
3·3	1·8	3·4	3·5	5·4	4·1	3·6	Tyrosine
42·1	43·6	28·6	36·9	27·8	39·0	41·0	Serine
22·0	17·9	19·1	18·3	19·1	20·1	24·0	Threonine
—	—	—	tr.	—	—	—	Cystine (½)
6·5	6·1	6·2	7·0	5·8	6·6	4·7	Methionine
49·5	49·9	44·8	49·0	46·5	51·1	50·1	Arginine
4·7	4·7	4·5	5·4	3·3	5·1	5·7	Histidine
25·3	27·6	19·0	26·6	35·6	24·6	25·9	Lysine
45·5	48·0	48·1	47·2	47·1	49·3	46·3	Aspartic acid
72·8	62·4	74·1	77·7	73·7	73·0	69·6	Glutamic acid
92·8	102·0	98·5	90·9	90·4	92·8	89·1	Hydroxyproline
4·9	4·0	9·6	5·9	—	8·0	5·8	Hydroxylysine
220·7	221·4	227·5	226	213·4	211·9	217·3	Total imino acids
59°C	57–59°C	—	60–67°C	59°C	—	59°C	T_S
—	—	—	36–39°C	37–38°C	—	—	T_D

1936). Fitton-Jackson (1968) has described the formation of elastoidin in regenerating tails of teleosts. Needle-like rods of elastoidin with precise lateral alignment of the constituent molecules are aggregated into filaments and the filaments formed into concentric sheets, to form rods with the component filaments precisely aligned in relation to each other, and each sheet being equivalently aligned in relation to its neighbours. Elastoidin gives collagen-like X-ray photographs (Champetier and Fauré-Fremiet, 1938b), with a highly ordered structure (McGavin, 1962), and it contracts on heating in water, but at a temperature (64°C) above that of ordinary shark collagen (53°C); however, the total imino acid content is low—about the same as in collagen from sharks. Elastoidin does not dissolve to give a gelatin but forms a rubber-like mass. Chemical analysis (Tables 18.3 and 18.4) has separated from it three components, a gelatin, an insoluble residue containing little glycine and hydroxyproline, but a high concentration of tyrosine (Gross and Dumsha, 1958) and a protein rich in tryptophan (Ramachandran, 1962). Electron microscopy shows only ribbon-like fibrils of collagen with a 600–700 Å banding. Presumably there must be some interaction between the collagen and the other components to produce the

Table 18.2 Amino acid composition of oxbone and dentine collagens, compared with tendon collagen. Residues per 1000 total residues. (Piez and Likens, 1960.)

	Ox tendon collagen	Ox bone collagen Cortical	Ox bone collagen Cancellous	Ox dentine collagen
Alanine	108	109	101	112
Glycine	344	337	333	327
Valine	22	20	22	22
Leucine	27	25	26	25
Isoleucine	11	11	11	11
Proline	117	123	118	116
Phenylalanine	13	13	14	16
Tyrosine	4·9	4·3	4·0	3·2
Serine	34	34	36	39
Threonine	17	16	18	16
Cystine	—	—	—	—
Methionine	5·9	5·0	5·3	4·1
Arginine	47	50	49	47
Histidine	4·6	4·1	4·0	3·7
Lysine	20	26	28	19
Aspartic acid	46	45	46	52
Glutamic acid	74	74	78	72
Hydroxyproline	94	98	102	103
Hydroxylysine	11	5·7	5·0	13

Table 18.3 Amino acid composition of elastoidin. Residues per 10^5 g of protein. (Damodaran *et al.*, 1956.)

Alanine	128·0
Glycine	338·0
Valine	23·2
Leucine	20·0
Isoleucine	20·5
Proline	115·3
Phenylalanine	12·7
Tyrosine	39·5
Serine	31·5
Threonine	20·3
Cystine	1·5
Methionine	12·0
Arginine	49·5
Histidine	11·1
Lysine	25·6
Aspartic acid	48·1
Glutamic acid	74·9
Hydroxyproline	66·8
Hydroxylysine	5·4

Table 18.4 Elastoidin (a) Analysis of whole fibre, water-soluble (gelatin) and water-insoluble (residue) fractions. Grammes per 1000 g dry weight. (Gross and Dumsha, 1958.)

	Carcharias glaucus			*Mustelus vulgaris*			*Squalus acanthias*		
Species Fraction	Fibre	Gela-tin	Resi-due	Fibre	Gela-tin	Resi-due	Fibre	Gela-tin	Resi-due
Percent of whole fibre	100	73·3	18·3	100	74·6	14·3	100	80·6	18·4
Glycine	207	254	141	228	278	131	211	238	113
Hydroxyproline	78	100	34	63	90	27	53	59	23
Proline	110	111	73	110	122	74	133	127	76
Tyrosine	70	24	220	67	38	205	82	45	237

(b) Partial chemical composition of elastoidin, and of three fractions extracted from it. Percentage content on moisture-free basis. (Ramachandran, 1962.)

	Elastoidin	Fraction *A* "Elastoin"	Fraction *B* "Elastrin"	Fraction *C* "Elastagen"
Carbohydrate as glucose	0·41	0·41	0·79	0·97
Tyrosine	6·62	21·02	none	none
Tryptophan	1·62	none	10·2	0·04
Hydroxyproline	8·45	3·19	3·97	10·41

marked temperature stability. In spite of its highly ordered structure, elastoidin is more extensible (15 per cent) than collagen (10 per cent).

2. *Elastin and other connective tissue fibres*

Elastins from only a few mammals, and from the chicken, have been analysed, and these show little chemical variation (Table 18.5). The lysine content of chicken elastin decreases during development as the lysine is used to form desmosine and isodesmosine (Miller *et al.*, 1964). No analyses exist for elastin from other vertebrates. Cellulose fibres have been found in mammalian connective tissues but whether they occur generally in vertebrate dermis is not known. Isaacs *et al.* (1963) found cellulose-like material in dentine from a fossil ostracoderm, although whether it was an original component of the tissue it is impossible to say.

3. *Ichthylepidin*

While it is generally held that teleost scales are wholly mesodermal products it has been suggested that they, like the elasmobranch denticle, have epidermal components. The scales have generally an outer calcified layer from which a protein, ichthylepidin, can be extracted and an inner fibrous layer, part of which is calcified and which contains collagen (Moss and Jones, 1964), and it is the ichthylepidin layer which is held by various workers to be epidermal in origin. Ichthylepidin, while it gives a collagen-like X-ray picture and contains hydroxyproline (Table 18.6), does not dissolve to form a gelatin when boiled in water. It has, in the analysis of Seshaiya *et al.*

Table 18.5 Amino acid composition of elastins.

	Chicken[a] 12 day embryo Aorta	Chicken[a] 1 year Aorta	Pig[b] Aorta	Sheep[b] Aorta	Ox[c] Aorta
Alanine	172	177	201	200	199
Glycine	352	352	288	279	239
Valine	177	174	165	184	160
Leucine	62	58	79	81	88
Isoleucine	19	20	27	36	40
Proline	122	124	154	158	—
Phenylalanine	22	22	54	50	60
Tyrosine	11	12	29·1	21·0	23
Serine	5·4	4·1	8·2	—	11
Threonine	4·2	4·6	13·1	11·6	16
Cystine	0·5	0·6	1·5	—	—
Methionine	—	—	—	—	3
Arginine	4·9	4·5	9·3	9·0	17
Histidine	<0·2	<0·2	—	—	3
Lysine	5·7	1·6	5·1	3·7	12
Aspartic acid	1·9	1·8	4·4	6·1	20
Glutamic acid	12	12	28	—	34
Hydroxyproline	24	23	15	14·6	15
Tryptophan	—	—	—	—	5

[a] Miller *et al.*, 1964. Residues per 1000 total residues.
[b] Neuman, 1949. } Grammes amino acid per 1000 g protein.
[c] Gotte *et al.*, 1963. }

(1963), a slightly higher content of imino groups than collagen from the same source but this could hardly explain its marked resistance to solution. It also contains more tyrosine and some cystine and these may contribute linkages that give it its stability.

4. *Matrix of enamel*

The matrix of tooth enamel is secreted by epidermal cells. It is composed almost entirely of protein, which is not evenly distributed through the enamel, but concentrated in the spaces between the enamel crystals. In embryo teeth the matrix forms from 20 to 30 per cent of the weight of the tooth but by the time the tooth is fully grown it forms only 0·C6 per cent, and this reduction is accompanied by chemical changes in the protein. X-ray photography of the matrix shows mainly diffuse halos, characteristic of unorientated molecules, although Glimcher *et al.* (1961) found evidence for a cross-β-type of structure. Graham and Pautard (1963) failed to detect such a structure but obtained an α-keratin configuration on stretching the matrix. It is possible to separate the protein from embryonic bovine enamel into a neutral-soluble fraction with a high proline and histidine and a low serine content which forms 75–90 per cent of the protein, according to the age of the material, and an acid-soluble fraction with a lower proline and histidine and higher serine content (Table 18.6). Very small quantities of hydroxyproline occur in both fractions (Glimcher *et al.*, 1964a), although Piez (1961) and

Table 18.6 Amino acid composition of proteins from bovine enamel, shark enamel, and carp ichthylepidin. Residues per 1000 total residues.

	Bovine embryonic enamel			Bovine erupted enamel			
	Decalcified[a] matrix (incisor)	Neutral[a] soluble fraction (incisor)	Acid-soluble fraction (incisor)	Decalcified[a] matrix (incisor)	High proline[a] fraction (incisor)	Shark scale enamel matrix	Carp scale ichthylepidin
Alanine	22	24	24	59	26	63	71
Glycine	70	52	83	195	70	248	263
Valine	37	37	34	36	37	38	31
Leucine	96	97	102	67	92	36	35
Isoleucine	30	35	23	28	30	28	32
Proline	213	265	173	90	210	80	103
Phenylalanine	36	23	47	39	31	18	20
Tyrosine	49	39	47	7	40	29	25
Serine	63	47	91	102	77	96	55
Threonine	29	27	34	48	30	27	34
Cystine ($\frac{1}{2}$)	1·0	3·0	0·5	11	3	0	4·2
Methionine	49	44	47	5	42	10	29
Arginine	27	16	38	29	23	33	24
Histidine	62	78	43	26	60	10	30
Lysine	19	14	23	35	20	27	16
Aspartic acid	37	34	41	94	55	107	46
Glutamic acid	156	161	144	128	153	73	80
Hydroxyproline	tr.	tr.	5·9	0	0	62	87
Hydroxylysine	2·5	2·7	3·5	~1	~1	17	15

[a] Glimcher *et al.*, 1964a,b.
[b] Moss and Jones, 1964.

Eastoe (1963) failed to find it in embryonic pig enamel and human enamel. Although the protein from erupted bovine enamel contains more serine, glutamic acid, and glycine than that from embryonic enamel, it is possible to separate from it a neutral-soluble, high proline and histidine fraction (Table 18.6) similar to that found in embryonic enamel, although only in small quantities. Maturation of the tooth enamel is accompanied by the loss of protein components rich in proline. No hydroxyproline was found in carefully cleaned samples of erupted enamel (Glimcher *et al.*, 1964b). The enamel proteins, particularly those from erupted teeth, are rich in cystine and in this they resemble intracellular keratin. One might expect enamel proteins, since they are extracellular, to resemble rather the extracellular cement between keratin containing epidermal cells. However, enamel proteins have no similarity to the protein extracted by Dedeurwaeder *et al.* (1964) from between the cells of wool with its high content of tyrosine and very low cystine content (Table 17.1). Enamel protein has high content of phosphorus with 59, 27, and 98 atoms of phosphorus per 1000 amino acid residues in neutral-soluble embryonic, acid-soluble embryonic, and adult matrix proteins, respectively.

Protein from shark enamel (Moss and Jones, 1964) differs from that of erupted bovine enamel in containing significant quantities of hydroxyproline and in a higher content of glycine, although it otherwise shows a marked resemblence to bovine enamel protein. Hydroxyproline in moderate quantities had been previously reported from bovine enamel proteins but when Gilmcher *et al.* (1964) took considerable trouble to obtain samples, uncontaminated by dentine or cementum, hydroxyproline was absent from the material. Whether the hydroxyproline of shark enamel is part of the structure or a contaminant has not yet been determined. The similarity in composition between shark enamel protein and carp ichthylepidin is taken as evidence by Moss *et al.* that ichthylepidin is indeed epidermally produced.

5. *The ground substance*

The chemical composition of the ground substance in which the collagen and elastin and cellulose fibres when present, are embedded is very complex. Various acid mucopolysaccharides (p. 9) have been obtained from mammalian dermis, in which they form about 5 per cent by weight of the tissue. In adult pig dermis the main components are DS (64 per cent) and hyaluronic acid (30 per cent). In embryo pig skin DS only forms from 5 to 12 per cent, while hyaluronic acid forms 78 per cent (Loewi and Meyer, 1958). The percentages of the different acid mucopolysaccharides in human aorta connective tissue in which CS—C is an important component differ quite markedly with age (Kaplan and Meyer, 1960).

In cartilage acid mucopolysaccharides may form as much as 50 per cent of the weight of the tissue. The proportions of the different acid mucopolysaccharides present in cartilage differ from that in soft connective tissue, but the composition of cartilage within the different vertebrate classes also shows quite a wide variation. *Myxine*-cartilage contains SCS—A with a molar ratio of sulphate to hexosamine exceeding 2. In this it recalls the highly sulphated mucopolysaccharides found in some invertebrates. The acid mucopolsaccharides of *Myxine* are also unique in resisting digestion by hyaluronidase. *Petromyzon*-cartilage on the other hand contains only CS—C, although CS—A is found in the notochord. Elasmobranch cartilage contains SCS—C and CS—C, with smaller amounts of CS—A and SKS. Although most elasmobranchs have calcium salts deposited in the spinal cartilage, this does not occur in the seven-gilled shark (Urist, 1961) and in this species no SCS is present in the cartilage. In the bony teleosts the ratio of CS—C to CS—A is approximately 2, but in the sturgeon, *Acipenser*, in which ossification of the skeleton is much reduced, the cartilage contains only CS—C. This suggests some correlation between a high concentration of CS—C and loss of ossification or calcification. In *Latimeria*, which has a mainly cartilaginous skeleton, although descended from well-ossified, freshwater, bony fish, the cartilage is mainly composed of SCS—C; in this it resembles the sulphated acid mucopolysaccharides of *Myxine* and elasmobranchs rather than that of the bony fish (Mathews, 1965, 1967).

The cartilage matrix of amphibians, reptiles, and birds, except for that of the mud puppy, *Necturus*, is mainly composed of CS—A and this is the generally predominant component of young adult cartilage in mammals, although in the cartilage of man CS—C and SKS are the most important components. The composition of embryonic cartilage varies with the age of the embryo and may differ quite considerably from that of the adult (Mathews, 1965, 1967). CS—C is characteristic of embryonic cartilage but this is gradually reduced as the percentage of CS—A increases to that found in the adult. Mucopolysaccharides only account for about 5 per cent by weight of the extracellular organic matrix of bone and the main component of this is CS—A.

Non-collagenous, extracellular proteins also are present in very small quantities in connective tissues. Some of them are plasma proteins but a distinct alkali-soluble protein (Table 18.7), believed to form part of the ground substance, has been extracted from calf-dermis (Bowes *et al.*, 1958). Other proteins (Table 18.7) firmly bound to mucopolysaccharides have been extracted from ox-cartilage (Partridge and Davis, 1958) and from ox-bone (Eastoe and Eastoe, 1954). The collagen fibres of connective tissue are believed not only to be embedded in this complex matrix but to establish linkages with various components of the matrix.

Undoubtedly the mechanical properties of connective tissue are in part determined by the mechanical properties of the acid mucopolysaccharides which form the main component of the ground substance, particularly in cartilage where the ground substance forms such an important part of the tissue. In what way they are responsible for the special properties of cartilage is not known in any detail. The water-holding capacity of mucopolysaccharides, especially sulphated ones (Needham, 1965), impedes the passage of interstitial water, providing the resistance to compression that is characteristic of cartilage. Acid mucopolysaccharides are long chain polymers. A submicroscopic organization of these polymers in ground substance, as has been suggested by Weiss and Ferris (1954), would certainly influence the

Table 18.7 Amino acid composition of non-collagenous proteins from connective tissues.

	Glycoprotein mucopoly-saccharide complex from ox-bone[a]	Mucoprotein from ox nasal septum cartilage[b]	Alkali-soluble protein from calf-skin[c]	Impurity from ox-hide gelatin[d]
Alanine	370	411	51·3	90
Glycine	265	346	63·9	125
Valine	450	454	47·6	63
Leucine	727	773	73·1	117 (Leucine + Isoleucine)
Isoleucine	365	350	38·1	
Proline	424	786	41·7	67
Phenylalanine	286	742	26·5	33
Tyrosine	196	454	23·3	42
Serine	361	265	47·7	25
Threonine	413	327	37·3	25
Cystine	113	—	0	62
Methionine	84	122	5·7	tr.
Arginine	387	23	141·6	75
Histidine	265	183	36·2	29
Lysine	426	350	77·5	50
Aspartic acid	966	73	84·8	100
Glutamic acid	116·7	124	80·1	152
Hydroxyproline	0	0	4·2	11
Hydroxylysine	0	—	?	—
Glucosamine	12·3	6·1	6·3	—
Galactosamine	76·7	5·6	10·2	—

[a] Eastoe and Eastoe, 1954.
[b] Partridge and Davis, 1958.
[c] Bowes *et al.*, 1958.
[d] Maron, 1958.
a, *b*, and *d*: grammes of amino acid per 1000 g of protein.
c: grammes of amino acid nitrogen per 1000 g total protein nitrogen.

mechanical properties of connective tissues but such an organization has yet to be demonstrated. The significance of the varying proportions of different acid mucopolysaccharides in connective tissue structures, both with age in the same species and between similar structures in different species, has yet to be determined.

6. *Mineral composition of bone*

As yet the mineral components, which form about 65 per cent of the dry weight of vertebrate bone, are known only in general terms. Calcium, magnesium, phosphate, carbonate, chloride, fluoride, and citrate are all found in bone and the main component is a calcium phosphate believed to be hydroxyapatite, $Ca_{10}(PO_4)_6(OH)_2$ in the form of small rod-shaped crystals of submicroscopic size (Finean and Engstrom, 1953). However, calcium phosphate with a Ca/P molar ratio lower than that in hydroxyapatite also has been found. When bone mineral is heated to 200–600°C some pyrophosphate is formed, which does not form on heating pure hydroxyapatite (Glimcher and Krane, 1968). Dallemagne (1964) has evidence that, in the early stages of bone-formation, a salt with a low Ca/P molar ratio is the main component but that in later stages it is partly replaced by hydroxyapatite. Magnesium may replace calcium in some crystals. The exact composition of the other mineral components and their relationship to the hydroxyapatite is not known. It seems likely that all these ions are sufficiently isomorphous to be deposited in the crystal lattice of bone mineral independently, but statistically in definite ratios. This is probably also revelant to the mineral components of the skeletons of other phyla.

D. Calcification

How the mineral salts come to be deposited in the organic matrix is as yet far from understood. Normally the ions, which form the mineral matter of bone, are held in a state of equilibrium in the blood and do not precipitate out spontaneously. It is now generally believed that calcification is initiated by a process of heterogeneous nucleation, i.e. many centres which promote the deposition of calcium phosphate are formed in the bone matrix.

Collagen is favoured as providing the nucleating centres. Collagen fibres, not only from demineralized bone but also from soft connective tissue such as rat-tail tendon, when soaked in metastable calcium phosphate solutions induce the deposition of calcium phosphate crystals. Of the various forms of collagen that can be reconstituted from tropocollagen, only native-type collagen fibres are able to initiate crystal-formation (Glimcher, 1959; Glimcher *et al.*, 1957).

There is increasing evidence both from *in vitro* experiments and from observation of the early stages of bone-formation (Robinson and Watson, 1955; Sheldon and Robinson, 1957; Fitton-Jackson Randall, 1956; Fitton-Jackson, 1957) that the nucleation sites occur within the collagen fibres in the gaps between the end of one macromolecule and the beginning of the next, in the $\frac{1}{4}$ staggered arrangement of the native-type fibril. Calculations based on the density of apatite and calcium and the size of the holes in the collagen fibrils indicate that approximately 50 per cent of the mineral phase can be accommodated in the holes.

When first formed, the crystals do not show any preferred orientation in the collagen fibrils. This initial stage is followed by a process of recrystallization and growth leading to the development of asymmetrical crystals with their *c*-axes approximately parallel to the collagen fibrils. When full mineralization of the tissues has been achieved crystals are probably deposited in the collagen fibrils in sites other than within the holes. Restriction of the crystals to sites within the fibrils possibly may be responsible for controlling the amount of calcification that a tissue can undergo, for in all types of bone it remains remarkably constant at about 65 per cent (Glimcher and Krane, 1968).

Glimcher and his co-workers have attempted, *in vitro*, to determine whether any particular side chains of the collagen molecules in the vicinity of the holes react with calcium or phosphate ions to initiate nucleation. They have so far failed to reach any definite conclusions since they have found it difficult to isolate the possible effects of the collagen side chains from the overall steric and electrostatic environment within the holes and the influence they also may have on the metastable solutions of calcium and phosphate (Glimcher and Krane, 1968).

Collagen-bound organic phosphate, which is in higher concentration in collagens from calcifying tissues than in that from soft connective tissues, is suspected of playing an important part in calcification, for it would be ideally suited for initiating, localizing and regulating the formation of inorganic crystals in a highly ordered fashion (Glimcher and Krane, 1968). The breakdown of organic phosphates might provide both phosphates and energy for calcification (Needham, 1964).

Cameron (1961), Decker (1966), and Hancock and Boothroyd (1967) have all failed to detect any relationship between the inorganic crystals and collagen-fibrils, in the earliest stages of calcification, and Cameron suggested that the crystals lying in or on the collagen fibrils in Fitton-Jackson and Randall's (1956) micrographs may have resulted from precipitation from their fixative, which contained calcium and phosphate ions. Hancock and Boothroyd suggest that the long, fine osteoblastic processes which they have shown to permeate the extracellular matrix of ossifying tissue and which are present in particularly high numbers in regions where crystallization is iminent, may be of importance in the initial seeding. Collagen cannot play a part in the calcification of enamel and where calcification occurs around blood vessels it is associated with elastin fibrils and not collagen.

Calcification is always associated with rapid mucopolysaccharide formation, although Weatherell *et al.* (1964) do not believe that this mucopolysaccharide plays a part in calcification. When crystals first appear, the mucopolysaccharide in the immediate area develops metachromatic staining with toluidine blue and prior treatment of the tissue with toluidine blue can hinder calcification (Miller *et al.*, 1952). From electron micrographs Hancock and Boothroyd (1966) describe, in the earliest stages of calcification, an increased contrast between the collagen fibrils and the matrix, as if the fibrils are surrounded and permeated by a material of greater electron density, and a similar observation was made by Fitton-Jackson and Randall (1956). The identity of the material is not known but may be a mucopolysaccharide—calcium complex, in some way concerned with nucleation. Weatherell *et al.* (1964) suggest that mucopolysaccharide could be a source of oxidizable carbohydrate to provide energy for calcification as no glycogen is found in the matrix.

Digby (1966) has suggested that bone matrix, like the periostracum of mollusc shells, has the properties of a semiconductor and that deposition of calcium results from electrochemical reactions between calcium in the body fluids and the alkalinity that prevails in the growing region of the bone, due to the establishment of an electrical potential gradient in the structure.

If collagen from soft tissues can, *in vitro*, provide nucleation sites for apatite deposition, then there must be some mechanism preventing this from happening *in vivo*. Fleisch and Bisaz (1964) have suggested that pyrophosphate, found in plasma and urine and which, *in vitro*, will inhibit apatite deposition, may act as a calcification inhibitor. In the tissue where calcification does take place, pyrophosphate could be destroyed by pyrophosphatase, which has been found in calcifying tissues (see Fleisch and Bisaz, 1964, for references), thus allowing the apatite crystals to be precipitated in these tissues; there is as yet no evidence to substantiate this suggestion. Glimcher (1960) has suggested that calcium-binding by the mucopolysaccharides, that occur in much higher concentrations in soft connective tissues and in cartilage than in bone, may restrain a high proportion of available calcium ions and prevent them from reaching the collagen fibrils.

Calcification is not controlled solely by physical factors however. It is not impossible that bone cells, besides elaborating the substances of the organic bone matrix also play an active part in calcification, but this has yet to be proved. Hormones influence calcification both directly and indirectly. The parathyroid gland, which has evolved as an endocrine organ in amphibia and higher vertebrates, controls the balance of calcium and phosphate in the blood by stimulating the resorption of these materials from bone when their level in the blood for any reason falls below normal. Thyrocalcitonin, secreted by the thyroid, not only inhibits the resorption of calcium from the bone, but is concerned also in promoting the deposition of calcium in bone (Hirsch *et al.*, 1963; Wase *et al.*, 1967; Raisz and Nieman, 1967). These two hormones between them determine the distribution of available calcium and phosphate between the skeleton and the blood system. Various hormones, like thyroxin that effect the general metabolic activities of vertebrates, influence indirectly the skeleton of these animals. Oestrogens have a specific influence on the bones of birds. At the beginning of the breeding cycle secondary calcification occurs in the marrow spaces of the bones and this calcium is later withdrawn to supply calcium for the eggshells, where it is deposited not as the phosphate but as carbonate. A somewhat similar temporary increased deposition of calcium occurs in mice during reproduction. Gonadotropins, in so far as they are growth-promoting substances, influence the growth of the skeleton. In young, castrated, male mice the skeleton remains small. Vitamins, too, influence bone, e.g. vitamin C through promoting the formation of collagen. Resorption of bone, which normally occurs in the remodelling of bone during growth, can be greatly increased by excessive doses of vitamin A. Vitamin D controls the absorption of calcium through the mucous membrane of the intestine and its subsequent availability for bone-formation. It also controls the citrate content of bone. Whether or not it directly influences the matrix of bone is a matter of dispute.

Calcification of connective tissue is not peculiar to vertebrates; it occurs also in various invertebrates, i.e. Porifera, Alcyonacea, echinoderms, molluscs, and as scattered calcareous bodies in some cestodes. In the invertebrates calcification is not intimately associated with collagen fibres, although these are present in the tissues. Invertebrate mesodermal calcification generally originates within cells as small spicules which later may become extracellular and aggregated to form massive structures. Although phosphates may be present in invertebrates, mesodermal calcareous structures are predominantly composed of calcium carbonate, in the form of either calcite or aragonite.

The basis for this difference in calcification between vertebrates and invertebrates is not yet known. Chemical differences do exist also between vertebrate and invertebrate collagens and possibly there are slight differences in the organization of the tropocollagen molecules. Mucopolysaccharides from cephalopod and arthropod cartilage are more heavily sulphated than those from cartilage in bony vertebrates, but resembling the heavily sulphated mucopolysaccharides that occur in some cartilaginous fishes. Whether similar heavily sulphated mucopolysaccharides occur generally in invertebrate connective tissue is not known, nor is it known whether they can inhibit phosphatic calcification of the embedded collagen fibres. Until the processes of calcification in both vertebrates and invertebrates are better understood, it is not possible to say where the differences really lie.

The most usual calcification of invertebrates occurs, not in the dermis but outside the epidermis in the formation of tubes, shells and cups. Although again phosphate may be present, it is the carbonate that forms the bulk of the structure, except in the shells of most inarticulate brachiopods; here the shells mineral is mainly calcium phosphate. Although the epidermis is able, in some phyla, to secrete collagen, yet generalizing from the few analyses that have been made, collagen rarely occurs in the matrix of these external, calcareous structures. Only in two of the 96 mollusc shells that Degens *et al.* (1967) analysed did they find hydroxyproline. It is therefore of interest that Jope (1965) reported hydroxyproline from the matrix of the phosphatic shells of *Lingula* and *Discinisca* but none from the three articulate, carbonate shells she analysed. Of course, if collagen should be present in these external calcareous structures this would not necessarily mean that it is intimately associated with calcium deposition.

Currey (1962) has pointed out how important it is, from a mechanical point of view, that bone is made up of very small crystals embedded in collagen. The crystals are brittle but if one of them cracks, the crack will run out of the crystal into the surrounding collagen, which will deform, but not itself crack. Only under the influence of very high stress will the other crystals become involved. A similar situation occurs in the calcified structures of invertebrates, both in external shells and tubes and in internal structures, where the crystals remain small and embedded in an organic matrix. Even in echinoderm skeletal plates, where each plate may be essentially a single crystal, the spongy nature of the plates, the spaces of which are filled with organic material, helps to prevent cracks spreading right across the plate.

CHAPTER 19
Vertebrate Egg Cases

The majority of free-living, marine animals have unspecialized methods of reproduction, with external fertilization and development from a relatively unprotected egg. Freshwater provides a much more exacting environment for reproduction, generally associated with internal fertilization, well-protected eggs or viviparity. It is therefore surprising, perhaps, that amongst the cyclostomes marine hag-fish lay large eggs surrounded by a tough, brownish case secreted round them by the follicle cells, while freshwater lampreys produce small eggs covered only by a delicate vitelline membrane. The eggs are laid in a nest and covered with sticky material to which sand grains adhere, giving them a measure of protection and camouflage. Nothing is known of the chemistry of the hag-fish egg shell.

All elasmobranchs, the majority of which are marine, produce, after internal fertilization, large eggs well protected by a tough casing, or are viviparous; even where viviparity occurs the eggs are generally surrounded initially by a rudimentary capsule, which is later resorbed. The form of the capsule, which varies in detail from species to species, is more or less rectangular, with the corners drawn out into hollow tubes which serve to anchor the capsule to weeds and through which a circulation of water round the developing embryo is said to take place. The wall of the capsule is made up of three or four distinct layers. Each layer is composed of a number of laminae of orientated fibrils. In all the laminae forming one layer the fibrils are parallel, while the fibrils in successive layers run at right angles to each other, either parallel to, or at right angles to the length of the capsule. The shell-gland that secretes the capsules is situated at the top of the oviduct, and has been described by Filhol and Garrault (1938), Brown (1955), and Threadgold (1957). Early chemical analysis showed that the capsule material contained 0·85–1·47 per cent sulphur (Fauré-Fremiet and Baudouy, 1938) and it was therefore called a keratin. When it is first formed and white it swells, and is dissolved like keratin by agents that break disulphide bonds but when it has turned brown and hard it no longer is attacked by these agents (Brown, 1952). It gives a collagen-type X-ray photograph (Champetier and Fauré-Fremiet, 1938a), and Gross *et al.* (1958) showed it contained hydroxyproline (2·8 per cent) but considerably less glycine, (11 per cent) than normal vertebrate collagen. Attempts to separate from it a hydroxyproline-rich protein met with no success and no other diffraction pattern appeared superimposed on the collagen X-ray pattern (Bear, 1952). Brown found indications of quinone-tanning and Threadgold, who found polyphenol oxidase and a phenol, probably a phenolic protein, in the gland, thinks the evidence favours some form of auto-tanning. He found the gland also produced two different proteins, suggesting that chemically the capsule is heterogeneous. Embryos develop within the capsule often for as long as nine months and the necessity for a resistant capsule is obvious.

The majority of teleosts have external fertilization and the eggs are never surrounded by more than a vitelline memebrane. Some cyprinodonts, however, have internal fertilization and development. Teleosts eggs are of two kinds, small pelagic eggs with thin vitelline membranes which float in the surface layers of water, and larger, demersal eggs that sink to the bottom and have thicker membranes, which are often adhesive, so that the eggs stick together or to weeds and stones. Eggs of freshwater teleosts are of the latter type. In addition, the development of nest-building, habits in freshwater teleosts helps prevent the eggs from being carried into unsuitable environments by river currents. Some small cyprinodonts, living in streams that dry up annually, bury their eggs in mud where they survive desiccation until water returns to the stream bed. The membrane of salmon eggs, formed round it in the ovary, is chemically very resistant (Young and Inman, 1938). Although cystine is present in the material (Young and Inman, 1938; Block, 1938; Brown, 1955), it does not dissolve in reagents that break disulphur bonds. It has not been subjected to X-ray analysis.

Amphibia normally breed in water and have external fertilization, the eggs being protected only by a vitelline membrane and an outer layer of jelly. They have become independent of water for reproduction only where they resort to internal fertilization and viviparity. Reptiles, on the other hand, which have become independent of water for reproduction, although they have internal fertilization, still lay eggs.

Shells of reptile eggs are generally hard and consist of a calcareous layer covered inside and out by organic membranes (Gadow, 1901; Young, 1950), but in chameleons the shell is soft and parchment-like. In the tortoise, *Amyda euphratica*, the calcareous layer is formed from fine-grained fibrous crystals of aragonite radiating from nuclei lying towards the base of the layer and there is little indication of organic matrix being present. The outer membrane is perforated by pores which pass through the crystalline layer to the inner, two-layered membrane. Pores are absent in *Testudo graeca* and *T. radiata* (Young, 1950). In alligator eggs the membranes are fibrous with the fibres arranged spirally around the egg but mutually at right angles in the two membranes (Gadow, 1901). When the shell is soft the egg may increase in size during development due, it is said, to the absorption by the embryo of water through the shell. The incubation of reptile eggs may sometimes last as long as a year.

All birds are oviparous, although there is no particular reason why this should be associated with flight. Pregnant bats are able to fly and, on the other hand, the weight of eggs in the female bird just before laying must be considerable. That oviparity persisted in birds must, in part, have depended on the evolution of complex behaviour patterns related to nest building, incubation of the eggs, and care of the young (Bellairs, 1960). Bird egg shells, although composed of a calcareous layer between an outer and inner organic membrane, differ from reptile shells in being proportionately thinner and more pervious to water and this can be correlated with the shorter incubation time of the bird egg (Young, 1950). They also differ in having the calcium in the form of calcite. It is from the shell that the embryo obtains calcium for skeleton-formation. Apatite is withdrawn from the maternal skeleton and the Ca changed into calcite to form the shell and then calcite is withdrawn from the shell and changed into apatite again to be deposited in the skeleton of the developing chick!

The outer membrane or cuticle of the hen's egg shell is 90 per cent protein, containing appreciable quantities of cystine and tyrosine (Table 19.1). The organic matrix of the calcite layer contains 70 per cent protein and 11 per cent polysaccharide. Chemically the matrix protein resembles the non-collagenous protein from mammalian cartilage (Table 19.1). Of the polysaccharide, 35 per cent is chondroitin A and C. Some galactosamine also is present, in a form other than chondroitin sulphate, together with glucosamine galactose, mannose, and fructose. The inner membrane again contains

Table 19.1 Organic matter of hen's egg-shell. Grammes of amino acid per 1000 g of organic matter.

	Matrix[a]	Cuticle[a]	Inner[a] membranes	Non-collagenous protein (Pig-cartilage)
Alanine	50	36	26	564
Glycine	54	93	37	707
Valine	53	—	—	446
Leucine } Isoleucine }	87	42	33	626
		45	15	299
Proline	38	43	66	107
Phenylalanine	40	30	39	331
Tyrosine	6	46	29	187
Serine	44	58	54	638
Threonine	52	58	48	335
Cystine	18	42	82	tr.
Methionine		69*	60*	—
Arginine	53	62	46	973
Histidine	6	8	28	287
Lysine	23	52	24	327
Aspartic acid	67	90	94	558
Glutamic acid	115	128	148	757
Hydroxyproline	0	0	0	0
Hydroxylysine	—	—	—	—
Total	706	902	823	722·4

* Methionine + Valine.
[a] Baker and Balch, 1962.
[b] Muir, 1958.

appreciable quantities of cystine (Table 19.1). Neither the cuticle, the matrix, nor the inner membrane contain hydroxyproline (Baker and Balch, 1962). The inner membrane, besides lacking hydroxyproline, has an amino acid composition quite different from collagen but gives a collagen-like X-ray photograph (Champetier and Fauré-Fremiet, 1938b). Nematinid silk, which also lacks hydroxyproline, similarly gives a collagen-like X-ray photograph,

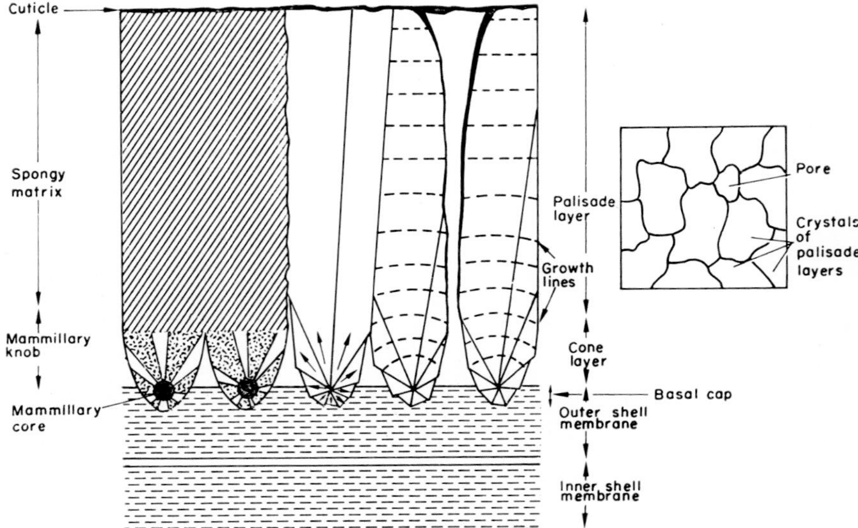

Fig. 19.1. Radial section of a generalized bird egg shell. On the left a decalcified shell showing the distribution of organic matrix. On the right the mineral structure of the shell. The arrows indicate the crystal axes. Inset is a tangential section of the outer part of the shell. (Wilbur and Simkiss, 1968, after Schmidt, 1962.)

but here the amino acid composition has some resemblance to that of collagen.

The detailed structure of the calcite layer differs from species to species (Tyler and Simkiss, 1959; Terepka, 1963). In the hen's egg, the calcite forms columnar masses stretching from the inner membrane to the cuticle, the inner part of each column being cone-shaped and embedded in the inner membrane (Fig. 19.1). A recent review of theories of egg-shell formation has been given by Wilbur and Simkiss (1968). Pores pass though the mineral between the columns of calcite and these are of great importance in the respiratory exchanges of the developing embryo.

In the hen's egg the vitelline membrane separating yolk from white is thin, transparent, and made up of two layers. The inner layer is formed in the ovary before the egg is fertilized and is composed of a three dimensional network of protein fibres that chemically most resemble the non-collagen protein component of connective tissue (Table 19.2). The outer membrane, which is added in the oviduct after fertilization, is made up of a varying number of sub-layers, each a lattice of fine fibrils. In certain regions of the outer sub-layer, particularly near the chalazae, the fibrils run together to form bundles thick enough to be seen by the light microscope. Chemically the membrane material resembles lysozyme and conalbumen, proteins which are concerned in preventing bacterial invasion (Table 19.2). Between

Table 19.2 Amino acid compositions of inner and outer layers of vitelline membrane of hen's egg, compared with those of other relevant proteins. Residues per 1000 total residues.

	Inner layer of vit. memb.[a]	Non-collagenous protein from dermis[b]	Wool keratin[c]	Outer layer of vit. memb.[a]	Lysozyme[c]	Conalbumen[c]
Alanine	75	72	55	77	84	58
Glycine	97	90	99	89	92	89
Valine	70	67	46	56	50	82
Leucine	113	102}	98	75	67	78
Isoleucine	34	53}		46	50	45
Proline	97	59	95	46	17	50
Phenylalanine	29	38	25	32	25	41
Tyrosine	16	32	30	18	25	30
Serine	79	67	108	85	80	70
Threonine	53	52	61	77	59	58
Cysteine	16	0	112	47	84	19
Methionine	5	6	6	7	17	16
Arginine	62	52	68	75	92	51
Histidine	36	17	8	16	8	19
Lysine	17	55	22	50	50	81
Aspartic acid	84	120	61	131	167	118
Glutamic acid	117	113	109	73	33	95
Hydroxyproline	—	5	—	—	—	—
Hydroxylysine	—	—	—	—	—	—

[a] Bellairs *et al.*, 1963.
[b] Bowes *et al.*, 1958.
[c] Tristram, 1953.

the inner and outer parts of the membrane is a continuous, fine granular layer. After about the first 4 days of incubation the yolk becomes enclosed by the yolk sac and the vitelline membrane disintegrates (Bellairs *et al.*, 1963).

The eggs of monotremes, although very small, are covered by shells which appear essentially similar to those of reptile eggs. There are four layers, and the layer below the outer cuticular layer contains a limited amount of calcium carbonate (Hill and Hill, 1933). Shells also occur round the eggs of the marsupials, *Dasyurus* and *Phascolarctos*. The shells are broken when the eggs are transferred to the pouch (Young, 1962).

Where shells occur round vertebrate eggs they resemble invertebrate exoskeletal structures in various ways, although exoskeletons are secreted by the epidermis while vertebrate egg shells are secreted by the oviducts, which are mesodermal in origin. Vertebrate egg shells may be stabilized by quinone-tanning like some invertebrate exoskeletons, e.g. those of some coelenterates some nematodes and molluscs, and arthropods, and may be strengthened by calcium in the form of calcite or aragonite, as in the shells, tubes, etc. of many invertebrates. While the quinone-tanned protein of elasmobranch eggs resemble collagen, as does the quinone-tanned protein of the byssus of *Mytilus*, the matrix of calcified vertebrate egg shells, like that of the majority of invertebrate shells, contains no collagen. Since the epithelium of the oviduct of vertebrates can, in the elasmobranchs, produce collagen while brachiopods demonstrate that calcium phosphate can be deposited in external membranes, there is no apparent reason why vertebrates could not use the same material for their endoskeletons and for their egg shells, but presumably there must be some reason why they do not.

In using different materials for skeletons and egg shells vertebrates resemble many invertebrates. Trematodes have quinone-tanned protein egg cases. Nematodes have a chitin-free, collagenous cuticle but often produce egg shells containing considerable quantities of chitin, although these eggs, on the face of it, are subjected to approximately the same hazards as are those of trematodes. Arthropods have chitin in the cuticle but do not often employ it in making egg shells and oothecae. In the latter there is more emphasis on quinone-tanned proteins. These do occur also in the cuticle but it may be significant that in one insect at least, the robin moth, tryptophan and not tyrosine is used as the tanning agent for the cocoon. Whether these variations are essentially functional in significance, or rather represent the most economical deployment of materials that may be in short supply for the animal, is not known. Not all the calcium of the bird's egg shell is used in forming the embryo's skeleton. However, by forming the shell of carbonate which is in abundant supply this ensures that no phosphate, which is less easily obtainable, is lost to the animal. In this case, therefore, the phosphate anion, but not the calcium cation, may be the subject of an economy mechanism.

CHAPTER 20
The Decay of Structural Materials

Eventually an animal dies, through the attack of a predator or parasites, accident, disease, or old age, and at once processes are set in motion that disintegrate and dispose of the body. If this were not so evolution of life would have been impossible, since it has depended upon the availability of the elements of life which can only be maintained by the return to circulation of the material contained in organisms that have died.

When an animal is eaten whole by a predator most of the soft tissues including collagen can be digested and used as food by the predator. Most of the hardened tissues, because they are more slowly digested or immune to attack by digestive enzymes, pass undigested through the body. Faecal pellets of owls contain the undigested bones and fur of the mice they have eaten. However, although the owls cannot digest the fur, other organisms can and the faecal pellets become colonized by taenid moths, larvae, and fungi able to digest keratin.

What a predator leaves is soon eaten away by a succession of other animals that feed on carrion, and by bacteria and fungi, and the same fate befalls animals that die from other causes. The actual sequence of events depends upon the size of the animal, and the environment and other conditions under which it dies. Details of the disposal of bodies under the sea or in fresh water have not been worked out because of the difficulties in observing the process, but the disposal of large bodies on land have been quite fully studied (see Elton, 1966 for references). The sequence of scavengers and actual species involved varies not only with geographical locality but also with the season of the year. In England the main large carrion feeders are foxes, badgers, crows, and gulls while insects, particularly beetles and flies, are important as they both feed on the carcass and lay eggs in it so that their larvae have an abundant food supply. Soon all the soft tissues of the body are disposed of, while again the hardened tissues are more slowly broken down. Enzyme systems are known that will break down chitin, keratin, and the organic components of bones and shells. The breakdown of calcareous structures is further aided by various organisms that are able to bore into them (Yonge, 1963b), not only thus breaking them down mechanically, but also opening up paths by which other organisms can penetrate into the material and by increasing the surface that is exposed to destructive agents.

As yet, however, no enzyme capable of breaking down quinone-tanned protein has been described. Quinone-tanned proteins often occur in egg cases making them resistant to the same organisms that destroy dead bodies, so that the shell can protect the developing embryo, or the parasite's egg until it reaches the right host. Escape from such egg cases is generally by a special micropyle or operculum. In spite of the resistance of quinone-tanned proteins to chemical disintegration, quinone-tanned skeletal structures are not greatly in evidence in the environment, although crab exoskeletons are a common feature of the sea shore and so are empty elasmobranch egg cases. Although these structures may not be broken down enzymatically, they tend to be brittle and eventually are probably reduced to small particles by abrasion. Fossil insects are not very numerous and there are only some 31 known specimens of fossil elasmobranch egg cases.

Under certain conditions the processes leading to the complete decomposition of the body are arrested and the body may become embedded in sediment and preserved as a fossil. The conditions that arrest decomposition vary enormously, from desiccation in hot, arid regions to death by falling into highly saline water which serves to pickle the body, or being trapped in a bog where the humic acid from decaying vegetation may tan the body proteins and protect them from decay. Rolfe and Brett (1969) have reviewed the literature dealing with conditions that prevent decay and aid the preservation of the remains to form fossils. They also consider the chemical changes that are likely to occur, with time, to the fossils embedded in the sediments.

The greater the amount of hardened skeletal material an animal possesses, particularly material hardened by inorganic salts, the more likely it is to occur as a fossil, because this material not only decays slowly but also is sufficiently robust to withstand being broken up by mechanical forces. Under exceptional circumstances, however, soft tissues also become fossilized, preserving even cellular details. Occasionally animals that lack any hard parts may leave surprisingly detailed casts of their bodies, when they become trapped in fine-grained, rapidly deposited sediments. It is hardly surprising, however, that there is virtually no fossil record of such animals as platyhelminthes, nemertines, nematodes, annelids, and tunicates, which either totally lack hardened structures, or possess only small or microscopic hardened teeth or bristles; even if these survive in some form as fossils, they can neither be detected with ease nor identified as belonging to specific animals. Simpson (1960) has estimated that only about 1–10 per cent of all species that have lived have left a fossil record. He has further considered in detail the factors that have determined both the nature of the fossil record of life on earth and the availability of this record to the palaeontologist.

It is possible from the study of fossils to get some insight into the evolutionary history of animals but this is necessarily almost entirely concerned with the evolution of anatomical structure. It has been shown recently, however, that organic material occurs within certain fossils that were formerly thought to be entirely inorganic, and this would appear to offer possibilities of obtaining a record of certain aspects of biochemical evolution as well.

Calcified structures seem to offer good protection to enclosed organic material, for Grégoire (1958, 1959) has obtained, from decalcified fossil mollusc shells, conchiolin that resembles morphologically that from related modern shells, and a number of workers (see Armstrong and Tarlo, 1966, for references) have obtained from fossil vertebrate bone and teeth banded fibres that resemble collagen.

Since the animals generally have undergone at least some decomposition before fossilization sets in it is not surprising that the amount of organic matter present in fossils is usually considerably less than in comparable modern structures. Ho (1965) found that bone from late Pleistocene bones (up to 1 million years old) contained from a sixth to four-fifths the amount present in recent bone, while fossil teeth of the same age contained between a sixteenth and two-thirds of that of their recent counterparts. The amino acid composition of the protein remaining also shows considerable differences from that of modern related species. This is most clearly seen in organic remains of fossil bone where one can assume that the original material was mainly collagen with a chemical composition similar to modern collagen. Armstrong and Tarlo (1966) found no hydroxyproline or hydroxylysine in pliosaur bone and tooth from upper Jurassic rocks (150 million years old), although both did contain relatively large amounts of glycine and alanine (Fig. 20.1). The amino acid composition of late Pleistocene bone is much

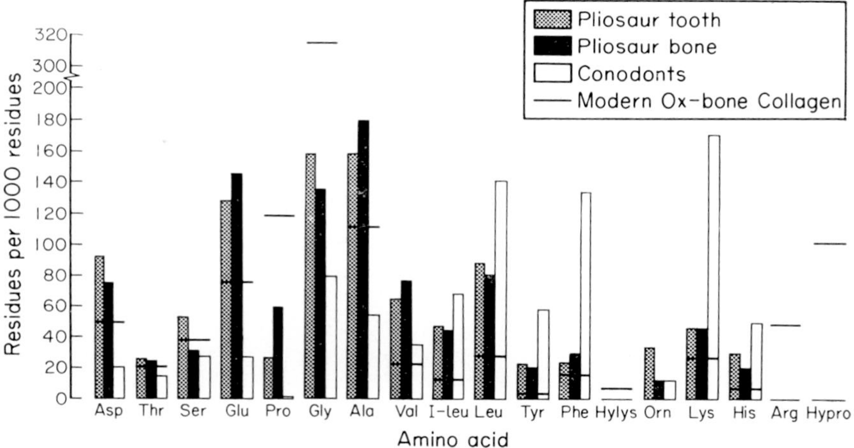

Fig. 20.1. Comparison between the amino acid composition of various fossils and modern ox-bone collagen (Armstrong and Tarlo, 1966).

closer to that of modern bone, and both hydroxyproline and hydroxylysine are present in concentrations not much below that in modern bone (Table 20.1), although glycine occurs in somewhat higher concentrations (Ho, 1966).

Comparison between the organic material of fossil mollusc shells and that of modern species is made difficult by the variations that are known to occur in amino acid content between recent species and, even to a limited

Table 20.1 Amino acid composition of proteins from Late Pleistocene bones and modern bone. Residues per 1000 total residues.

	Fossil bones			Recent
	Horse	Ungulate	Direwolf*	Ox femur
Alanine	110·8	97·8	62·1	105·6
Glycine	371·5	359·3	370·7	325·3
Valine	8·6	14·6	5·2	15·8
Leucine	23·4	30·0	17·0	26·3
Isoleucine	9·4	8·2	5·2	7·0
Proline	83·9	104·4	107·6	107·8
Phenylalanine	9·4	15·9	14·6	15·6
Tyrosine	1·2	0·7	0·8	6·4
Serine	31·0	34·6	12·9	36·9
Threonine	15·3	27·8	8·1	24·0
Cystine	—	—	—	5·7
Methionine	21·3	7·6	6·0	11·8
Arginine	48·1	59·6	103·4	47·8
Histidine	4·5	9·4	16·1	7·6
Lysine	25·5	35·4	69·3	27·5
Aspartic acid	69·6	45·6	33·1	62·2
Glutamic acid	71·4	63·8	73·3	71·1
Hydroxyproline	72·9	77·0	83·1	86·7
Hydroxylysine	4·1	8·2	10·2	7·5

* *Aenocyon dirus.*

Table 20.2 Amino acid composition of conchiolin from fossil and recent nautiloids. Residues per 1000 total residues. (Florkin *et al.*, 1961.)

	Nautilus Eocene	*Aturia* Oligocene	*Nautilus* Modern
Alanine	97	152	272
Glycine	208	233	357
Valine	31	57	22
Leucine	56	62	19
Isoleucine	31	33	18
Proline	37	48	18
Phenylalanine	—	—	} 24
Tyrosine	—	—	
Serine	240	167	109
Threonine	56	38	15
Cystine	—	—	—
Methionine	—	—	—
Arginine	—	—	—
Histidine	—	—	—
Lysine	—	—	—
Aspartic acid	87	90	90
Glutamic acid	156	119	55
Hydroxyproline	—	—	—
Hydroxylysine	—	—	—

extent, between individuals of the same species according to the conditions under which they live. Some genera, like, *Nucula*, *Mytilus*, and *Lithodomus*, extend from Palaeozoic times to the present but no single species has such an extended range in time so that there is no absolute standard for comparison. Florkin *et al.* (1961) found that conchiolin from a Eocene (70 million years old) and an Oligocene (40 million years old) fossil nautiloid contained less glycine and alanine but more serine and glutamic acid than that from modern *Nautilus* (Table 20.2). Not all the differences were sequential with time, indicating that *Aturia* had diverged from *Nautilus*.

Jope (1967a, b) has demonstrated, in recent brachiopods, consistent systematic differences in the amino acid composition of the shell matrix in the different orders of both articulates and inarticulates, and has used these as standards against which to test the amino acid composition of fossil shells from these orders. The fossils showed raised glutamic acid and serine and lowered glycine content (Table 20.3) the differences increasing with the age

Table 20.3 Amino acid composition of fossil and recent brachiopod shells. Residues per 1000 total residues.

	Articulata				
	Terebratulida		Rhynchonellida		
	Epithyris oxonica Jurassic	*Terebratulina* sp. Recent	*Camarotoechia* sp. Mid Silurian	*Globirhynchia subobsoleta* Jurassic	*Nodosaria nigricans* Recent
Alanine	69	51	77	65	86
Glycine	171	258	132	208	332
Valine	55	99	71	57	31
Leucine	79	52	81	72	23
Isoleucine	44	55	35	39	14
Proline	24	29	29	21	88
Phenylalanine	34	24	36	28	83
Tyrosine	18	—	22	17	2
Serine	119	86	114	132	69
Threonine	33	48	53	29	27
Cystine	—	1	—	—	2
Methionine	11	11	7	5	9
Arginine	53	83	47	47	39
Histidine	14	6	17	17	15
Lysine	50	37	58	45	13
Aspartic acid	92	121	92	95	119
Glutamic acid	134	79	129	122	46
Hydroxyproline	—	—	—	—	—
Hydroxylysine	—	—	—	—	—

of the fossil. No other genus of animal has been known to exist over such a long period as the inarticulate brachiopod, *Lingula*, which is found in Ordovician rocks (500 million years old) and is still living today round the shores of Japan. The amino acid content of *Lingula* shells covering this span of time should make an interesting study.

It is also possible to obtain organic material from structures that were not initially calcified. Abderhalden and Heyns (1933) obtained chitin from the wing sheaths of fossil coleopterans and Carlisle (1964) obtained protein and chitin from the 550-million-year-old fossil, *Hyolitellus*, which gave support to his suggestion that it represents a pogonophore tube.

The differences in amino acid composition between fossil and modern types not only result from the initial partial decay of the animal before fossilization but appear to be also dependent on the age of the fossil. Abelson (1957, 1963), in a series of "accelerated ageing experiments", has attempted to determine how the composition and subsequent history of the embedding rock are likely to have affected the organic content of the contained fossils. Proteins in solution progressively hydrolyse but at room temperature complete hydrolysis would take 10^4–10^5 years. It is speeded up by heat, by the presence of certain salts in solution, by changes in pH, and by the presence of autolytic enzymes. The individual amino acids show different stabilities. At room temperature, solutions of alanine might persist for billions of years but would disintegrate in about 1000 years at 120°C and in a few hours at 250°C; rocks during their history may often be subjected to temperatures as high as 200°C for long periods of time. Glutamic acid, glycine, isoleucine, leucine, proline, and valine have about the same stability as alanine, while threonine, tyrosine, lysine, and phenylalanine are less stable; serine and aspartic acid are relatively easily destroyed and tend to be in reduced amounts, or lacking in the oldest fossils. However, Armstrong (Tarlo, 1967) has now found, in similar material, some of the amino acids Abelson failed

Table 20.4 Amino acid content of various vertebrate fossils. (Abelson, 1963.)

Name	Approximate age (years)	Formation	Amino acid content (μM/g)	Principal constituents
Plesippus (prehistoric horse)	Late Pliocene 5×10^6	Hagerman Lake Beds, Idaho	0·6	Ala, gly
Plesippus (prehistoric horse, tooth)	Late Pliocene 5×10^6	Hagerman Lake Beds, Idaho	1·5	Gly, ala, leu, val, glu
Mesohippus (prehistoric horse, tooth)	Oligocene 40×10^6	White River, Nebraska	0·31	Ala, gly
Mosasaurus (dinosaur)	Cretaceous 100×10^6	Lance, Lance Creek, Wyoming	2·8	Ala, gly, glu, leu, val, asp
Stegosaurus (dinosaur)	Jurassic	Morrison, Como Bluff, Wyoming	0·26	Ala, gly, glu
Dinichthys (fish)	Devonian 360×10^6	Ohio Black Shale	3·0	Gly, ala, glu, leu, val, asp

to find in his oldest fossils. Abelson (1963) himself has shown, in one series of analyses of fossil bone, that the oldest material contained the largest variety and the highest concentration of amino acids (Table 20.4).

Erdman *et al.* (1956) have shown, however, that the same amino acids that have been found in fossils also occur free in the surrounding rock. Armstrong and Tarlo (1966) have determined the concentration of the different amino acids at two distances from an ichthyosaur bone of upper Jurassic age. Glutamic acid, valine, isoleucine, tyrosine, phenylalanine, and lysine are found in the highest concentration nearest the fossil suggesting that these amino acids may have diffused into the rock from the fossil. The other amino acids are in higher concentrations further out from the fossil indicating that these, derived from the decay of other organisms, have diffused from the rock into the fossil (Fig. 20.2). The amino acid content of

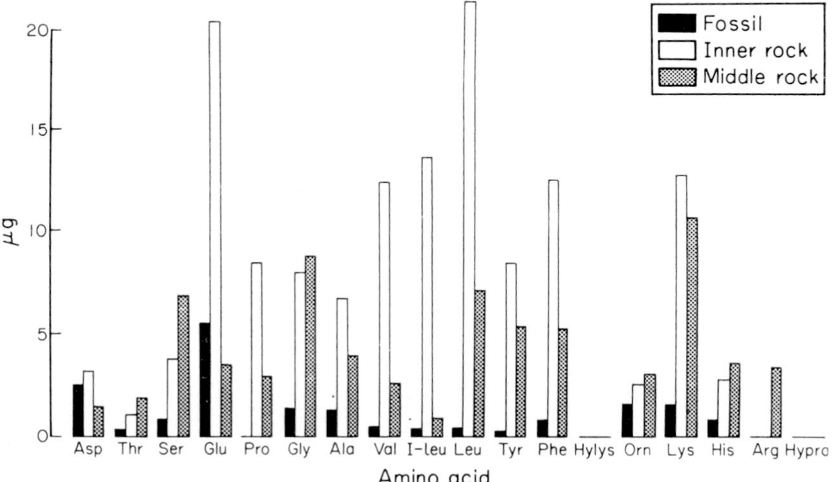

Fig. 20.2. Comparison between the amino acid content (μg/g of material) of a fossil and of the rock at two distances from the fossil (Armstrong and Tarlo, 1966).

fossils is influenced, therefore, not only by the initial amino content of the material and by time, but also by a variety of other factors whose precise effect it is impossible to estimate precisely. Certainly, it hardly seems likely, as suggested by Ho (1967), that the initial imino acid content of fossil bone collagen can be estimated with sufficient accuracy to indicate the body temperature of extinct animals.

In spite of these difficulties analyses of organic material from two types of fossil provide information that deserves to be given some consideration. Foucart *et al.* (1965) found that graptolite remains contained no chitin as they might be expected to do if graptolites were related to coelenterates. The chemical composition of the skeletal material resembles rather that of the periderm of recent *Rhabdopleura*, supporting Kozlowski's (1947) contention that graptolites are related to the Pterobranchia. Again the dissimilarity between the amino acid composition of bone and teeth from a Triassic pliosaur on the one hand, and an upper Devonian conodont on the other, do not on their face value substantiate Tarlo's (1960) suggestion that conodonts which are calcium phosphate-containing structures formed parts of primitive vertebrates (Armstrong and Tarlo, 1966).

Eglington and Calvin (1967) have called "chemical fossils" the resistant, complex, organic molecules which can have been formed only by living organisms and which occur in sedimentary rocks unconnected with fossil structures. It is hoped that the knowledge of the occurrence and variety of these substances in rocks of different ages will not only throw light on the evolution of these complex molecules, but also help to establish whether life existed at the time the earliest known rocks were laid down. Techniques used for looking for chemical fossils may also help to establish if life has ever existed on the moon or when rock specimens become available, on the planets. Components of structural materials have not yet been looked for amongst these chemical fossils but perhaps this is where the remains of quinone-tanned skeletal structures will be found.

CHAPTER 21

Conclusions

The mechanical properties of skeletal materials almost invariably depend upon the combination in them of two or more different structural elements. Fibrous proteins or fibres of chitin provide coherence and pliability, the particular orientation of the fibres determining the relative strength of the structure in any particular direction. The parallel arrangement of collagen fibres in tendon provide maximum strength in a direction parallel to the fibres, a more random arrangement of collagen fibres, as in mammalian connective tissue, making the structure more equally strong in all directions. The fibres are always embedded in a matrix, although the relative proportions of fibres to matrix shows considerable variation. The matrix helps to hold the fibres in place and protects them from damage by compression (Needham, 1965). They may also lubricate the relative movements of the fibres that must take place in certain structures, such as connective tissue, skin, and cuticles, with movements of the animal. These matrices may be mainly protein, as in arthropod cuticle and vertebrate keratin, or mainly acid mucopolysaccharide, as in connective tissue and cartilage of both invertebrates and vertebrates. Not only are the fibrous units embedded in the matrix, but almost certainly linkages of various sorts are established between the different components so that the structure can be considered as a chemical unit.

While strength and hardness are essential for some skeletal materials, elasticity is important in others. Elastin and the elastic fibres found in various invertebrates by Owen and Elder are not the only materials in animals that are markedly elastic. Resilin, like elastin, owes its extensibility to a randomly arranged network of protein chains, linked together by covalent bonds, it is unusual in that, even when these networks are stretched, they do not become sufficiently orientated to give an X-ray diffraction picture. Hair which has a well-defined X-ray picture owes its elasticity partly to the chains becoming pulled out to the extended β-form, and partly to straightening out of disorientated regions of the chains; the same mechanism may also account for the elasticity of the egg cases of *Buccinum*, although here the material is very highly organized. Most silk fibroins, the organized regions of which are already in the extended β-form, owe their elasticity to the chains in the disorganized regions, which may form as much as 70 per cent of the material, becoming reversibly pulled out straight. The byssus threads of

Mytilus and the hinge material of bivalves, both of which give a collagen type X-ray photograph, are elastic but on what the elasticity depends is not yet known, nor is it known in the case of the spermatophore of cephalopods.

Animal structural materials are remarkably stable in the normal environment of the animal, a necessary characteristic in materials that are relied on for support and protection. This has been achieved by the establishment of intermolecular linkages that are uninfluenced by the sort of changes in their environment that they are likely to encounter. When the environment is particularly severe, as on dry land or in the gut of a parasite's host, the structural materials at the surface of the animal and forming the egg shells are virtually immune, in the case of terrestrial animals, to the effects of changes in temperature and humidity, to the attacks of bacteria and fungi, and generally to abrasion, while in parasites they are also resistant to the ionic conditions of the host's gut and to digestive enzymes. The stability of structural materials has often been increased, also by the elimination of water from the material, making it dry and hard.

Over and above the van der Waals' forces and electrovalent bonds that exist in all structural materials, the formation of disulphur bonds as in keratin, the conversion of tyrosine or lysine residues to form intermolecular bridges, as in resilin and elastin, the addition of quinones to tan the protein, and the deposition of inorganic salts are ways in which stability is achieved and by which strength and hardness are added to the materials, so little is really known about invertebrate structural materials that it is more than likely that other types of linkages influencing the properties of the materials also occur. Even in vertebrate materials there is undoubtedly much still to be discovered.

It is not possible to see any clear evolutionary trends in the type of structural material used or in the types of linkages employed. Quinone-tanning occurs in the walls of bacteria (Clutterbuck, 1968), in Radiolaria, possibly in sponges, certainly in coelenterates, platyhelminthes, nematodes, annelids, arthropods, molluscs, and vertebrates. Forms of auto-tanning involving tyrosine may turn out to be more primitive than tanning employing an independent tanning agent, for the tyrosine of mammalian collagen is believed to be located in atypical (low imino acid content) peptide sequences at the end of monomers and these sequences may control aggregation of the monomer units (Rubin *et al.*, 1963, 1965). Quinone-tanning is not restricted to one type of protein: both collagen-like protein, as in the byssus of *Mytilus* and the egg case of elasmobranchs, and non-collagenous proteins in the arthropod cuticle, may become quinone-tanned.

Even keratin, which formerly was considered as essentially a vertebrate structural material, appears to have a more general distribution. Disulphur-bonded proteins occur in the spindle fibres of all cells. They also occur in coelenterate nematocyst walls, in the sclerites of trematodes, in cestode embryophores, and in echinoderm epidermal cells. It may be that the epidermis of all animals has the capacity to produce either keratin or mucus according to conditions, as has the epidermis of birds and mammals, but so far this does not seem likely. The epidermis of turbellarians, which produces mucus in aquatic forms, continues to produce mucus in those forms that have become terrestrial, although the epidermis of related trematodes is able to produce keratinized structures. Moreover, vitamin A does not appear to be essential for the majority of invertebrates, so that some other control of the switch mechanism would be necessary. Disulphur bonds are also widely distributed in extracellular proteins, in both invertebrates and vertebrates, and in both they may be associated with quinone-tanning in stabilizing structural materials.

Chitin occurs widely in invertebrates, including protozoans, although it has not yet been found in vertebrates. It is found in external secretions of the epidermis; only rarely and without certainty, has it been reported from internal connective tissue. Purified chitin fibres are stronger than silk, and hair, and slightly stronger than collagen, and so chitin must confer considerable strength on structures like arthropod cuticle. However, it is not always apparent why some chitin-containing structures need great strength. *Pocillopora* coral matrix is 98 per cent chitin, while other related species contain no chitin at all, although they all live under approximately the same conditions. The tubes of pogonophorans contain chitin although they are embedded in mud and are probably rarely subjected to tensile stress. It may be that sometimes chitin is used to save protein rather than for the provision of strength and this idea is further supported by its occurrence in egg cases (see page 152). Cellulose occurs sporadically right through the animal kingdom and it may occur more widely than it has yet been shown to do.

Inorganic salts of various kinds are also of wide occurrence, particularly calcium salts, although the deposition of calcium phosphate in close association with collagen fibres appears to be restricted to vertebrates. Whether this indicates a fundamental difference between vertebrate and invertebrate collagen is not known, although neither X-ray analysis nor electron microscopy indicate such a difference.

Perhaps this paucity of well-marked evolutionary trends in the type of structural materials used in animals indicates that even the earliest animals had wide potentialities for the manufacture of different kinds of structural materials and that evolution has not been dependent on the evolution of new structural materials, but rather on the adaptation and combination of basic materials to new purposes. However, once first bone and then a keratinized epithelium had evolved in the vertebrates, these conditioned the whole evolution of this phylum, and this is paralleled in the invertebrates by the part the hardened cuticle has played in the evolution of the arthropods.

There is little doubt that of all the structural materials the most important is collagen. While it has not yet been found definitely in protozoans, it occurs in all multicellular animals, forming the framework on which they are constructed, and the basis for the complex endoskeleton of vertebrates. It is secreted and adapted to form cuticles, threads, and egg-cases. From X-ray and electron microscopic evidence, its organization is remarkably constant in all groups of animals, although its chemical composition shows a certain amount of variation. This is shown mainly in the total content of imino groups which in vertebrates determines the thermal stability, and this in turn is related to the environmental temperature of the collagen. However collagens from mammals and from invertebrate parasites of mammals, which all have approximately the same T_D, do vary in their content of imino acids (Fig. 5.6(*b*)). In a number of invertebrates, while T_D is related to the environmental temperature, there is no relationship between the T_D and the total imino acid content of the collagen. In these invertebrate collagens the T_D is higher than would be expected from the imino acid content and so factors other than the imino acid content must contribute to their stability. Invertebrate collagens tend to contain a few cystine residues which are lacking in vertebrate collagens and slightly more tyrosine (Table 21.1). Both these residues could contribute covalent links that would increase the stability but undoubtedly there is still a very great deal to be found out about invertebrate collagens.

Table 21.1 Tyrosine and cystine content of collagens. Residues per 1000 total residues

	Tyrosine	Cystine half
Internal collagens		
Spongin *A*	4·7	3·3
Spongin *B*	4·0	6·0
Metridium	7·9	3·2 (cysteine)
Physalia float	5·6	1·6 (cysteine)
Actinia equina	0·4	—
Fasciola hepatica	5	—
Ascaris lumbricoides	17	0
Macracanthorhynchus	10	tr.
Digaster longmani	9	3
Lumbricus sp.	4	0
Allolobophora sp.	8	4
Paracentrotus	4·5	—
Thyone	7·9	2·5
Holothuria forskali	11	0
Strongylocentrotus	6	—
Lytechinus	2	—
Helix aspersa	9	—
Insect corpus cardiacum	24·8	5·0
Insect carcass	31·7	7·9
Human skin	3·5	tr.
Rat tendon	5·4	—
External collagens		
Nematode cuticle	4	16
Rhinodrilus fafneri	19	7
Digaster longmani	6	0
Lumbricus sp.	6·3	tr.
Lumbricus sp.	0	n.d.
Lumbricus sp.	6	0
Lumbricus terrestris	2·3	0
Pheretima megascolidioides	4	4
Allolobophora sp.	3	1
(*Nematus* silk)	25	—

Recently, evidence has begun to accumulate suggesting that in collagen from invertebrates and the most primitive vertebrates, the triple helix has three identical α-chains (Pikkarainen and Kulonen, 1967) and this is certainly the case in *Actinia equina* (Nordwig and Hayduk, 1969) and in cyclostomes (Pikkarainen, 1968). It is only in elasmobranchs (Lewis and Piez, 1964) and other gnathostomes above that that there is evidence for chemical differences in the chains of triple helix. While most vertebrates have at least two different chains, cod-skin collagen has all three chains different (Piez, 1964; Olsen, 1967) and this is now believed to be the case also in some other vertebrate collagens. Pikkarainen suggests that possibly the appearance of different α-chains may be related to the appearance of ossification in bony fishes and this in turn might be associated with the concentration of phosphorus in the α_2 chains of mammalian bone and dentine collagen, but this does not recognize the evidence for ossification already in the ancestors of present day cyclostomes.

Only a very small quantity of carbohydrate is firmly associated with vertebrate collagens but invertebrate collagens have considerably more. These include amino sugars, hexoses, pentoses, and methylpentoses (Table 21.2). When vertebrate collagen is autoclaved the sugars are removed from

Table 21.2 Sugar content of vertebrate and invertebrate collagens: Grammes per 1000 g dry, ash free weight.

	Hexosamine	Hexose	Pentose
Vertebrate			
Cow[a]	2	8	1
Calf[a]	4	15	1
Carp[a]	4	10	2
Skate egg case[a]	2	20	4
Invertebrate			
Earthworm cuticle[b]	176	140	24
Helix gelatin[c]	6·5	120	—
Ascaris cuticle[d]	6·5	1·9	15
Thyone[a]	11	33	6
Metridium[a]	10	89	9
Physalia float[a]	5	57	8
Spongin *A*[a]	18	110	27
Spongin *B*[a]	2	49	5

[a] Gross *et al.*, 1958.
[b] Watson, 1958.
[c] Williams, 1960.
[d] Watson and Silvester, 1959.

association with the collagen but this is not the case in invertebrate collagen and the sugars remain attached to the gelatin (Gross *et al.*, 1958).

Now that the genetic code for the production of the different amino acids is being worked out it is possible to explore not only the genetic basis for changes in structural proteins but also the genetic relationships between

Table 21.3 Serine, proline, hydroxyproline, threonine, and alanine content of various structural proteins. Residues per 1000 total residues

	Serine	Proline	Hydroxy-proline	Threo-nine	Alanine	Total
Internal collagens						
Spongin *A*	38	78	108	43	56	323
Spongin *B*	24	73	94	21	93	305
Metridium	54	63	49	39	113	318
Physalia float	47	63	61	33	66	270
Actinia equina	38	75	93	38	66	310
Fasciola hepatica	32	111	95	25	57	320
Ascaris lumbricoides	40	129	59	31	66	325
Macracanthorhynchus	38	109	58	26	69	300
Digaster longmani	59	59	115	33	54	320
Allolobophora	52	53	95	36	59	295
Paracentrotus	63	87	67	41	83	341
Thyone	54	109	60	39	71	333
Holothuria forskali	55	81	54	51	114	355
Strongylocentrotus	88	90	73	33	72	348
Lytechinus	52	95	100	32	92	371
Helix aspersa	61	104	100	28	72	365
Insect corpus cardiacum	50	64	9	45	81	245
Insect carcass	63	88	41	40	105	337
Human dermis	37	125	91	18	114	385
Elastin	8	109	—	8	223	348
External collagens						
Ascaris lumbricoides	17	291	19	16	78	421
Rhinodrilus fafneri	89	21	99	68	89	366
Digaster longmani	98	10	116	77	92	393
Lumbricus sp.	85	11	153	57	99	403
Lumbricus sp.	105	13	165	52	103	438
Lumbricus sp.	121	7	155	53	97	429
Lumbricus terrestris	83	8	165	49	100	405
Pheretima	86	8	148	54	92	388
Allolobophora	85	22	137	59	100	403
Brachiopoda						
Periostracum						
Lingula } Inarticulata phosphatic	89	45	21	45	132	330
Discinisca } Inarticulata phosphatic	47	6	12	54	231	349
Crania Inarticulata calcite	93	9	—	74	81	259
Laqueous } Articulata calcite	97	33	—	32	16	178
Terebratalia } Articulata calcite	70	70	—	37	41	159
Shell matrix						
Lingula	53	67	43	34	215	412
Discinisca	35	55	137	27	188	442
Crania	140	49	—	58	61	308
Laqueous	85	40	—	48	49	222
Terebratalia	90	58	—	51	74	273
Mollusca						
Periostracum						
**Akera* } Gastropods	112	45	19	33	101	310
**Hydactina* } Gastropods	73	38	12	72	89	284
**Dolabella* } Gastropods	105	31	—	55	79	270
**Aplysia*	120	37	—	52	89	289

Table 21.3 *(continued)*

	Serine	Proline	Hydroxy-proline	Threo-nine	Alanine	Total
Mollusca *(continued)*						
**Nautilus*	67	74	65	55	79	300
**Mytilus edulis* Bivalva	60	24	—	19	38	141
**Artica islandica* Bivalva	23	25	—	5	21	74
Shell matrix						
Mytilus edulis Inner aragonite	114	44	—	24	259	341
outer calcite	96	31	—	14	302	443
Abductin	36	7	—	7	27	77
Mytilus edulis byssus	42	83	71	34	84	314
Arthropod silks						
Nematus	110	97	—	35	138	380
Pachypasa otus	93	—	—	10	272	375
Pachymeta flavia	105	—	—	12	342	459
Braura truncata	86	—	—	10	293	389
Galleria mellonella	167	47	—	18	249	481
Resilin	79	76	—	30	109	294

* Degens *et al.*, 1967.

different types of structural protein. Rudall (1968a and b) has pointed out that collagen, elastin, resilin, and many arthropod silks all contain a high proportion of glycine residues but show considerable variation in their content of alanine, proline plus hydroxyproline (since hydroxyproline is probably formed *in situ* in the collagen molecule from proline), serine, and threonine, but that if the sum of all these residues present in collagens, elastin, resilin, and silks is considered, the totals show a marked similarity (Table 21.3). The genetic code words representing the combinations of uracil, cytosine, guanine, and adenine which specify these four amino acids differ only in the initial letters of the code (Fig. 21.1) and a change in the

SERINE	UCU	UCC	UCA	UCG
PROLINE	CCU	CCC	CCA	CCG
THREONINE	ACU	ACC	ACA	ACG
ALANINE	GCU	GCC	GCA	GCG

A, adenine; C, cytosine; G, guanine; U, uracil.

Fig. 21.1. "Code words" for serine, proline, threonine, and alanine.

initial letter results in one of the three other amino acids being produced instead (Nirenberg *et al.*, 1965). Such changes have been brought about chemically, for instance, by treating cytosine with nitrous acid which causes the amino group (NH_2) to be replaced by a hydroxyl group (OH) converting it to uracil and there is evidence that this type of change has taken place in the production of mutant strains of viruses (Fraenkel-Conrat, 1964).

In collagen itself, although the sum of serine, alanine, imino acids, and threonine remains approximately constant throughout both invertebrate and vertebrate material, there is a slight tendency for the total of these amino acids to increase with the evolutionary status of the animal source (Table 21.3). The total for external collagens is markedly higher than for internal collagens. While there is a certain amount of variation in the amount of each amino acid present, there is a tendency for alanine and imino acids to increase (although the amount of proline converted to hydroxyproline is very variable) and for threonine to decrease.

Rudall suggests that similar but more extensive changes have brought about the evolution of various different structural materials from a basic structural protein. He takes as his basic material fibroins from Warwicker's Group III silks (Table 12.2), with a glycine content approximately similar to that of collagen, but with a much higher alanine and lower proline content and lacking hydroxyproline. By changing some of the code words for alanine to those for proline, some of which is then converted to hydroxyproline, a collagen-like protein would be obtained. Since as far as is known silk fibroins are restricted to arthropods where they fulfil a specialized role, while collagens are universally present in multicellular animals, it would seem more reasonable, if these proteins are related in this way, to take collagen as the basic material.

The sum total of the amino acids under consideration in fibroins is above that in internal collagens generally, and considerably above those in the two insect collagens that have been analysed (Table 21.3), although these are from an orthopteran and not from a lepidopteran. Since fibroins are epidermal secretions it might be more reasonable to compare them with the cuticular collagens of nematodes and annelids. These cuticular collagens do have a total of alanine, serine, imino acids, and threonine nearer to that of Group III fibroins than do internal collagens.

Resilin differs from collagen in lacking hydroxyproline and in its higher content of aspartic acid and tyrosine, from which the di- and tri-tyrosine linkages are formed. Its content of the amino acids under consideration is lower than that in the majority of collagens but above that in corpus cardiacum and corpus allatum collagens. However, if the mean of these amino acids in these two insect collagens is taken, then this is practically identical with the content of these amino acids in resilin (mean for insect collagen 292/1000; Resilin 294/1000). Moreover, the mean content for tyrosine also is virtually the same as that of resilin (mean for insect collagen 28/1000; resilin 27/1000), but as in the case of fibroins it might be more logical to look for similarities with external collagens. Resilin has little similarity on the basis of alanine, serine, imino acids, and threonine-content, with collagen of either nematode or annelid cuticle, and also no similarity with the fibroins, except that they also have a high tyrosine content. Perhaps a material intermediate between resilin and fibroins will be found amongst the proteins of insect cuticle when this has been fully described.

Elastin differs from collagen, not only in virtually lacking hydroxyproline, but also in containing considerably more alanine. Elastin also contains less lysine but this may be due to the conversion of some of the lysine into desmosine and isodesmosine. As far as is known, true elastin does not occur in invertebrates, although elastic fibres somewhat resembling elastin and digested by elastase do occur in many of them. Assuming the evolution of elastin from collagen, did elastin evolve from a primitive collagen through an elastic fibre stage, to become a true elastin in vertebrates, or did it evolve initially in the vertebrates? Unfortunately, the chemical composition of invertebrate elastic fibres is not yet known. The total of relevant amino acids in bovine elastin (348/1000) is above that of collagens from the more primitive invertebrates, but somewhat similar to that in echinoderm collagens. Since elastin occurs in all vertebrates, did it evolve first in primitive vertebrates or even in the protochordates and then undergo evolution independent of collagen, or is it in each animal directly related to the collagen of that animal? While the content of these four amino acids in bovine elastin is below that of bovine collagen (374/1000), there is other evidence that would relate an animal's elastin to its collagen. In the first place, it is thought that elastin is secreted by the same cells that secrete collagen. Secondly, Burton *et al.* (1957) have converted collagen into typically elastin-like material by treating it with alkaline buffers and certain pancreatic enzymes. These experiments suggest a relationship between an animal's collagen and elastin, although not a direct genetic relationship; it is one based on enzymatic conversion of collagen to elastin by splitting off an hydroxyproline fraction which, in turn, leads to the loss of the collagen helix-structure.

The totals for the relevant residues in the periostracums of inarticulate brachiopods with phosphatic shells (Table 12.3) lie within the range of the totals for invertebrate collagens but are somewhat lower than those for external collagens. The totals for the periostracums of articulate brachiopods with calcite shells are very considerably lower, while that of *Crania* is intermediate. Perhaps it is significant that the totals for the articulates approach those of the periostracums of bivalve molluscs that have a similar shell mineral composition. The totals for the bivalve periostracums are reflected also in the total for abductin. The totals for *Nautilus* and gastropod periostracums, some of which contain hydroxyproline are nearer those of the inarticulates. The totals for the shell matrices of inarticulates approach those for invertebrate external collagens: so do those for *Mytilus edulis* shell matrix, while those for articulate brachiopods approach those of invertebrate internal collagens. Therefore, from the point of view of these particular residues, the periostracums of inarticulates, gastropods, and *Nautilus*, and the shell matrices of brachiopods and bivalves may possibly have some relationship to collagen, but the periostracums of articulates and bivalves and abductin could hardly have any close evolutionary connection with it.

Undoubtedly, it is far too simple-minded to hope to point out evolutionary relationships between structural proteins by only considering the serine, alanine, proline, and threonine-content of these different materials. Serine can be produced by other code-words as well. From the evidence of chemically induced mutations in tobacco-mosaic virus the code for aspartic acid can mutate to produce alanine, that for serine to produce phenylalanine, glycine, and leucine, while the proline code can change to give leucine (Fraenkel-Conrat, 1964). While it would appear reasonable that the various structural proteins have evolved from a primitive type of fibrous protein, and that most probably a collagen, it is not yet possible to substantiate this theory.

Definite cuticles may have evolved from specialization of the mucus coverings so often secreted by primitive animals that lack a cuticle, although mucus is secreted by special cells within the epidermis whereas all epidermal cells are generally concerned in the secretion of cuticles. Mucus contains both protein and polysaccharide and has the potential to evolve in various directions. Where the cuticle is mainly protein as in annelids and nematodes, polysaccharides, probably similar to those in mucus, do occur in the cuticle matrix. In arthropod cuticles the polysaccharide is represented by the more specialized chitin, with protein in the matrix.

The animal body, organized on a framework of connective tissue containing collagen fibres, had some requirement to develop internal spaces virtually free from connective tissue. This is implied by the evolution of coeloms, pseudocoels, and haemocoels. Solid animals were unable to develop local movements of the body without also involving the whole body, more or less, in the movements. Of course, this had its uses, contractions of the muscular body wall not only produced locomotive movements but also helped to circulate food and respiratory gases through the body and to move food through the gut, making a muscular wall to the alimentary canal unnecessary. However, once internal spaces had evolved, movements of the body wall no longer necessarily involved movements of the gut, which then evolved a muscular coat of its own, making the passage of food through the animal independent of its general activity. The spaces also reduced the amount of shearing to which the animal's internal tissues were subjected, and made easier the evolution of limbs and their carefully controlled, independent movements. Since the transport of substances through the body no longer depended upon muscular contraction of the body surface there was now nothing to hinder the evolution of hardened body surfaces as in arthropods.

A field as yet unexplored is the possible influence of food supplies and the capacity to manufacture particular amino acids on the structural materials employed and on their detailed composition. It is known that the food of silk worms influences the quality of the silk, and that of sheep the quality of wool, and nutrition has been shown to influence the composition of collagen also. Vitamins play an important part in the metabolism of structural materials in vertebrates but their possible importance in the formation of invertebrate structural materials is completely unexplored. Only one reference to the possible effect of a vitamin on an invertebrate material has been found: *Schistosoma mansoni* bred in mice lacking vitamin C in the diet failed to produce proper egg shells (Krakower *et al.*, 1944).

The relationship between T_S and T_D of an animal's collagen and the general environment of the collagen indicates a further relationship between structural materials and ecology. As yet, this is an almost unexplored field but relationships probably exist between the materials of egg shells and oothecae and conditions during the development of the egg, between the structure of cysts and the conditions against which they have to protect, between the properties of surface layers and resistance to desiccation. At the physiological level there are relationships between processes of growth and the organization of the skeleton and between size and the mechanical properties of the skeleton. Not only does the skeleton largely determine the types of movement an animal can make, but there is also a relationship between the elastic properties of skeletal, and particularly surface and connective tissues, and the types of movement an animal can make. It is therefore not an exaggeration to say that the way of life of an animal is an expression of the properties and organization of its structural materials.

Nor can structural materials, chemically stable as they are, be looked on as being, once formed, independent of the animal's metabolic activity and only of mechanical significance. Although the integrity of the animal depends upon the permanence of these structures, and while the structure as a whole is permanent, the material of which it is made is being constantly broken down and replaced, although generally the turnover rate is rather slow. This allows for skeletons to be adapted to varying stresses as growth and development proceed. It also allows for making good wear and tear and more extensive damage. Skeletal materials also form reserves of material, that can be called upon in times of need. All this means that structural materials are an intimate and integral part of an animal's metabolic processes. Structural materials, therefore, must be critically appraised in considering the whole organism in its environment, which is the true unit of biology.

APPENDIX

Classification of Animals

PHYLUM PROTOZOA

Class Mastigophora (= Flagellata)
- Sub-class Phytomastigina
  - Order Phytomonodina
  - Order Xanthomonadina
  - Order Chloromonadina
  - Order Euglenoidina
  - Order Cryptomonadina
  - Order Dinoflagellata
  - Order Ebriideae
  - Order Silicoflagellata
  - Order Coccolithophorida
  - Order Chrysomonadina
- Sub-class Zoomastigina (= Zooflagellata)
  - Order Protomonadina
  - Order Metamonadina (= Polymastigina + Hypermastigina)

Class Rhizopoda
  - Order Rhizomastigina
  - Order Amoebina
  - Order Testacea
  - Order Foraminifera

Class Actinopoda
  - Order Radiolaria
  - Order Heliozoa

Class Sporozoa
- Sub-class Gregarinomorpha
  - Order Archigregarina
  - Order Eugregarina
  - Order Schizogregarina
- Sub-class Coccidiomorpha
  - Order Prococcidia
  - Order Eucoccidia

Class Cnidosporida

Class Ciliata
- Sub-class Holotricha
  - Order Gymnostomatida
  - Order Suctorida
  - Order Chonotrichida
  - Order Trichostomatida
  - Order Hymenostomatida
  - Order Astomatida
  - Order Apostomatida
  - Order Thigmotrichida
  - Order Peritrichida
- Sub-class Spirotricha
  - Order Heterotrichida
  - Order Oligotrichida
  - Order Tintinnida
  - Order Entodiniomorphida
  - Order Ctenostomatida
  - Order Hypotrichida

PHYLUM PORIFERA*

Class Calcarea

Class Hexactinellida

Class Demospongiae
- Sub-class Tetractinellida
- Sub-class Monaxonida
- Sub-class Keratosa

* Classification, Hyman 1940.

PHYLUM CNIDARIA (= COELENTERATA)

Class Hydrozoa
  - Order Athecata (= Gymnoblastea)

Order Thecata (= Calypyo-blastea)
Order Limnomedusae
Order Trachymedusae
Order Narcomedusae
Order Siphonophora
Class Scyphozoa
Order Stauromedusae
Order Cubomedusae
Order Coronatae
Order Semaeostomae
Order Rhizostomae
Class Anthozoa
Sub-class Ceriantipatharia
Order Antipatharia
Order Ceriantharia
Sub-class Octocorallia
Order Alcyonacea
Order Gorgonacea
Order Pennatulacea
Sub-class Zoantharia
Order Zoanthiniaria
Order Corallimorpharia
Order Actinaria
Order Ptychodactiaria
Order Scleractinia

PHYLUM CTENOPHORA

PHYLUM PLATYHELMINTHES

Class Turbellaria
Order Acoela
Order Rhabdocoela
Order Alloeocoela
Order Tricladia
Order Polycladia
Class Trematoda
Order Monogenea
Order Aspidogastrea
Order Digenea
Class Cestoda
Sub-class Cestodaria
Sub-class Eucestoda
Order Proteocephala
Order Tetraphyllidea
Order Lecanicephala
Order Disculicepitidea
Order Diphyllidea
Order Trypanorhyncha
Order Cyclophyllidea
Order Caryophyllidea
Order Caryophyllidea
Order Nippotaeniidea
Order Pseudophyllidea

PHYLUM NEMERTINA

Class Anopla
Order Palaeonemertina
Order Heteronemertina
Class Enopla
Order Hoplonemertina
Order Bdellonemertina

PHYLUM ASCHELMINTHES

Class Rotifera
Class Gastrotricha
Class Echinoderida
Class Priapulida
Class Nematomorpha
Class Nematoda
Order Chromadoroidea
Order Dorylaimoidea
Order Mermithoidea
Order Enoploidea
Order Araeolaimoidea
Order Monhysteroidea
Order Desmoscolecoidea
Order Rhabditoidea
Order Rhabdiasoidea
Order Oxyuroidea
Order Ascaroidea
Order Strongyloidea
Order Spiruroidea
Order Dracunculoidea
Order Filarioidea
Order Trichuroidea
Order Dioctophymoidea

PHYLUM ACANTHOCEPHALA

PHYLUM ENTOPROCTA

PHYLUM POLYZOA

PHYLUM PHORONIDA

PHYLUM BRACHIOPODA

Class Inarticulata
Order Atremata
Order Neotremata
Class Articulata

PHYLUM MOLLUSCA

Class Aplacophora
Class Polyplacophora
Class Monoplacophora
Class Gastropoda
 Sub-class Prosobranchia
 Sub-class Opisthobranchia
 Sub-class Pulmonata
Class Scaphopoda
Class Bivalvia
Class Cephalopoda

PHYLUM SIPUNCULA

PHYLUM ECHIURA

PHYLUM ANNELIDA

Class Polychaeta
Class Myzostomaria
Class Oligochaeta
Class Hirudinea
Class Archiannelida

PHYLUM ARTHROPODA

Class Onychophora
Class Pauropoda
Class Diplopoda
Class Chilopoda
Class Symphyta
Class Insecta
Class Crustacea
Class Merostomata }
Class Arachnida } Chelicerata
Class Pycnogonida }
Class Pentastomida
Class Tardigrada

PHYLUM CHAETOGNATHA

PHYLUM POGONOPHORA

PHYLUM ECHINODERMATA

Sub-phylum Pelmatozoa
Class Crinoidea
Sub-phylum Eleutherozoa
Class Holothuroidea
Class Echinoidea
Class Asteroidea
Class Ophiuroidea

PHYLUM CHORDATA

Sub-phylum Hemichordata
Class Enteropneusta
Class Pterobranchia
  Order Rhabdopleurida
  Order Cephalediscida
Class Planctosphaeroidea
Sub-phylum Urochordata
Class Ascidiacea
Class Thaliacea
Class Larvacea
Sub-phylum Cephalochordata
Sub-phylum Vertebrata
Class Agnatha
Class Selachii Elasmobranchii
Class Bradyodonti
 Sub-class Holocephali
Class Pisces Osteichthyes
 Sub-class Palaeopterygii
  Order Chondrosrei
  Order Cladistia
 Sub-class Neopterygii Teleostei
  Holostei
 Sub-class Crossopterygii
Class Amphibia
Class Reptilia
Class Aves
Class Mammalia
 Sub-class Prototheria
  Order Monotremata
 Sub-class Theria
 Infra-class Metatheria
  Order Marsupialia
 Infra-class Eutheria

References

ABDERHALDEN, E. and HEYNS, K. (1933). Nachweis von Chitin in Flügelresten von Coleopteran des oberen Mitteleochäus (Fundstelle Geiseltal). *Biochem. Z.* **259,** 320–21.

ABELSON, P. H. (1957). Some aspects of paleobiochemistry. *Ann. N.Y. Acad. Sci.*, **69,** 276–85.

ABELSON, P. H. (1963). Geochemistry of amino acids. *In* "Organic Geochemistry" (ed. I. A. Breger) pp. 451–5. Monograph No. 16, Earth Sciences Series. Pergamon Press, Oxford.

ABOLINŠ-KROGIS, A. (1963a). On the protein stabilizing substances in the isolated *b*-granules and in the regenerating membranes of the shell of *Helix pomatia* (L). *Ark. Zool.* **15,** 475–84.

ABOLINŠ-KROGIS, A. (1963b). The morphological and chemical basis of the initiation of calcification in the regenerating shell of *Helix pomatia* (L). *Acta Univ. Upsaliensis*, (20) 22 pp.

ABOLINŠ-KROGIS, A. (1968). Shell regeneration in *Helix pomatia* with special reference to the elementary calcifying particles. *In* "Studies on the Structure, Physiology and Ecology of Molluscs" (ed. V. Fretter) Symposium of the Zoological Society, London, No. 22, pp. 75–92. Academic Press, London.

ADKINS, E. D. T., FLOWER, N. E., and KENCHINGTON, W. (1966). Studies on the oöthecal protein of the tortoise beetle, *Aspidomorpha*. *Jl R. Microsc. Soc.* **86,** 123–35.

AIKAWA, M. (1966). The fine structure of the erythrocytic stages of three malarial parasites, *Plasmodium fallax, P. lophurae* and *P. cathemerium*. *Am. J. trop. Med. Hyg.* **15,** 449–71.

AIKAWA, M., HEPLER, P. K., HUFF, G. C., and SPRINZ, H. (1966). The feeding mechanism of avian malarial parasites. *J. Cell. Biol.* **28,** 355–73.

AKERMANN, D. and MULLER, E. (1941). Uber das vorkommen von Dibromtyrosin neben Dijodtyrosin in Spongin. *Hoppe-Seyler's Z. physiol. Chem.* **269,** 146–57.

ALEXANDER, R. McN. (1962). Visco-elastic properties of the body-wall of sea-anemones. *J. exp. Biol.* **39,** 373–86.

ALEXANDER, R. McN. (1964). Visco-elastic properties of the mesogloea of jellyfish. *J. exp. Biol.* **41,** 363–70.

ALEXANDER, R. McN. (1966). Rubber-like properties of the inner hinge-ligament of Pectinidae. *J. exp. Biol.* **44,** 119–130.

AMOS, W. B. (1968). Personal communication.

ANDERSEN, S. O. (1963). Characterization of a new type of cross-linkage in resilin, a rubber-like protein. *Biochim. biophys. Acta.*, **69,** 249–62.

ANDERSEN, S. O. (1964). The cross-links in resilin identified as dityrosine and trityrosine. *Biochim. biophys. Acta*, **93,** 213–15.

ANDERSEN, S. O. (1966). Covalent cross-links in a structural protein, resilin. *Acta Physiol. scand.* **66,** (Suppl. 263), 1–81.

ANDERSEN, S. O. (1967). Isolation of a new type of cross link from the hinge ligament protein of molluscs. *Nature, Lond.* **216,** 1029–30.

ANDERSEN, S. O. (1968a). Personal communication.

ANDERSEN, S. O. (1968b). Personal communication.

ANDERSEN, S. O. and WEIS-FOGH, T. (1964). Resilin. A rubber-like protein in arthropod cuticle. *In* "Advances in Insect Physiology" (ed. J. W. L. Beament *et al.*) Vol. 2, pp. 1–65. Academic Press, London and New York.

ANDERSON, E. and BEAMS, H. W. (1959). The cytology of *Tritrichomonas* as revealed by the electron microscope. *J. Morph.* **104,** 205–35.

ANDERSON, E. and BEAMS, H. W. (1960). The fine structure of the Heliozoan, *Actinosphaerium nucleofilum*. *J. Protozool.* **7,** 190–9.

ANDERSON, E. and BEAMS, H. W. (1961). The ultrastructure of *Tritrichomonas* with special reference to the blepharoplast complex. *J. Protozool.* **8,** 71–5.

ANGELL, R. W. (1967). The test structure and composition of the Foraminifer, *Rosalina floridana*. *J. Protozool.* **14,** 299–307.

ANYA, A. O. (1964). Studies on the structure of the female reproductive system and egg-shell formation in *Aspicularis tetraptera* Schulz, (Nematoda: Oxyuroidea). *Parasitology*, **54,** 699–719.

ANYA, A. O. (1966a). The structure and chemical composition of the nematode cuticle, observations on some oxyurids and *Ascaris*. *Parasitology*, **56,** 179–98.

ANYA, A. O. (1966b). Experimental studies on the physiology of hatching of eggs of *Aspicularus tetraptera* Schulz (Oxyuroidea: Nematoda). *Parasitology*, **56,** 733–44.

ANYA, A. O. (1966c). Localization of ribonucleic acid in the cuticle of nematodes. *Nature, Lond.* **209,** 827–8.

APPELLOF, A. (1893). Die schallen von *Sepia*, *Spirula* und *Nautilus*. *K. svensk. Akad. Handl.* **25,** No. 7, 106 pp.

APPLEGATE, V. C. (1950). Natural history of the sea-lamprey, *Petromyzon marinus* in Michigan. *Spec. scient. Rep. Wildl. Serv.* (*Fisheries*) No. 55, 237 pp.

APPLEGATE, S. P. (1967). A survey of shark hard parts. *In* "Sharks, Skates and Rays" (ed. P. W. Gilbert, R. F. Mathewson and D. P. Rall) pp. 37–67. Johns Hopkins Press, Baltimore.

ARMSTRONG, J. A. and HAWKINS, F. (1964). Electron microscope sections of filarial worms. *Trans. R. Soc. trop. Med. Hyg.* **58,** 9.

ARMSTRONG, W. G. and TARLO, L. B. H. (1966). Amino-acid components in fossil calcified tissue. *Nature, Lond.* **210,** 481–2.

ARVY, L. (1955). Mise en évidence des fibres élastiques chez quelque invertebrés. *Bull. Soc. zool. Fr.* **80,** 103.

ASHHURST, D. E. (1964). Fibrillogenesis in the wax-moth, *Galleria mellonella*. *Q. Jl microsc. Sci.* **105,** 391–403.

ASHHURST, D. E. (1965). The connective tissue sheath of the locust nervous system: its development in the embryo. *Q. Jl microsc. Sci.* **106,** 61–73.

ASTBURY, W. T. (1938). X-ray adventures amongst the proteins. *Trans. Faraday Soc.* **34,** 378–88.

ASTBURY, W. T. and BELL, F. O. (1939). X-ray data on the structure of natural fibres and other bodies of high molecular weight. *Tabul. biol.* **17,** 90.

ATKINS, E. D. T. (1967). A four-strand coil coil model for some insect fibrous proteins. *J. molec. Biol.* **24,** 139–41.

AUBER, L. (1956). The anatomy of follicles producing wool fibres, with special reference to keratinization. *Trans. R. Soc., Edinb.* **62,** 191–254.

AYER, J. P. (1964). Elastic Tissue. *In* "International Review of Connective Tissue Research", (ed. D. A. Hall) Vol. 2, pp. 33–100. Academic Press, New York and London.

BACCETTI, B. (1961). Indagini comparative sulla ultra struttura della fibrilla collagene nei diversi ordini degli insetti. *Redia*, **46,** 1–7.

BAILEY, A. J. (1968). The nature of collagen. *In* "Comprehensive Biochemistry", (eds. M. Florkin and E. H. Stotz) Vol. 26B. pp. 297–423 Elsevier, Amsterdam.

BAILEY, K. and WEIS-FOGH, T. (1961). Amino acid composition of a new rubber-like protein, resilin. *Biochim. biophys. Acta*, **48,** 452–9.

BAILEY, S. W. (1954). Hardness of arthropod mouth parts. *Nature, Lond.* **173,** 503.

BAKER, J. R. and BALCH, D. A. (1962). A study of the organic material of hen's egg shell. *Biochem. J.* **82,** 352–362.

BAKER, J. R. and LAINSON, R. (1967). The fine structure of the gametocytes of an adeleine haemogregarine. *J. Protozool.* **14,** 232–8.

BALDWIN, E. (1937). "An Introduction to Comparative Biochemistry". Cambridge University Press, London.

BARKER, R. M. (1964). Microtextural variation in pelecypod shells. *Malacologica*, **2,** 69–86.

BARRAS, D. R. and STONE, B. A. (1965). The chemical composition of the pellicle of *Euglena gracilis* var. *bacillaris. J. Biochem.* **47,** 14–15.

BARRETT, J. M. (1958). Some observations on *Actinosphaerium nucleofilum* n.sp., a fresh water actinophryid. *J. Protozool.* **5,** 205–09.

BARRINGTON, E. J. W. (1967). *Invertebrate Structure and Function*. Nelson, London.

BARRINGTON, E. J. W. and BARRON, N. (1960). On the organic binding of iodine in the tunic of *Ciona intestinalis. J. mar. biol. Ass. U.K.* **39,** 513–23.

BARRINGTON, E. J. W. and THORPE, A. (1968). Histochemical and biochemical aspects of iodine binding in the tunic of the ascidian, *Dendrodoa grossularia* (van Beneden). *Proc. R. Soc.* **B171,** 91–109.

BARRNETT, R. J. and SOGNNAES, R. F. (1955). Comparative observations on certain histochemical characteristics of keratinous tissue of fish, reptiles, birds and mammals. *J. dent. Res.* **34,** 670–1.

BARRNETT, R. J. and SOGNNAES, R. F. (1962). Histochemical distribution of protein bound sulphydryl and disulphide groups in vertebrate keratins. *In* "Fundamentals of Keratinization" (ed. E. O. Butcher and R. F. Sognnaes) pp. 27–43. American Association for the Advancement of Science, Washington.

BATESON, W. (1885). Later stages in the development of *Balanoglossus. Q. Jl microsc. Sci.* **25,** (Suppl.), 81–122.

BATHAM, E. J. (1960). The fine structure of epithelium and mesogloea in a sea-anemone. *Q. Jl microsc. Sci.* **101,** 481–85.

BATHAM, E. J. and PANTIN, C. F. A. (1950). Muscular and hydrostatic action in the sea anemone *Metridium senile* (L). *J. exp. Biol.* **27,** 264–89.

BEAMENT, J. W. L. (1946). The formation and structure of the chorion of the egg of an Hemipteran, *Rhodnius prolixus. Q. Jl microsc. Sci.* **87,** 393–439.

BEAMENT, J. W. L. (1955). Wax-secretion in the cockroach. *J. exp. Biol.* **32,** 514–38.

BEAMS, H. W., KING, R. L., TAHMISIAN, T. N. and DEVINE, R. (1960). Electron microscope studies on *Lophomonas striata* with special reference to the nature and position of the striations. *J. Protozool.* **7,** 91–101.

BEAMS, H. W. and SEKHON, S. S. (1951). Fine structure of body wall and cells in pseudocoelom of the nematode *Rhabditis pellio. J. Ultrastruct. Res.* **18,** 580–94.

BEAMS, H. W., TAHMISIAN, T. N., ANDERSON, E. and WRIGHT, W. (1961). Studies on the fine structure of *Lophomonas blattarum* with special reference to the so-called parabasal apparatus. *J. Ultrastruct. Res.* **5,** 166–83.

BEAMS, H. W., TAHMISIAN, T. N., DEVINE, R. L., and ANDERSON, E. (1959). Studies in the fine structure of a gregarine parasitic in the gut of the grasshopper, *Melanoplus differentialis. J. Protozool.* **6,** 136–46.

BEAR, R. S. (1952). The structure of collagen fibrils. Advances in Protein Chemistry, **7,** 109–62.

BECKET, E. B. and BOOTHROYD, B. (1961). Some observations on the fine structure of the mature larva of the nematode *Trichinella spiralis. Ann. trop. Med. Parasit.* **55,** 116–24.

BEEDHAM, G. E. (1954). Properties of the non-calcareous material in the shell of *Anodonta cygnaea. Nature, Lond.* **174,** 750.

BEEDHAM, G. E. (1958). Observations on the non calcareous component of the shell of the Lamellibranchia. *Q. Jl microsc. Sci.* **99,** 341–57.

BEEDHAM, G. E. (1965). Repair of the shell in species of *Anodonta. Proc. zool. Soc., Lond.* **145,** 107–24.

BEEDHAM, G. E. and OWEN, G. (1965). The mantle and shell of *Solemya parkinsoni* (Protobranchia: Bivalvia). *Proc. zool. Soc., Lond.* **145,** 405–30.

BEEDHAM, G. E. and TRUEMAN, E. R. (1958). The utilization of I^{131} by certain lamellibranchs with particular reference to shell secretion. *Q. Jl microsc. Sci.* **99,** 199–204.

BEEDHAM, G. E. and TRUEMAN, E. R. (1967). The relationship of the mantle and shell of the Polyplacophora in comparison with that of other Mollusca. *J. Zool., Lond.* **151,** 215–31.

BEEDHAM, G. E. and TRUEMAN, E. R. (1968). The cuticle of the Aplacophora and its evolutionary significance. *J. Zool., Lond.* **154,** 443–51.

BEHNKE, O. and ZEALANDER, T. (1967). Filamentous structure of microtubules of the marginal bundle of mammalian blood platelets. *J. Ultrastruct. Res.* **19,** 147–65.

BELL, L. G. E. (1962). Polysaccharides and cell membranes. *J. theor. Biol.* **3,** 132–3.

BELLAIRS, R. (1960). Development of Birds. *In* "Biology and Comparative Physiology of Birds" (ed. A. J. Marshall). Academic Press, New York and London.

BELLAIRS, R., HARKNESS, M., and HARKNESS, R. D. (1963). The vitelline membrane of the hen's egg: a chemical and electron microscope study. *J. Ultrastruct. Res.* **8,** 339–59.

BENNET-CLARK, H. C. (1962). Active control of the molecular properties of insect endocuticle. *J. Insect Physiol.* **8,** 627–33.

BENNET-CLARK, H. C. (1963). The relation between epicuticular folding and the subsequent size of an insect. *J. Insect Physiol.* **9,** 43–6.

BENTLEY, P. J. and BLUMER, W. F. C. (1962). Uptake of water by lizard *Moloch horridus. Nature, Lond.* **194,** 699–700.

BENTLEY, P. J. and SCHMIDT-NEILSON, K. (1966). Cutaneous water loss in reptiles. *Science, N.Y.* **151,** 1547–9.

BENUSAN, H. B. (1965). A novel hypothesis for the mechanism of cross-linking in collagen. *In* "Structure and Function of Connective and Skeletal Tissue" (ed. S. Fitton-Jackson *et al.*) pp. 42–6. Proceedings of N.A.T.O. Advanced Study Group, St. Andrews, 1964. Butterworth, London.

BERGEL, D. H. (1961). The static elastic properties of the arterial wall. *J. Physiol.* **156,** 445–57.

BERGENHAYN, J. R. M. (1930). Kurze Bemerkungen zur Kenntnis der Schalenstruktur und Systematik der Loricaten. *K. Svensk. Vetensk-Akad. Handl.* (3) **9,** 1–54.

BERGMANN, W. and LESTER, D. (1940). Coral reefs and the formation of petroleum. *Science, N.Y.* **92,** 452–3.

BERN, H. A., HARKNESS, R. D., and BLAIR, S. M. (1954). Radioautographic studies of keratin formation. *Proc. nat. Acad. Sci. Wash.* **41,** 55–60.

BERRILL, N. J. (1949). The polymorphic transformations of *Obelia. Q. Jl microsc. Sci.* **90,** 235–64.

BERRILL, N. J. (1955). "The Origin of the Vertebrates". Oxford University Press, London.

BERTHELOT, M. P. E. (1858). Sur la transformation en sucre de divers principes immédiats contenus dans les tissus des animaux invertébrés. *C.r. hebd. Séanc. Acad. Sci., Paris,* **47,** 227–30.

BEVELANDER, G. and BENZER, P. (1948). Observations on the non calcareous component of the shell of the Lamellibranchia. *Biol. Bull. mar. biol. Lab., Woods Hole,* **94,** 176–83.

BEVELANDER, G. and NAKAHARA, H. (1960). Development of the skeleton of the sand dollar (*Echinarachnius parma*). *Publs. Am. Ass. Adv. Sci.* **64,** 41–56.

BEYER, H. (1886). Structure of glottidia. *Stud. Biol. Lab. Johns Hopkins Univ.* **3.**

BIDDER, A. M. (1950). The digestive mechanism of the European squids *Loligo vulgaris, Loligo forbesii, Alloteuthis media* and *Alloteuthis subulata. Q. Jl microsc. Sci.* **91,** 1–43.

BIDDER, A. M. (1962). Use of the tentacles, swimming and buoyancy control in the pearly *Nautilus. Nature, Lond.* **196,** 451–4.

BIDDER, G. P. (1898). The skeleton and classification of calcareous sponges. *Proc. R. Soc.* B. **64,** 61–76.

BIRD, A. F. (1956). Chemical composition of the nematode cuticle. Observations on the whole cuticle. *Expl Parasitol.* **5,** 350–8.

BIRD, A. F. (1957). Chemical composition of the nematode cuticle. Observations on individual layers and extracts from these layers in *Ascaris lumbricoides* cuticle. *Expl Parasitol.* **6,** 383–403.

BIRD, A. F. (1958a). Further observations on the structure of nematode cuticle. *Parasitology,* **48,** 32–9.

BIRD, A. F. (1958b). The adult female cuticle and egg sac of the genus *Meloidogyne* Goeldi, 1887. *Nematologica,* **3,** 205–12.

BIRD, A. F. (1959). Development of the root-knot nematodes *Meloidogyne javanica* (Treub) and *Meloidogyne hapla* Chitwood in tomato. *Nematologica,* **4,** 31–42.

BIRD, A. F. (1968). Changes associated with parasitism in nematodes III. Ultrastructure of egg shell. *J. Parasitol.* **54,** 475–89.

BIRD, A. F. and DEUTSCH, K. (1957). The structure of the cuticle of *Ascaris lumbricoides* var. *suis*. *Parasitology*, **47,** 319–28.

BIRD, A. F. and ROGERS, G. E. (1965). Ultrastructure of the cuticle and its formation in *Meloidogyne javanica*. *Nematologica*, **11,** 224–30.

BIRD, A. F. and ROGERS, W. P. (1956). Chemical composition of the cuticle of third stage nematode larvae. *Expl Parasitol.* **5,** 449–57.

BIRKBECK, M. S. C. and MERCER, E. H. (1957). The electron-microscopy of the human hair follicle. I. Introduction and the hair cortex. *J. biophys. biochem. Cytol.* **3,** 203–13.

BLACKMORE, P. D. and TOD, R. (1957). Mineralogy of some Foraminifera as related to their classification and ecology. *J. Paleont.* **33,** 1–15.

BLACKSTAD, T. W. (1963). The skin and its derivatives. *In* "The Biology of Myxine" (ed. A. Brodal and R. Fänge) pp. 195–230. Universitetsforlaget, Oslo.

BLACKWELL, J., PARKER, K. D., and RUDALL, K. M. (1965). Chitin in pogonophore tubes. *J. mar. biol. Ass. U.K.* **45,** 659–61.

BLAKEY, P. R., EARLAND, C., and STELL, J. G. P. (1963). Calcification of keratin. *Nature Lond.* **198,** 481.

BLANK, I. H. (1952). Factors which influence the water-content of the stratum corneum. *J. invest. Derm.* **18,** 433–9.

BLANK, I. H. (1953). Further observations on factors which influence the water content of the stratum corneum. *J. invest. Derm.* **21,** 259–71.

BLANK, I. H. and SHAPPIRIO, E. B. (1955). The water content of the stratum corneum. *J. invest. Derm.* **25,** 391–401.

BLOCK, R. J. (1938). Comparative biochemistry of the proteins. *Cold Spring Harb. Symp. quant. Biol.* **6,** 79–90.

BLOWER, G. (1951). A comparative study of the chilopod and diplopod cuticle. *Q. Jl microsc. Sci.* **92,** 141–61.

BLOWER, J. G. (1955). Millipedes and centipedes as soil animals. *In* "Soil Zoology" (ed. D. K. McE. Kevan). Butterworth, London.

BOBIN, G. (1944). Morphogénèse des soies chez les annelides polychètes. *Annls Inst. océanogr., Monaco.* **22,** 1–106.

BOBIN, G. and MAZONÉ, H. (1944). Topographie, histologie, charactères physiques et chemique des soie *d'Aphrodite aculeata*. *Bull. Soc. zool. Fr.* **69,** 125–34.

BØGGILD, O. B. (1930). The shell structure of the mollusks. *K. dansk. Vidensk. Selsk. Skr.* **2,** 233–326.

BOGITSCH, B. J. (1962). The chemical nature of metacercarial cysts. I. Histological and histochemical observations on the cyst of *Posthodiplostomum minimum*. *J. Parasitol.* **48,** 55–60.

BOILLY, B. (1967). Contribution a l'étude ultrastructurale de la cuticle epidermique et pharyngienne chez une annelide polychète (*Syllis amica* Quadrefages). *J. Microscopie,* **6,** 469–84.

BONNER, J. T. (1955). A note concerning the distribution of polysaccharides in the early development of the hydromedusan *Phialidium gregarium*. *Biol. Bull. mar. biol. Lab., Woods Hole,* **108** 18–20.

BOOTH, H. and SAUNDERS, B. C. (1956). Studies in peroxidase action. Part X. The oxidation of phenols. *J. chem. Soc.* **19,** 940–8.

BORG, F. (1926). Studies on recent cyclostomatous Bryozoa. *Zool. Bidrag,* **10,** pp. 181–505.

BORNSTEIN, P. and PIEZ, K. A. (1964). A biochemical study of human skin collagen and the relation between intra- and intermolecular cross-linking. *J. clin. Invest.* **43,** 1813–23.

BORNSTEIN, P. and PIEZ, K. A. (1965). Collagen: Structural studies based on the cleavage of methionyl bonds. *Science, N.Y.* **148,** 1353–5.

BOUILLION, J. (1956). Étude monographique du genre *Limnocnida* (Limnomedusae). *Annls Soc. r. zool. Belg.* **87,** 254–500.

BOUILLION, J. (1960). Sur la nature de la coquille chez les mollusques. *Annls Soc. r. zool. Belg.* **89,** 229–37.

BOUILLION, J. and VANDERMEERSSCHE, G. (1957). Structure et nature de la mésoglée des Hydro et Scyphoméduses. *Annls Soc. r. zool. Belg.* **87,** 9–25.

BOULIGAND, Y. (1965a). Sur une architecture torsadée répandue dans de nombreuses cuticles d'Arthropodes. *C.r. hebd. Séanc. Acad. Sci., Paris,* **261,** 3665–8.

25

BOULIGAND, Y. (1965b). Sur une disposition fibrillaire torsadée commune à plusiers structures biologiques. *C.r. hebd. Séanc. Acad. Sci., Paris*, **261,** 4864–7.

BOWES, J. H., ELLIOT, R. G., and MOSS, J. A. (1958). The extraction of soluble protein from skin by alkaline solutions. *In* "Recent Advances in Gelatin and Glue Research" (ed. G. Stainsby) p. 71. Pergamon Press, London.

BRADBURY, P. C. (1966). The Fine Structure of the mature tomite of *Hyalophysa chattoni. J. Protozool.* **13,** 591–607.

BRADBURY, S. (1957). A histochemical study of the connective tissue fibres in the leech, *Glossisiphonia companata. Q. Jl microsc. Sci.* **98,** 29–45.

BRADBURY, S. and MEEK, G. A. (1958). A study of fibrogenesis in the leech, *Hirudo medicinalis. Q. Jl microsc. Sci.* **99,** 143–8.

BRANDENBURGER, E. and FREY-WYSSLING, A. (1947). Über die Membransubstanzen von *Chlorochytridion tuberculatum* W. Vischer. *Experientia,* **3,** 492–3.

BRANDT, T. VON (1940). Further observations upon the composition of Acanthocephala *J. Parasitol.* **26,** 301–7.

BRANDT, T. VON (1952). "Chemical Physiology of Endoparasitic Animals" p. 176. Academic Press, New York.

BRIEN, P. (1930). Contribution a l'etude de la regeneration naturelle et experimentale chez les Clavelinidae. *Annls Soc. r. zool. Belg.* **61,** 19–112.

BRIEN, P. (1948). Les Appendiculaires. *In* "Traité de Zoologie" (ed. P-P. Grassé) Tome XI, pp. 867–94. Masson et Cie, Paris.

BRODY, I. (1959). An ultrastructural study on the role of the keratohyalin granules in the keratinizing process. *J. Ultrastruct. Res.* **3,** 84–104.

BRODY, I. (1960). The ultrastructure of the tonofibrils in the keratinization process of normal human epidermis. *J. Ultrastruct. Res.* **4,** 264–97.

BRÖKELMANN, Y. and FISCHER, A. (1966). Uber die Cuticula von *Platynereis dumerilli* (Polychaeta). *Z. Zellforsch.* **70,** 131–5.

BRØNDSTED, H. V. and CARLSEN, F. E. (1951). A cortical cytoskeleton in expanded epithelium cells of sponge gemmula. *Expl Cell. Res.* **2,** 90–6.

BROUSSY, J. (1932). Thesis No. 52. Université de Montpellier, Faculté de Medicine.

BROWN, C. H. (1950a). Structural Proteins in the Invertebrata and Vertebrata: a histochemical study. Ph.D. Thesis. Cambridge University.

BROWN, C. H. (1950b). Quinone-tanning in the animal kingdom. *Nature, Lond.* **165,** 275.

BROWN, C. H. (1950c). Keratins in invertebrates. *Nature, Lond.* **166,** 439.

BROWN, C. H. (1950d). A review of the methods available for the determination of the types of forces stabilizing structural proteins in animals. *Q. Jl microsc. Sci.* **91,** 331–9.

BROWN, C. H. (1952). Some structural proteins of *Mytilus edulis. Q. Jl microsc. Sci.* **93,** 487–502.

BROWN, C. H. (1955). Egg-capsule proteins of selachians and trout. *Q. Jl microsc. Sci.* **96,** 483–8.

BRUNET, P. C. J. (1963). Synthesis of an aromatic ring in insects. *Nature, Lond.* **199,** 492.

BRUNET, P. C. J. (1966). The metabolism of aromatic compounds. *In* "Aspects of Insect Biochemistry" (ed. T. W. Goodwin) Academic Press, London and New York.

BRUNET, P. C. J. (1967). Sclerotins. *Endeavour,* **26,** 68–74.

BRUNET, P. C. J. and CARLISLE, D. B. (1958). Chitin in Pogonophora. *Nature, Lond.* **182,** 1689.

BRUNET, P. C. J. and KENT, P. W. (1955). Mechanism of sclerotin formation: the participation of a Beta-glucoside. *Nature, Lond.* **175,** 819.

BUCHANAN, J. B. and HEDLEY, R. H. (1960). A contribution to the biology of *Astrorhiza limicola* (Foraminifera). *J. mar. biol. Ass. U.K.* **39,** 549–60.

BULLOUGH, W. S. (1962). The control of mitotic activity in adult mammalian tissues. *Biol. Rev.* **37,** 307–42.

BUONOCORE, C. (1958). Un nuovo strado nel secreto serico del *Bombyx mori. Annali Sper. Agr. Rome,* **12,** 681–5.

BURFIELD, S. (1927). *Sagitta. Liverpool Mar. Biol. comm. mem.*

BURGE, R. E. and HYNES, R. D. (1959). The thermal denaturation of collagen in solution and its structural implications. *J. molec. Biol.* **1,** 155–64.

BURGE, W. E. and BURGE, G. L. (1915). The protection of parasites in the digestive tract against the action of digestive enzymes. *J. Parasitol.* **1,** 179–83.

Burgess, G. H. O. (1956). Absence of keratin in teleost epidermis. *Nature, Lond.* **178,** 93–4.

Burton, A. C. (1954). Relation of structure to the function of the tissues of the wall of blood vessels. *Physiol. Rev.* **34,** 619–40.

Burton, D., Hall, D. A., Keech, M. F., Reed, R., Saxl, H., Tonbridge, R. E. and Wood, M. J. (1955). Apparent transformation of collagen fibrils into elastin. *Nature, Lond.* **176,** 966–9.

Burton, P. R. (1963). A histochemical study of vitelline cells, egg capsules and Mehlis gland in the frog lung-fluke, *Haematoloechus medioplexus. J. exp. Zool.* **154,** 247.

Burton, P. R. (1964). The ultrastructure of the integument of the frog lung-fluke, *Haematoloechus medioplexus* (Trematoda: Plagiorchiidae). *J. Morph.* **115,** 305–18.

Bychowsky, B. E. (1937). Monogenetic trematodes, their classification and phylogeny. 509 pp. Academy of Sciences. U.S.S.R. Moscow, Leningrad.

Calgren, O. (1956). Actiniaria from depths exceeding 6000 metres. *Galathea Rep.* **2,** 9–16.

Calman, W. T. (1909). Crustacea. *In* "A Treatise on Zoology" Pt. VII. (ed. Ray Lankester). A. and C. Black, London.

Calvery, H. O. (1933). Some analyses of egg-shell keratin. *J. biol. Chem.* **100,** 183–6.

Cameron, A. T. (1914). Contributions to the biochemistry of iodine. I. The distribution of iodine in plant and animal tissues. *J. biol. Chem.* **18,** 335–80.

Cameron, D. A. (1961). The fine structure of osteoblasts in the metaphysis of the tibia of the young rat. *J. biophys. biochem. Cytol.* **9,** 583–94.

Campion, M., Lawn, A. M., and Nisbet, R. H. (1964). The fine structure of the glands in the collar of *Archachatina. J. Physiol.* **170,** 46 pp.

Cannon, H. G. (1947). On the anatomy of the pedunculate barnacle *Lithotrya. Phil. Trans. R. Soc.* B. **233,** 89–136.

Carbonell, L. H. and Apitz, R. (1960). Sulphydryl and disulphide groups in the cuticle of *Ascaris lumbricoides. Expl Parasitol.* **10,** 263–7.

Carlgren, O. (1956). Actiniaria from depths exceeding 6000 metres. *Galathea Rep.* **2,** 9–16.

Carlisle, D. B. (1958). On the exuvia of *Priapulus caudatus* Lamark. *Ark. Zool.* **12,** 79–81.

Carlisle, D. B. (1964). Chitin in a Cambrian fossil, *Hyolitellus. Biochem. J.* **90,** 1c.

Cerfontaine, P. (1890). Recherches sur la système cultané et sur le système musculaire du *Lumbricus terrestris. Archs Biol., Paris,* **10,** 327–428.

Champetier, G. and Fauré-Fremiet, E. (1938a). Étude roentgénographique des kératines sécrétees. *C.r. hebd. Séanc. Acad. Sci., Paris,* **207,** 1133–5.

Champetier, G. and Fauré-Fremiet, E. (1938b). Étude roentgénographique de quelques collagènes. *J. Chim. Phys.* **35,** 223–32.

Champetier, G. and Fauré-Fremiet, E. (1942). Étude roentgénographique de quelques corneines d'Anthozroaires. *C.r. hebd. Séanc. Acad. Sci., Paris,* **215,** 94–6.

Chapman, D. M. (1966). Evolution of the scyphistoma. *In* "The Cnidaria and their Evolution" (ed. W. J. Rees) pp. 51–72. Symposia of the Zoological Society of London, No. 16. Academic Press, London and New York.

Chapman, D. M. (1968). Structure, histochemistry and formation of the podocyst and cuticle of *Aurelia aurita. J. mar. biol. Ass. U.K.* **48,** 187–208.

Chapman, D. M., Pantin, C. F. A., and Robson, E. A. (1962). Muscle in coelenterates. *Revue can. Biol.* **21,** 267–76.

Chapman, G. (1953a). Studies on the mesogloea of coelenterates. I. Histology and chemical properties. *Q. Jl microsc. Sci.* **94,** 155–76.

Chapman, G. (1953b). Studies on the mesogloea of coelenterates. II. Physiological properties. *J. exp. Biol.* **30,** 440–451.

Chapman, G. (1959). The mesogloea of *Pelagia noctiluca. Q. Jl microsc. Sci.* **100,** 599–610.

Chapman, G. (1966). The structure and functions of the mesogloea. *In* "The Cnidaria and their Evolution" (ed. W. J. Rees) pp. 147–65. Symposia of the Zoological Society of London, No. 16. Academic Press, London and New York.

Chave, K. E. (1952). A solid solution between calcite and dolomite. *J. Geol.* **60,** 190–2.

Chen, Y. T. (1950). Investigations into the biology of *Peranema trichophorum* (Euglenineae) *Q. Jl microsc. Sci.* **91,** 279–308.

Chessin, E. M. (1965). The significance of ultra-structures in the taxonomy of protozoa. Progress in Protozoology (Abstract of papers) Second International Conference on Protozoology, London, 1965.

CHETVERIKOV, S. S. (1918). *A. Rep. Smithsonian Inst.* No. 441.

CHEUNG, T. S. (1966). The development of egg membranes and egg attachment in the shore crab, *Carcinus maenas*, and some related decapods. *J. mar. biol. Ass. U.K.* **46,** 373–400.

CHITWOOD, B. G. (1936). Observations on the chemical nature of the cuticle of *Ascaris lumbricoides* var. *suis*. *Proc. Helminth. Soc. Wash.* **3,** 39–49.

CHITWOOD, B. G. (1938). Further studies on nemic skeletoids and their significance in the chemical control of nemic pests. *Proc. Helminth. Soc. Wash.* **5,** 68–75.

CHITWOOD, B. G. and CHITWOOD, M. B. (1950). An introduction to Nematology (Second Edit.) Monumental Printing Co., Baltimore.

CHVAPIL, M. and JENŠOVSKÝ, L. (1963). The shrinkage temperature of collagen fibres isolated from the tail tendons of rats of various ages and from different places of the same tendon. *Gerontologia*, **7,** 18–29.

CLARK, A. H. (1911). On the inorganic constituents of the skeletons of two recent crinoids. *Proc. U.S. natn. Mus.* **39,** 487–8.

CLARK, M. E. and CLARK, R. B. (1960a). The fine structure and histochemistry of the ligaments of *Nephthys*. *Q. Jl microsc. Sci.* **101,** 133–48.

CLARK, R. B. and CLARK, M. E. (1960b). Ligamentary system and the segmental musculature of *Nephthys*. *Q. Jl microsc. Sci.* **101,** 149–76.

CLARK, R. B. and COWEY, J. B. (1958). Factors controlling the change of shape of certain nemertine and turbellarian worms. *J. exp. Biol.* **35,** 731–48.

CLARKE, A. J., COX, P., and SHEPHERD, A. M. (1967). Chemical composition of the egg shells of the potato cyst-nematode, *Heterodera rostochiensis* Woll. *Biochem. J.* **104,** 1056–60.

CLARKE, F. W. and WHEELER, W. C. (1922). The inorganic constituents of marine invertebrates. *Prof. Pap. U.S. geol. Surv.*, No. 124, 1–62.

CLEGG, J. A. (1958). Unpublished work.

CLEGG, J. A. (1965). Secretion of lipoprotein by Mehlis gland in *Fasciola hepatica*. *Ann. N.Y. Acad. Sci.* **118,** 969–86.

CLEGG, J. A. and SMYTH, J. D. (1968). Growth, development, and culture methods: parasitic platyhelminths. *In* "Chemical Zoology, Vol. 2, Porifera, Coelenterata and Platyhelminthes" (ed. M. Florkin and B. T. Scheer). Academic Press, New York and London.

CLOUDSLEY-THOMPSON, J. L. (1950). The water relations and cuticle of *Paradesmus gracilis*. (Diplopoda. Strongylosomidae). *Q. Jl microsc. Sci.* **91,** 453–64.

CLUTTERBUCK, A. J. (1968). Personal communication.

COE, M. J. (1967). Local migration of *Tilapia grahami* Boulenger in Lake Magadi, Kenya in response to diurnal temperature changes. *E.Af. Wildl. J.* **5,** 171–4.

COGGESHALL, R. E. (1966). A fine structural analysis of the epidermis of the earthworm, *Lumbricus terrestris. L. J. Cell. Biol.* **28,** 95–108.

COHEN, N. W. (1952). Comparative rates of dehydration and hydration of some californian salamanders. *Ecology*, **33,** 462–79.

COIL, W. H. and KUNTZ, R. E. (1963). Observations on the histochemistry of *Synocoelium spathulatum* n.sp. *Proc. Helminth. Soc. Wash.* **30,** 60–5.

COLLEY, F. C. (1967). Fine structure of sporozoits of *Eimeria neischulzi*. *J. Protozool.* **14,** 217–20.

COMFORT, A. (1951). The pigmentation of molluscan shells. *Biol. Rev.* **26,** 285–301.

CONGDON, E. D. (1937). The primary types of extra-organic gross connective tissue structures. *Anat. Rec.* **67,** 193–203.

CORFIELD, M. C., ROBSON, A., and SKINNER, B. (1958). The amino acid composition of three fractions from oxidized wool. *Biochem. J.* **68,** 348–52.

COTT, H. B. (1940). "Adaptive Coloration in Animals". Methuen, London.

COWAN, P. M., NORTH, A. C. T., and RANDALL, J. T. (1955). X-ray diffraction studies of collagen fibres. *In* "Fibrous Proteins and their Biological Significance". pp. 115–26. S.E.B. Symposium IX.

COWEY, J. B. (1952). The structure and function of the basement membrane system in *Amphiporus lactifloreus* (Nemertinea). *Q. Jl microsc. Sci.* **93,** 1–15.

COX, R. W., GRANT, R. A., and HORN, R. W. (1967). The structure and assembly of collagen fibrils I. Native collagen fibrils and their formation from tropocollagen. *Jl R. Microsc. Soc.* **87,** 123–42.

CREWTHER, W. G. and DOWLING, L. M. (1960). Effects of chemical modifications on the physical properties of wool: A model of the wool fibre. *J. Text. Inst.* **51,** T775–91.

Crewther, W. G., Fraser, R. D. B., Lennox, F. G., and Lidley, H. (1965). The chemistry of keratins. *Adv. Protein Chem.* **20,** 191–346.

Crick, F. H. C. (1952). Is α-keratin a coiled coil? *Nature, Lond.* **170,** 882–3.

Crites, J. L. (1958). The chemistry of the membranes of the egg envelope of *Cruzia americana* Maplestone, 1930. *Ohio J. Sci.* **58,** 343–6.

Crompton, D. W. T. (1963). Morphological and histochemical observations on *Polymorphus minutus* (Goeze, 1782), with special reference to the body wall. *Parasitology*, **53,** 663–85.

Crompton, D. W. T. and Lee, D. L. (1965). The fine structure of the body wall of *Polymorphus minutus* (Goeze, 1782) (Acanthocephala). *Parasitology*, **55,** 357–64.

Crounse, R. G. (1965). An approach to a common keratin sub unit. *In* "Biology of the Skin and Hair Growth" (ed. A. G. Lyne and B. F. Short) pp. 307–12. Angus and Robertson, Sydney.

Cruz, H. (1948). Further studies on the development of *Cysticercus fasciolaris* and *Cysticercus pisiformis* with special reference to the growth and sclerotization of the rostellar hooks. *J. Helminth. Soc. Wash.* **22,** 79–98.

Currey, J. D. (1962). Strength of bone. *Nature, Lond.* **195,** 513–4.

Currey, J. D. (1967). The failure of exoskeletons and endoskeletons. *J. Morph.* **123,** 1–16.

Currey, J. D. and Nichols, D. (1967). Absence of organic phase in echinoderm calcite. *Nature, Lond.* **214,** 81–3.

Curtis, A. S. G. (1967). The Cell Surface: its molecular role in morphogenesis. Logos Press, London. Academic Press, New York.

Dallemagne, M. G. (1964). Phosphate and carbonate in bone and teeth. *In* "Bone and Tooth", Proceedings First European Symposium (ed. H. J. J. Blackwood) pp. 171–74. Pergamon Press, Oxford.

Damas, D. (1968). Origine et structure du spermatophore de *Glossiphonia complanata* (Hirudinée, Rhynchobdelle). *Archs. Zool. exp. gén.* **109,** 79–85.

Damodaran, M., Sivaraman, C., and Dhavalikar, R. S. (1956). Amino acid composition of elastoidin. *Biochem. J.* **62,** 621–5.

Darmon, S. and Rudall, K. M. (1950). Infra-red and X-ray studies of chitin. *Discuss. Faraday Soc.* **9,** 251–60.

Das, S. M. (1936). On the structure and function of the ascidian test. *J. Morph.* **59,** 539–601.

Davson, H. and Danielli, J. F. (1943). "The Permeability of Natural Membranes". Cambridge University Press, London.

Dawes, B. (1940). Notes on the formation of the egg capsules in the monogenetic trematode, *Hexacotyle extensicauda* Dawes, 1940. *Parasitology*, **32,** 287–95.

Dawson, A. B. (1920). Integument of *Necturus maculosus*. *J. Morph.* **34,** 487–577.

Dawson, B. (1960). Use of collagenase in the characterisation of pseudocoelomic membranes of *Ascaris lumbricoides*. *Nature, Lond.* **187,** 799.

Dawson, J. A. (1963). The oral cavity, the jaws and the horny teeth of *Myxine*. *In* "The Biology of Myxine" (ed. A. Brodal and R. Fänge) pp. 231–55. Universitetsforhget, Oslo.

Decker, J. D. (1966). An electron microscopic investigation of osteogenesis in the embryonic duck. *Am. J. Anat.* **118,** 591–613.

Dedeurwaeder, R. A., Dobb, M. G., and Sweetman, B. J. (1964). Selective extraction of a protein fraction from wool keratin. *Nature, Lond.* **203,** 48–9.

Degens, E. T. and Love, S. (1965). Comparative studies in amino acids in shell structures of *Gyraulus trochiformis* Stahl, from the Tertiary of Steinheim, Germany. *Nature, Lond.* **205,** 876–8.

Degens, E. T., Spencer, D. W., and Parker, R. H. (1967). Paleobiochemistry of molluscan shell proteins. *Comp. Biochem. Physiol.* **20,** 553–79.

Denison, R. H. (1956). A review of the habitat of the earliest vertebrates. *Fieldiana, Geol. Mem.* **11,** 359–457.

Dennell, R. (1947). The occurrence and significance of phenolic hardening in the newly formed cuticle of Crustacea Decapoda. *Proc. R. Soc. Lond.* B. **134,** 485–502.

Dennell, R. (1949). Earthworm chaetae. *Nature, Lond.* **164,** 370.

Dennell, R. (1960). Integument and exoskeleton. *In* "The Physiology of Crustacea", (ed. T. H. Waterman) Vol. 1, pp. 449–69. Academic Press, New York and London.

DENTON, E. J. and GILPIN-BROWN, J. B. (1961). The buoyancy of the cuttle fish, *Sepia officinalis* L. *J. mar. biol. Ass. U.K.* **41,** 319–42.

DENTON, E. J. and GILPIN-BROWN, J. B. (1966). On the buoyancy of the pearly *Nautilus*. *J. mar. biol. Ass. U.K.* **46,** 723–59.

DEVANESEN, D. W. (1922). The development of the calcareous parts of the lantern of Aristotle in *Echinus miliaris*. *Proc. R. Soc. Lond.* B. **93,** 468–85.

DEYL, Z., EVERITT, A., and ROSMUS, J. (1968). Changes in collagen cross-linking in rats with reduced food intake, *Nature, Lond.* **217,** 670–1.

DEYRUP, I. J. (1964). Water balance and kidney. *In* "Physiology of the Amphibia" (ed. J. A. Moore) pp. 251–328. Academic Press, New York and London.

DHAVALIKAR, R. S. (1962). Amino acid composition of indian silk fibroins and sericins: Part II—Sericins. *J. scient. indust. Res.* **21** C, 303.

DICKSON, M. R. and MERCER, E. H. (1962). Fine structural changes accompanying desiccation in *Philodina roseola*. *J. Miscroscopie*, **6,** 331–48.

DIGBY, P. (1966). Mechanism of calcification in mammalian bone. *Nature, Lond.* **212,** 1250–2.

DIGBY, P. S. B. (1968). The mechanism of calcification in the molluscan shell. *In* "Studies in the Structure, Physiology and Ecology of Molluscs" (ed. V. Fretter). Symposia of the Zoological Society of London and the Malacological Society, Vol. 22. Academic Fress, London.

DINGLE, J. T. (1961). Studies on the mode of action of excess vitamin A. 3. Release of a bound protease by the action of vitamin A. *Biochem. J.* **79,** 509–12.

DISSANAIKE, A. J. and CANNING, E. V. (1957). The mode of emergence of the sporoplasm in microsporidia and its relation to the structure of the spores. *Parasitology*, **47,** 92–9.

DITMARS, R. L. (1933). "Reptiles of the World". John Lane, London.

DIXON, K. E. (1964). Excystment of metacercariae of *Fasciola hepatica* L. in vitro. *Nature, Lond.* **202,** 1240–1.

DIXON, K. E. (1965). The structure and histochemistry of the cyst wall of the metacercaria of *Fasciola hepatica* L. *Parasitology*, **55,** 215–26.

DIXON, K. E. and MERCER, E. H. (1964). The fine structure of the cyst wall of the metacercaria of *Fasciola hepatica*. *Q. Jl microsc. Sci.* **105,** 385–9.

DODD, J. R. (1963). Paleoecological implications of shell mineralogy in two pelecypod species. *J. Geol.* **71,** 1–11.

DODGE, J. D. (1965). Thecal fine-structure in the dinoflagellate genera, *Prorocentrum* and *Exuviaella*. *J. mar. biol. Ass. U.K.* **45,** 607–14.

DODGE, J. D. (1967). The fine structure of the dinoflagellate *Aureodinium pigmentosum* gen. et sp. nov. *Br. phycol. Bull.* **3,** 327–36.

DODSON, J. W. (1963). On the nature of tissue interactions in embryonic skin. *Expl Cell. Res.* **31,** 233–5.

DOREY, A. E. (1965). The organisation and replacement of the epidermis in acoelous turbellarians. *Q. Jl microsc. Sci.* **106,** 147–72.

DOWNE, A. E. R. (1962). Serology of insect proteins. I. Preliminary studies on the proteins of insect cuticle. *Can. J. Zool.* **40,** 957–67.

DRUM, R. W. (1968). Electron microscopy of siliceous spicules from the freshwater sponge *Heteromyenia*. *J. Ultrastruct. Res.* **22,** 12–21.

DUCROS, C. (1967). Contribution a l'étude du tannage de la radula chez les gasteropodes. *Ann. Histochim.* **12,** 243–73.

DUDICH, E. (1931). Systematische und biologische Untersuchungen über die Kalkeinlagerungen des Crustaceenpanzers in polarisiertem Lichte. *Zoologica, Stuttg.* **30,** 1–154.

DUMONT, J. N., ANDERSON, E., and CHOMYN, E. (1964). The fine structure of the peripheral nerve and its ensheathing artery in the horse-shoe crab, *Limulus polyphemus*. *Am. Zool.* **4,** 314.

DUNACHIE, J. F. (1963). The periostracum of *Mytilus edulis*. *Trans. R. Soc. Edinb.* **65,** 383–411.

EARLAND, C., BLAKEY, P. R., and STELL, J. G. P. (1962). Studies on the structure of keratin, IV. The molecular structure of some morphological components of keratins. *Biochim. biophys. Acta*, **56,** 268–74.

EARLAND, C. and RAVEN, D. J. (1961). Isolation of a cystine derivative from silk. *Nature, Lond.* **192,** 1185–6.

Eastoe, J. E. (1957). The amino acid composition of fish collagen and gelatin. *Biochem. J.* **65,** 363–8.

Eastoe, J. E. (1960). Organic matrix of tooth enamel. *Nature, Lond.* **187,** 411–2.

Eastoe, J. E. (1963). The amino acid composition of proteins from the oral tissues II. The matrix proteins in dentine and enamel from developing human deciduous teeth. *Archs. Oral Biol.* **8,** 633–52.

Eastoe, J. E. (1967). Composition of collagen and allied proteins. *In* "Treatise on Collagen", (ed. G. N. Ramachandran) Vol. 1, pp. 1–67. Academic Press, London and New York.

Eastoe, J. E. and Eastoe, B. (1954). The organic constituents of mammalian compact bone. *Biochem. J.* **57,** 453–9.

Eastoe, J. E. and Leach, A. A. (1958). A survey of recent work on the amino acid composition of vertebrate collagen and gelatin. *In* "Recent Advances in Gelatin and Glue Research" (ed. G. Stainsby) pp. 173–8. Pergamon Press, London.

Ebling, F. J. (1945). Formation and nature of opercular chaetae of *Sabellaria alveolata. Q. Jl microsc. Sci.* **85,** 153–76.

Ebling, F. J. (1965). Comparative and evolutionary aspects of hair replacement. *In* "Comparative Physiology and Pathology of the Skin" (ed. A. J. Rook and G. S. Walton) pp. 87–102. Blackwell, Oxford.

Ebling, W. (1964). The permeability of insect cuticle. *In* M. Rockstein, "The Physiology of Insecta", Vol. 3, pp. 507-556. Academic Press, New York and London.

Ebner, V. von (1887). Ueber den feineren Bau der Skelettheile der Kalkschwamme, nebst Bemerkungen über Kalkskelete überhaupt. *Sber. Akad. Wiss., Wien* Abt. I. **95,** 55–149.

Echlin, P. (1966). Origins of photosynthesis. *Science Jl* **2,** April, 42–7.

Echlin, P. (1969). The origins of plants. *New Scientist* **42,** 286–9.

Eckert, J. and Schwarz, R. (1965). Zur Struktur der Cuticula invasionsfähiger Larven einiger Nematoden. *Z. ParasitKde.* **26,** 116–42.

Edkin, R. M. and Westfall, J. A. (1962). Fine structure of the notochord of *Amphioxus. J. Cell. Biol.* **12,** 646–51.

Edwards, F. S. (1964). (See Andersen and Weis-Fogh, 1964).

Eglinton, G. and Calvin, M. (1967). Chemical Fossils. *Scient. Am.* **216,** (1) 32–43.

Ehret, C. F. and Powers, E. L. (1959). The cell surface of *Paramecium. Int. Rev. Cytol.* **8,** 87–133.

Eisen, A. Z. and Gross, J. (1965). The role of epithelium and mesenchyme in the production of a collagenolytic enzyme and a hyaluronidase in the anurian tadpole. *Devl. Biol.* **12,** 408–18.

Elder, H. Y. (1966). The fine structure of some invertebrate fibrillar and lamellar elastica. *Proc. R. Microscop. Soc.* **1,** 99–100.

Elder, H. Y. and Owen, G. (1967). Occurrence of "elastic" fibres in invertebrates. *J. Zool., Lond.* **152,** 1–8.

Elkan, E. (1968). Mucopolysaccharides in the anuran defence against desiccation. *J. Zool., Lond.* **155,** 19–53.

Ellenby, C. (1946). Nature of the cyst wall of the potato-root eelworm, *Heterodera rostochiensis* Wollenweber and its permeability to water. *Nature, Lond.* **157,** 302.

Ellenby, C. (1963). Masked polyphenols in the cuticle of a cyst-forming nematode. *Experientia,* **19,** 256–7.

Ellenby, C. (1968). Desiccation survival in the plant parasitic nematodes, *Heterodera rostochiensis* Wollenweber and *Ditylenchus dipsaci* (Kuhn) Filipjev. *Proc. R. Soc. Lond.* B. **169,** 203–13.

Elliot, D. H. (1965). Structure and function of mammalian tendon. *Biol. Rev.* **40,** 392–421.

Elton, C. S. (1966). *The pattern of Animal Communities.* Methuen, London.

Endean, R. (1955). Studies of the blood and tests of some Australian ascidians III. The formation of the test of *Pyura stolonifera* (Heller). *Aust. J. mar. Freshwat. Res.* **6,** 157–64.

Endean, R. (1961). The test of the ascidian, *Phallusia mammillata. Q. Jl microsc. Sci.* **102,** 107–17.

Erdman, J. G., Marlett, E. M., and Hanson, W. E. (1956). Survival of amino acids in marine sediments. *Science, N.Y.* **124,** 1026.

ERNST, V. and RUIBAL, R. (1966). The structure and development of the digital lamellae of lizards. *J. Morph.* **120,** 233–42.

'ESPINASSE, P. G. (1939). Developmental anatomy of the Brown Leghorn breast feather and its reaction to oestrone. *Proc. zool. Soc., Lond.* (Ser. A) **109,** 247–86.

'ESPINASSE, P. G. (1964) Feathers. *In* "A New Dictionary of Birds" (ed. A. Landsborough Thompson) pp. 272–7. Nelson, London.

ESPRING, U. (1957a). A factor inhibiting fertilization of sea urchin eggs from extracts of the alga *Fucus vesiculosus* 1. The preparation of the factor inhibiting fertilization. *Arkiv. f. Kemi, Stockh.* **11,** 107–15.

ESPRING, U. (1957b). A factor inhibiting fertilization of sea urchin eggs from extracts of the alga *Fucus vesiculosus.* 2. The effect of the factor inhibiting fertilization on some enzymes. *Arkiv. f. Kemi, Stockh.* **11,** 117–27.

EVANS, F. G. (1957). *Stress and Strain in Bone.* Thomas, Springfield, Illinois.

FAIRBAIRN, D. (1957). The biochemistry of *Ascaris. Expl Parasitol.* **6,** 491–554.

FAIRBAIRN, D. and PASSEY, R. F. (1957). Occurrence and distribution of trehalose and glycogen in the eggs and tissues of *Ascaris lumbricoides. Expl Parasitol.* **6,** 566–74.

FAURÉ-FREMIET, E. (1912). Graisse et glycogène dans le développement de l'oeuf d'*Ascaris megalocephala. Bull. Soc. zool. Fr.* **37**(b), 233–4.

FAURÉ-FREMIET, E. (1913). La formation de la membrane interne de l'oeuf d'*Ascaris megacephata. C.r. Séanc. Soc. Biol.*, **74,** (20) 1183–4.

FAURÉ-FREMIET, E. (1931). Étude histologique de *Ficulina ficus. Arch. Anat. Microsc.* **28,** 421–48.

FAURÉ-FREMIET, E. and BAUDOUY, C. (1938). Sur l'ovokératine des selaciens. *Bull. Soc. Chim. biol.* **20,** 14–23.

FAURÉ-FREMIET, E. and GARRAULT, H. (1944). Proprietes physiques de l'ascarocollagene. *Bull. Biol. Fr. Belg.* **78,** 206–14.

FAURÉ-FREMIET, E., ROUILLER, C., and GAUCHERY, M. (1956a). Structure et origine du peduncle chez *Chilodochona. J. Protozool.* **3,** 188–93.

FAURÉ-FREMIET, E., ROUILLER, C., and GAUCHERY, M. (1956b). Les structures myoides chez les cilies. Étude au microscope électronique. *Archs. Anat. Microsc.* **45,** 139–61.

FAURÉ-FREMIET, E., ROUILLER, C., and GAUCHERY, M. (1956c). L'appareil squelettique et myoide des urcéolaires: étude au microscope électronique. *Bull. Soc. zool. Fr.* **81,** 77–84.

FAURÉ-FREMIET, E., STOLKOWSKI, J., and DUCORNET, J. (1948). Étude experimentale de la calcification tegumentaire chez un infusoire cilie, *Coleps hirtus. Biochim. biophys. Acta* **2,** 668–73.

FAURÉ-FREMIET, E. and THAUREAU, J. (1944). Proteines de structure et cytosquelette chez les Urcéolarides. *Bull. Biol. Fr. Belg.* **78,** 143–56.

FELL, H. B. (1964). The experimental study of keratinization in organ culture. *In* "The Epidermis" (ed. W. Montagna and W. C. Lobitz) pp. 61–79. Academic Press, New York and London.

FELL, H. B. and MELLANBY, E. (1953). Metaplasia produced in cultures of chick ectoderm by high vitamin A. *J. Physiol.* **119,** 470–88.

FERRY, J. D. (1941). A fibrous protein from the slime of the hagfish. *J. biol. Chem.* **138,** 263–8.

FILHOL, J. and GARRAULT, H. (1938). La secretion de la prokeratine et la formation de la capsule ovulaire chez les selaciens. *Archs. Anat. Microsc.* **34,** 105–45.

FILSHIE, B. K. and ROGERS, G. E. (1961). The fine structure of a keratin. *J. molec. Biol.* **3,** 784–6.

FILSHIE, B. K. and ROGERS, G. E. (1962). An election microscope study of the fine structure of feather keratin. *J. Cell. Biol.* **13,** 1–12.

FINEAN, J. B. and ENGSTROM, A. (1953). The low-angle scatter of X-rays from bone tissue. *Biochim. biophys. Acta,* **11,** 178–89.

FINLEY, H. E. and BACON, A. L. (1965). The morphology and biology of *Pyxicola nolandi* (Ciliate, Peritrichida, Vaginicolidae). *J. Protozool.* **12,** 123–31.

FISCHER, F. G. and BRANDER, J. (1960). Eine Analyse der Gespinste der Kreusspinne. *Hoppe-Seyler's Z. physiol Chem.* **320,** 92–102.

FISCHER, F. G. and NEBEL, H. J. (1955). Nachweis und Bestimmung von Glucosamin und Galactosamin auf Paper chromatogrammen. *Hoppe-Seyler's Z. physiol. Chem.* **302,** 10–19.

Fischer, P. H. (1940). Structure et evolution de l'epithelium de l'operculum chez *Purpura lapillus* L. *Bull. Soc. zool. Fr.* **65,** 199–204.

Fisher, L. R. and Kon, S. K. (1959). Vitamin A in the invertebrates. *Biol. Rev.* **34,** 1–36.

Fitton-Jackson, S. (1957). The fine structure of developing bone in the embryonic fowl. *Proc. R. Soc. Lond.* B. **146,** 270–80.

Fitton-Jackson, S. (1964). Connective Tissue Cells. *In* "The Cell" (ed. J. Brachet and A. E. Mirsky) Vol. VI, pp. 387–520. Academic Press, New York and London.

Fitton-Jackson, S. (1965). Macromolecular order in the ground substance. *In* "Structure and Function of Connective and Skeletal Tissue" (ed. S. Fitton-Jackson *et al.*) pp. 156–60. Butterworth, London.

Fitton-Jackson, S. (1968). The morphogenesis of collagen. *In* "Treatise on Collagen" (ed. B. S. Gould) Vol. 2, Pt. B., pp. 1–60. Academic Press, London and New York.

Fitton-Jackson, S., Kelly, F. C., North, A. C. T., Randall, J. T., Seeds, W. E., Watson, M., and Wilkinson, G. W. (1965). The byssus threads of *Mytilus edulis* and *Pinna nobilis*. *In* "Nature and Structure of Collagen" (ed. J. T. Randall and S. Fitton-Jackson) pp. 106–16. Butterworth, London.

Fitton-Jackson, S. and Randall, J. T. (1956). Fibrogenesis and the formation of matrix in developing bone. *In* Ciba Foundation Symposium on "Pone Structure and Metabolism" (ed. G. E. W. Wolstenholme and C. M. O'Connor) pp. 47–62. Churchill, London.

Fjerdingstad, E. J. (1961). Ultrastructure of the collar of the choanoflagellate, *Codonosiga botrytis*. *Z. Zellforsch. mikrosk. Anat.* **54,** 499–510.

Fleisch, H. and Bisaz, S. (1964). Role of collagen, pyrophosphate and pyrophosphatase in calcification. *In* "Bone and Tooth", Proceedings First European Symposium of the Bone and Tooth Society. (ed. H. J. J. Blackwood) pp. 249–56. Pergamon Press, Oxford.

Fleischmajer, R. and Fishman, L. (1965). Amino-acid composition of human dermal collagen. *Nature, Lond.* **205,** 264–6.

Flesch, P. (1958). Chemical data on human epidermal keratinization and differentiation. *J. invest. Derm.* **31,** 63–73.

Florkin, M., Grégoire, C., Bricteux-Grégoire, S., and Schoffeniels, E. (1961). Conchiolines de nacres fossilles. *C.r. hebd. Séanc. Acad. Sci., Paris,* **252,** 440–2.

Flower, N. E., Geddes, A. J., and Rudall, K. M. (1969). Ultra-structure of the fibrous protein from the egg capsules of the whelk, *Buccinum undatum*. *J. Ultrastruct. Res.* **26,** 262–73.

Flower, N. E. and Kenchington, W. (1967). Studies on insect fibrous proteins: the larval silk of *Apis*, *Bombus* and *Vespa* (Hymenoptera: Aculeata). *Jl R. Microsc. Soc.* **86,** 297–310.

Fott, B. and Ludvik, J. (1956). Uber den submicroskopischen Bau des Panzers von *Ceratium hirundinella*. *Prestia,* **28,** 276–8.

Foucart, M. F. (1966a). Localisation du collagène dans le test d'un oursin (Echinoderme). *Bull. Acad r. Belg. Cl. Sci.* **52,** 316–9.

Foucart, M. F. (1966b). Composition chimique de la matrice organique du test d'un echinoderm *Paracentrotus lividus* Lamarek. *Bull. Acad. r. Belg. Cl. Sci.* **52,** 1155–62.

Foucart, M., Bricteux-Grégoire, S., and Jeuniaux, C. (1965). Composition chimique du tube d'un pogonophore (*Siboglinum* sp.) et des formations squelettique de deux pterobranches. *Sarsia,* **20,** 35–41.

Foucart, M. F., Bricteux-Grégoire, S., Jeuniaux, C., and Florkin, M. (1965). Fossil proteins of graptotites. *Life Sciences,* **4,** 467–71.

Foucart, M. F. and Jeuniaux, C. (1966). Paléobiochimie et position systématique des Graptolithes. *Annls. Soc. r. zool. Belg.* **95,** 39–45.

Fox, D. L. (1966). Pigmentation of Molluscs. *In* "Physiology of Mollusca" (ed. K. M. Wilbur and C. M. Yonge) Vol. 2. Academic Press, New York and London.

Fox, H. M. and Ramage, H. (1931). A spectroscopic analysis of animal tissues. *Proc. R. Soc.* B. **108,** 157–73.

Fraenkel, G. and Rudall, K. M. (1940). A study of the physical and chemical properties of the insect cuticle. *Proc. R. Soc.* B. **129,** 1–34.

Fraenkel-Conrat, H. (1964). The genetic code of a virus. *Sci. Ann.* **211,** 47–54.

Francoise, J. (1968). Nature conjonctive du "tentorium" des Diploures (Insectes, Apterygotes). Étude ultrastructurale. *C. r. hebd. Séanc. Acad. Sci., Paris,* **267,** 1976–8.

Fredericq, L. (1878). Sur la digestion des albuminoides chez quelque invertébrés. *Bull. Acad. r. Belg. Cl. Sci.* (2) **46,** 213–28.

Fretter, V. (1937). The structure and function of the alimentary canal of some species of Polyplacophora (Mollusca). *Trans R. Soc., Edinb.* **59,** 119–64.

Fretter, V. (1941). The genital ducts of some British stenoglossan prosobranchs. *J. mar. biol. Ass. U.K.* **25,** 173–211.

Friedmann, H. (1935). Notes on the differential threshold of reaction to vitamin D—deficiency in the house sparrow and the chick. *Biol. Bull. mar. biol. lab., Woods Hole,* **69,** 71–4.

Friend, D. S. (1960). The fine structure of *Giardia muris*, *J. Cell. Biol.* **29,** 317–32.

Fritsch, F. G. (1935). "The Structure and Reproduction of the Algae", Vol. 1. Cambridge University Press, London.

Fujimoto, D. and Adams, E. (1964). Intra species composition in collagen from cuticle and body of *Ascaris* and *Lumbricus*. *Biochem. biophys. Res. Commun.* **17,** 437–42.

Furneaux, P. J. S. and MacFarlane, J. E. (1965). Identification, estimation and localization of catecholamines in eggs of the house cricket, *Acheta domesticus,* (L). *J. Insect Physiol.* **11,** 591–600.

Fyffe, A. (1819). Account of some experiments made with the view of ascertaining the different substances from which iodine can be procured. *Edinb. Phil. J.* **1,** 254–8.

Gabe, M. and Prennant, M. (1949). Donnes histologiques sur le tissue conjonctif des Polyplacophores. *Archs. Anat. microsc. Morph. exp.* **38,** 65–78.

Gabe, M. and Prennant, M. (1957). Particularités histochimiques du ruban radulaire et des dents de la radula chez quelques mollusques. *Bull. Soc. zool. Fr.* **82,** 195–6.

Gabe, M. and Prennant, M. (1958). Particularités histochimiques de l'appareil radulaire chez quelques mollusques. *Ann. Histochim.* **3,** 95–112.

Gadow, H. (1901). Amphibia and Reptiles. *In* "The Cambridge Natural History" (eds. S. F. Harmer and A. E. Shipley) Vol. VIII. Macmillan, London.

Gallacher, I. H. C. (1964). Chemical composition of hooks isolated from hydatid scolices. *Expl Parasitol.* **15,** 110–7.

Gansen-Semal, P. van (1960). Occurrence of a non-fibrillar elastin in the earthworm. *Nature, Lond.* **186,** 654–5.

Garnham, P. C. C., Baker, J. R., and Bird, R. G. (1962). The fine structure of *Lankesterella garnhami*. *J. Protozool.* **9,** 107–14.

Garnham, P. C. C., Bird, R. G., and Baker, J. R. (1961). Electron microscope studies of motile stages of malaria parasites II. The fine structure of the sporozoite of *Laverania* (= *Plasmodium falcipara*). *Trans. R. Soc. trop. Med. Hyg.* **55,** 98–102.

Garnham, P. C. C., Bird, R. G., and Baker, J. R. (1963). Electron microscope studies of malarial parasites IV. The fine structure of the sporozoites of four species of *Plasmodium*. *Trans. R. Soc. trop. Med. Hyg.* **57,** 27–31.

Garrault, H. (1934). Le tissu élastique du pédoneule de *Pollicipes cornucopiae* Leach. *Archs Anat. microsc.,* **30,** 199–215.

Garrault, H. (1936). Developpement des fibres d'elastoidine (actinotrichia) chez les salmonides. *Archs Anat. microsc.* **32,** 105–37.

Garrault, H. (1937). Structure de la membrane basale sous-epidermique chez les embryons de selaciens. *Archs Anat. microsc.* **33,** 167–76.

Garrido, I. and Blanco, J. (1947). Structure cristalline des piquants d'oursin. *C. r. hebd. Séanc. Acad. Sci., Paris,* **224,** 485.

Gaupner, H. and Fischer, I. (1933). Beiträge zur Kenntnis der Goldfischhaut II. Über die Bildung der Perlorgane bei *Carassius auratus*. *Zool. Anz.* **103,** 279–85.

Gegenbaur, C. (1858). Anatomische Untersuchungen eines *Limulus* mit besonderer Berucksichtigung der Gewebe. *Abh. naturforsch. Ges. Halle,* **4.**

Gelei, J. von (1936). Das erregungsleitende System der Ciliaten. C.r. XII Congr. Int. Zool. Lisbonne, pp. 174–209. Conférences et Communications Scientifiques.

Gibbons, I. R. (1965). An effect of adenosine triphosphate on the light scattered by suspensions of cilia. *J. Cell. Biol.* **26,** 707–12.

Gibbons, I. R. and Grimstone, A. V. (1960). On flagellar structure in certain flagellates. *J. Cell. Biol.* **7,** 697–716.

GILLESPIE, J. M., O'DONNEL, I. J., THOMPSON, E. O. P., and WOODS, E. F. (1960). Preparation and properties of wool proteins. *J. Text. Inst.* **51,** *T*703–15.

GILMOUR, D. (1961). *The Biochemistry of Insects*. Academic Press, New York and London.

GLIMCHER, M. J. (1959). Molecular biology of mineralised tissues with particular reference to bone. *Rev. mod. Phys.* **31**, 359–93.

GLIMCHER, M. J. (1960). Specificity of the molecular structure of organic matrices in mineralisation. *In* "Calcification in Biological Systems" (ed. R. F. Sognnaes) pp. 421–87. American Association for the Advancement of Science, Washington, Publ. No. 64.

GLIMCHER, M. J., BONAR, L. C. and DANIEL, E. J. (1961). The molecular structure of the protein matrix of bovine dental enamel. *J. molec. Biol.* **3,** 541–6.

GLIMCHER, M. J., FRIBERG, U. A. and LEVINE, P. T. (1964). The isolation and amino acid composition of the enamel proteins of erupted bovine teeth. *Biochem. J.* **93,** 202–10.

GLIMCHER, M. J., HODGE, A. J., and SCHMITT, F. O. (1957). Macromolecular aggregation states in relation to mineralisation: The collagen-hydroxyapatite system as studied *in vitro*. *Proc. natn. Acad. Sci. U.S.A.* **43**. 860–7.

GLIMCHER, M. J. and KRANE, S. M. (1964). The identification of serine phosphate in enamel proteins. *Biochim. biophys. Acta*, **90,** 477–83.

GLIMCHER, M. J. and KRANE, S. M. (1968). The organisation and structure of bone and the mechanism of calcification. *In* "Treatise on Collagen" (ed. B. S. Gould) Vol. 2, Pt. B, pp. 68–241. Academic Press, London and New York.

GLIMCHER, M. J., MECHANIC, G. L. and FRIBERG, U. A. (1964). The amino acid composition of the organic matrix and neutral-soluble and acid-soluble components of embryonic bovine enamel. *Biochem. J.* **93,** 198–202.

GODEAUX, J. (1964). Le revêtement cutané des tuniciers. *Studium Generale* **17,** *Heft* **3,** 176–90.

GOFFINET, G. (1965). Conchioline, nacroïne et chitine dans la coquille des mollusques. *Mém. Lic. Sci. Zool. Fac. Sci. Univ. Liège*. 100 pp. (dactyl., inedit.)

GOMOT, L. (1959). Contribution a l'étude du developpement embryonnaire de la glande uropygienne chez le canard. *Archs Anat. microsc. Morphol. exp.* **48,** 63–141.

GÖNNERT, R. (1955). Schistosomiasis-Studien II. Über die Eibildung bei *Schistosoma mansoni* und das Schicksal der Eier im Wirts organismus. *Z. Tropen. Med. Parasit.* **6,** 33–52.

GÖNNERT, R. (1962). Histologische untersuchungen uber den Feinbau die Eibildungsstatte (Oogenotop) von *Fasciola hepatica*. *Z. ParasitKde*, **21,** 475–92.

GOODEY, T. (1913). The excystment of *Colpoda cucullus* from its resting cysts and the nature and properties of the cyst membrane. *Proc. R. Soc.* B. **86,** 427–39.

GOODRICH, E. S. (1896). Notes on the oligochaetes, with the description of a new species. *Q. Jl microsc. Sci.* **39,** 51–69.

GORBMAN, A., CLEMENTS, M., and O'BRIEN, R. (1954). Utilization of radio-active iodine by invertebrates with a special study of several Annelida and Mollusca. *J. exp. Zool.* **127,** 75–89.

GOREAU, T. F. (1959). The physiology of skeleton-formation in corals I. A method of measuring the rate of calcium deposition by corals under different conditions. *Biol. Bull. mar. biol. Lab., Woods Hole*, **116,** 59–73.

GOREAU, T. F. (1961). Problems of growth and calcium deposition in reef corals. *Endeavour*, **20,** 32–9.

GOTTE, L., MENEGHELLI, V. and CASTELLANI, A. (1965). Electron microscope observations and chemical analyses of human elastin. *In* "The Structure and Function of Connective Tissue" (ed. S. Fitton-Jackson *et al.*) pp. 93–100. Proceedings of N.A.T.O. Advanced Study Group, St. Andrews, 1964. Butterworths, London.

GOTTE, L., STERN, P., ELDSDEN, D. F., and PARTRIDGE, S. M. (1963). The chemistry of connective tissues 8. The composition of elastin from three bovine tissues. *Biochem. J.* **87,** 344–51.

GOULD, B. S. (1968). The role of certain vitamins in collagen formation. *In* "Treatise on Collagen" (ed. B. S. Gould). Vol. 2, Part A, pp. 323–66. Academic Press, London and New York.

GRAFF, L. VON (1883). Über *Rhodope veranii*. Kölliker. *Morph. Jb.* **8,** 73–83.

GRAHAM, G. N. and PAUTARD, F. G. E. (1963). Mature enamel matrix. *J. dent. Res.* **42,** 1100.

Grant, R. A., Cox, R. W. and Horne, R. W. (1967). The structure and assembly of collagen fibrils. *Jl R. Microsc. Soc.* **87,** 143–55.

Grassé, P. P. (1956). L'ultrastructure de *Pyrsonympha vertens* (Zooflagellata Pyrsonymphina): les flagelles et leur coaptation avec le corps, l'axostyle contractile, le paraxostyle et le cytoplasme. *Archs Biol., Paris,* **67,** 595–611.

Green, J. C. and Jennings, D. H. (1967). A physical and chemical investigation of the scales produced by the Golgi apparatus within, and found on the surface of cells of *Chrysochromulina chiton* Parke and Manton. *J. exp. Bot.* **18,** 359–70.

Greenlee, T. K., Ross, R. and Hartman, J. L. (1966). The fine-structure of elastic fibres. *J. Cell. Biol.* **30,** 59.

Grégoire, Ch. (1958). Essai de détection au microscope électronique des dentelles organiques dans les nacres fossiles (ammonites, céphalopodes, gastéropodes et pélécypodes). *Arch. int. Physiol. Biochim.* **66,** 674–6.

Grégoire, Ch. (1959). A study on the remains of organic components in fossil mother-of-pearl. *Bull. Inst. r. Sci. natn. Belg.* **35,** 1–14.

Grégoire, Ch. (1961a). Structure of the conchiolin cases of the prisms in *Mytilus edulis* Linné. *J. biophys. biochem. Cytol.* **9,** 395–400.

Grégoire, Ch. (1961b). Sur la structure submicroscopique de la conchioline associée aux prismes des coquilles des mollusques. *Bull. inst. r. Sci. nat. Belg.,* **31,** 1–34.

Grégoire, Ch. (1967). Sur la structure des matrices organiques des coquilles de mollusques. *Biol. Rev.* **42,** 653–88.

Grégoire, Ch., Duchateau, G. and Florkin, M., (1955). La trame protidique des nacres et des perles. *Ann. Inst. Océanogr.* **31,** 1–36.

Grell, K. G. V. and Wohlfarth-Botterman, K. E. (1957). Licht und electronmikroscopische untersuchungen an den Dinoflagellaten *Amphidinium elegans.* n.sp. *Z. Zellforsch.* **47,** 7–11.

Grimstone, A. V. (1959). Cytoplasmic membranes and the nuclear membrane in the flagellate *Trichonympha. J. biophys. biochem. Cytol.* **6,** 369–78.

Grimstone, A. V. (1961). Fine structure and morphogenesis in protozoa. *Biol. Rev.* **36,** 97–150.

Grimstone, A. V. (1962). Cilia and Flagella. *Br. med. Bull.* **18,** 238–41.

Grimstone, A. V. (1963). Fine structure of some polymastigote flagellates. *Proc. Linn. Soc., Lond.* **174,** 49–52.

Grimstone, A. V. (1966). Structure and function in Protozoa. *A. Rev. Microbiol.* **20,** 131–50.

Grimstone, A. V. and Cleveland, L. R. (1965). The fine structure and function of the contractile axostyle of certain flagellates. *J. Cell. Biol.* **24,** 387–466.

Grimstone, A. V. and Gibbons, I. R. (1966). The fine structure of the centriolar apparatus and associated structures in the complex flagellates *Trichonympha* and *Pseudotrichonympha. Phil. Trans. R. Soc.* B. **250,** 215–42.

Grimstone, A. V., Horne, R. W., Pantin, C. F. A., and Robson, E. A. (1958). The fine structure of the mesenteries of the sea anemone, *Metridium senile. Q. Jl microsc. Sci.* **99,** 523–40.

Gross, J. (1956). The behaviour of collagen units as a model in morphogenesis. *J. biophys. biochem. Cytol.* **2,** 261–74.

Gross, J. (1963). Comparative biochemistry of collagen. *In* "Comparative Biochemistry" (eds. M. Florkin and H. S. Mason) Vol. V, pp. 307–42. Academic Press, New York and London.

Gross, J. and Dumsha, B. (1958). Elastoidin: a two component member of the collagen class. *Biochim. biophys. Acta,* **28,** 268–70.

Gross, J., Dumsha, B., and Glazer, N. (1958). Comparative biochemistry of collagen. Some amino acids and carbohydrates. *Biochim. biophys. Acta,* **30,** 293–7.

Gross, J. and Lapier, C. M. (1962). Collagenolytic activity in amphibian tissues: A tissue-culture assay. *Proc. natn. Acad. Sci. U.S.A.* **48,** 1014–22.

Gross, J., Sokal, Z., and Rougvie, M. (1956). Structural and chemical studies on the connective tissue of marine sponges. *J. Histochem. Cytochem.* **4,** 227–44.

Grove, A. J. (1925). On the reproductive processes of the earthworm *Lumbricus terrestris. Q. Jl microsc. Sci.* **69,** 245–90.

GROVE, A. J. and COWLEY, L. F. (1926). On the reproductive processes of the brandling worm, *Eisenia foetida* (Sav.) *Q. Jl microsc. Sci.* **70,** 559–81.

GUPTA, B. L. and LITTLE, C. (1969). Studies on Pogonophora II. Ultrastructure of the tentacular crown of *Siphonobrachia*. *J. mar. biol. Ass., U.K.* **49,** 717–41.

GUPTA, B. L., LITTLE, C., and PHILIP, A. M. (1966). Studies on Pogonophora: Fine structure of the tentacles. *J. mar. biol. Ass. U.K.* **46,** 351–72.

GUSTAVSON, K. H. (1955). The function of hydroxyproline in collagens. *Nature, Lond.* **175,** 70–4.

HAAS, F. (1935). Bivalvia, Teil 1. *In* "Klassen und Ordnungen des Tierreichs" Band III(3), Mollusca, Abt. 3 (ed. H. G. Bronn). Akad. Verlagsges., Leipzig.

HACKMAN, R. H. (1958). Biochemistry of the insect cuticle. *In* Proceedings of the Fourth International Congress on Biochemistry, Vienna, 1958, Vol. 12, pp. 48–57.

HACKMAN, R. H. (1960). Studies on Chitin. IV. The occurrence of complexes in which chitin and protein are covalently linked. *Aust. J. biol. Sci.* **13,** 568.

HACKMAN, R. H. (1964). Chemistry of the Insect cuticle. *In* "The Physiology of Insecta" (ed. M. Rockstein) Vol. III, pp. 471–502. Academic Press, New York and London.

HACKMAN, R. H. and GOLDBERG, M. (1958). Proteins of the larval cuticle of *Agrianome spinicollis* (Coleoptera). *J. Insect. Physiol.* **2,** 221–31.

HAEKEL, E. (1886). Report on the Scientific Results of the Exploring Voyage of H.M.S. Challenger 1873–76. *Zoology*, Vol. XVIII Report on the Radiolaria. Plates.

HAGGIS, G. H., MICHIE, D., MUIR, A. R., ROBERTS, K. B., and WALKER, P. M. B. (1964). *Introduction to Molecular Biology*. Longmans, London.

HALDANE, J. B. S. (1930). On being the right size. *In* "Possible Worlds". Phoenix Library, Chatto and Windus, London.

HALIBURTON, W. D. (1885). On the occurrence of chitin as a constituent of the cartilages of *Limulus* and *Sepia*. *Q. Jl microsc. Sci.* (N.S.) **25,** 173–81.

HALL, C. E. and SLAYTER, H. S. (1959). The fibrinogen molecule: its size, shape and mode of polymerization. *J. biophys. biochem. Cytol.* **5,** 11–15.

HALL, D. A., LLOYD, P. F., SAXL, H., and HAPPEY, F. (1958). Mammalian cellulose. *Nature, Lond.* **181,** 470–2.

HALL, D. A. and SAXL, H. (1961). Studies of human and tunicate cellulose and of their relationship to reticulin. *Proc. R. Soc.* B. **155,** 202–17.

HALSTEAD, L. B. (Tarlo, L. B. H.) (1969). *The Pattern of Vertebrate Evolution*. Oliver and Boyd, Edinburgh.

HAMMOND, R. A. (1967). The fine structure of the trunk and praesoma wall of *Acanthocephalus ranae* (Schrank 1788) Lühe, 1911. *Parasitology*, **57,** 475–86.

HAMMOND, R. A. (1968a). Some observations on the role of the body wall of *Acanthocephalus ranae* in liquid uptake. *J. exp. Biol.* **48,** 217–25.

HAMMOND, R. A. (1968b). Observations on the body surface of some acanthocephalans. *Nature, Lond*, **218,** 872–3.

HANCOCK, N. M. and BOOTHROYD, B. (1966). Electron microscope observations of osteogenesis. *In* "Fourth European Symposium on Calcified Tissues" (eds. P. J. Gaillard, A. van den Hooff, and R. Steendijk). International Congress Series 120, Excerpta Medica Foundation, Amsterdam.

HARDING, J. J. (1963). The amino acid composition of human collagens from adult dura mater and post-menopausal uterus. *Biochem. J.* **86,** 574–6.

HARE, P. E. (1963). Amino acids in the proteins from aragonite and calcite in the shells of *Mytilus californianus*. *Science, N.Y.* **139,** 216–17.

HARE, P. E. and ABELSON, P. H. (1965). Amino acid composition of some calcified proteins. Yb. Carnegie Inst. Wash. **64,** 223–31.

HARKNESS, R. D. (1961). Biological functions of collagen. *Biol. Rev.* **36,** 399–463.

HARKNESS, R. D. (1964). The physiology of the connective tissues of the reproductive tract. *Int. Rev. Connect. Tissue Res.* **2,** 155–211.

HARKNESS, R. D. (1968). Mechanical properties of collagenous tissues. *In* "Treatise on Collagen" (ed. B. S. Gould) Vol. 2, pt. A. Academic Press, London and New York.

HARLEY, J. C. (1961). The shell of acridid eggs. *Q. Jl microsc. Sci.* **102,** 249–55.

HARMS, J. W. (1929). Die Realisation von Genen und die consecutive Adoption I. Phasen in der Differenzierung der Anlagenkomplexe und die Frage Landtierwerdung. *Z. wiss. Zool.* **133,** 212–397.

Harper, E., Seifter, S., and Scharrer, B. (1967). Electron microscopic and biochemical characterization of collagen in blattarian insects. *J. Cell. Biol.* **33,** 385–93.

Harris, J. E. and Crofton, H. D. (1957). Structure and function in the nematodes: internal pressure and cuticular structure in *Ascaris. J. exp. Biol.* **34,** 116–30.

Hay, E. D. (1964). Secretion of a connective tissue protein by developing epidermis. *In* "The Epidermis" (eds. W. Montagna and W. C. Lobitz. Jr.) pp. 97–114. Academic Press, New York and London.

Hecht, S. (1918). The physiology of *Ascidia atra* Lesueur I. General Physiology. *J. exp. Zool.* **25,** 229–60.

Hedley, R. H. (1956). Studies of serpulid tube formation I. The secretion of calcareous and organic components of the tube of *Pomatoceros triqueter. Q. Jl microsc. Sci.* **97,** 411–19.

Hedley, R. H. (1960). The iron containing shell of *Gromia oviformis. Q. Jl microsc. Sci.* **101,** 279–94.

Hedley, R. H. (1962). The significance of an inner "chitinous lining" in saccaminid organisation with special reference to a new species of *Saccamina* (Foraminifera) from New Zealand. *N.Z. Jl Sci.* **5,** 375–89.

Hedley, R. H. (1963). Cement and iron in the arenaceous foraminifera. *Micropaleontology,* **9,** 433–41.

Hedley, R. H. and Bertaud, W. D. (1962). Electron microscope observations on *Gromia oviformis* (Scarcodina). *J. Protozool.* **9,** 79–87.

Hedley, R. H., Parry, D. M., and Wakefield, J. St. J. (1967). Fine structure of *Shepheardella taeniformis* (Foraminifera: Protozoa). *J. R. Microsc. Soc.* **87,** 445–56.

Hedley, R. H. and Wakefield, J. St. J. (1967). A collagen-like sheath in the arenaceous foraminifer *Haliphysema* (Protozoa). *J. R. Microsc. Soc.* **87,** 475–81.

Hedley, R. H. and Wakefield, J. St. J. (1969). Fine structure of *Gromia oviformis. Bull. Br. Mus. nat. Hist.* (*Zool.*) **18,** (2), 69–89.

Henrikson, R. C. and Matoltsy, A. G. (1968a). The fine structure of teleost epidermis I. Introduction and filament-containing cells. *J. Ultrastruct. Res.* **21,** 194–212.

Henrikson, R. C. and Matoltsy, A. G. (1968b). The fine structure of teleost epidermis II. Mucus cells. *J. Ultrastruct. Res.* **21,** 213–21.

Henrikson, R. A. and Matoltsy, A. G. (1968c). The fine structure of teleost epidermis III. Club cells and other cell-types. *J. Ultrastruct. Res.* **21,** 222–32.

Henry, S. M. (1962). The significance of micro-organisms in the nutrition of insects. *Trans. N.Y. Acad. Sci.* **24,** 676–83.

Hepler, K. P., Huff, C. G., and Sprinz, H. (1966). The fine structure of the exoerythrocytic stages of *Plasmodium fallax. J. Cell. Biol.* **30,** 333–58.

Herber, E. C. (1950). Studies on the biochemistry of cyst envelopes of the fluke, *Notocotylus urbanensis. Proc. Penn. Acad. Sci.* **24,** 140–3.

Herlant-Meewis, H. (1948). Contribution à l'étude histologique des spongiares. *Annls. Soc. r. zool. Belg.* **79,** 5–36.

Hertwig, O. (1873). Untersuchungen über den Bau und die Entwickelung des Cellulose-Mantels der Tunicaten. *Jena Z. Naturw.* **7,** 46–73.

Hertwig, O. and Hertwig, C. W. T. R. (1878). *Der organismus der Medusen und seine stellung zur Keimblättertheorie.* Jena, Denkschr 2 (Heft 1), Gustav Fischer.

Hess, A. (1961). The fine structure of cells in *Hydra. In* "Biology of Hydra" (eds. H. M. Lenhoff and W. F. Loomis) pp. 1–8. University of Miami Press.

Hildemann, W. H. (1959). A cichlid fish, *Symphysodon discus,* with unique nurture habits. *Am. Nat.* **93,** 27–34.

Hill, C. J. and Hill, J. P. (1933). The development of the Monotremata. I The histology of the oviduct during gestation. II The structure of the egg shell. *Trans. zool. Soc. Lond.* **21,** 413–76.

Hillman, R. E. (1961). Formation of the periostracum in *Mercenaria mercenaria. Science, N.Y.* **134,** 1754–5.

Hinz, E. (1963). Electronenmikroskopische Untersuchungen an *Parascaris equorum. Protoplasma,* **56,** 202–41.

Hirsch, P. F., Gauthier, G. F., and Manson, P. L. (1963). Thyroid hypocalcemic principle and recurrent laryngeal nerve-injury as factors affecting the response to parathyroidectomy in rats. *Endocrinology*, **73,** 244–52.

Hirschman, H. (1959). Histological studies on the anterior region of *Heterodera glycines* and *Hoplolaimus tylenchiformis* (Nematoda, Tylenchida). *Proc. Helminth. Soc. Wash.* **26,** 73–90.

Hirschman, H. (1960). External characters and body wall of nematodes. *In* "Nematology, Fundamentals and Recent Advances with Emphasis on Plant Parasitic and Soil Forms" (eds. J. N. Sasser and W. R. Jenkins). University of N. Carolina Press, Chapel Hill.

Ho, T-Y. (1965). The amino acid composition of bone and tooth proteins in late pleistocene mammals. *Proc. U.S. Acad. Sci.* **54,** 26–31.

Ho, T-Y. (1966). The isolation and amino acid composition of the collagen in pleistocene mammals. *Comp. Biochem. Physiol.* **18,** 353–8.

Ho, T-Y. (1967). Relationship between amino acid contents of mammalian bone collagen and body temperature as a basis for estimating the body temperature of prehistoric mammals. *Comp. Biochem. Physiol.* **22,** 113–19.

Hodge, A. J. (1967). Structure at the electron microscopic level. *In* "Treatise on Collagen" (ed. G. N. Ramachandran) Vol. 1, pp. 185–205. Academic Press, London and New York.

Hodge, A. J., Highberger, J. H., Deffner, G. G. J., and Schmitt, F. O. (1960). The effect of proteases on the tropocollagen macromolecule and on its aggregation-properties. *Proc. natn. Acad. Sci. U.S.A.* **46,** 197–206.

Hodge, A. J., Petruska, J. A., and Bailey, A. J. (1965). The sub-unit structure of the tropocollagen macromolecule and its relation to various ordered aggregation-states. *In* "Structure and Function of Connective and Skeletal Tissue" (ed. M. S. Fitton-Jackson *et al.*) pp. 31–41. Butterworth, London.

Hodgman, C. D. (ed.) (1965). *Handbook of Chemistry and Physics*. Edition 46. Chemical Rubber Co., Cleveland, Ohio.

Hosker, A. (1936). V. Studies on the Epidermal structures of birds. *Phil. Trans. R. Soc.* **B226,** 143–88.

Hou, H. C. (1928). Studies on the glandula uropygialis of birds. *Chin. J. Physiol.* **2,** 345–80.

Hou, H. C. (1930). Further observations on the relation of the preen gland of birds to rickets. *Chin. J. Physiol.* **4,** 79–92.

Hovasse, R. (1932). Note préliminaire sur les Ebriacées. *Bull. Soc. zool. Fr.* **57,** 118–31.

Hoyes, A. D. (1967). Acid mucopolysaccharide in human fetal epidermis. *J. invest. Derm.* **48,** 598–601.

Hubendick, B. (1948). Über den Bau und das Wachstum des Konzentrischen Operculartypus bei Gastropoden. *Ark. Zool.* **40A,** (10), 1–28.

Huggins, M. L. (1967). The structure of alpha keratin. *Proc. natn. Acad. Sci. U.S.A.* **43,** 204–9.

Hulme, A. C. and Arthington, W. (1952). New amino acids in young apple fruits. *Nature, Lond.* **170,** 659–60.

Hunt, S. (1966). Carbohydrate and amino acid composition of the egg capsule of the whelk *Buccinum undatum*. *Nature, Lond.* **210,** 436.

Hyman, L. H. (1940). *The Invertebrates. Vol. I. Protozoa through Ctenophora*. McGraw-Hill, New York.

Hyman, L. H. (1951a). *The Invertebrates. Vol. II. Platyhelminthes and Rhynchocoela*. McGraw-Hill, New York.

Hyman, L. M. (1951b). *The Invertebrates. Vol. III. Acanthocephala, Aschelminthes and Entoprocta*. McGraw-Hill, New York.

Hyman, L. (1955). *The Invertebrates. IV. Echinodermata*. McGraw-Hill, New York.

Hyman, L. (1958). The occurrence of chitin in the lophophorate phyla. *Biol. Bull. mar. biol. Lab., Woods Hole* **114,** 106–12.

Hyman, L. (1959). *The Invertebrates. Volume V. Smaller Coelomate Groups*. McGraw-Hill, New York.

Hyman, L. (1966). Further notes on the occurrence of chitin in invertebrates. *Biol. Bull. mar. biol. Lab., Woods Hole* **130,** 94–5.

Hyman, L. (1967). *The Invertebrata. Vol. VI. Mollusca I.* McGraw-Hill, New York.

Ibrahim, I. K. A. and Hollis, J. P. (1967). Cuticle ultrastructure of *Meloidogyne Napla*. *Proc. Helminth. Soc. Wash.* **34,** 137–9.

INATOMI, S., SAKUMOTO, D., ITANO, K., and TANAKA, H. (1963). Studies on the submicroscopic structure of body surface of larval nematodes. *Jap. J. Parasit.* **12,** 16–39.

INGLIS, W. G. (1964a). The structure of the nematode cuticle. *Proc. zool. Soc.* **143,** 465–502.

INGLIS, W. G. (1964b). The comparative anatomy of the ascaridoid cuticle (Nematoda). *Bull. Soc. zool. Fr.* **89,** 317–38.

INGLIS, W. G. (1965). Patterns of evolution in parasitic nematodes. *In* Third Symposium of the British Society for Parasitology, pp. 79–124. Blackwell Scientific Publications, Oxford.

ISAACS, W. A., LITTLE, K., CURREY, J. D., and TARLO, L. B. H. (1963). Collagen and a cellulose-like substance in fossil dentine and bone. *Nature, Lond.* **197,** 192.

ISENBERG, H. D., DOUGLAS, S. D., LAVINE, L. S., SPICER, S. S., and WEISSFELLNER, H. (1966). A protozoan model of hard tissue formation. *Ann. N.Y. Acad. Sci.* **136,** 155–90.

ISENBERG, H. D., LAVINE, L. S., MANDD, C., and WEISSFELLNER, H. (1965). Qualitative chemical composition of the calcifying organic matrix obtained from cell-free coccoliths. *Nature, Lond.* **206,** 1153–4.

IVANOV, A. V. (1963). *Pogonophora.* Translated from the Russian and edited by D. B. Carlisle. Academic Press, London and New York.

IZARD, J. and BROUSSY, J. (1964). Acid mucopolysaccharides in the cuticle of the gizzard of earthworms. *Nature, Lond.* **201,** 1338.

JACKSON, D. S. and BAILEY, J. P. (1968). Collagen-glycosaminoglycan interactions. *In* "Treatise on Collagen" (ed. B. S. Gould) Vol. 2a, pp. 189–211. Academic Press, London and New York.

JACOBS, L. and JONES, M. P. (1939). Studies on oxyuriasis XXI. The chemistry of the membranes of the pinworm egg. *Proc. Helminth. Soc. Wash.* **6,** 57–60.

JAKUS, M. A. (1945). The structure and properties of the trichocysts of *Paramecium*. *J. exp. Zool.* **100,** 457–86.

JAKUS, M. A. (1956). Studies on the cornea II. The fine structure of Descemets' membrane. *J. biophys. biochem. Cytol.* **2,** (Suppl.) 243.

JAMUR, M. P. (1966). Electron microscope studies on the body wall of the nematode *Nippostronglylus brasiliensis*. *J. Parasitol.* **52,** 209–32.

JARRETT, A. and SPEARMAN, R. I. C. (1964). *Histochemistry of the skin: Psoriasis.* A monograph on normal and parakeratotic epidermal keratinization. British Universities Press, London.

JARRETT, A., SPEARMAN, R. I., and HARDY, J. A. (1959). Histochemistry of keratinization. *Br. J. Derm.* **71,** 277–95.

JAROSCH, R. (1959). Zur Gleitbewegung der niederne Organismen. *Protoplasma* **50,** 277–89.

JAYLE, M. F. (1939). Étude comparative de l'action catalytiques des peroxydases vegetales et de l'hemoglobine. *Bull. Soc. Chim. biol.* **21,** 14–47.

JENSEN, M. and WEIS-FOGH, T. (1962). Biology and physics of locust flight V. Strength and elasticity of locust cuticle. *Phil. Trans. R. Soc.* B. **245,** 137–69.

JEUNIAUX, C. (1963). *Chitine et Chitinolyse.* Masson et Cie, Paris.

JEZYK, P. F. and FAIRBAIRN, D. (1967). Ascarosides and ascaroside esters in *Ascaris lumbricoides* (Nematoda). *Comp. Biochem. Physiol.* **23,** 691–705.

JOHNSON, D. J. and SIKORSKI, J. (1962). Molecular and fine structure of alpha-keratin. *Nature, Lond.* **194,** 31–4.

JOHRI, L. N. (1957). A morphological and histochemical study of egg-formation in a cyclophyllidean cestode. *Parasitology,* **47,** 21–8.

JOHRI, L. N. and SMYTH, J. D. (1956). A histochemical approach to the study of helminth morphology. *Parasitology,* **46,** 107–16.

JONES, B. M. (1954). On the role of the integument in acarine development and its bearing on pupa-formation. *Q. Jl microsc. Sci.* **95,** 169–81.

JONES, E. I., MCCANCE, R. A., and SHACKLETON, L. R. B. (1935). The role of iron in the structure of the radula teeth of certain marine molluscs. *J. exp. Biol.* **12,** 59–64.

JONES, W. C. (1954a). The orientation of the optical axis of spicules of *Leucosolenia complicata*. *Q. Jl microsc. Sci.* **95,** 33–48.

JONES, W. C. (1954b). Spicule form in *Leucosolenia complicata*. *Q. Jl microsc. Sci.* **95,** 191–203.

JONES, W. C. (1955a). Crystalline properties of spicules of *Leucosolenia complicata*. *Q. Jl microsc. Sci.* **96,** 129–49.

JONES, W. C. (1955b). The sheath of spicules of *Leucosolenia complicata*. *Q. Jl microsc. Sci.* **96,** 411–21.

JONES, W. C. (1956). Colloidal properties of the mesogloea in species of *Leucosolenia*. *Q. Jl microsc. Sci.* **97,** 269–85.

JONES, W. C. (1961). Properties of the wall of *Leucosolenia variabilis* I. The skeletal layer. *Q. Jl microsc. Sci.* **102,** 531–50.

JONES, W. C. (1967). Sheath and axial filament of calcareous sponge spicules. *Nature, Lond.* **214,** 365–8.

JOPE, M. (1965). Composition of brachiopod shell. *In* "Treatise on Invertebrate Paleontology" (ed. R. C. Moore) Part H. Brachiopoda, Vol. 1, pp. H.156–H.163. The Geological Society of America Inc. and University of Kansas Press.

JOPE, M. (1967a). The protein of brachiopod shell I. Amino acid composition and implied protein taxonomy. *Comp. Biochem. Physiol.* **20,** 593–600.

JOPE, M. (1967b). The protein of brachiopod shell II. Shell protein from fossil articulates: Amino acid composition. *Comp. Biochem. Physiol.* **20,** 601–5.

JØRGENSEN, C. B. (1944). On the spicule formation of *Spongilla lacustris*. *Biol. Meddr. Kjobenhavn,* **19,** (7), 1–44.

JØRGENSEN, C. B. and LARSEN, L. O. (1960). Hormonal control of moulting in amphibians. *Nature, Lond.* **185,** 244–5.

JOSSE, J. and HARRINGTON, W. F. (1964). Role of pyrrolidine residues in the structure and stabilization of collagen. *J. molec. Biol.* **9,** 269.

KADO, Y. (1960). Studies on shell-formation in Mollusca. *J. Sci. Hiroshima Univ.* (Ser. B1) **19,** 163–210.

KAFATOS, F. C., TARTAKOFF, A. M., and LAW, J. H. (1967). Cocoonase I. Preliminary characterization of a proteolytic enzyme from silk moths. *J. biol. Chem.* **242,** 1477–87.

KAGEI, N. (1960). Morphological studies on thread worms, Filaroidea. Report 1. Morphological structure of *Setaria cervi*. *Acta. Med. Univ. Kagoshima.* **2,** 142–9.

KAN, S. P. and DAVEY, K. G. (1968a). Moulting in a parasitic nematode III. The histochemistry of cuticle deposition and protein synthesis. *Can. J. Zool.* **46,** 723–7.

KAN, S. P. and DAVEY, K. G. (1968b). Moulting of a parasitic nematode, *Phocanema decipiens* II. Histochemical study of the larval and adult cuticle. *Can. J. Zool.* **46,** 235–41.

KAPLAN, D. and MEYER, K. (1960). Mucopolysaccharides of aorta at various ages. *Proc. Soc. exp. Biol. Med.* **105,** 78–81.

KAPUR, S. P. and GIBSON, M. A. (1967). A histological study of the development of the mantle edge and shell in the fresh water gastropod, *Helisoma duryi eudiscus* (Pilsbry). *Can. J. Zool.* **45,** 1169–82.

KAPUR, S. P. and GIBSON, M. A. (1968a). Histochemical studies of dopa oxidase and peroxidase in the mantle edge of the freshwater gastropod, *Helisoma duryi eudiscus* (Pilsbry). *Can. J. Zool.* **46,** 165–7.

KAPUR, S. P. and GIBSON, M. A. (1968b). A histochemical study of the development of the mantle edge and shell in the freshwater gastropod, *Helisoma duryi eudiscus* (Pilsbry). *Can. J. Zool.* **46,** 481–91.

KARLSON, P. and SEKERIS, C. E. (1962). N-acetyl-dopamine as sclerotizing agent of the insect cuticle. *Nature, Lond.* **195,** 183.

KAWAI, Y., SENO, M., and ANNO, K. (1966). Chondroitin polysulphate of squid cartilage. *J. Biochem.* **60,** 317–21.

KAYE, G. W. C. and LABY, T. H. (1948). *Physical and Chemical Constants*. Longmans, London.

KELLY, D. E. (1966). Fine structure of desmosomes, hemidesmosomes and an adepidermal globular layer in developing newt epidermis. *J. Cell. Biol.* **28,** 51–72.

KELLY, R. E. and RICE, R. V. (1967). Abductin: a rubber-like protein from the internal triangular hinge lagement of *Pecten*. *Science, N.Y.* **155,** 208–10.

KENCHINGTON, W. (1969a). Silk secretion in sawflies. *J. Morph.* **127,** 355–62.

KENCHINGTON, W. (1969b). The hatching thread of preying mantids: an unusual chitinous structure. *J. Morph.* **129,** 307–16.

KENNAUGH, J. (1959). An examination of the cuticles of two scorpions *Pandinus imperator* and *Scorpiops hardwickii*. *Q. Jl microsc. Sci.* **100,** 41–50.

Kennedy, C. H. (1927). The exoskeleton as a factor limiting and directing the evolution o insects. *J. Morph.* **44,** 267–312.

Kennedy, W. J., Taylor, J. D., and Hall, A. (1969). Environmental and biological controls on bivalve shell mineralogy. *Biol. Rev.* **44,** 499–530.

Kent, P. W. (1964). Chitin and mucosubstances. *In* "Comparative Biochemistry" (eds. M. Florkin and H. S. Mason) Vol. VII, pp. 93–136. Academic Press, New York and London.

Kessel, E. (1941). Ban und Bildung des Prosobranchien-Deckels. *Z. Morph. Ökol. Tiere,* **38,** 197–250.

Khayatt, R. M. and Chamberlain, N. H. (1948). *J. Text. Inst.* **39,** T.185.

King, R. C. and Koch, E. A. (1963). Studies on the ovarian follicke cells of *Drosophila. Q. J. microsc. Sci.* **104,** 297–320.

Kirk, J. T. O. (1964). The effect of trypsin on the pellicle of *Euglena gracilis. J. R. Microsc. Soc.* **82,** 205–10.

Kitching, J. A. (1952). Observations on the mechanism of feeding in the suctorian *Podophyra. J. exp. Biol.* **29,** 255–66.

Kitching, J. A. (1967). Contractile vacuoles, ionic regulation and excretion. *In* "Research in Protozoology" (ed. T. T. Cheng) Vol. 1, pp. 307–36, Pergamon Press, Oxford.

Kitzan, S. M. and Sweeny, P. R. (1968). A light and electron microscope study of the structure of *Protopterus annectens* epidermis. *Can. J. Zool.* **46,** 767–72.

Kligman, A. M. (1964). The biology of the stratum corneum. *In* "The Epidermis" (eds. W. Montagna and W. C. Lobitz) pp. 387–430. Academic Press, New York and London.

Knight, D. P. (1968). Cellular basis for quinone-tanning of the periscarc in the thecate hydroid *Campanularia* (≡ *Obelia*) *flexuosa*. Hincks. *Nature, Lond,* **218,** 584–6.

Knight-Jones, E. W. (1953). Feeding in *Saccoglossus. Proc. Zool. Soc.* **123,** 637–54.

Knowles, H. R., Hart, E. B. and Halpin, J. G. (1935). The relation of the preen gland to rickets in the domestic fowl. *Poult. Sci.* **14,** 33–6.

Koch, J. C. (1917). The laws of bone architecture. *Am. J. Anat.* **21,** 177–298.

Koczy, F. F. and Titze, H. (1958). Radium content of carbonate shells. *J. mar. Res.* (*Sears Fdn.*) **17,** 302–11.

Koehler, J. K. (1965). A fine structure-study of the rotifer integument. *J. Ultrastruct. Res.* **12,** 113–34.

Koehler, J. K. (1966). Some comparative fine structure relationships of the rotifer integument. *J. exp. Zool.* **162,** 231–5.

Koehler, J. K. and Hayes, T. L. (1969). The rotifer jaw: A scanning and transmission electron microscope study I. The trophi of *Philodina acuticornis odiosa. J. Ultrastruct. Res.* **21,** 402–18.

Kofoid, C. A., McNeil, E., and Kopac, M. J. (1931). Chemical nature of the cyst-wall in human intestinal protozoa. *Proc. Soc. exp. Biol. Med., N.Y.,* **29,** 100–2.

Koidsumi, K. (1957). Antifungal action of cuticular lipids. *J. Insect Physiol.* **1,** 40–51.

Kölliker, A. von (1864). Icones Histiologicae oder Atlas der vergleichenden Gewebelchre. Wilhelm Engleman, Leipzig.

Kozlowki, R. (1947). Les affinités des graptolites. *Biol. Rev.* **22,** 93–108.

Kraepelin, K. (1887). Die deutschen süsswasserbryozoen I. Anatomischer-systematischer Teil. *Abh. Geb. Naturw. Hamburg,* **10,** 168 pp.

Krakower, C., Hoffman, W. A., and Axtmayer, J. H. (1944). Defective granular eggshell formation by *Schistosoma mansoni* in experimentally infected guinea pigs, on a vitamin C deficient diet. *J. Infect. Dis.* **74,** 178–83.

Krall, J. F. (1968). The cuticle and epidermal cells of *Dero obtusa* (Family Naididae). *J. Ultrastruct. Res.* **25,** 84–93.

Krishnan, G. (1953). On the cuticle of the scopion *Palamneus swammerdami. Q. Jl microsc. Sci.* **94,** 11–21.

Krishnan, G. (1954). The epicuticle of an arachnid, *Palamneus swammerdami. Q. Jl microsc. Sci.* **95,** 371–81.

Krishnan, G. (1958). Some aspects of cuticular organisation of the branchiopod *Streptocephalus dichotomus. Q. Jl microsc. Sci.* **99,** 359–71.

Krishnan, G., Ramachandran, G. N., and Santaman, M. S. (1955). Occurrence of chitin in the epicuticle of an arachnid *Palamneus swammerdami*. *Nature, Lond.* **176,** 557–8.

Kroon, D. B., Veerkamp, T. A., and Loeven, W. A. (1952). X-ray analysis of the process of extension of the wing of the butterfly. *K. Nederl. Akad. Wetenschappen Proc. Ser.* C. **55,** 209–14.

Krukenberg, C. F. W. (1885a). "Vergleichend-Physiologische Vorträge 4. Vergleichenden Physiologie der Thierischen Gerüstsubstanzen". Carl Winter Universitätbuchhandlung, Heidelberg.

Krukenberg, C. F. W. (1885b). Über das Vorkommen des Chitins. *Zool. Anz.* **8,** 412–5.

Kruyt, H. R. (1952). "Colloid Science". Vol. I. Elsevier, Amsterdam.

Kudo, R. R. (1951). Observations on *Pelomyxa Illinoisensis*. *J. Morph.* **88,** 145–73.

Kükenthal, W. (1916). System und stammesgeschichte der Scleraxonier und der Ursprung der Holaxonier. *Zool. Anz.* **47,** 170–6.

Kümmel, G. (1957). Die Gleitbewegung der Gregarinen. *Arch. Protistenk.* **102,** 501–22.

Kunike, G. (1925). Nachweis und Verbreitung organischer skelet-substanzen vei Tieren. *Z. vergl. Physiol.* **2,** 233–53.

Kusakabe, D. and Kitamori, R. (1948). Anatomical structure of byssus of *Anadara* and *Barbatia*. *Contr. cent. Fish. Stn Japan,* **80,** 253–63.

Labbe, A. (1929). Les organes palléaux et le tissue conjonctif du manteau de *Rostanga*. *Arch. anat. Microsc.* **25,** 87–103.

Labbe. A. (1933). Sur la présence de spicules siliceux dans les téguments des oncidiadés. *C.r. hebd. Séanc. Acad. Sci., Paris,* **197,** 533–5.

Lackey, J. B. (1940). Some new flagellates from the Woods Hole area. *Am. Midl. Nat.* **23,** 463–71.

Lacy, D. and Miles, H. B. (1959). Observations by electron microscopy on the structure of an acephaline gregarine *Apolocystis elongata* (Phillips and MacKinnon). *Nature, Lond.* **183,** 1456–7.

Lafon, M. (1943). Sur la structure et la composition chimique du tégument de la limule (*Xiphosura polyphemus* L.) *Bull. Inst. Oceanogr.* No. 850.

Lagler, K. F., Bardach, J. E., and Miller, R. R. (1962). "Ichthyology". J. Wiley, New York.

Lamport, D. T. A. and Northcote, D. H. (1960). Hydroxyproline in primary cell walls of higher plants. *Nature, Lond.* **188,** 665–6.

Landucci, J. M., Pouradier, J., and Durante, M. (1958). Sur la position des aldehydes dans la molecule de collagen. *In* "Recent Advances in Gelatin and Glue Research" (ed. G. Stainsby) pp. 62–7. Pergamon Press, London.

Lane, C. E. and Dodge, E. (1958). The toxicity of *Physalia* nematocysts. *Biol. Bull. mar. biol. Lab., Woods Hole,* **115,** 219–26.

Lane, N. J. (1963). Microvilli on the external surfaces of gastropod tentacles and body walls. *Q. Jl microsc. Sci.* **104,** 495–504.

Lang, K. (1948). Contribution to the ecology of *Priapulus caudatus* Lam. *Ark. Zool.* **41a** (9) 1–8.

Langner, E. (1937). Untersuchungen an Tegument und Epidermis bei Diplopoden. *Zool. Jb.* (*abh. Anat.*) **63,** 483–541.

Lash, J. W. and Whitehouse, M. W. (1960). An unusual polysaccharide in chondroid tissue of the snail, *Busycon:* polyglucose sulphate. *Biochem. J.* **74,** 351–5.

Laubenfels, M. W. de (1932). Physiology and morphology of Porifera exemplified by *Iotrochota birotulata* Higgin. *Publ. Carnegie Inst., Wash.* **435,** 37–66.

Lawerence, D. J. and Bern, H. A. (1960). Mucus metaplasia and mucous gland formation in keratinized adult epithelium *in situ* treated with vitamin A. *Exp. Cell. Res.* **21,** 443–6.

Leach, A. A. (1957). The amino acid composition of amphibian, reptilian, and avian gelatins. *Biochem. J.* **67,** 83–7.

Leadbeater, B. and Dodge, J. D. (1966). The fine structure of *Wolosynskia micra* sp.nova. new marine Dinoflagellate. *Br. Phycol. Bull.* **3,** 1–17.

Leblond, C. P., Puchtler, H., and Clermont, Y. (1960). Structures corresponding to terminal bars and terminal web in many types of cells. *Nature, Lond.* **186,** 784–8.

Lee, D. L. (1962). Studies on the function of the pseudosuckers and holdfast organ of *Diplostomun phoxini* Faust (Strigeida, Trematoda). *Parasitology,* **52,** 103–12.

LEE, D. L. (1965). The cuticle of adult *Nippostrongylus brasiliensis*. *Parasitilogy*, **55,** 173–81.

LEE, D. L. (1966a). An electron microscope study of the body wall of the third-stage larva of *Nippostrongylus brasiliensis*. *Parasitology*, **56,** 127–34.

LEE, D. L. (1966b). Structure and composition of Helminth cuticle. *In* "Advances in Parasitology," (ed. Ben Dawes) Vol. 4, pp. 187–254. Academic Press, New York and London.

LEE, H. H. K., JONES, A. W., and WYANT, K. D. (1959). Development of the taenid embryophore. *Trans. Am. microsc. Soc.* **78,** 335–57.

LEEDALE, G. F. (1964). Pellicle structure in *Euglena*. *Br. phycol. Bull.* **2,** 291–306.

LEEDALE, G. F. (1967). *Euglenoid Flagellates*. Prentice-Hall, London and New York.

LEES, A. D. (1947). Transpiration and the structure of the epicuticle in ticks. *J. exp. Biol.* **23,** 379–410.

LEESON, C. R. and THREADGOLD, L. T. (1961). The differentiation of the epidermis in *Rana pipiens*. *Acta Anat.* **44,** 159–73.

LEGHISSA, S. and MAZZI, S. (1959). Contributo ad una migliora conoscenza sulla struttura e composizione della mesodermide negli Antozoi. *Riv. Biol. Perugia*, **51,** 293–325.

LEGROS CLARK, W. E. (1936). The problem of the claw in primates. *Proc. zool. Soc.* **106,** 1–24.

LEGROS CLARK, W. E. (1945). *The Tissues of the Body*. 2nd edition. Clarendon Press, Oxford.

LEHMANN, F. E., MANNI, E., and BAIRATI, A. (1956). Der feinbau von Plasmalemma und Kotraktiler vacuole bei *Amoeba proteus* in Schnitt und Fragment präparaten. *Revue suisse Zool.* **63,** 246–55.

LENDENFELD, R. VON (1889). *A monograph of the horny sponges*. Published for the Royal Society by Trübner, London.

LENHOFF, H., KLINE, E. S. and HURLEY, R. (1957). An hydroxyproline-rich, intracellular, collagen-like protein of *Hydra* nematocysts. *Biochem. Biophys. Acta*, **26,** 204–5.

LENHOFF, H. M., SCHROEDER, R., and LEIGH, W. H. (1960). The collagen-like nature of metacercarial cysts of a new species of *Ascocotyle*. *J. Parasitol.* **46,** Supplement, 36.

LEONARDI, G. (1965). Ricerche sulla struttura del corion di uova di Ascidic. *Atti Accad. naz. Lincei Rc.* **39** (8), 118–22.

LERNER, H. (1954). Zur Kenntnis des Feinbaues der Flottoblastenschalen von *Plumatella repens* (L.) und *Cristatella mucedo* Cuvier. *Ber. oberhess. Ges. Nat.- u. Heilk.* (New Ser.) **27,** 111–22.

LEVI, C. (1956). Étude des Halisarca de Roscoff: embryologie et systématique des Démosponges. *Arch. Zool. exp. gén.* **93,** 1–184.

LEVI, C. (1963). Scleroblastes et spiculogenèse chez une éponge siliceuse. *C. r. hebd. Séanc. Acad. Sci., Paris*, **256,** 497–8.

LEWIN, R. A. (1958). The cell wall of *Platymonas*. *J. gen. Microbiol.* **19,** 87–90.

LEWIN, R. A., OWEN, M. J., and MELNICK, L. L. (1951). Cell wall structure in *Chlamydomonas*. *Expl Cell. Res.* **2,** 708–10.

LEWIS, M. S. and PIEZ, K. A. (1964). The characterisation of collagen from the skin of the dogfish shark, *Squalus acanthias*. *J. biol. Chem.* **239,** 3336–40.

LIM, C. F. (1965). Functional morphology of the byssus and associated glands in the bivalve genus *Anadara*. *J. amin. Morph. Physiol.* **12,** 113–31.

LINDER, H. J. (1960). Egg shell formation in *Chirocephalopsis bundyi* II. Histochemistry of egg-shell formation. *J. Morph.* **107,** 259–84.

LISON, L. (1953). *Histochimie et Cytochimie Animale*. (Deuxieme édition). Gauthier-Villars, Paris.

LISTER, J. J. (1900). *Astrosclera willeyana*, the type of a new family sponges. *Willey's zool. Res.* **4,** 459–82.

LITTLEFORD, R. A., KELLER, W. F., and PHILLIPS, N. E. (1947). Studies on the vital limits of water loss in the plethodont salamanders. *Ecology*, **28,** 440–7.

LOCKE, M. (1964). The structure and formation of the integument of insects. *In* "The Physiology of Insecta" (ed. M. Rockstein) Vol. III, pp. 379–470. Academic Press, New York and London.

LOEBLICH, A. R. and TAPPAN, H. (1964). *Treatise on Invertebrate Palaeontology*. C. Geological Society of America and University of Kansas Press.

LOEWI, G. and MEYER, K. (1958). The acid mucopolysaccharides of embryonic skin. *Biochim. biophys. Acta*, **27,** 453–6.

LOISEL, G. (1898). Contribution a l'histo-physiologie des eponges I. Les fibres des *Reniera*. *J. Anat. Physiol., Paris*, **34,** 1–43.

LONGLEY, J. B. (1950). A cytological study of the formation of keratin. Ph.D. Thesis, Cambridge University.

LÖSER, E. (1965). Die eibildung bei cestoden. A. *Parasitenk.* **25,** 556–80.

LOTMAR, W. and PICKEN, L. E. R. (1950). A new crystallographic modification of chitin and its distribution. *Experienta*, **6,** 58–9.

LOW, E. M. (1951). Halogenated amino acids of the bath sponge. *J. Mar. Res.* **10,** 239–245.

LOWENSTAM, H. A. (1954). Factors affecting the aragonite/calcite ratios in carbonate-secreting marine organisms. *J. Geol.* **62,** 284–324.

LOWENSTAM, H. A. (1967). Lepidocrocite, an apatite mineral in the teeth of chiton. *Science, N.Y.* **156,** 1373–5.

LOWEY, S., GOLDSTEIN, L., COHEN, C., and LUCK, S. M. (1967). Proteolytic degradation of myosin and the meromyosins by a water-insoluble polyanionic derivative of trypsin. *J. molec. Biol.* **23,** 287–304.

LUCAS, F. (1964). Spiders and their silks. *Discovery* **25** (1), 20–5.

LUCAS, F. (1966). Cystine content of silk fibroin (*Bombyx mori*). *Nature, Lond.* **210,** 952–3.

LUCAS, F. and RUDALL, K. M. (1968). Extracellular fibrous proteins: The silks. *In* "Comprehensive Biochemistry" (eds. M. Florkin and E. H. Stotz) Vol. 26B, pp. 475–558. Elsevier, Amsterdam.

LUCAS, F., SHAW, J. T. B., and SMITH, S. G. (1955). The chemical composition of some silk fibroins and its bearing on their physical properties. *J. Text. Inst.* T.440.

LUCAS, F., SHAW, J. T. B., and SMITH, S. G. (1957). Amino acid composition of the silk of *Chrysopa* egg-stalks. *Nature, Lond.* **179,** 905–7.

LUCAS, F., SHAW, J. T. B., and SMITH, S. G. (1958). The silk fibroins. *Adv. Protein Chem.* **13,** 107–242.

LUCAS, F., SHAW, J. T. B. and SMITH, S. G. (1960). Comparative studies of fibroins I. The amino acid composition of various fibroins and its significance in relation to their crystal structure and taxonomy. *J. molec. Biol.* **2,** 339–49.

LYNCH, D. L. and BOGITSH, B. J. (1962). The chemical nature of metacercarian cysts II. Biochemical investigations of the cyst of *Posthodiplostomum minimum*. *J. Parasitol.* **48,** 241–3.

LYONS, K. M. (1966). The chemical nature and evolutionary significance of monogenean attachment sclerites. *Parasitology*, **56,** 63–100.

MCBRIDE, O. W. and HARRINGTON, W. F. (1967). *Ascaris* cuticle collagen: on the disulphide cross linkages and the molecular properties of the subunits. *Biochemistry*, **6,** 1484–98.

MCCONNELL, D. (1963). Inorganic constituents in the shell of the living brachiopod, *Lingula*. *Bull. geol. Soc. Am.* **73,** 363–4.

MCGAVIN, S. (1962). The structure of elastoidin in relation to that of tendon collagen. *J. molec. Biol.* **5,** 275–83.

MCGAVIN, S. (1964). Optical rotation in protein fibres. *J. molec. Biol.* **9,** 601–4.

MACGREGOR, H. C. and THOMASSON, P. A. (1965). The fine structure of two archigregarines, *Selenidium fallax* and *Ditrypanocysitis cirratuli*. *J. Protozool.* **12,** 438–43.

MACKIE, G. (1960). Studies on *Physalia physalia*. *Discovery Rep.* **30,** 371–408.

MACKIE, G. O. and MACKIE, G. V. (1967). Mesogoeal ultrastructure and reversible opacity in a transparent siphonophora. *Vie Milieu* (Ser. A) **18** (Fasc. 1), 47–65.

MACLENNAN, R. F. (1937). Growth in the ciliate *Ichthyophthirius* I. Maturity and encystment. *J. exp. Zool.* **76,** 423–40.

MCLOUGHLIN, C. B. (1961a). The importance of mesenchymal factors in the differentiation of chick epidermis I. The differentation in culture of the isolated epidermis of the embryonic chick and its response to excess vitamin A. *J. Embryol. exp. Morph.* **9,** 370–84.

MCLOUGHLIN, C. B. (1961b). The importance of mesenchymal factors in the differentiation of chick epidermis II. Modification of epidermal differentiation by contact with different types of mesenchyme. *J. Embryol. exp. Morphl.* **9,** 385–408.

MCMANUS, M. A. and ROTH, L. E. (1967). Microtubular structure in myxomycete plasmodia. *J. Ultrastruct. Res.* **20,** 260–6.

MADERSON, P. F. A. (1965a). Histological changes in the epidermis of snakes during the sloughing cycle. *J. Zool. Lond.* **146,** 98–113.

MADERSON, P. F. A. (1965b). The structure and development of the squamate epidermis. *In* "Biology of the skin and hair growth" (eds. A. G. Lyne and B. F. Short). Angus and Robertson, Sydney.

MADERSON, P. F. A. (1966). Histological changes in the epidermis of the tokay (*Gekko gecko*) during the sloughing cycle. *J. Morph.* **119,** 39–50.

MANGIN, L. (1907). Observations sur la constitution de la membrane des péridiniens. *C. r. hebd. Séanc. Acad. Sci., Paris,* **144,** 1055–7.

MANN, T., MARTIN, A. W., and THIERSCH, J. B. (1966). Spermatophores and spermatophoric reactions in the giant octopus of the North Pacific, *Octopus dofleine martini. Nature, Lond.* **211,** 1279–82.

MANSARD, M. (1954). Inclusions cytoplasmiques ovulaires chex quelques nématodes parasites des vertébrés. *C. r. hebd. Séanc. Soc. Biol., Paris,* **148,** 2014–17.

MANTON, I. (1959). Electron microscopical observations on a very small flagellate: the problem of *Chromulina pusilla,* Butcher. *J. mar. biol. Assoc. U.K.* **38,** 319–33.

MANTON, I. (1964). Further observations on the fine structure of the haptoneme in *Prymnesium parvum. Arch. mikrobiol.* **49,** 315–30.

MANTON, I. (1966). Observations on scale production in *Prymnesium parvum. J. cell. Sci.* **1,** 375–80.

MANTON I. (1967). Further observations on scale formation in *Chrysochromulina chiton. J. cell. Sci.,* **2,** 411–8.

MANTON, I. and LEEDALE, G. F. (1961a). Observations on the fine structure of *Paraphysomonas vestita,* with special reference to the Golgi apparatus and the origin of scales. *Phycologica,* **1,** 37–57.

MANTON, I. and LEEDALE, G. T. (1961b). Further observations on the fine structure of *Chrysochromulina ericina* (Parke and Mantone). *J. mar. biol. Ass. U.K.* **41,** 145–55.

MANTON, I. and LEEDALE, G. F. (1969). Observations on the microanatomy of *Coccolithus* and *Cricosphaera carterae,* with special reference to the origin of coccoliths and scales. *J. mar. biol. Ass. U.K.* **49,** 1–16.

MANTON, I. and PARKE, M. (1960). Further observations on small green flagellates with special reference to possible relations of *Chromulina pusilla* Butcher. *J. mar. biol. Ass. U.K.* **39,** 275–98.

MANTON, I. and PARKE, M. (1965). Observations on the fine structure of two species of *Platymonas* with special reference to flagellar scales and the mode of origin of the theca. *J. mar. biol. Ass. U.K.* **45,** 743–54.

MANTON, I. and PETERFI, L. S. (1969). Observations on the fine structure of coccoliths, scales and the protoplast of a freshwater coccolithophorid, *Hymenomonas roseola* Stein, with supplementary observations on the protoplast of *Cricosphaera carterae. Proc. R. Soc.* B. **172,** 1–15.

MANTON, I., RAYNS, G. D., ETTL, H., and PARKE, M. (1965). Further observations on green flagellates with scaly flagella. *J. mar. biol. Ass. U.K.* **45,** 241–56.

MANTON, S. M. (1940). On two new species of the hydroid *Myriothela. Scient. Rep. Br. Graham Ld Exped. 1934–1937.* (4) 255–94.

MANTON, S. M. (1941). On the hydrorhiza and claspers of the hydroid *Myriothela cocksi* (Vigurs). *J. mar. biol. Ass. U.K.* **25,** 143–50.

MANTON, S. M. (1953). The evolution of arthropodan locomtory mechanisms IV. The structure, habits, and evolution of the Diplopoda. *J. Linn. Soc.* (*Zool.*) **42,** 299–368.

MANTON, S. M. (1958). Habits of life and evolution of body design in Arthropoda. *J. Linn. Soc.* (*Zool.*) **44,** 58–72.

MANTON, S. M. (1961). The evolution of arthropodan locomotory mechanisms VII. Functional requirements and body design in Colobognatha (Diplopoda) together with a comparative account of diplopod burrowing techniques, trunk musculature and segmentation. *J. Linn. Soc.* (*Zool.*) **44,** 383–461.

MARKS, M. H., BEAR, R. S., and BLAKE, C. H. (1949). X-ray diffraction evidence of collagen type protein fibres in the Echinodermata, Coelenterata, and Porifera. *J. exp. Zool.* **111,** 55–77.

MARON, N. (1958). A polarographic investigation of a polypeptide impurity from gelatin. *In* "Recent Advances in Gelatin and Glue Research" (ed. G. Stainby) pp. 221–4. Pergamon Press, Oxford.

MARRINAN, H. J. and MANN, J. (1956). Infrared spectra of the crystalline modifications of cellulose. *J. Polym. Sci.* **21,** 301–11.

MARTOJA, R. and BASSOT, J. M. (1965). Existence d'un tissue conjonctif de type cartilagineux chez certains insectes orthoptères. *C. r. hebd. Séanc. Acad. Sci. Paris,* **261,** 2954–7.

MASER, D. M. and RICE, R. V. (1962). Biophysical and biochemical properties of earthworm cuticle collagen. *Biochim. biophys. Acta,* **63,** 255–65.

MASER, D. M. and RICE, R. V. (1963a). The denaturation and renaturation of earthworm cuticle collagen. *Biochim. biophys. Acta,* **74,** 283–94.

MASER, D. M. and RICE, R. V. (1963b). Soluble earthworm cuticle collagen: a possible dimer of tropocollagen. *J. Cell. Biol.* **18,** 569–77.

MASON, H. S. (1955). Comparative biochemistry of the phenolasc complex. *Adv. Enzymol.* **16,** 105–48.

MASON, H. S. (1957). Mechanisms of oxygen metabolism. *Adv. Enzymol.* **19,** 79–233.

MASON, P. (1966). Threads of our existence. *Ormond Pap.* **1,** 9–16.

MATHEWS, M. B. (1965). A comparative study of acid mucopolysaccharide protein complexes. *In* "Structure and Function of Connective and Skeletal Tissue" (eds. S. Fitton-Jackson, S. M. Partridge, R. D. Harkness, and G. R. Tristram) pp. 181–91, Butterworth, London.

MATHEWS, M. B. (1967). Marcomolecular evolution of connective tissue. *Biol. Rev.* **42,** 499–549.

MATHEWS, M. B., DUH, J. and PERSON, P. (1962). Acid mucopolysaccharides of invertebrate cartilage. *Nature, Lond.* **193,** 378–9.

MATOLTSY, A. G. (1962a) Structural and Chemical properties of keratin-forming tissues. *In* "Comparative Biochemistry" (eds. M. Florkin and H. S. Mason) Vol. IV, pp. 343–69. Academic Press, New York and London.

MATOLTSY A. G. (1962b) Mechanism of keratinization. *In* "Fundamentals of Keratinization" (eds. E. O. Butcher and R. F. Sognnaes) pp. 1–25, Publ. No. 70. American Association for the Advancement of Science.

MATOLTSY, A. G. (1964). Prekeratin. Nature, Lond. **201,** 1130–1.

MATOLTSY, A. G. (1965). Soluble prekeratin. *In* "Biology of the Skin and Hair Growth" (eds. A. G. Lyne and B. F. Short) pp. 291–305. Angus and Robertson, Sydney.

MATOLTSY, A. G., DOWNES, A. M., and SWEENEY, T. M. (1968). Studies of the epidermal water barrier II. Investigation of the chemical nature of the water barrier. *J. invest. Derm.* **50,** 19–26.

MATOLTSY, A. G. and MATOLTSY, M. N. (1966). The membrane protein of horny cells *J. invest. Derm.* **46,** 127–9.

MATOLTSY, A. G. and PARAKKAL, P. F. (1965). Membrane-coated granules of keratinizing epithelia. *J. Cell. Biol.* **24,** 297–307.

MEEK, G. A. (1966). Intracellular collagen fibres. *J. Physiol. Lond.* **182,** 3 pp.

MEENAKSHI, V. E. (1963). "The organic matrix of the gastropod shell (apple snail *Pila globosa*)". Proceedings of the Sixteenth International Congress of Zoology, Washington, Vol. II, p. 79.

MERCER, E. H. (1951). Formation of silk fibre by the silk-worm. *Nature, Lond.* **168,** 792–3.

MERCER, E. H. (1952). Observations on the molecular structure of byssus fibres. *Aust. J. mar. Freshwat. Res.* **3,** 199–204.

MERCER, E. H. (1958). The election microscopy of keratinized tisses. *In* "The Biology of Hair Growth" (eds. W. Montagna and R. A. Ellis) pp. 91–111. Academic Press, New York and London.

MERCER, E. H. (1959). An election microscopic study of *Amoeba proteus. Proc. R. Soc.* B. **150,** 216.

MERCER, E. H. (1961). *Keratin and Keratinization.* Pergamon Press, Oxford.

MERCER, E. H., MUNGER, B. L., ROGERS, G. E., and ROTH, S. I. (1964). A suggested nomenclature for fine structural components of keratin and keratin-like products of cells. *Nature, London.* **201,** 367–8.

MEREDITH, R. (ed.) (1956). *Mechanical Properties of Textile Materials*. North Holland Publishing, Amsterdam.

MEYER, F. K. (1931). Röntgenographische Untersuchungen an Gastropodenschalen. *Jena Z. Naturw.* **65,** 487–512.

MEYER, F. K. and WEINECK, E. (1932). Die Verbreitung des Kalzium Karbonates im Tierreich unter besonderer Berücksichtigung der Wirbellosen. *Jena* A. *Naturwiss.*, **66,** 149–222.

MEYMARIAN, E. (1961). Host-parasite relationships in *Echinococcus* IV. Hatching and activation of *Echinococcus granulosus* ova *in vitro*. *Am. J. trop. Med. Hyg.* **10,** 719–26.

MIGNOT, J. P. (1965). Ultrastructure des Eugléniens I. Étudé de la cuticle chez differentes espèces. *Protistologica*, **1,** 5–15.

MIGNOT, J. P. (1966). Structure et ultra-structure de quelques Euglenomonadines. *Protistologica*, **2,** 51–117.

MILLAR, R. H. (1951). The stolonic vessels of the Didemnidae. *Q. Jl microsc. Sci.* **92,** 249–54.

MILLARD, A. and RUDALL, K. M. (1960). Light and electron microscope studies of fibres. *Jl R. Microsc. Soc.* **79,** 227–31.

MILLARD, A. and RUDALL, K. M. (1962). Ribbons of a helical fibrous protein. *In* "Fifth international Congress for Election Microscopy" (ed. S. S. Breese) Vol. 2, pp. 1–10. Academic Press, New York and London.

MILLER, E. J., MARTIN, G. R. and PIEZ, K. A. (1964). The utilization of lysine in the biosynthesis of elastin cross links. *Biochem. biophys. Res. Commun.* **17,** 248–53.

MILLER, P. L. (1960). Respiration in the desert locust II. The control of the spiracles. *J. exp. Biol.* **37,** 237–63.

MILLER, P. L. (1964). Respiration—Aerial gas transport. *In* "The Physiology of Insects" (ed. M. Rockstein) Vol. III, pp. 557–615. Academic Press, New York and London.

MILLER, Z. B., WALDMAN, J., and MCLEAN, F. C. (1952). The effect of dyes on the calcification of hypertrophic rachitic cartilage *in vitro*. *J. exp. Med.* **95,** 497–508.

MILLOTT, N. (1954). Sensitivity to light and the reactions to changes in light intensity of the echinoid *Diadema antillarum* Philippi. *Phil. Trans. R. Soc.* B. **238,** 187–220.

MINCHIN, E. A. (1900). *A Treatise on Zoology. Vol. II. The Porifera and Coelenterata* (ed. E. Ray Lankester). A. and C. Black. London.

MINCHIN, E. A. (1908). Materials for a monograph of the Ascons II. The formation of spicules in the genus *Leucosolenia* with some notes on the histology of the sponges. *Q. Jl microsc. Sci.* **52,** 301–55.

MONNÉ, L. (1955). On the histochemical properties of the egg envelopes and external cuticle of some parasitic nematodes. *Ark. Zool.* **9,** 93–113.

MONNÉ, L. (1959). On the external cuticles of various helminths and their role in the host-parasite relationship. A histochemical study. *Ark. Zool.* **12,** 343–58.

MONNÉ, L. (1962). On the formation of the egg shells of the Ascaroidea particularly *Toxascaris leonina* Linst. *Ark. Zool.* **15,** 277–84.

MONNÉ, L. and BORG, K. (1954). On the gram-staining of the egg-envelopes of parasitic worms. *Ark. Zool.* **6,** (2), 555–7.

MONNÉ, L. and HÖNIG, G. (1954a). On the properties of the egg envelopes of the parasitic nematodes *Trichuris* and *Capillaria*. *Ark. Zool.* **6,** (2), 559–67.

MONNÉ, L. and HÖNIG, G. (1954b). On the properties of the shells of the coccidian oocysts, *Ark. Zool.* **7,** (2), 251–6.

MONNÉ, L. and HÖNIG, G. (1954c). On the embryonic envelopes of *Polymorphus botulus* and *P. minutus* (Acanthocephala). *Ark. Zool.* **7,** (2) 559–75.

MOORE, R. C. and HARRINGTON, H. J. (1956). Scyphozoa. *In* "Treatise on Invertebrate Paleontology" Part F. (ed. by R. C. Moore) pp. 27–38. Geological Society of America and University of Kansas Press.

MOORE, T. (1957). *Vitamin A*. Elsevier, Amsterdam.

MORGAN, T. H. (1942). Cross- and self-fertilization in the ascidian *Styela*. *Biol. Bull. mar. biol. Lab., Woods Hole*, **82,** 161–71.

MORI, S. (1922). The changes in the para-ocular glands which follow the administration of diets low in fat soluble A: with notes of the effect of the same diets on the salivary glands and the mucosa of the larynx and trachae. *Bull. Johns Hopkins Hosp.* **33,** 357–9.

Moritz, K. and Storch, V. (1970). Über den aufbau des integumentes der priapuliden und der sipunculiden (*Priapulus caudalis* Lamarck, *Phascolion strombi* Montagu). *Z. Zellforsch.* **105,** 55–64.

Mornin, L. and Francis, D. (1967). The fine structure of *Nematodinium armatum*, a naked Dinoflagellate. *J. Microscopie*, **6,** 759–72.

Morris, J. E. and Afzelius, B. A. (1967). The structure of the shell and outer membranes in encysted *Artemia salina* embryos during cryptobiosis and development. *J. Ultrastruct. Res.* **20,** 244–59.

Morrison, P. R. (1946). Physiological observations on water loss and oxygen consumption in *Peripatus*. *Biol. Bull. mar. biol. Lab., Woods Hole,* **91,** 181–8.

Morseth, D. (1965). Ultra-structure of developing taeniid embryophores and associated structures. *Expl. Parasitol.* **16,** 207–16.

Morseth, D. (1966). Chemical composition of embyophoric blocks of *Taenia hydatigena, Taenia ovis* and *Taenia pisiformis* eggs. *Expl. Parasitol.* **18,** 347–54.

Morton, J. E. (1967). *Molluscs* (4th Edition). Hutchinson's University Library, London.

Moscona, A. A. (1950). Studies of the eggs of *Bacillus libanicus* (Orthoptera, Phasmidae) I. The egg envelopes. *Q. Jl Microsc. Sci.* **91,** 183–93.

Moscona, A. A. (1961). Environmental factors in experimental studies on histogenesis. *Colloques int. Cent. natn. Rech. Scient.* **101,** 154–68.

Moss, M. L. (1962). Studies of the acellular bone of teleost fish II. Response to fracture under normal and acalcemic conditions. *Acta Anat.* **48,** 46–60.

Moss, M. L. (1964). The phylogeny of mineralised tissues. *Int. Rev. Gen. exp. Zool.* **1,** 297–331.

Moss, M. L., Bé, A. W. H., and Ericson, D. B. (1963). Aspects of calcification in planktonic Foraminifera (Sarcodina). *Ann. N.Y. Acad. Sci.* **109,** 65–81.

Moss, L. M. and Freilich, M. (1963). Studies of the acellular bone of teleost fish IV. Inorganic content. *Acta Anat.* **55,** 1–8.

Moss, L. M. and Jones, S. J. (1964). Calcified ectodermal collagens of shark tooth enamel and teleost scale. *Science, N.Y.* **145,** 940–1.

Moss, M. L. and Meehan, M. M. (1967). Sutural connective tissues in the test of an echinoid, *Arbacia punctulata*. *Acta Anat.* **66,** 279–304.

Moulins, M. (1968). Étude ultra-structurale d'une formation de soutien épidermo-conjunctive inédite chez les insectes. *Z. Zellforsch.* **91,** 112–34.

Muir, H. (1958). The nature of the link between protein and carbohydrate of a chondroitin sulphate complex from hyaline cartilage. *Biochem. J.* **69,** 195–204.

Mullinger, A. M. (1964). The fine structure of ampullary electric receptors in *Amiurus*. *Proc. R. Soc.* B. **160,** 345–57.

Murphy, W. H. and Gottschalk, A. (1961). Studies on mucoproteins VII. The linkage of the prosthetic group to aspartic and glutamic and residues in bovine submaxillary gland mucoprotein. *Biochim. biophys. Acta.* **52,** 349–60.

Murray, P. D. F. (1936). *Bones: Study of the development and structure of the Vertebrate Skeleton*. Cambridge University Press, London.

Mutvei, H. (1964). On the shells of *Nautilus* and *Spirula* with notes on the shell secretion in non-cephalopod molluscs. *Ark. Zool.* **16,** 223–78.

Muzii, E. O. and Skinner, C. H. W. (1966). Calcite deposition during shell repair by the aragonitic gastropod *Murex fulvescens*. *Science, N.Y.* **151,** 201–3.

Myall, L. C. (1895). *The Natural History of Aquatic Insects*. Macmillan, London.

Nachmias, V. T. (1964). Fibrillar structures in the cytoplasm of *Chaos chaos*. *J. Cell. Biol.* **23,** 183–8.

Nadler, J. E. (1929). Notes on the loss and regeneration of the pellicle in *Blepharisma undulans*. *Biol. Bull. mar. biol. Lab., Woods Hole,* **56,** 327–30.

Napper, D. H. (1967). Modern theories of colloid stability. *Sci. Prog., Oxford* **55,** 91–109.

Needham, A. E. (1947). Excessive hydration in an animal with an open type of circulation. *Nature, Lond.* **160,** 755.

Needham, A. E. (1964). *The Growth Process in Animals*. Pitman, London.

Needham, A. E. (1965). *The Uniqueness of Biological Materials*. Pergamon Press, Oxford.

Needham, A. E. (1968). Distribution of protoporphyrin, ferrihaem and indoles in the body-wall of *Lumbricus terrestris* L. *Comp. Biochem. Physiol.* **26,** 429–42.

NEFF, R. J. and BENTON, W. F. (1962). Localisation of cellulose in the cysts of *Acanthamoeba* sp. *J. Protozool.* **9,** (Suppl. 11).

NEFF, R. J., BENTON, W. F., and NEFF, R. H. (1964). The composition of the mature cyst-wall of the soil amoeba *Acanthamoeba* sp. *J. Cell. Biol.* **23** (Abstract No. 133).

NESBIT, R. H. (1961). Some aspects of the structure and function of the nervous system of *Archachatina* (*Calachatina*) *marginata* (Swainson). *Proc. R. Soc.* B. **154,** 267–87.

NEUMAN, R. E. (1949). The amino acid composition of gelatins, collagens and elastins from different sources. *Arch. Biochem.* **24,** 289–98.

NEVILLE, A. C. (1963a). Daily growth layers in locust rubber-like cuticle influenced by an external rhythm. *J. Insect Physiol.* **9,** 177–86.

NEVILLE, A. C. (1963b). Growth and deposition of resilin and chitin in locust rubber-like cuticle. *J. Insect Physiol.* **9,** 265–78.

NEWBY, W. W. (1941). The development and structure of the slime-net glands of *Urechis*. *J. Morph.* **69,** 303–16.

NEWBY, W. W. (1946). The slime-glands and thread-cells of the hagfish, *Polistrotrema stouti*. *J. Morph.* **78,** 397–409.

NICHOLAS, W. L. and HYNES, H. B. N. (1963). "Embryology, post-embryonic development and phylogeny of the Acanthocephala". *In* "Lower Metazoa" (ed. E. C. Dougherty) p. 385–402. University of California Press, Berkeley.

NICHOLAS, W. L. and MERCER, E. H. (1965). The ultra structure of the tegument of *Moniliformis dubius* (Acanthocephala). *Q. Jl microsc. Sci.* **106,** 137–46.

NICHOLS, D. (1966). *Echinoderms* (revised ed.). Hutchinson University Library, London.

NICHOLS, D. (1967). Pentamerism and the calcite skeleton in echinoderms. *Nature, Lond.* **215,** 665–6.

NIIZIMA, M. (1960). Comparative histology of the arrangement of collagenous fibres of the corium of vertebrates. *Collagen Symp.* **3,** 51–9.

NIIZIMA, M., MIZUHIRA, A., and OKADA, S. (1954). Comparative anatomical studies on the arrangement of connective tissue fibres of the corium of vertebrates. *Bull. Tokyo med. dental Univ.* **1,** 15–20.

NIRENBERG, M. W., LEDER, P., BERNFIELD, M., BRIMACOMBE, R., TRUPIN, J. and ROTHMAN, F. (1965). RNA code words and protein synthesis VII. On the general nature of the RNA code. *Proc. natn. Acad. Sci. U.S.A.* **53,** 1161–8.

NISSEN, H. (1963). Röntgengefugeanalyse am Kalzit von Echinoderm skeletten. *Neues. Jb. Geol. Abh.* **117,** 230–4.

NOIROT-TIMOTHÉE, C. (1958). L'ultrastructure de la limite ectoplasme-endoplasme de des fibres formant le caryophore chez les cilies du genre *Isotricha*. Stein (Holotriches Trichostomes). *C. r. hebd. Séanc. Acad. Sci., Paris*, **247,** 692–5.

NOIROT-TIMOTHÉE, C. (1959). Recherches sur l'ultrastructure *d'Opalina ranarum*. *Annls. Sci. nat.* **12,** 265–81.

NOIROT-TIMOTHÉE, C. (1960). Étude d'une famille de Cilies: les Ophryoscolecidae, structures et ultrastructures. *Annls Sci. nat.* **12,** 527–718.

NORDWIG, A. and HAYDUK, U. (1967). A contribution to the evolution of collagen. *J. molec. Biol.* **26,** 351–2.

NORDWIG, A. and HAYDUK, U. (1969). Invertebrate collagens: Isolation, characterization and phylogenetic aspects. *J. molec. Biol.* **44,** 161–72.

NORDWIG, A., HÖRMAN, H., KÜHN, K. and GRASSMAN, W. (1961). Weitere Versuche zum Abbau des Kollagens durch Kollagenase. *Hoppe-Seyler's Z. physiol. Chem.* **325,** 242–50.

NORMAN, J. R. (1963). *A History of Fishes* (2nd Edition) (ed. P. H. Greenwood). Ernest Benn, London.

NORTHROP, J. W. (1926). The resistance of living organisms to digestion by pepsin or trypsin. *J. gen. Physiol.* **9,** 497–502.

NUÑEZ, J. A. (1963). Central nervous control of the mechanical properties of the cuticle in *Rhodnius prolixus*. *Nature, Lond.* **199,** 621–2.

NURSE, F. R. (1950). Quinone-tanning in the cocoon-shell of *Dendrocoelum lacteum*. *Nature, Lond.* **165,** 570.

ODHNER, N. (1952). Petites opisthobranches peu connus de la côte Méditerranéenne de France. *Vie Milieu*, **3,** 136–47.

Odland, G. F. (1960). A submicroscopic granular component in human epidermis. *J. invest. Derm.* **34,** 11–15.

Ogren, R. E. (1956). Development and morphology of the oncosphere of *Mesocestoides corti*, a tape worm of mammals. *J. Parasitol.* **42,** 414–28.

Ogren, R. E. (1959). The hexacanth embryo of a dilepid tapeworm III. The formation of shell and inner capsule around the oncosphere. *J. Parasitol.* **45,** 580–5.

Ogren, R. E. and Magill, R. M. (1962). Demonstration of protein in the protective envelopes and embryonic muscle in oncospheres of *Hymenolepis diminuta*, a tape worm of mammals. *Proc. Penn. Acad. Sci.* **36,** 160–7.

Okazaki, K. (1960). Skeleton formation of sea urchin larvae II. Organic matrix of the spicule. *Embryologia*, **5,** 283–320.

Oldfield, E. (1955). Observations on the anatomy and mode of life of *Lasaea rubra* and *Turtonia minuta*. *Proc. malac. Soc. Lond.* **31,** 226–49.

Oldfield, E. (1961). The functional morphology of *Kellia suborbicularis* (Montagu), *Montacuta ferruginosa* (Montagu) and *M. substriata* (Montagu) Mollusca, Lamellibranchiata. *Proc. malac. Soc. Lond.* **34,** 255–95.

Olsen, B. R. (1967). Electron microscope studies on collagen VI. The structure of segment-long spacing aggregates consisting of molecules renatured from isolated α-fractions of codfish skin collagen. *J. Ultrastruct. Res.* **19,** 446–73.

Olsson, R. (1961). The skin of *Amphioxus*. *Z. Zellforsch.* **54,** 90–104.

O'Neil, C. H. (1964). Isolation and properties of the cell surface membrane of *Amoeba proteus*. *Expl. Cell. Res.* **35,** 477–96.

Orton, I. H. and Amirthalingan, C. (1926). Notes on shell-deposition in oysters. With a note on the chemical composition of 'chalky' deposits in shells of *O. edulis*. *J. mar. biol. Ass. U.K.* **14,** 935–54.

Ørvig, T. (1967). Phylogeny of tooth-tissues. Evolution of some calcified tissues in early vertebrates. *In* "Structural and Chemical Organisation of Teeth" (ed. A. E. W. Miles) Vol. 1, pp. 45–110. Academic Press, New York and London.

Ørvig, T. (1968). The dermal skeleton; general considerations. *In* "Current Problems of Lower Vertebrate Phylogeny" (ed. T. Ørvig). Proceedings of the Fourth Nobel Symposium. Interscience Publishers, New York.

Owen, G. (1959). A new method of staining connective tissue fibres, with a note on Liang's method for nerve fibres. *Q. Jl microsc. Sci.* **100,** 421–4.

Paasche, E. (1968). Biology and physiology of coccolithophorids. *A. Rev. Microbiol.* **22,** 71–86.

Parakkal, P. F. and Matoltsy, G. A. (1964). A study of the fine structure of the epidermis of *Rana pipiens*. *J. Cell. Biol.* **20,** 85–94.

Parakkal, P., Montagna, W., and Ellis, R. A. (1962). The skin of primates XI. The skin of the white-browed gibbon. *Anat. Rec.* **143,** 169–73.

Park, H. D., Greenblatt, C. L., Mattern, C. F. T., and Merril, C. R. (1967). Some relations between *Chlorohydra*, its symbionts and some other chlorophyllous forms. *J. exp. Zool.* **164,** 141–62.

Parke, K. D. and Rudall, K. M. (1947). The silk of the egg-stalk of the green lace-wing fly. *Nature, Lond.* **179,** 905–6.

Parke, M. and Adams, I. (1960). The motile *Crystallolithus hyalinus* (Gardar and Markali) and non-motile phases in the life history of *Coccolithus pelagicus* (Wallich) Schiller. *J. mar. biol. Ass. U.K.* **39,** 263–74.

Parke, M. and Manton, I. (1965). Preliminary observations on the fine structure of *Prasinocladus marinus*. *J. mar. biol. Ass. U.K.* **45,** 525–36.

Parke, M. and Manton, I. (1967). The specific identity of the algal symbiont in *Convoluta roscoffensis*. *J. mar. biol. Ass. U.K.* **47,** 445–64.

Parke, M., Manton, I., and Clarke, B. (1955). Studies on marine flagellates II. Three new species of *Chrysochromulina*. *J. mar. biol. Ass. U.K.* **34,** 579–609.

Parke, M., Manton, I., and Clarke, B. (1959). Studies on marine flagellates V. Morphology and microanatomy of *Chrysochromulina strobilus*. sp. nov. *J. mar. biol. Ass. U.K.* **38,** 169–88.

Parry, D. A. and Brown, R. H. J. (1959). The hydraulic mechanism of the spider leg. *J. exp. Biol.* **36,** 423–33.

PARTRIDGE, S. M. and DAVIS, H. F. (1958). The chemistry of connective tissues IV. The presence of a non-collagenous protein in cartilage. *Biochem. J.* **68,** 298–305.

PASPALEFF, G. W. (1938a). Über die Entwicklung von *Rhizostoma pulmo* Agass. *Trud. Chernomorsk. biol. Sta. Varna,* **7,** 1–25.

PASPALEFF, G. W. (1938b). Über die Polypen-und Medusenform von *Ostroumovia inkermanica* (Pal-ostr). I. Einteilung II. Die Polypen *Ostroumovia inkermanica* (Pal-ostr.). *Trud. Chernomorsk. biol. Sta. Varna,* **7,** 27–44.

PAULIN, J. J. (1967). The fine structure of *Nyctotherus cordiformis*. *J. Protozool.* **14,** 183–96.

PAULING, L. and COREY, R. B. (1953). Compound helical configurations of polypeptide chains: Structure of proteins of the α-keratin type. *Nature, Lond.* **171,** 59–61.

PAUTARD, F. G. E. (1959a). Adventitious formation of bone salts by algal flagella and the question of apatite-formation in Protozoa generally. Proceedings of the Fifteenth International Congress on Zoology, London, 1958, pp. 478–9.

PAUTARD, F. G. E. (1959b). Hydroxyapatite as a developmental feature of *Spirostomum ambiguum*. *Biochim. biophys. Acta.* **35,** 33–46.

PAUTARD, F. G. E. (1961a). Calcium, phosphorus and the origin of back-bones. *New Scientist,* **12,** 364–6.

PAUTARD, F. G. E. (1961b). Studies of enamel and baleen. *J. dent. Res.* **40,** 1285–91.

PAUTARD, F. G. E. (1962). The molecular-biological background to the evolution of bone. *Clin. Orthopaed.* **24,** 230–44.

PAVANS DE CECCATTY, M. and THINEY, Y. (1963). Microscopie électronique de la fibrogenèse cellulaire du collagène chez l'eponge siliceuse, *Tethya lyncurium* L.K. *C. r. hebd. Séanc. Acad. Sci., Paris* **256,** 5406–8.

PAVLOVA, L. I. (1963). Materialy Nauchnoi Konf. Vsesoyuz obshchest. *Gelm. Mat. Sci. Conf. Soc. Helminth. Mosc.* 1963, **2,** 34–6.

PEAKALL, D. B. (1964). Composition, function and glandular origin of the silk fibroins of the spider, *Araneus diadematus* Cl. *J. exp. Zool.* **156,** 52.

PEEBLES, C. R. (1957). Ultra-structure of *Rhabditis strongyloides*. *J. Parasitol.* **43,** (Suppl.), 45.

PELC, S. R. and FELL, H. B. (1960). The effect of excess vitamin A on the uptake of labelled compounds by embryonic skin in organ culture. *Exp. Cell. Res.* **19,** 99–113.

PERSON, P. and MATHEWS, M. B. (1967). Endoskeletal cartilage in a marine polychaete, *Eudistyla polymorpha*. *Biol. Bull. mar. biol. Lab., Woods Hole,* **132,** 244–52.

PERSON, P. and PHILPOTT, D. E. (1963). Invertebrate cartilage. *Ann. N.Y. Acad. Sci.* **109,** 113–26.

PERSON, P. and PHILPOTT, D. E. (1969). The nature and significance of invertebrate cartilage. *Biol. Rev.* **44,** 1–16.

PETERSEN, J. B. and HANSEN, J. B. (1960). On some neuston organisms. *Bot. Tidsskr.* **56,** 197–234.

PETRUSKA, J. A. and HODGE, A. J. (1964). A subunit model for the tropocollagen macromolecule. *Proc. natn. Acad. Sci. U.S.A.* **51,** 871–6.

PETTUS, D. (1958). Water relationships in *Natrix sipedon*. *Copeia* (1958) 207–11.

PFEIFFER, W. and PLETCHER, T. F. (1964). Club-cells and granular cells in the skin of lamprey. *J. Fish. Res. Bd. Can.* **21,** 1083–7.

PHILLIPS, J. H. (1956). Isolation of active nematocysts of *Metridium senile* and their chemical composition. *Nature, Lond.* **78,** 932.

PHILPOTT, C. W. (1963). Halide localisation in the teleost chloride cell and its identification by selected area electron diffraction. *Protoplasma,* **60,** 7–23.

PHILPOTT, C. W. and COPELAND, D. E. (1965). Fine structure of chloride cells from three species of *Fundulus*. *J. Cell. Biol.* **18,** 389–404.

PHILPOTT, D. E., CHAET, A. B. and BARNETT, A. L. (1966). A study of the secretory granules of the basal disc of *Hydra*. *J. Ultrastruct. Res.* **14,** 74–84.

PICKEN, L. E. R. (1941). On the Bicoecidae, a family of colourless flagellates. *Phil. Trans. R. Soc.* B. **230,** 451–71.

PICKEN, L. E. R. (1960). *The organisation of cells and other organisms.* Clarendon Press, Oxford.

PICKEN, L. E. R. and LOTMAR, W. (1950). Orientated protein in chitinous structures. *Nature, Lond.* **165,** 599–600.

PICKEN, L. E. R., PRYOR, M. G. M. and SWANN, M. M. (1947). Orientation of fibrils in natural membranes. *Nature, Lond.* **159,** 434.

PICKEN, L. E. R. and SKAER, R. J. (1966). A review of researches on nematocysts. *In* "The Cnidaria and their Evolution" (ed. J. W. Rees) pp. 19–42. Symposia of the Zoological Society of London Vol. 16. Academic Press, London and New York.

PIERCE, S., KUSSOV, V., VALENTI, R., and SMETANA, D. G. (1968). Cytochemical studies on the test of *Allogromia laticollare. Micropaleontology,* **14,** 242–6.

PIEZ, K. A. (1961). Amino acid composition of some calcified proteins. *Science, N.Y.* **134,** 841.

PIEZ, K. A. (1964). Non-identity of the three α-chains in codfish skin collagen. *J. biol. Chem.* **239**, P.C. 4315–6.

PIEZ, K. A. (1967). Soluble collagen and its components resulting from its denaturation. *In* "Treatise on Collagen" (ed. G. N. Ramachandran) Vol. I, pp. 207–48. Academic Press, London and New York.

PIEZ, K. A. and GROSS, J. (1959). The amino acid composition and morphology of some invertebrate and vertebrate collagens. *Biochem. biophys. Acta,* **34,** 24–39.

PIEZ, K. A. and GROSS, J. (1960). The amino acid composition of some fish collagens: Relation between composition and structure. *J. biol. Chem.* **235,** 995–8.

PIEZ, K. A., IRREVERRE, F. and WOLFF, H. L. (1956). The separation and determination of cyclic imino acids. *J. biol. Chem.* **223,** 687–97.

PIEZ, K. A. and LIKINS, R. C. (1960). The nature of collagen II. Vertebrate collagens. *In* "Calcification in Biological Systems" (ed. R. F. Sognnaes) pp. 411–20. American Association for the Advancement of Science, Washington.

PIKKARAINEN, J. (1968). The molecular structures of vertebrate skin collagens. *Acta physiol. scand.* (Suppl.) **309,** 1–72.

PIKKARAINEN, J. and KULONEN, E. (1967). On the evolution and development of collagen. *Scand. J. Clin. Lab. Invest.* (Suppl.) **95,** 40.

PIPA, R. L. and WOOLEVER, P. S. (1965). Insect neurometamorphosis II. The fine structure of perineural connective tissue, adipohemocytes and the shortening central nerve cord of a moth, *Galleria mellonella. Z. Zellforsch. mikrosk. Anat.* **68,** 80–101.

PITELKA, D. R. (1961). Observations on the kinetoplast-mitochondrion and the cytostome of *Bodo. Exp. Cell. Res.* **25,** 87–93.

PITELKA, D. R. (1963). *Electron microscope study of Protozoa.* Pergamon Press, Oxford.

PITELKA, D. R. and SCHOOLEY, C. N. (1955). Comparative morphology of some protistan flagella. *Univ. Calif. Publ. Zool.* **61,** 79–128.

PLUMMER, J. M. (1966). Collagen formation in Achatinidae associated with a specific cell type. *Proc. malacol. Soc. Lond.* **37,** 189–97.

PORTER, K. R. (1956). Observations on the fine structure of animal epidermis. *In* "Proceedings of the International Conference on Electron Microscopy, 1954" (ed. R. Ross) pp. 539–546. Royal Microscopical Society, London.

PORTER, K. R. (1966). Cytoplasmic microtubules and their function. *In* Ciba Foundation Symposium. "Principles of Biomolecular Organisation" (eds. G. E. W. Wolstenholme and M. O'Connor). Churchill, London.

POURADIER, J. and ACCARRY, A. M. (1962). The nucleic acids present in gelatin. *In* "Collagen" (ed. N. Ramanathan) pp. 411–417. Interscience, New York.

POURADIER, J. and VENET, A. M. (1952a). Recherches sur les aldehydes existant dans les gelatines. *Bull. Soc. chim. Fr.* **19,** 347–50.

POURADIER, J. and VENET, A. M. (1952b). Contribution a l'ètude de la structure des gélatines V. Dégradation de la gélatine en solution isoélectrique. *J. Chim. phys.* **49,** 238–44.

PRENANT, M. (1919). Recherches sur les rhabdites des turbellaries. *Arch. Zool. exp. gen.* **58,** 219–50.

PRENANT, M. (1924). Contributions a l'étude cytologique du calcaire. *Bull. biol. Fr. Belg.* **58,** 331–80.

PRENANT, M. (1927). Les formes minéralogiques du calcaire chez les êtres vivants, et le problème de leur determinisme. *Biol. Rev.* **2,** 365–93.

PRENANT, M. (1928). Notes histologiques sur *Terebratulina caput-serpentis* L. *Bull. Soc. zool. Fr.* **53,** 113–25.

PRINGSHEIM, E. G. (1946). On Iron Flagellates. *Phil. Trans. R. Soc.* **232,** 311–42.

Propst, A. and Leb, D. (1964). Vergleichende elektronenmikroskopische studien an Glaskörper-, zonula- und Kollagenfibrillen. *Z. Zellforsch. mikrosk. Anat.* **61,** 829.

Pryor, M. G. M. (1940a). On the hardening of the ootheca of *Blatta orientalis. Proc. R. Soc.* B. **128,** 378–93.

Pryor, M. G. M. (1940b). On the hardening of the cuticle of insects. *Proc. R. Soc.* B. **128,** 393–406.

Pryor, M. G. M. (1955). Tanning of blowfly puparia. *Nature, Lond.* **175,** 600.

Pryor, M. G. M. (1962). Sclerotization. *In* "Comparative Biochemistry" (eds. M. Florkin and H. S. Mason) Vol. IV. Academic Press, New York and London.

Pryor, M. G. M., Russell, P. B. and Todd, A. R. (1946). Protocatechuic acid, the substance responsible for the hardening of the cockroach ootheca. *Biochem. J.* **40,** 627–8.

Puytorac, P. de (1959). Le cytosquelette et les systemes fibrillaires du cilie *Metaradiophyra gigas* de Puytorac, d'après étude au microscope électronique. *Arch. Anat. Microsccp.* **48,** 49–62.

Puytorac, P. de. (1965). Fibrillary ultra structure and cytoplasmic skeletal ultrastructure in ciliates. *In* "Progress in Protozoology" (Abstract of papers) Second International Conference of Protozoology, London, 1965.

Pyefinch, K. A. (1945). Unpublished results.

Pyefinch, K. A. and Downing, F. S. (1949). Notes on the general biology of *Tubularia larynx. J. mar. biol. Ass. U.K.* **28,** 21–43.

Pyne, C. K. (1959). L'ultrastructure de *Crystobia helicis* (Flagellé, Fam. Bodonidae). *C. r. hebd. Séanc. Acad. Sci., Paris,* **248,** 1410–13.

Pyne, C. K. (1960). L'ultrastructure de l'appareil basal des flagelles chez *Cryptobia helicis* (Flagellé, Bodonidae). *C. r. hebd. Séanc. Acad. Sci., Paris,* **250,** 1912–4.

Radhakrishnan, A. M. and Girri, K. V. (1954). The isolation of *allo* hydroxy-L-proline from sandal (*Santalum album* L.) *Biochem. J.* **58,** 57–61.

Raiz, L. G. and Nieman, I. (1967). Early effects of parathyroid hormone and thyrocalcitonin on bone in organ culture. *Nature, Lond.* **214,** 486–7.

Ramachandran, G. N. (1967). Structure of collagen at the molecular level. *In* "Treatise on Collagen" (ed. G. N. Ramachandran) Vol. I, pp. 103–179. Academic Press, London and New York.

Ramachandran, L. K. (1962). Elastoidin—a mixture of three proteins. *Biochem. biophys. Res. Commun.* **6,** 443–8.

Randall, J. T. (1953). Physical and chemical problems of fibre-formation and structure. *In* "Nature and Structure of Collagen" (eds. J. T. Randall and S. Fitton-Jackson). Butterworth, London.

Randall, J. T. (1956). Fine structure of some ciliate protozoa. *Nature, Lond.* **178,** 9–14.

Randall, J. T. (1957). Aspects of macromolecular orientation in collagenous tissues. *J. cell. comp. Physiol.* **49** (Suppl. 1), 113.

Randall, J. T. (1959a). Contractility in the stalks of Vorticellidae. *J. Protozool.* **6,** (Suppl.), 30.

Randall, J. T. (1959b). The stalks of Epistylidae. *J. Protozool.* **6,** (Suppl.), 30–1.

Randall, J. T. and Fitton-Jackson, S. F. (1958). Fine structure and function in *Stentor polymorphus. J. biophys. biochem. Cytol.* **4,** 807–30.

Randall, J. T., Fraser, R. D. B., Jackson, S. F., Martin, A. V. W. and North, A. C. T. (1952). Aspects of collagen structure. *Nature, Lond.* **169,** 1029–33.

Rasmont, R. (1956). La gemmulation des spongillides. *Ann. Soc. r. Zool. Belg.* **86,** 349–87.

Ratnayake, W. E. (1960). The chemical composition of gregarine hard parts. *J. Parasitol.* **46,** 22.

Raup, D. M. (1960). Ontogenetic variation in the crystallography of echinoid calcite. *J. Paleont.* **34,** 1041–50.

Raup, D. M. (1962). The phylogeny of calcite crystallography in echinoids. *J. Paleont.* **36,** 793–810.

Raup, D. M. (1965). The Endoskeleton. *In* "Physiology of Echinodermata" (ed. R. A. Boolootian) pp. 379–96. Interscience, New York.

Rawles, M. E. (1960). The integumentary system. *In* "Biology and Comparative Physiology of Birds" (ed. A. J. Marshall) Vol. 1, pp. 189–240. Academic Press, New York and London.

Ray, C. (1958). Vital limits and rates of desiccation in salamanders. *Ecology*, **39,** 75–83.

Reed, R. and Rudall, K. M. (1948). Electron microscope studies on the structure of earthworm cuticle. *Biochem. biophys. Acta*, **2,** 7–18.

Rees, W. J. (1957). Evolutionary trends in the classification of capitate hydroids and medusae. *Bull. Br. Mus.* (*Nat. Hist.*) *Zool.* **4,** 453–534.

Reger, J. F. (1967). The fine structure of the gregarine *Pyxinoides balani* parasitic in the barnacle, *Balanus tintinnabulum. J. Protozool.* **14,** 488–97.

Reichard, A. (1902). Über cuticular und Gerüst-substanzen bei Wirbellosan Tieren. Diss. Frankfurt. Quoted from Chitwood and Chitwood, 1950.

Reid, D. M. (1943). Occurrence of crystals in the skin of Amphipoda. *Nature, Lond.* **151,** 504–5.

Reid, W. M., Nice, S. J. and MacIntrye, R. C. (1949). Certain factors which influence activation of the hexacanth embryo of the fowl tapeworm, *Baillientina cesticillus. Trans. Ill. Acad. Sci.* **42,** 165–8.

Renaud, F. L., Rowe, A. J. and Gibbons, I. R. (1968). Some properties of the protein forming the outer fibres of cilia. *J. Cell. Biol.* **36,** 79–90.

Rhodin, J. A. G. and Reith, E. J. (1962). Ultra-structure of keratin in oral mucosa, skin, esophagus, claw and hair. *In* "Fundamentals of Keratinization" (eds. E. O. Butcher and R. F. Sognnaes) pp. 61–94. Publ. 70. American Association for the Advancement of Science.

Rich, A. and Crick, F. H. C. (1961). The molecular structure of collagen. *J. molec. Biol.* **3,** 483–506.

Richards, A. G. (1951). *The Integument of Arthropods*. University of Minnesota Press, Minneapolis.

Ridewood, W. G. (1921). On the calcification of the vertebral centra in sharks and rays. *Phil. Trans. R. Soc.* B. **210,** 311–407.

Rigby, B. J. (1967a). Relation between the shrinkage of native collagen in acid solution and the melting temperature of the tropocollagen molecule. *Biochim. biophys. Acta*, **133,** 272–7.

Rigby, B. J. (1967b). Correlation between serine and thermal stability of collagen. *Nature, Lond.* **214,** 87.

Rigby, B. J. (1968). Thermal transitions in some invertebrate collagens and their relation to amino acid content and environmental temperature. *In* "Symposium on Fibrous Proteins, Australia 1967" (ed. W. G. Crewther). Butterworth, Sidney.

Ris, H. and Plaut, W. (1962). Ultrastructure of DNA-containing areas in the chloroplasts of *Chlamydomonas. J. Cell. Biol.* **13,** 383–91.

Robertson, J. D. (1957). The habitat of the early vertebrates. *Biol. Rev.* **32,** 156–87.

Robertson, J. D. and Pantin, C. F. A. (1938). Tube formation in *Pomotoceros triqueter* (L). *Nature, Lond.* **141,** 648–9.

Robinow, C. F. (1956). Observations on vase-shaped, iron containing houses of two colourless flagellates of the family Bicoecidae. *J. biophys. biochem. Cytol.* **2** (4), Suppl., 233–4.

Robinson, R. A. and Watson, M. L. (1955). Crystal-collagen relationship in bone as observed in the electron microscope III. Crystal and collagen morphology as a function of age. *Ann. N.Y. Acad. Sci.* **60,** 596–628.

Robson, E. A. (1964). The cuticle of *Peripatopsis moseleyi. Q. Jl microsc. Sci.* **105,** 281–99.

Roche, J. (1952). Biochemie comparée des seléroprotéines iodées des Anthozoaires et des Spongiaires. *Experientia*, **8,** 45–54.

Roche, J., Andre, S. and Covelli, I. (1963). Sur la fixation de l'iode (I^{131}) par la moule (*Mytilus gallo provincialis* L.) et la nature des combinaisons iodees elabororees. *Comp. Biochim. Physiol.* **9,** 291–300.

Roche, J., Andre, S. and Covelli, I. (1964). Sur la fixation et l'utilisation de l'iode (I^{131}) par deux bryozoaires, *Bugula neritina* L. et *Schizoporella errata* (Waters, 1878). *Comp. Biochem. Physiol.* **11,** 215–21.

Roche, J., Ranson, G. and Eysserie-Lafon, M. (1951). Sur la composition des scleroproteines des coquilles des mollasques (conchiolines). *C. r. hebd. Séanc. Soc. Biol. Paris*, **145,** 1474–7.

Roe, D. A. (1956). A fibrous keratin processed from the human epidermis I. The extraction and physical properties of a fibrous protein found in the human epidermis. *J. invest Derm.* **27,** 1–8.

ROGERS, G. E. (1958). Some observations on the proteins of the inner root sheath cells of hair follicles. *Biochim. biophys. Acta*, **29,** 33–43.

ROGERS, G. E. (1959a). Newer findings on the enzymes and proteins of hair follicles. *Ann. N.Y. Acad. Sci.* **83,** 408–28.

ROGERS, G. E. (1959b). Electron microscopy of wool. *J. Ultrastruct. Res.* **2,** 309–30.

ROGERS, G. E. (1964a). Structural and biochemical features of the hair follicle. *In* "The Epidermis" (eds. W. Montagna and W. C. Lobitz) pp. 179–235. Academic Press, New York and London.

ROGERS, G. E. (1964b). Isolation and properties of inner sheath-cells of hair follicles. *Expl Cell. Res.* **33,** 264–76.

ROGERS, W. P. (1962). *The Nature of Parasitism.* Academic Press, New York and London.

ROGERS, W. P. and SOMMERVILLE, R. I. (1963). The infective stage of nematode parasites and its significance in parasitism. *In* "Advances in Parasitology" (ed. B. Dawes) Vol. I. Academic Press, London and New York.

ROGGEN, D. R., RASKI, D. J. and JONES, N. O. (1967). Further electron microscope observations of *Xiphinema index*. *Nematologica*, **13,** 1–16.

ROLFE, W. D. I. and BRETT, D. W. (1969). Fossilization Processes. *In* "Organic Geochemistry" (eds. G. E. Eglinton and M. T. J. Murphy). Longman, London.

ROLFE, W. D. I. and INGHAM, J. K. (1967). Limb structure, affinity and diet of the carboniferous 'centipede', *Arthropleura*. *Scott. J. Geol.* **3,** 118–24.

ROMER, A. S. (1933). Eurypterid influence on vertebrate history. *Science, N.Y.* **78,** 114–17.

ROMER, A. S. (1942). Cartilage as an embryonic adaptation. *Am. Nat.* **76,** 394–404.

ROMER, A. S. (1945). *Vertebrate Paleontology*. University of Chicago Press, Chicago.

ROMER, A. S. (1954). *Man and the Vertebrates*. Penguin, Harmondsworth, Middx.

ROSIN, S. (1946). Über Bau und Wachstum der Grenzlamelle der Epidermis bei Amphibienlarven: Analyse einer Orthogonalen Fibrillenstruktur', *Revue suisse Zool.* **53,** 133–201.

ROSS, R. (1968). The connective tissue fibre forming cell. *In* "Treatise on Collagen" (ed. B. S. Gould) Vol. 2. Academic Press, London and New York.

ROSS, R. and BENDITT, E. P. (1961). Wound-healing and collagen formation I. Sequential changes in components of guinea-pig skin wounds observed in the electron microscope. *J. biophys. biochem. Cytol.* **11,** 677–700.

ROSS, R. and BORNSTEIN, P. (1969). The elastic fibre I. The separation and partial characterisation of its macromolecular components. *J. Cell. Biol.* **40,** 366–81.

ROTH, L. E. (1956). Aspects of ciliary fine structure in *Euplotes patella*. *J. biophys. biochem. Cytol.* **2,** (Suppl.), 235–42.

ROTH, L. E. (1959). An electron-microscope study of the cytology of the protozoan *Perinema trichophorum*. *J. Protozool.* **6,** 107–16.

ROTH, S. I. (1965). The cytology of the murine resting (teleogen) hair follicle. *In* "Biology of the Skin and Hair Growth" (eds. A. G. Lyne and B. F. Short) pp. 233–50. Angus and Robertson, Sydney.

ROTH, S. I. and CLARK, W. H. (1964). Ultrastructed evidence related to the mechanism of keratin synthesis. *In* "The Epidermis" (eds. W. Montagna and W. C. Lobitz) pp. 303–37. Academic Press, New York and London.

ROTH, S. I. and JONES, W. A. (1967). The ultra-structure and enzymatic activity of the boa constrictor (*Constrictor constrictor*) skin during the resting phase. *J. Ultrastruct. Res.* **18,** 304–23.

ROTHMAN, S. (1964). Keratinization in historical perspective. *In* "The Epidermis" (eds W. Montagna and W. C. Lobitz) pp. 1–12. Academic Press, New York and London.

ROTHSCHILD, LORD (1965). *A Classification of Living Animals.* (2nd Edition) Longmans, London.

ROUILLER, C. and FAURÉ-FREMIET, E. (1958). Structure fine d'un flagelle chrysomonadien, *Chromulina psammobia*. *Expl Cell. Res.* **1,** 1–13.

ROUILLER, C., FAURÉ-FREMIET, E., and GAUCHERY, M. (1956a). Les tentacules d'*Ephelota*: étude au microscope electionique. *J. Protozool.* **3,** 194–200.

ROUILLER, C., FAURÉ-FREMIET, E., and GAUCHERY, M. (1956b). Fibres scleroproteique d'origine ciliaire chez les infusoires peritriches. *C. r. hebd. Séanc. Acad. Sci., Paris*, **242,** 180–2.

ROUILLER, C., FAURÉ-FREMIET, E., and GAUCHERY, M. (1956c). Origine ciliaire des fibrilles scléroprotéiques chez les ciliés péritriches. *Exp. Cell. Res.* **11,** 527–41.

ROWAN, W. B. (1956). The mode of hatching of the egg of *Fasciola hepatica*. *Expl. Parasitol.* **5,** 118–37.

ROWAN, W. B. (1957). The mode of hatching of the eggs of *Fasciola hepatica* II. Colloidal nature of the viscous cushion. *Expl Parasitol.* **6,** 131–42.

ROWAN, W. B. (1962). Permeability of the egg shell of *Fasciola hepatica*. *J. Parasitol.* **48,** 499.

RUBIN, A. L., DRAKE, M. P., DAVISON, P. F., PFAHL, D., SPEAKMAN, P. T., and SCHMITT, F. O. (1965). Effects of pepsin treatment on the interaction properties of tropocollagen molecules. *Biochemistry*, **4,** 181–90.

RUBIN, A. L., PFAHL, D., SPEAKMAN, P. T., DAVISON, P. F., and SCHMITT, F. O. (1963). Tropocollagen: Significance of protease-induced alterations. *Science, N.Y.* **139,** 37–9.

RUDALL, K. M. (1947). X-ray studies of the distribution of protein chain types in the vertebrate epidermis. *Biochim. biophys. Acta*, **1,** 549–62.

RUDALL, K. M. (1950). Fundamental structures in biological systems. *Prog. Biophys.* **1,** 39–72.

RUDALL, K. M. (1952). The proteins of the mammalian epidermis. *Adv. Protein Chem.* **7,** 253–90.

RUDALL, K. M. (1955). The distribution of collagen and chitin. *Symp. Soc. exp. Biol.* **9,** 49–71.

RUDALL, K. M. (1965). Protein polysaccharide interactions in chitinous structures. *In* "Structure and Function of Connective and Skeletal Tissue" (eds. S. Fitton-Jackson *et al.*) pp. 191–6. Butterworth, London.

RUDALL, K. M. (1967). Conformation in chitin-protein complexes. *In* "Conformation of Biopolymers" (ed. G. N. Ramachandran) Vol. 2, pp. 751–65. Academic Press, London and New York.

RUDALL, K. M. (1968a). Comparative biology and biochemistry of collagen. *In* "Treatise on Collagen" (ed. B. S. Gould) Vol. II, Part A, pp. 83–138. Academic Press, London and New York.

RUDALL, K. M. (1968b). Intracellular fibrous proteins and the keratins. *In* "Comprehensive Biochemistry" (eds. M. Florkin and E. H. Stotz) Vol. 26b. Elsevier, Amsterdam.

RUDZINSKA, M. A. (1965). The fine structure and function of the tentacles in *Tokophya infusiorum*. *J. Cell. Biol.* **25,** 459–77.

RUDZINSKA, M. A. and TRAGER, W. (1957). Intracellular phagotrophy by malarial parasites: an electron microscope study of *Plasmodium lophurae*. *J. Protozool.* **4,** 190–9.

RUNHAM, N. W. (1961). The histochemistry of the radula of *Patella vulgata*. *Q. Jl microsc. Sci.* **102,** 371–9.

RUNHAM, N. W. (1963a). The histochemistry of the radula of *Acanthochitona communis*, *Lymnaea stagnalis*, *Helix pomatia*, *Scaphander lignarius* and *Archidoris pseudoargus*. *Ann. Histochim.* **8,** 433–42.

RUNHAM, N. W. (1963b). Rate of replacement of the molluscan radula. *Nature, Lond.* **194,** 992–3.

RUNHAM, N. W. (1963c). A study of the replacement mechanism of the pulmonate radula. *Q. Jl microsc. Sci.* **104,** 271–8.

RUSKA, C. and RUSKA, H. Z. (1961). Die cuticula der epidermis des regenwurms (*Lumbricus terrestris L.*). *Z. Zellforsch.* **53,** 759–64.

RYBICKA, K. (1964). Attempt of general approach to the embryology of cyclophyllidean cestodes. *Acta Parasitol. Pol.* **12,** 327–30.

RYBICKA, K. (1966). Embryogenesis in Cestodes. *In* "Advances in Parasitology" (ed. B. Dawes) Vol. 4, pp. 107–86. Academic Press, London and New York.

RYDER, M. L. (1958). Investigations into the distribution of protein chain types in the skin follicles of mice and sheep and the entry of labelled sulphur compounds. *Proc. R. Soc.* B. **67,** 65–82.

SACHS, I. B. (1956). The chemical nature of the cyst membrane of *Pelomyxa illinoisensis*. *Trans. Am. microsc. Soc.* **75,** 307–13.

SAGAN, L. (1967). The origin of mitosing cells. *J. theor. Biol.* **14,** 225–74.

SAINT-HILAIRE, K. (1931). Morphogenetische Untersuchungen des Ascidienmantels. *Zool. Jb., Jena Anat.* **54,** 435–608.

SALENSKY, W. (1892). Beiträge zur Embryonalentwicklung der Pyrosomen III. Bildung des tetrazoiden Embryos und Entwicklung der Ascidiozoiden. *Zool. Jb., Anat. Ontog.* **5,** 1–98.

Satir, P. (1965). Studies on Cilia. Examination of the distal region of the ciliary shaft and the role of the filament in motility. *J. Cell. Biol.* **26,** 805–34.

Savel, J. (1955). Études sur la constitution et le metabolisme proteiques *d'Ascaris lumbricoides* Linné, 1758. *Path. gén. comp.* **55,** 52–121, 213–79.

Savory, T. (1964). *Spiders and other Arachnids.* English Universities Press, London.

Schepotieff, A. (1907). Knospungsprozess und Gehäuse von *Rhabdopleura. Zool. Jb., Abt. Anat.* **24,** 193–238.

Schmeideberg, O. (1882). Über die chemische Zusammensetzung der Wohröhren von *Onuphis tabicola* Mäll. *Mitt. Zool. Stat. Neapel.* **3,** 373–392.

Schmidt, W. J. (1924). *Die Bausteine des Tierkorpers in polarisiertem Lichte.* F. Cohen, Bonn.

Schmidt, W. J. (1940). Zur Morphologie, Polarisationsoptik und Chemie der Greifhaken von *Sagitta. Z. Morphol. Okol. Tier* **37,** 63–82.

Schmidt, W. J. (1962). Leigt der Eischalen kalk der Vögel als submikroskopische Kristallite vor? *Z. Zellforsch.* **57,** 848–80.

Schmitt, F. O. (1959). Intermolecular properties of elongate protein macromolecules with particular reference to collagen (tropocollagen). *In* "Biophysical Science" (eds. J. L. Oncley *et al.*). Wiley, New York.

Schneider, D. (1963). Normal and phototropic growth reactions in the marine bryozoan *Bugula avicularia. In* "Lower Metazoa" (ed. T. C. Dougherty) pp. 357–71. University of California Press, Berkeley.

Schneider, K. C. (1902). *Lehrbuch der Vergleich enden Histologie der Tiere.* Gustav Fischer, Jena.

Scholander, P. F., Hock, R., Watters, V., Johnson, F., and Irving, L. (1950). Heat regulation in some arctic and tropical mammals and birds. *Biol. Bull. mar. biol. Lab., Woods Hole,* **99,** 225–36.

Scholtyseck, E., Hammond, D. M., and Ernst, J. V. (1966). Fine structure of the macrogametes of *Eimeria perforans, E. stiedae, E. bovis* and *E. auburnesis. J. Parasitol.* **52,** 975–87.

Schour, I. (1932). The teeth. *In* "Special Cytology", 2nd edition, (ed. by E. V. Cowdry). Hoeber, New York.

Schröder, K. (1936). Beitrage zur kenntnis der Spiculabildung, der larvenspiculation und der Variations-breite der Gerüstnadeln von Susswasser-Schwammen. *Z. Morph. Ökol. Tiere,* **31,** 245–67.

Schroeder, W. A., Kay, L. M., Lewis, B., and Munger, N. (1955). The amino acid composition of certain morphologically distinct parts of white turkey feathers and of goose feather barbs, and goose down. *J. Am. chem. Soc.* **77,** 3801–8.

Schulze, F. E. (1863). Über die Structur des Tunicatenmantels und sein Verhalten im polarisirten Lichte. *Z. wiss. Zool.* **12,** 175–88.

Schulze, F. E. (1871). Uber den Bau und die Entwicklung von *Cordylophora locustris* (Allman). Engelmann, Leipzig.

Schulze, F. E. (1878a). Untersuchungen über den Bau and die Entwicklung der Spongien. Die Gattung *Spongelia. Z. wiss. Zool.* **32,** 117–57.

Schulze, F. E. (1878b). Untersuchungen über den Bau und die Entwicklung der Spongien. Die Familie Aplysinidae. *Z. wiss. Zool.* **30,** 379–420.

Schulze, F. E. (1879). Untersuchungen über den Bau und die Entwicklung der Spongien. Die Gattung *Hircinia, Nardo* und *Oligoceras,* n.g. *Z. wiss. Zool.* **33,** 1–38.

Schulze, F. E. (1887). Hexactinellida. *Rep. Scient. Res. Voy. Challenger,* 1873–6. Vol. XXI. Plates.

Schulze, P. (1925). Zum morphologischen Feinbau der Kieselschwammnadeln. *Z. Morph. Ökol. Tiere,* **4,** 615–25.

Schuster, F. (1963). An electron microscopic study of *Naegleria gruberi* II. The cyst stage. *J. Protozool.* **10,** 313–20.

Sebestyen, O. (1935). Studies on *Diplopsalis acuta* (Apstein) Entz with remarks on the question of *Kolkwitziella salebrosa. Arch. protistenk.* **85,** 20–31.

Sengel, P. (1958). Recherches expérimentales sur la différenciation des germes plumaires et du pigment de la peau de l'embryon de poulet en culture *in vitro. Annls Sci. natn. Zool.* **20,** 431–514.

Seshaiya, R. V., Ambujabai, P., and Kalyani, M. (1963). Amino acid composition of ichthylepidin from fish scales. *In* "Aspects of Protein Structure" (ed. W. G. N. Ramachandran) pp. 343–9. Academic Press, London and New York.

Severaid, J. H. S. (1945). Pelage changes in the snowshoe hare, *Lepus Americanus Strathopus* Bangs. *J. Mammal.* **26,** 41–63.

Sewell, M. T. (1955). The histology and histochemistry of the cuticle of a spider, *Tegenaria domestica* (L). *Ann. ent. Soc. Am.* **48,** 107–18.

Sheffield, H. G. (1963). Electron microscopy of the bacillary band and stichosome of *Trichuris muris and T. vulpis*. *J. Parasitol.* **49,** 998–1009.

Sheffield, H. G. and Hammond, D. M. (1966). Fine structure of first generation merozoites of *Eimeria bovis*. *J. Parasitol.* **52,** 595–606.

Shelanski, M. L. and Taylor, E. W. (1967). Isolation of a protein subunit from microtubules. *J. Cell. Biol.* **34,** 549–54.

Sheldon, H. and Robinson, R. A. (1957). Electron microscope studies of crystal collagen relationships in bone IV. The occurrence of crystal within collagen fibrils. *J. biophys. biochem. Cytol.* **3,** 1011–5.

Shimizu, M. (1958). Études chimiques sur de "duvet originel" ou exfoliation de la soie. *C. r. hebd. Séanc. Soc. Biol., Paris,* **149,** 853.

Shostak, S., Patel, N. G. and Barnet, A. L. (1965). The role of mesogloea in mass cell-movements in *Hydra*. *Devl. Biol.* **12,** 434–50.

Shrivastava, S. C. (1957). Studies on the cuticle of some indian scorpions. Sulphur linkages and purines. *Proc. natn. Acad. Sci.* India, B. **27,** 74–7.

Silliman, B. (1846). On the chemical composition of the calcareous corals. *Am. J. Sci. Arts.* (2nd Series) **1,** 189–99.

Silverman, P. H. (1954a). Studies on the biology of some tapeworms of the genus *Taenia* I. Factors affecting hatching and activation of taenid ova, and some criteria of their viability. *Ann. trop. Med. Parasitol.* **48,** 207–15.

Silverman, P. H. (1954b). Studies on the biology of some tapeworms of the genus *Taenia* II. The morphology and development of the taeniid hexacanth embryo and its enclosing membranes, with some notes on the state of development and propagation of gravid segments. *Ann. trop. Med. Parasitol.* **48,** 356–65.

Silverman, P. H. (1956). The longevity of eggs of *Taenia pisiformis* and *T. Saginata* under various conditions. *Trans. R. Soc. trop. med. Hyg.* **50,** 8.

Simmonds, D. H. (1955). Variations in the amino acid composition of merino wool. Proceedings of the International Wool Textile Research Conference, Australia 1955, Vol. C. pp. 65–74.

Simmonds, R. A. (1958). Studies on the sheath of fourth stage larvae of the nematode parasite *Nippostrongylus muris*. *Exp. Parasitol.* **7,** 14–22.

Simpson, C. F. and White, F. H. (1964). Structure of *Trichomonas foetus* as revealed by electron microscopy. *Am. J. Vet. Res.* **25,** 815–24.

Simpson, G. G. (1960). The History of Life. *In* "Evolution after Darwin" (ed. S. Tax) Vol. I, pp. 117–80. Chicago University Press, Ann Arbor.

Simpson, T. L. (1963). The biology of the marine sponge *Microciona prolifera* (Ellis and Solander) I. A study of cellular function and differentiation. *J. exp. Zool.* **154,** 135–47.

Sinex, F. M. (1968). The role of collagen in aging. *In* "Treatise on Collagen" (ed. B. S. Gould) Vol. 2, Part B, pp. 410–48. Academic Press, London and New York.

Singh, K. S. and Lewert, R. M. (1959). Observations on the formation and chemical nature of metacercarial cysts of *Notocotylus urbanensis*. *J. Infect. Dis.* **104,** 138–41.

Singleton, L. (1957). The chemical structure of earthworm cuticle. *Biochim. biophys. Acta* **24,** 67–72.

Skaer, R. J. (1961). Some aspects of the cytology of *Polycelis nigra*. *Q. Jl microsc. Sci.* **102,** 295–317.

Skaer, R. J. (1965). The origin and continuous replacement of epidermal cells in the planarian *Polycelis tenuis* (Iijima). *J. Embryol. exp. Morph.* **13,** 129–39.

Smith, D. and Treherne, J. E. (1963). Functional aspects of the organisation of the insect nervous system. *In* "Advances in Insect Physiology" (eds. J. W. L. Beament, J. E. Treherne and V. B. Wigglesworth) Vol. 1, pp. 401–478. Academic Press, London and New York.

SMITH, G. M. (1950). *Freshwater Algae of the United States*. McGraw-Hill, New York.

SMITH, G. M. and COATES, C. W. (1936). On the histology of the skin of the lung fish, *Protopterus annectens*, after experimentally induced aestivation. *Q. Jl microsc. Sci.* **79,** 487–491.

SMITH, H. W. (1939). *Studies in the physiology of the kidney*. University of Kansas, Lawrence.

SMITH, H. W. (1953). *From Fish to Philosopher*. Little Brown, Boston.

SMITH, J. E. (1937). On the nervous system of the starfish *Marthasterias glacialis* (L). *Phil. Trans. R. Soc.* B. **227,** 111–73.

SMITH, J. W. (1965). Packing arrangements of tropocollagen molecules. *Nature, Lond.* **205,** 356–8.

SMITH J. W. and WARMSLEY, R. (1959). Factors affecting the elasticity of bone. *J. Anat.* **93,** 209–25.

SMYTH, J. D. (1953). Standardization of methyl green for specific staining of egg-shell material in a trematode. *Q. Jl microsc. Sci.* **94,** 243–6.

SMYTH, J. D. (1954). A technique for the histochemical demonstration of polyphenol oxidase and its application to egg-shell formation in helminths and byssus formation. *Q. Jl microsc. Sci.* **95,** 139–52.

SMYTH, J. D. (1956). Studies on tapeworm physiology IX. A histochemical study of egg-shell formation in *Schistocephalus solidus* (Pseudophyllidea). *Expl. Parasitol.* **5,** 519–40.

SMYTH, J. D. (1962). Lysis of *Echinococcus granulosus* by surface active agents in bile, and the role of this phenomenon in determining host specificity in helminthes. *Proc. R. Soc.* B. **156,** 553–72.

SMYTH, J. D. (1969). *The physiology of Cestodes*. Oliver and Boyd, Edinburgh.

SMYTH, J. D. and CLEGG, J. A. (1959). Eggshell formation in trematodes and cestodes. *Expl. Parasitol.* **8,** 286–323.

SNODGRASS, R. E. (1924). Anatomy and metamorphosis of the apple maggot, *Rhagoletis pomonella* Walsh. *J. Agr. Res.* **28,** 1–36.

SOGNNAES, R. F. and LUSTIG, L. (1955). Histochemical reactions of the Lamprey mouth. *J. dent. Res.* **34,** 132–43.

SOLLAS, I. B. (1907). The molluscan radula, its chemical composition and some points in its development. *Q. Jl microsc. Sci.* **51,** 115–36.

SOLLAS, W. J. (1888). Tetractinellids. *Rep. Scient. Res. Voy. Challenger, 1873–6* **25.**

SOTELO, J. R. and TRUJILLO-CENOZ, O. (1959). The fine structure of an elementary contractile system. *J. biophys. biochem. Cytol.* **6,** 126–8.

SOUTHWARD, E. C. and SOUTHWARD, A. J. (1966). A preliminary account of the general and enzyme histochemistry of *Siboglinum atlanticum* and other Pogonophora. *J. mar. biol. Ass. U.K.* **46,** 579–616.

SPEARMAN, R. I. C. (1960). The structure of epidermal keratinized cells. *J. Anat.* **94,** 293–4.

SPEARMAN, R. I. C. (1964). The evolution of mammalian keratinized structures. *In* "The Mammalian Epidermis and its Derivatives" (ed. F. J. Ebling) Symposia of the Zoological Society of London, No. 12, pp. 67–81. Academic Press, London and New York.

SPEARMAN, R. I. C. (1966). The keratinization of epidermal scales, feathers and hairs. *Biol. Rev.* **41,** 59–96.

STANACH, F. R., WOODHOUSE, M. A., and GRIFFIN, F. L. (1966). Preliminary observations on the ultrastructure of the body wall of *Pomphorhynchus laevis* (Acanthocephala). *J. Helminth.* **40,** 395–402.

STAUFFER, H. (1924). Die Lokomotion der Nematoden. *Zool. Jb., Abt. Syst.* **49,** 1–118.

STEGEMAN, H. (1963). Protein (Conchagen und Chitin) im Stützgewebe von Tintenfischen. *Z. phys. Chem.* **331,** 269–79.

STEIN, F. (1854). Die infusionsthiere auf ihre Entwickelungsgeschichte Untersucht. *Trans. microsc. Soc. Lond.* **2,** 272–6.

STEINER, G. (1928). Proceedings of the Helminthological Society of Washington. *J. Parasitol.* **14,** 65–6.

STEINER, G. (1930). Die Nematoden I. Tiel. *Dt. Süpol.-Exped., 1901–1903.* (*Zool.*) **12,** 170–215.

STEINER, G. and HOEPPLI, R. (1926). Studies on the exoskeleton of some Japanese marine nemas. *Arch. Schiffs- u. Tropenhyg.* **30,** 597–676.

STEINERT, M. and NOVIKOFF, A. B. (1960). The existence of a cytostome and the occurrence of pinocytosis in the trypanosome, *Trypanosoma mega*. *J. biophys. biochem. Cytol.* **8,** 563–70.

STENZEL, H. B. (1962). Aragonite in the resilium of oysters. *Science, N.Y.* **136,** 1121–2.

STEPHENSON, W. (1947). Physiological and histochemical observations on the adult liver fluke, *Fasciola hepatica* III. Eggshell formation. *Parasitology,* **38,** 128–44.

STRANACK, F. R., WOODHOUSE, M. A. and GRIFFIN, R. L. (1966). Preliminary observations on the ultrastructure of the bodywall of *Pomphorhynchus laevis* (Acanthocephala). *J. Helminth.* **40,** 395–402.

SULLIVAN, C. E. (1961). Functional morphology, microanatomy and histology of the "Sydney cockle", *Anadara trapezia* (Deshayes). *Aust. J. Zool.* **9,** 219–57.

SUMMERS, N. M. (1967). Cuticle sclerotization and blood phenol oxidase in the fiddler crab, *Uca pugnax. Comp. Biochem. Physiol.* **23,** 129–38.

SUMMERS, N. M. (1968). The conversion of tyrosine to catecholamines and the biogenesis of n-acetyl-dopamine in isolated epidermis of the fiddler crab, *Uca pugilator. Comp. Biochem. Physiol.* **26,** 259–69.

SUNDARA RAJULU, G. and GOWRI, N. (1967). Nature of chitin of the coenosteum of *Millepora* sp. *Indian J. exp. Biol.* **5,** 180–81.

SWANBECK, G. (1964). A theory for the structure of α-Keratin. *In* "The Epidermis" (eds. W. Montagna and W. C. Lobitz) pp. 339–49. Academic Press, New York and London.

SWANBECK, G. and THYRESSON, N. (1962). A study of the state of aggregation of the lipids in normal and psoriatic horny layer. *Acta derm.-vener. Stockh.* **42,** 445–7.

SWANBECK, G. and THYRESSON, N. (1965). The role of keratohyalin material in the keratinization process and its importance for the barrier function. *Acta derm.-vener.* **45,** 21–5.

SZAKALL, A. (1957). Physiologische Prizipien bei der Entwicklung Wirksamer Preparate zur, Gesunderhaltung der Haut. *Arzneimittel-Forsch.* **7,** 408–11.

SZENT-GYORGYI, A. G., COHEN, C., and PHILPOT, D. E. (1960). Light meromyosin fraction I. A helical molecule from myosin. *J. molec. Biol.* **2,** 133–42.

TAMARIN, A. and CARRICKER, M. R. (1967). The egg capsule of the muricid gastropod *Urosalpinx cinerea:* An integrated study of the wall by ordinary light, polarised light and electron microscopy. *J. Ultrastruct. Res.* **21,** 26–40.

TARLO, L. B. H. (1960). The invertebrate origins of the vertebrates. *Rep. Int. Geol. Congr.* (21st Sess.), **22,** 113–23.

TARLO, L. B. H. (1962). The earliest vertebrates. *New Scientist,* **14,** 151–3.

TARLO, L. B. H. (1964). The origin of bone. *In* "Eone and Tooth". Proceedings of the First European Symposium of the Eone and Tooth Society. (ed. H. J. J. Blackwood) pp. 3–17. Pergamon Press, Oxford.

TARLO, L. B. H. (1967). Biochemical evolution and the fossil record. *In* "The Fossil Record" (eds. W. B. Harland *et al.*) pp. 119–32. Geological Society, London.

TAYLOR, D. L. (1968). *In situ* studies on the cytochemistry and ultrastructure of a symbiotic marine dinoflagellate. *J. mar. biol. Ass. U.K.* **48,** 349–66.

TCHEON-TAI-CHUIN (1930). Le cycle evolutif du scyphistome de *Chrysaora. Trav. Stn biol. Roscoff,* **8,** 1–179.

TEREPKA, A. R. (1963). Structure and calcification in avian egg-shells. *Expl Cell. Res.* **30,** 171–82.

TETRY, A. (1959). Classe des Sipunculiens. *In* "Traite de Zoologie" (ed. P-P. Grasse) Vol. V. Masson et Cie, Paris.

THOMAS, H. J. (1944). Tegumental glands in the Cirripedia Thoracica. *Q. Jl microsc. Sci.* **84,** 257–82.

THOMAS, J., ELSDEN, D. F., and PARTRIDGE, S. M. (1963). Partial structure of two major degradation products from the cross-linkages in elastin. *Nature, Lond.* **200,** 651–2.

THOMPSON, D'A. W. (1942). *On Growth and Form.* Cambridge University Press, London.

THOR, C. J. B. and HENDERSON, W. F. (1940). The preparation of alkali chitin. *Am. Dye stuff Reporter,* **29,** 461–4.

THORSON, T. B. (1955). The relationship of water-economy to terrestrialism in amphibians. *Ecology,* **36,** 100–16.

THREADGOLD, L. T. (1957). A histochemical study of the shell-gland of *Scyliorhinus caniculus. J. Histochem. Cytochem.* **5,** 159–66.

THREADGOLD, L. T. (1962). An electron microscope study of the tegument and associated structures of *Dipylidium caninum. Q. Jl microsc. Sci.* **103,** 135–40.

THREADGOLD, L. T. (1963a). The tegument and associated structures of *Fasciola hepatica. Q. Jl microsc. Sci.* **104,** 505–12.

THREADGOLD, L. T. (1963b). The ultrastructure of the "cuticle" of *Fasciola hepatica. Exp. Cell. Res.* **30,** 238–42.

THREADGOLD, L. T. (1967). Electron-microscope studies of *Fasciola hepatica* III. Further observations on the tegument and associated structures. *Parasitology*, **57,** 633–7.

THREADGOLD, L. T. and GALLACHER, S. S. E. (1966). Electron microscope studies of *Fasciola hepatica* I. The ultrastructure and interrelationship of the parenchymal cells. *Parasitology*, **56,** 299–304.

THURM, U. (1963). Die Beziehungen zwischen mechanischen Reizgrössen und stationären Erregungs-zuständen bei Borstenfeld-Sensillen von Bienen. *Z. vergl. Physiol.* **46,** 351–82.

TILNEY, L. G., HIROMOTO, Y., and MARSLAND, D. (1966). Studies on the microtubules in Heliozoa III. A pressure analysis of the role of these structures in the formation and maintenance of the axopodia of *Actinosphaerium nucleofilum* (Barrett). *J. Cell. Biol.* **29,** 77.

TIMM, R. W. (1953). Observations on the morphology and histological anatomy of a marine nematode, *Leptosomatum acephalatum* Chitwood, 1936, new combination (Enoplidae: Leptrosomatinae). *Am. Midl. Nat.* **49,** 229–48.

TIXIER, R. (1945). Contribution a l'étude de quelque pigments pyrroliques naturels des coquilles de mollusques, de l'oeuf a'émeu et du squelette du corail bleu. *Annls Inst. océanogr., Monaco*, **22,** 343–97.

TOD, R. and BLACKMORE, P. D. (1956). Calcite and aragonite in Foraminifera. *J. Paleont.* **30,** 217–9.

TONER, P. G. and CARR, K. E. (1968). *Cell structure.* E. and S. Livingstone, Edinburgh.

TONG, W. and CHAIKOFF, I. L. (1961a). Stimulatory effects of cytochrome C and quinones on the I^{131} utilization by cell-free sheep thyroid preparations. *Biochim. biophys. Acta*, **46,** 259–70.

TONG, W. and CHAIKOFF, I. L. (1961b). I^{131} utilization by the aquarium snail and the cockroach. *Biochim. biophys. Acta*, **48,** 347–51.

TOPSENT, E. (1887). Contribution a l'étude des Clionides. *Archs. Zool. exp. gén.* (2 Sér) **5** (suppl.), 1–165.

TOPSENT, E. (1893). Contribution à l'histologie des spongiares. *C. r. hebd. Séanc. Acad. Sci., Paris*, **117,** 444–6.

TOWE, K. M. (1967). Echinoderm calcite: single crystal or polycrystalline aggregate. *Science, N.Y.* **157,** 1048–50.

TRACEY, M. V. (1958). Cellulose and chitinase in plant nematodes. *Nematologica*, **3,** 179–83.

TRAVIS, D. F. (1960). Matrix and mineral deposition in skeletal structures of the decapod Crustacea. *In* "Calcification in Biological Systems" (ed. R. F. Sognnaes) pp. 57–116. American Association for the Advancement of Science, Pub. No. 64.

TRAVIS, D. F., FRANCOIS, C. J., BONAR, L. C., and GLIMCHER, M. J. (1967). Comparative studies of the organic matrices of invertebrate mineralised tissues. *J. Ultrastruct. Res.* **18,** 519–48.

TRIM, A. R. (1949). The kinetics of the penetration of some representative anthelmintics and related compounds into *Ascaris lumbricoides* var. *suis. Parasitology*, **39,** 281–90.

TRISTRAM, G. R. (1953). The amino acid composition of proteins. *In* "The Proteins" (eds. H. Neurath and K. Bailey) Vol. 1A, pp. 181–233. Academic Press, New York and London.

TRUEMAN, E. R. (1942). The structure and deposition of the shell of *Tellina tenuis. J. R. microsc. Soc.* **62,** 69–92.

TRUEMAN, E. R. (1949). The ligament of *Tellina tenuis. Proc. zool. Soc. Lond.* **119,** 717–42.

TRUEMAN, E. R. (1950). Observations on the ligament of *Mytilus edulis. Q. Jl microsc. Sci.* **91,** 225–35.

TRUEMAN, E. R. (1951). The structure, development and operation of the hinge ligament of *Ostrea edulis. Q. Jl microsc. Sci.* **92,** 129–40.

TRUEMAN, E. R. (1953a). Observations on certain mechanical properties of the ligament of *Pecten. J. exp. Biol.* **30,** 453–67.

TRUEMAN, E. R. (1953b). The ligament of *Pecten. Q. Jl microsc. Sci.* **94,** 193–202.

TRUEMAN, E. R. (1966). Bivalve mollusks: fluid dynamics of burrowing. *Science, N.Y.* **152,** 523–5.

Tsujii, T., Sharp, D. G. and Wilbur, K. M. (1958). Studies on shell-formation VII. The submicroscopic structure of the shell of the oyster, *Crassostrea virginica*. *J. biophys. biochem. Cytol.* **4,** 275–84.

Tucker, J. B. (1968). Fine structure and function of the cytopharyngeal basket in the ciliate *Nassula*. *J. Cell. Sci.* **3,** 493–514.

Tuzet, O. (1932). Recherches sur l'histologie des eponges *Reniera elegans* et *R. Simulans*. *Archs Zool. exp. gén.* **74,** 169–92.

Tuzet, O. and Connes, R. (1964). Sur la presence de lophocytes chez le *Sycon*. *C. r. hebd. Séanc. Acad. Sci., Paris,* **258,** 4142–3.

Tyler, C. and Simkiss, K. (1959). A study of the shell of ratite birds. *Proc. zool. Soc., Lond.* **133,** 201–42.

Urist, R. M. (1961). Calcium and phosphorous in the blood and skeleton of the Elasmobranchii. *Endocrinology,* **69,** 778–801.

Usuka, G. and Gross, J. (1965). Morphological studies of connective tissue resorption in the tail fin of metamorphosing bull frog tadpole. *Devl. Biol.* **11,** 352–70.

Valcurone, M. (1953). Ricerche istochimiche sui granuli vitellini dei Policladi. *Arch. Zool. ital.* **38,** 245–68.

Valli, M. (1950). Fenoli e tannazione chinonica delle proteine. *Monit. Zool. ital.* **58,** 83–7.

Van Oosten, J. (1957). The skin and scales. *In* "The Physiology of Fishes" (ed. E. M. Brown.) Vol. 1, pp. 207–44. Academic Press, New York and London.

Vervoort, W. (1966). Skeletal structure in the Solanderiidae and its bearing on hydroid classification. *In* "The Cnidaria and their Evolution" (ed. W. J. Rees) Symposia of the Zoological Society of London, No. 16. Academic Press, London and New York.

Verway, J. (1960). Annual report of the Zoological station of the Netherlands Zoological Society for the year 1958. *Archs néerl. Zool.* **13,** 556–71.

Vinogradov, A. P. (1953). The elementary chemical composition of marine organisms. *Mem. Sears Fdn mar. Res.* **2.**

Viswanatha, T. and Irreverre, F. (1960). Occurrence of hydroxylysine in trypsin. *Biochim. biophys. Acta,* **40,** 564–5.

Vivier, E. and Schrevel, J. (1964). Étude au microscope électronique d'une gregarine du genre *Selenidium* parasite de *Sabellaria alveolata* L. *J. Microsc.* **3,** 651–70.

Von Klement, R. (1938). Die anorganische Skeletsubstanz. *Naturwissenschaften,* **26,** 145–52.

Vovelle, J. (1965). La tube de *Sabellaria alveolata*. *Archs Zool. exp. gén.* **106,** 1–187.

Wada, K. (1961). Crystal growth of molluscan shells. *Bull. natn. Pearl Res. Lab.* **7,** 703–828.

Wada, K. (1964). Studies on the mineralisation of the calcified tissue in molluscs VII. Histological and histochemical studies of organic matrices in shells. *Bull. natn. Pearl Res. Lab.* **9,** 1078–86.

Waele, A. de (1933). Recherches sur les migrations des Cestodes. *Bull. Acad. r. Belg. Cl. Sci.* **19** (5), 649–60.

Wainwright, S. A. (1963). Skeletal organisation in the coral *Pocillopora damicornis*. *Q. Jl microsc. Sci.* **104,** 169–83.

Wald, G. (1963). Phylogeny and Ontogeny at the molecular level. Fifth International Congress on Biochemistry, Vol. 13, pp. 12–51.

Wallace, H. R. and Doncaster, C. C. (1964). A comparative study of the movement of some microphagous, plant-parasitic and animal-parasitic nematodes. *Parasitology,* **54,** 313–26.

Walshe, B. M. (1947). Feeding mechanisms of *Chironomus* larvae. *Nature, Lond.* **160,** 474.

Ward, W. H. and Lundgren, H. P. (1954). The formation, composition and properties of the keratins. *Adv. Protein Chem.* **9,** 243–97.

Warwicker, J. O. (1960). Comparative studies of fibroins II. The crystal structures of various fibroins. *J. molec. Biol.* **2,** 350–62.

Wase, A. W., Solewski, J., Rickes, E., and Seidenberg, J. (1967). Action of thyrocalcitonin on bone. *Nature, Lond.* **214,** 388–9.

Wasielewski, T. von (1924). Fortschritte der Coccidienforschung. *Ergebh. Hyg. Bakt.* **6,** 305–49.

Watabe, N. (1965). Studies on shell formation XI. Crystal-matrix relationships in the inner layers of mollusk shells. *J. Ultrastruct. Res.* **12,** 351–70.

WATABE, N., MEENAKSHI, V. R., HARE, P. E., MENZIES, R. J., and WILBUR, K. M. (1966). Ultrastructure, histochemistry and amino acid composition of the shell of *Neopilina sp. Physiologist*, **9,** 315.

WATABE, N., SHARP, D. G. and WILBUR, K. M. (1958). Studies on shell formation IX. Electron microscopy of crystal growth in the nacreous layer of the oyster, *rassostrea virginica. J. biophys. biochem. Cytol.* **4,** 281–6.

WATABE, N. and WILBUR, K. M. (1960). Influence of the organic matrix on crystal type in molluscs. *Nature, Lond.* **188,** 334.

WATABE, N. and WILBUR, K. M. (1963). Experimental studies on calcification in molluscs and the alga *Coccolithus huxleyi. Ann. N.Y. Acad. Sci.* **109,** 82–112.

WATSON, B. D. (1965a). The fine structure of the body-wall in a free living nematode, *Euchromadora vulgaris. Q. Jl microsc. Sci.* **106,** 75–81.

WATSON, B. D. (1965b). The fine structure of the body-wall and the growth of the cuticle in the adult nematode *Ascaris lumbricoides. Q. Jl microsc. Sci.* **106,** 83–91.

WATSON, M. R. (1958). The chemical composition of earthworm cuticle. *Biochem. J.* **68,** 416–9.

WATSON, M. R. and SILVESTER, N. R. (1959). Studies of invertebrate collagen preparations. *Biochem. J.* **71,** 578–84.

WATTERSON, R. L. (1942). The morphogenesis of down feathers with special reference to the developmental history of melanophores. *Physiol. Zool.* **15,** 234–59.

WAY, M. (1950). The structure and development of the larval cuticle of *Diataraxia oleracea* (Lepidoptera). *Q. Jl microsc. Sci.* **91,** 145–82.

WEATHERALL, J. A., BAILEY, P. J. and WEIDMANN, S. M. (1964). Sulphated mucopolysaccharides and calcification. *In* "Bone and Tooth" (ed. by H. J. J. Blackwood). Proceedings of First European Symposium of the Bone and Tooth Society. Pergamon Press, Oxford.

WEBB, M. (1965). Additional notes on the adult and larva of *Siboglinum fiordicum* and on the possible mode of tube formation. *Sarsia,* **20,** 21–34.

WEIS-FOGH, T. (1960). A rubber-like protein in insect cuticle. *J. exp. Biol.* **37,** 889–907.

WEIS-FOGH, T. (1961a). Thermodynamic properties of resilin, a rubber-like protein. *J. molec. Biol.* **3,** 520–31.

WEIS-FOGH, T. (1961b). Molecular interpretation of the elasticity of resilin, a rubber-like protein. *J. molec. Biol.* **3,** 648–67.

WEIS-FOGH, T. (1965). Resilin as compared with Elastin. *In* "Structure and Function of Connective and Skeletal Tissue" (eds. S. Fitton-Jackson *et al.*) pp. 101–105. Proceedings of N.A.T.O. Advanced Study Group, St. Andrews, 1964. Butterworth, London.

WEISS, P. (1957). Macromolecular fabrics and patterns. *J. Cell. comp. Physiol.* **49,** (Suppl. I), 105–12.

WEISS, P. and FERRIS, W. (1954). Electron-microscopic study of the texture of the basement membrane of larval amphibian skin. *Proc. natn. Acad. Sci. U.S.A.* **40,** 528–40.

WELLING, S. R., CHUINARD, R. G., and COOPER, R. A. (1967). Ultrastructural studies of normal skin and epidermal papillomas of the flathead sole. *Z. Zellforsch.* **78,** 370–87.

WESSELLS, N. K. (1962). Tissue interactions during skin histo-differentiation. *Devl. Biol.* **4,** 87–107.

WEST, A. J. (1963). A preliminary investigation of the embryonic layers surrounding the acanthos of *Acanthocephalus jackonsi* Bullock 1962 and *Echinorhynchus* gadi (Zoega) Müller 1776. *J. Parasitol.* **49,** (Suppl.), 42–43.

WESTER, D. H. (1910). Über die Verbreitung und Lokalisation des Chitins in Tierreiche. *Zool. Jb. Abt. Syst.* **28,** 531–58.

WHITEAR, M. (1962). The fine structure of crustacean proprioceptors I. The cordotonal organs in the legs of the shore crab, *Carcinus maenas. Phil. Trans. R. Soc.* B. **245,** 291–324.

WHITEAR, M. (1965). The fine structure of crustacean proprioceptors II. The thoracico-coxal organs in *Carcinus, Pagurus* and *Astacus. Phil. Trans. R. Soc.* B **248,** 437–56.

WHYTE, L. P. (1958). Melanin: a naturally occurring cation-exchange material. *Nature, Lond.* **182,** 1427–8.

WIGGLESWORTH, V. B. (1948). The insect cuticle. *Biol. Rev.* **23,** 408–51.

WIGGLESWORTH, V. B. (1956). The haemocytes and connective tissue formation in an insect, *Rhodnius prolixus* (Hemiptera). *Q. Jl microsc. Sci.* **97,** 89–98.

WIGGLESWORTH, V. B. (1965). *The Principles of Insect Physiology* (6th Edition). Methuen, London.

WILBUR, K. M. and SIMKISS, K. (1968). Calcified Shells. *In* "Comprehensive Biochemistry" Vol. 26, Part A. Extracellular and Supporting Structures (eds. M. Florkin and E. H. Stotz). Elsevier, Amsterdam.

WILBUR, K. M. and WATABE, N. (1963). Experimental studies on calcification in molluscs and the alga *Coccolithus huxleyi. Ann. N.Y. Acad. Sci.* **109,** 82–112.

WILLEY, A. (1894). *Amphioxus and the Ancestry of the Vertebrates.* Macmillan, New York.

WILLIAMS, A. (1956). The calcareous shell of the Brachiopoda and its importance to their classification. *Biol. Rev.* **31,** 243–87.

WILLIAMS, A. (1965). Brachiopod anatomy. *In* "Treatise on Invertebrate Paleontology" (ed. R. C. Moore) Part H, Brachiopoda, Vol. 1, ph.H. 6–H57. The Geological Society of America and University of Kansas Press.

WILLIAMS, A. (1966). Growth and structure of the shell of living articulate brachiopods. *Nature, Lond.* **211,** 1146–8.

WILLIAMS, A. P. (1960). The chemical composition of snail gelatin. *Biochem. J.* **74,** 304–7.

WILMER, E. N. (1956). Factors which influence the acquisition of flagella by the amoeba *Naegleria gruberi. J. exp. Biol.* **33,** 583–603.

WILSON, H. V. and PENNY, J. T. (1930). The regeneration of sponges (*Microciona*) from dissociated sponges. *J. exp. Zool.* **9,** 537–77.

WILSON, R. A. (1967). The structure and permeability of the shell and vitelline membrane of the egg of *Fasciola hepatica. Parasitology,* **57,** 47–58.

WISSE, E. and DAEMS, W. T. (1968). Electron microscope observations on second-stage larvae of the potato root eelworm, *Heterodera rostochiensis. J. Ultrastruct. Res.* **24,** 210–31.

WOHLFARTH-BOTTERMAN, E. K. and KRUGER, F. (1954). Protistenstudien VI. Die Feinstruktur der Axopodien und Skeletelnadeln von Helizoen. *Protoplasma,* **43,** 177–91.

WOLBACH, S. B. and HOWE, P. R. (1925). Tissue-changes following deprivation of fat-soluble A vitamin. *J. exp. Med.* **42,** 753–77.

WOLPERT, L. and GUSTAFSON, T. (1961). Studies on the cellular basis of morphogenesis of sea-urchin embryo. *Expl. cell. Res.* **25,** 311–25.

WOOD, G. C. (1960). The formation of fibrils from collagen solutions III. The effect of chondroitin sulphate and some other naturally occurring polyanions on the rate of formation. *Biochem. J.* **75,** 605–12.

WOODLAND, W. (1907). Studies in spicule formation VI. The scleroblastic development of the spicules in Ophiuroidea and Echinoidea, and in the genera *Antedon* and *Synapta. Q. Jl microsc. Sci.* **51,** 31–44.

WRIGHT, K. A. (1963). Cytology of the bacillary bands of the nematode *Capillaria hepatica* (Bancroft, 1893). *J. Morph.* **112,** 233–59.

WRIGHT, K. A. (1965). The histology of the oesophageal region of *Xiphinema index* Thorne and Allen, 1950, as seen with the electron microscope. *Can. J. Zool.* **43,** 689–700.

WRIGHT, K. A. (1968). The fine structure of the cuticle and intercordal hypodermis of the parasitic nematodes, *Capillaria hepatica* and *Trichuris myocastoris. Can. J. Zool.* **46,** 173–9.

WRIGHT, K. A. and HOPE, W. D. (1968). Elaborations of the cuticle of *Acanthonchus duplicatus. Can. J. Zool.* **46,** 1005–11.

YANAGITA, T. M. and WADA, T. (1954). Effects of trypsin and thioglycollate upon the nematocysts of the sea anemone. *Nature, Lond.* **173,** 171.

YONGE, C. M. (1935). Origin and nature of the egg case in the crustacea. *Nature, Lond.* **136,** 67–8.

YONGE, C. M. (1962). On the primitive significance of the byssus in the Bivalvia and its effects in evolution. *J. mar. Biol. Ass. U.K.* **42,** 113–25.

YONGE, C. M. (1963a). The biology of coral reefs. *In* "Advances in Marine Biology", **1,** 209–60.

YONGE, C. M. (1963b). Rock-boring organisms. *In* "Mechanisms of Hard Tissue Destruction" (ed. R. F. Sognnaez) Publ. 75 of the Amer. Association for the Advancement of Science, Washington.

YONGE, C. M. (1963c). The biology of coral reefs. *In* "Advances in Marine Biology" **I,** 209–60. Academic Press, London and New York.

YONGE, C. M. and NICHOLLS, A. G. (1931). Studies in the physiology of corals II. Digestive enzymes with notes on the speed of digestion. *Scient. Rep. Barrier Reef Exped.* **1,** 59–81.

YOUNG, G. E. and INMAN, W. R. (1938). The protein of the casing of salmon eggs. *J. biol. Chem.* **124,** 189–93.

YOUNG, J. D. (1950). The structure and some physical properties of the testudinian egg-shell. *Proc. zool. Soc. Lond.* **120,** 455–69.

YOUNG, J. Z. (1962). *Life of the Vertebrates.* (2nd Edition) Clarendon Press, Oxford.

YUEN, P-H. (1967). Electron microscopical studies on *Ditylenchus dipsaci* (Kühn) I. Stomatal Region. *Can. J. Zool.* **45,** 1019–33.

Additional References

CHAPTER 1 THE COMPOSITION AND PROPERTIES OF STRUCTURAL MATERIALS

ATWOOD, M. M. and ZOLA, H. (1967). The association between chitin and protein in some chitinous tissues. *Comp. Biochem. Physiol.*, **20,** 993–98.
BARRETT, A. J. (1971). The biochemistry and function of mucosubstances. *Histochemical Journal*, **3,** 213–21.
DAVIS, N. R. and BAILEY, A. J. (1972). The chemistry of the collagen cross links. *Biochem. J.*, **129,** 91–96.
ELLIS, D. O. and McGAVIN, S. (1970). The structure of collagen. *J. Ultrastructure Research*, **32**, 191–211.
FRANZBLAU, C. (1971). Elastin in *Comprehensive Biochemistry*, Vol. 26C 659–709. (eds M. Florkin and E. H. Stotz) Elsevier, Amsterdam.
BODLEY, H. D. and WOOD, R. L. (1972). Ultrastructural studies on elastic fibres using enzymatic digestion of thin sections. *Anat. Rec.*, **172,** 71–88.
JONES, A. W. and BARSON, A. J. (1971). Elastogenesis in the developing chick lung: a light and electron microscopical study. *J. Anat.*, **110,** 1–15.
JOHN, R. and THOMAS, J. (1972). Chemical compositions of elastins isolated from aortas and pulmonary tissues of humans of different ages. *Biochem. J.*, **127,** 261–69.
KASHIWAR, H. K. (1970). Mineralised spherules in cartilage of bone revealed by cytochemical methods. *Am. J. Anat.*, **129,** 459–66.
MILLWARD, G. R. (1970). The substructure of α-keratin microfibrils. *J. Ultrastructure Research*, **31,** 349–55.
OAKES, B. W. and BIALKOWER, B. (1973). Loadstrain and structural studies of the elastic ligament of the wing of the domestic fowl. *J. Anat.*, **114,** 150.
PARTRIDGE, S. M. (1970). Isolation and characterisation of elastin. In *Chemistry and Molecular Biology of the intercellular Matrix*. Vol. 1. (ed. E. A. Balazes) Academic Press, New York.
PIEZ, K. A. (1968). Cross-linking of collagen and elastin. *Ann. Rev. Biochem.*, **37,** 547–70.
THOMAS, J. (1971). The isolation and composition of cross-linked peptides following enzymic hydrolysis of elastin. *Int. J. Biochem.*, **2,** 644–50.

CHAPTER 2 PROTOZOA

BABA, S. A. (1972). Flexural rigidity and elastic constant of cilia. *J. Exp. Biol.*, **56,** 459–67.
FERNAND, L. and WOLK, C. P. (1973). Structural studies on the glycolipids from the envelope of the heterocyst of *Anabaena cylindrica*. *Biochemistry*, **12,** 791–98.
MESSER, G. and BEN-SHAUL, Y. (1971). Fine structure of trichocyst fibrils of the dinoflagellate *Peridinium westii*. *J. Ultrastructure Research*, **37,** 94–104.
NETZEL, H. (1972a). Die Schlalenbildung bei *Difflugia oviformis* (Rhizopoda, Testacea). *Z. Zellforsch.*, **135,** 55–61.
NETZEL, H. (1972b). Morphogenese des Gehäuses von *Euglypha rotunda*. (Rhizopoda, Testacea). *Z. Zellforsch.*, **135,** 63–69.
NORDWIG, A., HAYDUK, U. and GERISH, G. (1969). Is collagen present in unicellular organism? *Hoppe-Seyler's Z. Physiol. Chem.*, **350,** 245–48.

ROTH, L. E., PIHLAJA, D. J. and SHIGENAKA, Y. (1970). Microtubules in the heliozoan axopodium. I. The gradion hypothesis of allosterism in structural proteins. *J. Ultrastructure Res.*, **30,** 7–37.

ROTH, L. E. and SHIGENAKA, Y. (1970). Microtubules in the heliozoan axopodium. II. Rapid degradation by cupric and nickelous ions. *J. Ultrastructure Res.*, **31,** 356–74.

SHIGENAKA, Y., ROTH, L. E. and PIHLAJA, D. J. (1971). Microtubules in the heliozoan axopodium. III. Degradation and reformation after dilute urea treatment. *J. Cell. Sci.*, **8,** 127–51.

CHAPTER 4 PORIFERA

PE, J. (1973). Etude quantitative de la régulation du squelette chez une éponge d'eau douce. *Arch. Biol.* (Bruxelles), **84,** 147–73.

STONE, A. R. (1970). Seasonal variations of spicule size in *Hymeniacidon perleve*. *J. Mar. biol. Ass. U.K.*, **50,** 343–48.

CHAPTER 5 COELENTERATA

BARGMANN, W. (1972). Zur Architektur der Mesoglea. Untersuchungen an der Rippenqualle *Pleurobranchia pileus*. *Z. Zellforsch*, **123,** 66–81.

BARNES, J. D. (1970). Coral skeletons: An explanation of their growth and structure. *Science*, **170,** 1305–08.

CHAPMAN, G. (1973). A note on the composition of some coelenterate exoskeletal materials. *Comp. Biochem. Physiol.*, **45B,** 279–82.

GOSLINE, J. M. (1971a). Connective tissue mechanics of *Metridium senile*. I. Structural and compositional aspects. *J. Exp. Biol.*, **55,** 763–74.

GOSLINE, J. M. (1971b). Connective tissue mechanics of *Metridium senile*. II. Visco-elastic properties and macromolecular model. *J. Exp. Biol.*, **55,** 775–95.

HAUSMAN, R. E. and BURNETT, A. L. (1969). The mesoglea of *Hydra*. I. Physical and histochemical properties. *J. Exp. Zool.*, **171,** 7–14.

KNIGHT, D. P. (1970a). Tanning cells in the thecate hydroid *Campanularia flexuosa*. *Proc. Chall. Soc.*, **IV,** 60–61.

KNIGHT, D. P. (1970b). Sclerotization of the perisarc of the calyptoblastic hydroid *Laomeda flexuosa*. The identification and localisation of dopamine in the hydroid. *Tissue and cell.*, **2,** 467–77.

WILFERT, M. and PETERS, W. (1969). Vorkommen von Chitin bei Coelenteraten. *Z. Morphol. Oekol. Tiere*, **64,** 77–84.

YOUNG, S. D. (1971). Organic material from scleractinian coral skeletons. I. Variation in composition between several species. *Comp. Biochem. Physiol.*, **40B,** 113–20.

YOUNG, S. D., O'CONNOR, J. D. and MUSCATINE, L. (1971). Organic material from scleractinian coral skeletons. II. Incorporation of ^{14}C into protein, chitin and lipid. *Comp. Biochem. Physiol.*, **40B,** 945–58.

CHAPTER 6 PLATYHELMINTHES AND NEMERTEA

ANANTARAMAN, S. and RAVINDRANATH, M. H. (1973). Chemical nature of the hooks of the cystacanth of *Moniliformis moniliformis*. *Acta histochem.*, **47,** 124–31.

LEE, D. L. (1972). The structure of the helminth cuticle in *Advances in Parasitology*, **10,** 347–79 (ed. B. Dawes) Academic Press. London.

BUNKE, D. (1972). Sklerotin-Komponenten in den Vitellocyten von *Microdalyellia fairchildi* (Turbellaria). *Z. Zellforsch*, **135,** 383–98.

BURTON, P. R. (1963). A histochemical study of vitelline cells, egg capsules and Mehlis' gland in the frog lung fluke *Haematolocchus medioplexus* *J. exp. Zool.*, **154,** 247–58.

HOCKLEY, D. and MCLAREN, D. J. (1971). The outer membrane of *Schistosoma mansoni*. *Trans. Roy. Soc. Trop. Med & Hyg.*, **65,** 432.

HOCKLEY, D. and MCLAREN, D. J. (1973). *Schistosoma mansoni:* Changes in the outer membrane of the tegument during development from cercaria to adult worm. *International Journal for Parasitology*, **3,** 13–25.

KILEJIAN, A. and SCHWABE, C. W. (1971). Studies on the polysaccharides of the *Echinococcus granulosus* cyst, with observations on a possible mechanism for laminated membrane formation. *Comp. Biochem. Physiol.*, **40B,** 25–36.

MOSSE, B. (1970). Honey-coloured, sessile, endogene spores. III. Wall structure. *Arch. Mikrobiol.*, **74,** 146–59.

PEDERSEN, K. J. (1968). Some morphological and histochemical aspects of nemertean connective tissue. *Z. Zellforsch.*, **90,** 570–95.

PENCE, D. B. (1967). The fine structure and histochemistry of infective eggs of *Dipylidium caninum*. *Journal of Parasitology*, **56,** 84–97.

RAMALINGAM, K. (1971). Studies on vitelline cells of Monogenea. IV. Presence of masked phenol and its significance. *Acta histochem.*, **41,** 72–78.

RAMALINGAM, K. (1972). Studies in vitelline cells of Monogenea. III. Nature of phenolic material and a possible alternate mechanism involved in hardening of eggshell in helminths. *Acta histochem.*, **44,** 71–76.

RAMLINGAM, K. (1973). Chemical nature of monogenean sclerites. I. Stabilization of clamp-protein by the formation of dityrosine. *Parasitology*, **66,** 1–7.

RAMALINGAM, K. (1973). The chemical nature of the eggshell of helminths. I. Absence of quinone tanning in the eggshell of the liver-fluke *Fasciola hepatica*. *International Journal for Parasitology*, **3,** 67–75.

SADANAND, A. V. (1971). Biochemical analyses of the cyst wall of *Echinococcus grannlosus*, Batsch. *Comp. Biochem. Physiol.* **40B,** 797–805.

SWIDERSKI, Z. (1973). Electron microscopy and histochemistry of oncospheral hook formation by the cestode *Catenotaenia pusilla*. *International Journal for Parasitology*, **3,** 27–33.

CHAPTER 7 ASCHELMINTHES

BONNER, T. P., MENEFEE, M. G. and ETGES, F. J. (1970). Ultrastructure of cuticle formation in a parasitic nematode, *Nematospiroides dubius*. *Z. Zellforsch.*, **104,** 193–204.

BONNER, T. P., WEINSTEIN, P. P. and SAZ, H. J. (1971). Synthesis of cuticular protein during the third moult in the nematode *Nippostrongylus brasiliensis*. *Comp. Biochem. Physiol.*, **40B,** 121–27.

BRODIE, A. E. (1970). Development of the cuticle in the rotifer *Asplanchna brightwelli*. *Z. Zellforsch.*, **105,** 515–25.

CLARKE, J. A., COX, P. M. and SHEPHERD, A. M. (1967). The chemical composition of the eggshells of the potato cyst-nematode, *Heterodera rostochiensis* Woll. *Biochem. J.* **104,** 1056–60.

CLARKE, A. J. (1968).The chemical composition of the cyst wall of the potato cyst-nematode, *Heterodera rostochiensis*. *Biochem. J.* **108,** 221–24.

CLARKE, A. J. (1970). The composition of the cyst wall of the beet cyst-nematode. *Heterodera schachtii*. *Biochem. J.*, **118,** 315–18.

CLÉMENT, P. (1969). Premières observations sur l'ultrastructure comparée des teguments de rotifères. *Vie et Milieu, Ser A. 20*, 461–80.

DE GRASS, A. T. (1972). Body wall ultrastructure of *Macroposthonia xenoplax* (Nematoda). *Nematologica*, **18,** 25–30.

HIRUMI, H., HUNG, C. L. and MARAMOROSCHI, K. (1970). Body wall ultrastructure of two plant parasite nematodes *Trichodorus christiei* and *Pratylenchus penetrans*. *Phytopathology*, **60,** 575.

JOHNSON, P. W., VAN GRUNDY, S. D. and THOMPSON, W. W. (1970). Cuticle ultrastructure of *Hemicycliophora arenaria*, *Aphelenchus arenae*, *Hirschmanniella gracilis* and *Hirschmanniella belli*. *J. Nemat.* **2,** 42–58.

KISIEL, M., HIMMELHOCH, S. and ZUCKERMAN, B. M. (1972). Fine structure of the bodywall and vulva area of *Pratylenchus penetrans*. *Nematologica*, **18,** 234–38.

LEE, D. A. (1970). The ultrastructure of the cuticle of adult female *Mermis nigrescens* (Nematoda). *J. Zool. Lond.*, **161,** 513–18.

LEE, D. L. and LEŠTAN, P. (1971). Oogenesis and eggshell formation in *Heterakis gallinarum* (Nematoda). *J. Zool. Lond.*, **164,** 189–96.

OZEROL, N. H. and SILVERMAN, P. H. (1972). Enzymatic studies on the exsheathment of *Haemonchus contortus* infective larvae: the role of leucine aminopeptidase. *Comp. Biochem. Physiol.*, **42B,** 109–21.

NICHOLAS, W. L. (1972). The fine structure of the cuticle of *Heterotylenchus*. *Nematologica*, **18,** 138–40.

SHEPHERD, A. M., CLARK, S. A. and DART, P. J. (1972). Cuticle structure in the genus *Heterodera*. *Nematologica*, **18,** 1–17.

STORCH, V. and WELSCH, U. (1969). Über den aufbau des Rotatorienintegumentes. *Z. Zellforsch.*, **95,** 405–14.

CHAPTER 8 ACANTHOCEPHALA

ROTHMAN, A. H. and ELDER, J. E. (1970). Histochemical nature of an acanthocephalan, a cestode and a trematode absorbing surface. *Comp. Biochem. Physiol.*, **33,** 745–62.

CHAPTER 9 ENDOPROCTA AND LOPHORATE PHYLA

HUNT, S. (1972). Scleroprotein and chitin in the exoskeleton of the ectoproct *Flustra foliacea*. *Comp. Biochem. Physiol.*, **43B,** 571–77.

JOPE, M. (1969). The protein of brachiopod shell. III. Comparison with structural protein of soft tissue. *Comp. Biochem. Physiol.*, **30,** 209–24.

JOPE, M. (1971). Constituents of brachiopod shells in *Comprehensive Biochemistry*, Vol. 26C, pp. 749–82. (eds M. Florkin and E. H. Stotz) Elsevier, Amsterdam.

JOPE, M. (1973). The protein of brachiopod shell. V. N-terminal end groups. *Comp. Biochem. Physiol.*, **45B,** 17–24.

CHAPTER 10 MOLLUSCA

ABOLINŠ-KROGIS, A. (1970). Electron microscope studies of the intracellular origin and formation of calcifying granules and calcium spherites in the hepatopancreas of the snail, *Helix pomatia*, L. *Z. Zellforsch.*, **108,** 501–15.

BACCETTI, B. and ROSATI, F. (1971). Electron microscopy on tardigrades: III. The integument. *J. Ultrastructure Res.*, **34,** 214–43.

BEVELANDER, G. and NAKAHARA, H. (1970). An electron microscope study of the formation and structure of the periostracum of a gastropod, *Littorina littorea*. *Cal. Tiss. Res.* **5,** 1–12.

BRICTEUX-GRÉGOIRE, S., FLORKIN, M. and GRÉGOIRE, CH. (1968). Prism conchiolin of modern or fossil molluscan shells. An example of protein paleization. *Comp. Biochem. Physiol.*, **24,** 567–72.

GRÉGOIRE, C. (1972). Structure of the molluscan shell in *Chemical Zoology*, Vol. VII, pp. 45–102, (eds M. Florkin and B. T. Scheer) Academic Press, New York.

HUNT, S. (1970). Invertebrate structural proteins. Characterization of the operculum of the gastropod mollusc. *Buccinum undatum*. *Biochim. biophys. Acta* **207,** 347–60.

HUNT, S. (1971). Comparison of three extracellular structural proteins in the gastropod mollusc, *Buccinum undatum* L., the periostracum, egg capsule and operculum. *Comp. Biochem. Physiol.*, **40B,** 37–46.

HUNT, S. (1973). Chemical and physical studies of the chitinous siphon sheath in the Lamellibranch *Lutraria lutraria* L and its relationship to the periostracum. *Comp. Biochem. Physiol.* **45B,** 311–23.

HUNT, S. and OATES, K. (1970). Fibrous protein ultrastructure of gastropod periostracum (*Buccinum undatum* L). *Experientia*, **26,** 1196–97.

MEENAKSHI, V. R. and SCHEER, B. T. (1970). Chemical studies on the internal shell of the slug, *Ariolimax columbianus* (Gould) with special reference to the organic matrix. *Comp. Biochem. Physiol.*, **34,** 953–57.

MEENAKSHI, V. R., HARE, P. E. and WILBUR, K. M. (1971). Amino-acids of the organic matrix of neogastropod shells. *Comp. Biochem. Physiol.*, **40B,** 1037–43.

MERCER, E. H. (1972). Byssus fibre-Mollusca, in *Chemical Zoology*, Vol. VII, Mollusca, pp. 147–54, (eds M. Florkin and B. T. Scheer) Academic Press, New York.

MOCZAR, M. and MOCZAR, E. (1973). Phylogenetic evolution of the cornea. Macromolecular composition of the corneal stroma of the squid (*Sepia officinalis*). *Comp. Biochem. Physiol.*, **45B,** 213–23.

PETERS, W. (1972). Occurrence of chitin in Mollusca. *Comp. Biochem. Physiol.*, **41B,** 541–50.

PIKKARAINEN, J., RANTENEN, J., VASTIMÄKI, M., KARI, A. and KULONEM, E. (1968). On collagens of invertebrates with special reference to *Mytilus edulis*. *European J. Biochem.*, **4,** 555–60.

PUJOL, J. P. (1970). Le collagène du byssus de *Mytilus edulis*. II. Etude autoradiographique de l'incorporation de ^{3}H-Proline. *Z. Zellforsch.*, **104,** 358–74.

PUJOL, J. P., HOUVENAGHEL, G. and BOUILLON, J. (1972). Le Collagène du byssus de *Mytilus edulis*. L. I. Ultrastructure des cellules sécrétrices. *Arch. Zool. exp. gén.*, **113,** 251–64.

PUJOL, J. P., ROLLAND, M., LASRY, S. and VINET, S. (1970). Comparative study of the amino-acid composition of the byssus in some common bivalve molluscs. *Comp. Biochem. Physiol.*, **34,** 193–201.

RAVINDRANATH, M. H. and RAMALINGAM, K. (1972). Histochemical identification of dopa, dopamine and catechol in phenol gland and mode of tanning of byssus threads in *Mytilus edulis*. *Acta Histochem.*, **42,** 87–94.

WILBUR, K. M. (1972). Shell formation in Molluscs, in *Chemical Zoology*, Vol. VII, Mollusca pp. 103–45, (eds M. Florkin and B. T. Scheer) Academic Press. New York.

CHAPTER 11 ANNELIDA, SIPUNCULA AND ECHIURA

BACCETTI, B. (1967). Collagen of the earthworm. *J. Cell. Biol.*, **34,** 885–91.

BANTZ, M. and MICHEL, C. (1971). Revêtement cuticulaire de la gaine de la trompe chez *Glycera convoluta* Keferstein (Annélide Polychète). *Z. Zellforsch.*, **118,** 221–42.

BOILLY, B. (1970). Présence de collagène dans le proventricule des Syllidiens (Annélides, Polychètes). *Z. Zellforsch.*, **103,** 263–81.

DEFRETIN, R. (1971). The tubes of polychaete annelids, in *Comprehensive Biochemistry*, Vol. **26C,** 713–47. (eds M. Florkin and E. H. Stotz) Elsevier, Amsterdam.

DAMAS, D. (1969). Données histochemiques sur la cuticle de *Glossiphonia complanata* (L) (Hirudinée, Rhychobdelle) *Arch. Zool. exp. gén.*, **110,** 417–33.

DAMAS, D. (1972). Durcissement de la cuticle des mâchoires chez *Hirudo medicinalis* (Annélide, Hirudinée) aboutissant aux structures dentaires étude histochemique et ultrastructurale. *Arch. Zool. exp. gén.*, **113,** 401–21.

FREYTAG, K. (1953). Untersuchungen über den Aufbau der Cuticula von *Hirudo medicinalis*, *Z. Zellforsch.*, **39,** 85–93.

HERMANS, C. O. (1970). The periodicity of collagen in the brain sheath of a polychaete. *J. Ultrastructure Res.*, **30,** 255–61.

KRALL, J. F. (1968). The cuticle and epidermal cells of *Dero obtusa* (Family Naididae). *J. Ultrastructure Res.*, **25,** 84–93.

KRISHNAN, N. and RAJULU, S. G. (1969). The integumentary mucus secretions of the earthworm, *Megascolex mauritii*, *Z. Naturforsch.*, **24B,** 1620–23.

MICHEL, C. (1969). Ultrastructure et histochemie de la cuticle pharyngienne chez *Eulalia viridis* (Müller) (Annélide Polychète Errante Phyllodicidae). Étude de ses rapports avec l'epithélium sus-jacent dans le cycle digestif. *Z. Zellforsch.*, **98,** 54–73.

MICHEL, C. (1971). Mise en évidence d'un système de tannage quinonique au niveau des mâchoires de *Nephtys hombergii* (Annélide Polychète) *Ann. Histochem.*, **16,** 273–82.

NEFF, J. M. (1971). Ultrastructural studies of the secretion of calcium carbonate by the serpulid polychaete worm *Pomatoceros caeruleus*. *Z. Zellforsch.*, **120,** 160–86.

PHILIPS-DALES, R. and PELL, J. S. (1970). The nature of the peritrophic membrane in the gut of the terebellid polychaete *Neoamphitrite figulus*. *Comp. Biochem. Physiol.*, **34,** 819–26.

RICHARDS, K. S. (1974). The ultrastructure of the cuticle of some British lumbricids (Annelida). *J. Zool. Lond.*, **172,** 303–16.

RUTSCHKE, E. (1970). Zur substructure der cuticula der Egel (Hirudinea). *Z. Morphol. Tiere*, **67,** 97–105.

STORCH, V. and MORITZ, K. (1970). Über die Regeneration des Integumentes aus Zellen im Ventralnervenstrang von *Phascolion stranbi* (Montagu) (Siphunculida). *Z. Zellforsch.*, **110,** 258–69.

CHAPTER 12 ARTHROPODA

ANDERSEN, S. O. (1970). Isolation of arterenone (2-amino-3′,4′-dihydroxyacetophenone) from hydrolysates of sclerotized insect cuticle. *J. Insect. Physiol.*, **16,** 1951–59.

Andersen, S. O. (1971a). Phenolic compounds isolated from insect hard cuticle and their relationship to the sclerotization process. *Insect Biochem.*, **1,** 157–70.

Andersen, S. O. (1971b). Resilin in *Comprehensive Biochemistry*, Vol. **26C,** pp. 633–56. (eds M. Florkin and E. H. Stotz) Elsevier, Amsterdam.

Andersen, S. O. (1972). An enzyme from locust cuticle involved in the formation of cross-links from N-acetyldopamine. *J. Insect Physiol.*, **18,** 527–40.

Andersen, S. O. and Barrett, F. M. (1970). The isolation of ketocatechols from insect cuticle and their possible role in sclerotization. *J. Insect Physiol.*, **17,** 69–83.

Armold, M. T., Blomquist, G. J. and Jackson, L. L. (1969). Cuticular lipids of insects III. The surface lipids of the aquatic and terrestrial life forms of the big stonefly, *Pteronarcys californica* Newport. *Comp. Biochem. Physiol.*, **31,** 685–92.

Ashhurst, D. E. and Costin, N. M. (1971a). Insect mucosubstances. I. The mucosubstances of developing connective tissues in the locust, *Locusta migratoria*. *Histochemical Journal*, **3,** 279–95.

Ashhurst, D. E. and Costin, N. M. (1971b). Insect mucosubstances. II. The mucosubstances of the central nervous system. *Histochemical Journal*, **3,** 297–310.

Ashhurst, D. E. and Costin, N. M. (1971c). Insect mucosubstances. III. Some mucosubstances of the nervous system of the wax-moth (*Galleria mellonella*) and the stick insect (*Carausius morosus*). *Histochemical Journal*, **3,** 379–87.

Barth, F. G. (1970). Die Feinstruktur des Spinnenintegumenets II. Die räumliche Anordnung der Mikrofasern in der lamellierten Cuticula und ihre Beziehung zur Gestalt der Porenkanäle. *Z. Zellforsch.*, **104,** 87–106.

Bernays, E. (1972). Changes in the first instar cubicle of *Schistocerca gregaria* before and associated with hatching. *J. Insect Physiol.*, **18,** 897–912.

Browning, T. O. (1969a). Permeability to water of the shell of the egg of the *Locusta migratoria migratorioides* with observations on the egg of *Teleogryllus commodus*. *J. Exp. Biol.*, **51,** 99–105.

Browning, T. O. (1969b). The permeability of the shell of the egg of *Teleogryllus commodus* measured with the aid of tritiated water. *J. Exp. Biol.*, **51,** 397–405.

Brück, E. and Stockem, W. (1972a). Morphologische Untersuchungen an der Cuticula von Insecten I. Die Feinstrucktur der larvalen Cuticula von *Blaberus trapezoideus* BURM. *Z. Zellforsch.*, **132,** 403–16.

Brück, E. and Stockem, W. (1972b). Morphologische Untersuchungen an der Cuticula von Insecten II. Die feinstruktur der larvalen Cuticula von *Periplanata americana* (L). *Z. Zellforsch.*, **132,** 417–30.

Chang, F. (1972). The effect of ligation on tanning in the larva of a tachinid parasite, *Lespesia archippivora*. *J. Insect Physiol.*, **18,** 729–35.

Cummings, M. R. (1972). Formation of the vitelline membrane and chorion in developing oocytes of *Ephestia kühniella*. *Z. Zellforsch.*, **127,** 175–88.

Earland, C. and Robins, S. P. (1973). A study of the cystine residues in *Bombyx mori* and other silks. *Int. J. Peptide Protein Res.*, **5,** 327–35.

Fox, F. R. and Mills, R. R. (1971). Cuticle sclerotization by the american cockroach: differential incorporation of leucine and tyrosine during post ecdysis. *J. Insect Physiol.*, **17,** 2363–74.

Fox, F. R., Seed, J. R. and Mills, R. R. (1972). Cuticle sclerotization by the american cockroach: immunological evidence for the incorporation of blood proteins into the cuticle. *J. Insect Physiol.*, **18,** 2065–70.

Greven, H. (1972). Vergleichende Untersuchungen am Integument von Hetero- and Eutardigraden. *Z. Zellforsch.*, **135,** 517–38.

Furneau, P. J. S., James, C. R. and Potter, S. A. (1969). The egg shell of the house cricket (*Acheta domesticus*), an electron microscope study. *J. Cell. Sci.*, **5,** 227–49.

Hackman, R. H. (1971a). The Integument of Arthropoda, in *Chemical Zoology*, Vol. VI. Arthropoda, pp. 1–53 (eds M. Florkin and B. T. Scheer) Academic Press, New York.

Hackman, R. H. (1971b). Distribution of cystine in a blowfly larval cuticle and stabilization of the cuticle by disulphide bonds. *J. Insect Physiol.*, **17,** 1065–71.

Hackman, R. H. and Goldberg, M. (1971). Studies on the hardening and darkening of insect cuticle. *J. Insect. Physiol.*, **17,** 335–47.

Hohnke, L. A. (1971). Enzymes of chitin metabolism in the decapod *Hemigrapsus nuclus*. *Comp. Biochem. Physiol.*, **40B,** 757–79.

Hunt, S. (1970a). Amino-acid composition of silk from the pseudoscorpion *Neobisium maritimum* (Leach) a possible link between the silk fibroins and the keratins. *Comp. Biochem. Physiol.*, **34,** 773–76.

Hunt, S. (1970b). Amino-acid composition of *Macrocentrus thoracicus* cocoon-wall protein. *Comp. Biochem. Physiol.*, **37,** 93–99.

Hunt, S. (1971). Molecular and ultrastructural characterisation of the silk from the larvae of the Large White butterfly *Pieris brassicae* (L) *Comp. Biochem. Physiol.*, **40B,** 715–21.

Hunt, S. and Legg, G. (1971). Characterization of the structural protein component in the spermatophore of the pseudoscorpion *Chthonius ischnocheles* (Herman). *Comp. Biochem. Physiol.*, **40B,** 475–79.

Jackson, L. L. (1972). Cuticular lipids of insects. IV. Hydrocarbons of the cockroaches *Periplaneta japonica* and *Periplaneta americana* compared to other cockroach hydrocarbons. *Comp. Biochem. Physiol.*, **41B,** 331–36.

Koeppe, J. K. and Mills, R. R. (1972). Hormonal control of tanning by the american cockroach: probable bursicon mediated translocation of protein-bound phenols. *J. Insect Physiol.*, **18,** 465–69.

Krall, J. F. (1968). The cuticle and epidermal cells of *Dero obtusa* (Family Naididae). *J. Ultrastructure Res.*, **25,** 84–93.

Krishnan, G. and Ravindranath, M. H. (1972). Mode of stabilization of prosclerotin in arthropod cuticles. *Acta histochem.*, **44,** 348–64.

Krishnan, G. (1969). Chemical components and mode of hardening of the cuticle of Collembola. *Acta histochem.*, **34,** 212–28.

Krishnan, G. (1970). Chemical nature of the cuticle and its mode of hardening in *Eoperipatus weldoni*. *Acta histochem.*, **37,** 1–17.

Lacombe, D. (1970). A comparative study of the cement glands in some balanid barnacles. (Cirripedia, Balanidae). *Biol. Bull.*, **139,** 164–79.

Lagarrigue, J. G. and Trilles, J.-P. (1969). Nouvelles recherches écologiques sur les Isopodes *Cymothoïdae méditerranéens*: I. L'importance, la calcification et les constituants organiques de la cuticle ses varations suivant les espèces. *Vie et Milieu, Ser. A 20*, 117–36.

Lensky, Y. and Rakover, Y. (1972). Resorption of moulting fluid proteins during the ecdysis of the honeybee. *Comp. Biochem. Physiol.*, **41B,** 521–31.

Lipke, H. and Geoghegan, T. (1971). Enzymolysis of sclerotized cuticle from *Periplaneta americana* and *Sarcophaga bullata*. *J. Insect Physiol.*, **17,** 415–25.

Maillard, Y. P. (1970). Étude comparée de la construction du cocon de ponte chez *Hydrophilus piceus* L. et *Hydrophilus caraboides* L. (Insecte, Coleopt. Hydrophilidae). *Bulletin de la Societé Zoologique de France*, **95,** 71–84.

McFarlane, J. E. (1970). The permeability of the cricket egg shell. *Comp. Biochem. Physiol.*, **37,** 133–41.

Neville, A. C. (1970). Cuticle ultrastructure in relation to the whole insect. In *Insect Ultrastructure* Symp. R. entomol. Soc. Lond. No. 5, pp. 17–39, (ed. A. C. Neville) Blackwell's Sci. Publications. Oxford.

Porcella, D. B., Rixford, C. E. and Slater, J. V. (1969). Moulting and calcification in *Daphnia magna*. *Physiological Zoology*, **42,** 148–59.

Richards, G. A. and Richards, P. A. (1971). Origin and composition of the peritrophic membrane of the mosquito, *Aedes aegypti*. *J. Insect Physiol.*, **17,** 2253–75.

Richter, C. J. J., Stolting, H. C. J. and Vlijon, L. (1971). Silk production in adult females of the wolf spider *Pardosa amentata* (Lycosidae, Araneae). *J. Zool. Lond.*, **165,** 285–90.

Srivastava, R. P. (1971). The amino-acid composition of cuticular proteins of different development stages of *Galleria mellonella*. *J. Insect. Physiol.*, **17,** 189–96.

Stevenson, J. R. (1969). Sclerotin in the crayfish cuticle. *Comp. Biochem. Physiol.*, **30,** 503–08.

Thompson, H. C. Jr. and Thompson, M. H. (1968). Isolation and amino-acid composition of the collagen of white shrimp. (*Penaeus setiferus*). I. *Comp. Biochem. Physiol.*, **27,** 127–32.

Thompson, H. C. Jr. and Thompson, M. H. (1970). Isolation and amino-acid composition of the collagen of white shrimp. (*Penaeus setiferus*). II. *Comp. Biochem. Physiol.*, **35,** 471–77.

Thompson, H. C. Jr. and Thompson, M. H. (1970). Amino-acid compositional relatedness between the protocollagen and insoluble collagen of white shrimp (*Penaeus setiferus*) and collagens of certain other invertebrates. *Comp. Biochem. Physiol.*, **36,** 189–93.

VINCENT, J. F. V. (1971). Effects of bursicon on cuticular properties in *Locusta migratorioides. J. Insect Physiol.*, **17,** 625–636.

WARD, C. W. (1972). Diversity of proteases in the keratinolytic larvae of the webbing clothes moth, *Tineola bisselliella. Comp. Biochem. Physiol.*, **42B,** 131–35.

WEIS-FOGH, T. (1970). Structure and formation of insect cuticle in: *Insect Ultrastructure* Symp. R. entomol. Soc. Lond. No. 5, pp. 165–85. (ed. A. C. Neville) Blackwell's Sci. Publications. Oxford.

WELINDER, B. S. (1974). The crustacean cuticle. I. Studies on the composition of the cuticle. *Comp. Biochem. Physiol.*, **47A,** 779–87.

YAMAZAKI, H. I. (1969). The cuticular phenol oxidase in *Drosophila viridis. J. Insect Physiol.*, **15,** 2203–11.

ZELAZNY, B. and NEVILLE, A. C. (1972a). Endocuticle layer formation controlled by non-circadian clocks in beetles. *J. Insect Physiol.*, **18,** 1967–79.

ZELAZNY, B. and NEVILLE, A. C. (1972b). Quantitative studies on fibril orientation in beetle endocuticle. *J. Insect Physiol.*, **18,** 2095–2121.

CHAPTER 15 HEMICHORDATA, GRAPTOLITOIDEA, UROCHORDATA AND CEPHALOCHORDATA

AZARIAH, J. (1973a). Studies on the cephalochordates of the Madras coast. 14. Polyphenols and phenoloxidase system in amphioxus *Branchiostoma lanceolatum. Acta histochem.*, **45,** 591–97.

AZARIAH, J. (1973b). Studies on the cephalochordates of the Madras coast. 15. The nature of the structural polysaccharide in amphioxus, *Branchiostoma lanceolatum. Acta histochem.*, **46,** S 10–17.

DILLY, P. N. (1971). Keratin-like fibres in the hemichordate *Rhabdopleura compacta. Z. Zellforsch.*, **117,** 502–15.

MISHRA, A. K. and COLVIN, J. R. (1969). The microscopic and submicroscopic structure of the tunic of two ascidians. *Can. J. Zool.*, **47,** 659–63.

MONNIOT, F. (1970). Les spicules chez les tuniciers aplousobranches. *Arch. Zool. exp. gen.*, **111,** 303–11.

SMITH, M. J. (1970a). The blood cells and tunic of the ascidian *Halocynthia aurantium* (Pallas). I. Haematology, tunic morphology and partition of cells between blood and tunic. *Biol. Bull.*, **138,** 354–78.

SMITH, M. J. (1970b). The blood cells and tunic of the ascidian *Halocynthia aurantium* (Pallas). II. Histochemistry of the blood cells and tunic. *Biol. Bull.*, **138,** 379–88.

SMITH, M. J. and DEHNEL, P. A. (1970). The chemical and enzymatic analyses of the tunic of the ascidian *Halocynthia aurantium*, (Pallas). *Comp. Biochem. Physiol.*, **35,** 17–30.

SMITH, M. J. and DEHNEL, P. A. (1971). The composition of tunic from four species of ascidians. *Comp. Biochem. Physiol.*, **40B,** 615–22.

WARDROP, A. B. (1970). The structure and formation of the test of *Pyura stolinfera*, (Tunicata). *Protoplasma*, **70,** 73–86.

CHAPTER 16 VERTEBRATA: EXTERNAL SKELETAL STRUCTURES

ALEXANDER, N. J. (1970). Comparison of α and β keratin in reptiles. *Z. Zellforsch.*, **110,** 153–65.

ALEXANDER, N. J. and PARAKKAL, P. F. (1969). Formation of α and β-type keratin in lizard epidermis during the moulting cycle. *Z. Zellforsch.*, **101,** 72–87.

BADEN, H. P. and MADERSON, P. F. A. (1970). Morphological and biophysical identification of fibrous proteins in the amniote epidermis. *J. Exp. Zool.*, **174,** 225–32.

BREATHNACH, A. S., GOODMAN, T., STOLINSKI, G. and GROSS, M. (1973). Freeze-fracture replication of cells of stratum corneum of human epidermis. *J. Anat.*, **114,** 65–81.

BROWN, G. A. and WELLINGS, S. R. (1970). Electron microscopy of the skin of the teleost *Hippoglossoides elassodon. Z. Zellforsch.*, **103,** 149–69.

CHIBON, P. (1972). Étude ultrastructurale et autoradiographique des dents chez les amphibiens. Relations entre la morphogenèse dentaire et l'activité thyroidienne. *Bulletin de la Société Zoologique de France*, **97,** 437–48.

DAWSON, J. (1969). The keratinised teeth of *Myxine glutinosa*. A histological, histochemical, ultrastructural and experimental study. *Acta Zoologica*, **50,** 35–68.

DOWNING, S. W. and NOVALES, R. R. (1971a). The fine structure of lamprey epidermis. I. Introduction and mucous cells. *J. Ultrastructure Res.*, **35,** 282–94.

DOWNING, S. W. and NOVALES, R. R. (1971b). The fine structure of lamprey epidermis. II. Club cells. *J. Ultrastructure Res.*, **35,** 295–303.

DOWNING, S. W. and NOVALES, R. R. (1971c). The fine structure of lamprey epidermis. III. Granula cells. *J. Ultrastructure Res.*, **35,** 304–13.

ELDEN, H. R. (1971) editor. *Biophysical Properties of the Skin*. Wiley Interscience, New York.

FARBMAN, A. I. (1970). The dual pattern of keratinization in the filiform papillae on rat tongue. *J. Anat.*, **106,** 233–42.

FLAXMAN, B. A. (1972). Cell differentiation and its control in vertebrate epidermis. *Am. Zoologist.*, **12,** 13–27.

FLAXMAN, B. A., MADERSON, P. F. A., SZABO, G. and ROTH, S. I. (1968). Control of cell differentiation in lizard epidermis *in vitro*. *Develop. Biol.*, **18,** 354–74.

JENSON, H. M. and MOTTET, N. K. (1970). Ultrastructural features of defective *in vitro* keratinization of chick embryonic skin. *J. Cell. Sci.*, **6,** 485–509.

JORGENSEN, C. B. and LARSEN, L. O. (1961). Moulting and hormonal control in toads. *Gen. Com. Endocrinol*, **1,** 145–53.

JORGENSEN, C. B. and LARSEN, L. O. (1960). Hormonal control of moulting in amphibians. *Nature, Lond.*, **185,** 244–45.

LAWRY, JAMES V. (1973). Dioptric modification of the scales overlying the photophores of the lantern fish. *Tarletonbeania crenularis* (Myctophidae). *J. Anat.* **114,** 54–63.

MATOLTSY, A. G. (1969). Keratinization of the avian epidermis. *J. Ultrastructure Res.*, **29,** 438–58.

MATOLTSY, A. G. and HUSZAR, T. (1972). Keratinization of the reptilian epidermis: an ultrastructural study of the turtle skin. *J. Ultrastructure Res.*, **38,** 87–101.

MADERSON, P. F. A., FLAXMAN, B. A., ROTH, S. I. and SZABO, G. (1972). Ultrastructural contributions to the identification of cell types in the lizard epidermal generation. *J. Morphol.*, **136,** 191–210.

MOSS, M. L. (1970). Comparative histology of dermal sclerifications in reptiles. *Acta Anat.*, **73,** 510–33.

NEW, D. A. T. (1965). Effects of excess vitamin A on cultures of skin from the tail and pads of the embryonic rat, and from the trunk, tail and pads of the embryonic rabbit. *Exp. Cell. Res.*, **39,** 178–83.

ROSEN, M. W. and CORNFORD, N. E. (1971). Fluid friction of fish slimes. *Nature, Lond.*, **234,** 49–51.

ROTH, S. I. and JONES, W. A. (1967). The ultrastructure and enzymatic activity of the boa constrictor (*Constrictor constrictor*) skin during the resting stage. *J. Ultrastructure Res.*, **18,** 304–23.

ROTH, S. I. and JONES, W. A. (1970). The ultrastructure of epidermal maturation in the skin of the boa constrictor (*Constrictor constrictor*). *J. Ultrastructure Res.*, **32,** 69–93.

SANTHANAKRISHNAN, G., MAHADEVAN, S. and RAJULU, G. S. (1973). Chemical nature of the gizzard lining in the domestic hen (*Gallus domesticus*). *Acta histochem.*, **47,** 254–65.

SASSE, D., PFEIFFER, W. and ARNOLD, M. (1970). Epidermal Organe am Kopf von *Morulius chrysophakedion* (Cyprinidae, Ostariophysi, Pisces). *Z. Zellforsch.*, **103,** 218–31.

SENGEL, P. (1971). The organogenesis and arrangement of cutaneous appendages in birds. *Advances in Morphogenesis*, **9,** 181–230.

SPEARMAN, R. I. C. (1968). Epidermal keratinization in the salamander and a comparison with other amphibia. *J. Morph.*, **125,** 129–44.

SPEARMAN, R. I. C. (1969). The epidermis of the gopher tortoise, *Testudo polyphemus*. *Acta Zoologica*, **50,** 1–9.

TREGEAR, R. T. (1966). *Physical functions of skin*. Academic Press. London.

URIST, M. R., UYENO, S., KING, E., OKADA, M. and APPLEGATE, S. (1972). Calcium and phosphorus in the skeleton and blood of the lung fish *Lepidosiren paradoxa* with comment on humoral factors on calcium homeostasis in the osteichthyes. *Comp. Biochem. Physiol.*, **42A,** 393–408.

WATERMAN, R. E. (1970). Fine structure of scale development in the teleost *Brachydanio verio*. *Anat. Rec.*, **168,** 361–80.

WELSCH, U. and Storch, V. (1971). Fine structure and enzyme histochemical observations on the notochord of *Ichthyophis glutinosus* and *Ichthyophis kohtaoensis* (Gymnophiona, Amphibia). *Z. Zellforsch.*, **117,** 443–50.

CHAPTER 17 GENERAL PROPERTIES OF KERATIN-CONTAINING STRUCTURES

MARTINEZ, I. R., JR. and PETERS, A. (1971). Membrane-coating granules and membrane modifications in keratinizing epithelia. *Am. J. Anat.*, **130,** 93–119.

MATOLTSY, A. G. and MATOLTSY, M. (1972). The amorphous component of keratohyalin granules. *J. Ultrastructure Res.*, **41,** 550–60.

ORWIN, D. F. G. and THOMPSON, R. W. (1972). An ultrastructural study of the membranes of keratinizing wool follicle cells. *J. Cell. Sci.*, **11,** 205–19.

WILSON, P. A., HENRIKSON, R. C. and DOWNES, A. M. (1971). Incorporation of [Me-^{3}H] methionine into wool follicle proteins: a biochemical and ultrastructural study. *J. Cell. Sci.*, **8,** 489–512.

WOODS, E. F. (1971). Chromatography of the soluble proteins from feathers. *Comp. Biochem. Physiol.*, **39A,** 325–31.

CHAPTER 18 VERTEBRATA: INTERNAL SKELETAL STRUCTURES

EASTOE, J. E. (1971). Dental enamel. In *Comprehensive Biochemistry*, Vol. **26C,** pp. 785–834 (eds M. Florkin and E. H. Stotz) Elsevier. Amsterdam.

ELDEN, H. R. (1970). The bone of reptiles. In *Biology of Reptiles* Vol. 1, 45–80 (eds G. Gans, A. d'A. BELLAIRS and T. S. PARSONS) Academic Press, New York.

ENGEL, M. B. and CATCHPOLE, H. R. (1972). Collagen distribution in the rat demonstrated by a specific anticollagen antiserum. *Am. J. Anat.*, **134,** 23–40.

JACKSON, D. S. (1970). Biological function of collagen in the dermis. *Advan. Biol. Skin.*, **10,** 39–48.

JARMAN, P. J. (1972). The development of a dermal shield in impala. *J. Zool. Lond.*, **166,** 349–56.

MOSS, M. (1972). The vertebrate dermis and the integumental skeleton. *Am. Zoologist*, **12,** 27–34.

O'CONNELL, J. J. and LOW, F. N. (1970). A histochemical and fine structural study of early extracellular connective tissue in the chick embryo. *Anat. Rec.*, **167,** 425–38.

CHAPTER 19 VERTEBRATA: EGG CASES

BUSTARD, H. R. (1968). The egg shell of gekkonid lizards: a taxonomic adjunct. *Copeia.*, 1968 (1), 162–64.

HAGENMAIER, H. E. (1973). The hatching process in fish embryos: II. The structure, polysaccharide and protein cytochemistry of the chorion of the trout egg, *Salmo gairdneri* (Rich). *Acta histochem.*, **47,** 61–69.

HASIAK, R. J., VADEHRA, D. V. and BAKER, R. C. (1970). Lipid composition of the egg exteriors of the chicken (*Gallus gallus*). *Comp. Biochem. Physiol.*, **37,** 429–35.

HOFFER, A. P. (1971). The ultrastructure and cytochemistry of the shell membrane-secreting region of the Japanese quail oviduct. *Am. J. Anat.*, **131,** 253–88.

WERNER, Y. L. (1972). Observations on eggs of eublepharid lizards, with comments on the evolution of the Gekkonoidea. *Zoologische Mededelingen*, Leiden, **47,** 211–24.

CHAPTER 21 CONCLUSIONS

ASHHURST, D. E. (1968). Fibroblasts—vertebrate and invertebrate. In *Cell structure and its interpretation* pp. 237–49 (eds S. M. McGee-Russell and K. P. A. Ross) Arnold, London.

BENTLEY, J. P. (1970). The biological role of ground substance. *Advan. Biol. Skin*, **10,** 103–21.

ELDER, H. Y. (1973). Distribution and functions of elastic fibres in the invertebrates. *Biol. Bull.*, **144,** 43–63.

HUNT, S. (1970). *Polysaccharide-Protein Complexes in Invertebrates*. Academic Press, London.

JEUNIAUX, C. (1971). Chitinous Structures in *Comprehensive Biochemistry*, Vol. **26C,** pp. 259–629 (eds M. Florkin and E. H. Stotz) Elsevier, Amsterdam.

MANLY, R. S. (1970). *Adhesion in Biological Systems*. Academic Press, London.

NORDWIG, A. and HAYDUK, U. (1971). A contribution to the evolution of structural proteins. In *Prebiotic and Biochemical Evolution* (eds A. P. Kimball and J. Oro) North-Holland Publishing Co., Amsterdam.

PIKKARAINEN, J. and KULONEN, E. (1969). Comparative chemistry of collagen *Nature, Lond.*, **223,** 839–41.

PIKKARAINEN, J. and KULONEN, E. (1972). Relations of various collagens, elastin, resilin and fibroin. *Comp. Biochem. Physiol.*, **41B,** 705–12.

RUDAL, K. M. and KENCHINGTON, W. (1973). The Chitin System. *Biol. Rev.*, **49,** 597–636.

SPEARMAN, R. I. C. (1973). *The Integument*. Cambridge University Press.

TOWE, K. M. (1972). Invertebrate shell structure and the organic matrix concept. *Biomineralisation*, **4,** 1–14.

TRACEY, M. V. (1968). The biochemistry of supporting materials in organisms. In *Advances in Comparative Physiology and Biochemistry*, **3,** 233–70. (ed. O. Lowenstein) Academic Press, New York.

Index